Environmental Engineering: Sustainable Practices

Environmental Engineering: Sustainable Practices

Edited by **Chuck Lancaster**

R CALLISTO
REFERENCE

New York

Published by Callisto Reference,
106 Park Avenue, Suite 200,
New York, NY 10016, USA
www.callistoreference.com

Environmental Engineering: Sustainable Practices
Edited by Chuck Lancaster

© 2016 Callisto Reference

International Standard Book Number: 978-1-63239-639-6 (Hardback)

Contents

Preface

Every book is initially just a concept; it takes months of research and hard work to give it the final shape in which the readers receive it. In its early stages, this book also went through rigorous reviewing. The notable contributions made by experts from across the globe were first molded into patterned chapters and then arranged in a sensibly sequential manner to bring out the best results.

This book is a valuable compilation of topics, ranging from the basic to the most complex advancements in the field of environmental engineering. It is the branch of engineering that uses the principles and practices of engineering to protect the environment and reduce the harmful effects of human activities on the environment. The various sub-fields of environmental engineering along with technological progress that have future implications are glanced at in this book. The objective of this text is to provide a comprehensive overview of this field and its applications. It also sheds light on some of the unexplored aspects of environmental engineering. It deals with recycling, waste disposal, waste water management, air pollution control, industrial hygiene, radiation protection and environmental sustainability. This book will help the readers in keeping pace with the rapid changes in this area of study.

It has been my immense pleasure to be a part of this project and to contribute my years of learning in such a meaningful form. I would like to take this opportunity to thank all the people who have been associated with the completion of this book at any step.

Editor

Simulating the yields of bioenergy and food crops with the crop modeling software BioSTAR: the carbon-based growth engine and the BioSTAR ET$_0$ method

Roland Bauböck

Abstract

Background: With a growing production and use of agricultural substrates in biogas facilities, the competition between food and energy production, environmental issues, and sustainability goals has seen an increase in the last decade and poses a challenge to policy makers. Statistical yield data has a low spatial resolution and only covers standard crops and makes no statement in regard to yields under climate change. To support policy makers and regional planners in an improved allocation of agricultural land use, a new crop model (BioSTAR) has been developed.

Results: Simulations with weather and yield data from 7 years and four regions in Lower Saxony have rendered overall good modeling results with prediction errors (RMSE and percentage) ranging from 1.6 t and 9.8% for winter wheat to 2.1 t and 11.9% for maize. The model-generated ET$_0$ and ET$_a$ values (mean of four locations) are lower than ET$_0$/ET$_a$ values calculated with the Penman-Monteith method but appear more realistic when compared to field trial data from northern and eastern Germany.

Conclusions: The model has proven to be a functioning tool for modeling site-specific biomass potentials at the farm level, and because of its Access® database interface, the model can also be used for calculating biomass yields of larger areas, like administration districts or states. Out of the seven crops modeled in this study, only limited yield and test site data was available for winter barley, winter rye, sorghum, and sunflower. For further improvement of model performance and model calibration, more trial data and data testing are required for these crops.

Keywords: Crop modeling; Energy crops; BioSTAR; Biomass potentials; Evapotranspiration modeling

Background

The demand for biomass from agricultural resources as an energy source is currently seeing a strong increase. This is particularly true for Germany, as the country is trying to double the share of bioenergy (agricultural, forest, and waste biomass combined) to the country's energy total by the year 2020 [1].

In 2011, 2.2 million ha of the total agricultural area (17 million ha) was already in use for either energy crop production or renewable primary products. Of this area, 800,000 ha was in use for biogas crops, mainly maize, 900,000 ha for oilseed rape (mainly for biodiesel production) and, the smallest share, 250,000 ha for starch and bioethanol production. By 2020, the agricultural area in use for renewable resource production in Germany is projected to be further expanded and will then have a share of around 20% of the country's total agricultural area. Even though Germany's food production is close to self-sufficient today, a growing competition between food production, environmental issues, sustainability goals, and the production of energy and renewable primary products is moving into the focus of policy makers and researchers. At present, the production of biogas from energy crops and agricultural wastes (manure and other residual materials) appears to be the most (land

Correspondence: rbauboe1@gwdg.de
Department of Cartography, GIS and Remote Sensing, Research Project 'BIS', University of Göttingen, Goldschmidtstraße 5, Göttingen 37077, Germany

resource) efficient way to use agricultural areas for energy production. This is due to the relatively high energy yield of biogas per hectare [2]. This advantage of biogas is even higher when power-heat cogeneration technology is applied.

In an intensively used agricultural landscape, as it is the case in Germany, good management and farming practices and diverse crop rotation cycles are of importance, and the introduction of new energy crops into the existing crop rotation cycles can be beneficial for ecological reasons [3,4]. One research project working on this interdisciplinary topic is the currently running bioenergy project of the University of Göttingen [5].

On the contrary, using mainly maize as a substrate in biogas facilities can lead to monocultures, soil erosion, and nitrate problems in the drinking water. This is even exacerbated in areas where a lot of maize is already grown for animal feed as is the case in the western part of Lower Saxony.

Using a crop modeling tool, yield differences of different crop rotations and crops can be approximated and optimized solutions, with economical as well as ecological perspectives in view, can be found out.

Crop models have been in existence for about four decades now [6]. Resource capture of an agricultural crop can be implemented in a model in different ways. Commonly used approaches are either carbon-based [7], radiation use efficiency (RUE)-based [8], water productivity-based (WP) [9], or transpiration-based (BTR) [10].

BioSTAR's primary growth engine is carbon-based, and it uses an asymptotic exponential light response curve [11]. Among the well-known crop models, the RUE approach is probably the one which is most often used. Examples for crop models with this type of growth engine are CropSyst [12], APSIM [13], CERES (DSSAT) [14], and LINTUL. Carbon-based growth engines are used in all of the older models from Wageningen such as WOFOST and in the model CROPGRO (DSSAT).

The water productivity approach is relatively new [15], and it has been implemented in the model AquaCrop [16]. The transpiration-based growth engine (BTR) is used as a second growth engine in the model CropSyst. Because the Tanner-Sinclair relationship becomes unstable at low VPD, the RUE method is used as a main growth engine in the model CropSyst.

Even though there are numerous crop models in existence today, no single model can claim to adequately cover all possible demands a user might put to such a model. One big advantage of developing a new model is the ability to structure and build the model according to user specifications and to be able to modify it and add on to it to suit future demands.

The crop model Biomass Simulation Tool for Agricultural Resources (BioSTAR) [17,18] has been developed to simulate climate and soil-dependent biomass yields for bioenergy crops, but obviously it can also be used to predict yields for food crops like wheat or rye. The model's software is built in such a way that, depending on the resolution of the input data, large-scale (single plots or farms) or small-scale (larger areas with many input datasets) yield predictions can be generated very easily. Novelties in the BioSTAR crop modeling software are a MS Access® database connection for fast data editing and organization and the possibility to choose between four different growth engines and four ET_0 methods. Validation runs for several agricultural crops grown in Lower Saxony have proven the models' capability to serve as a user-friendly biomass simulation tool for small- and large-scale agricultural planning.

Results and discussion
Biomass yields
To validate the model BioSTAR, yield, soil, and climate data from five different locations in Lower Saxony, Germany have been used. The first two locations are farm plots in Hedeper and in Troegen. The other two are field trial sites of the Chamber of Agriculture of Lower Saxony (LWK), situated in Poppenburg and in Werlte. Winter wheat and maize were grown at all four localities, sunflower, sorghum, winter rye, and winter barley only in Poppenburg and Werlte, and sugar beet only in Hedeper.

The overall simulation results (all have been performed with the carbon-based growth engine and the BioSTAR ET_0 method) have shown that the model predicts biomass yields at a good level of accuracy, though differences between cultures exist (Table 1). For the culture sugar beet (in the following referred to as beet), the analysis has been divided up into three parts: (1) all soil types, (2) clay soil types, and (3) no clay soils. This has been done to distinguish the unique reaction (overestimation) of beet to soils with high clay contents. The model produced the lowest error values (root-mean-square error (RMSE) and percentage error) for winter wheat (RMSE = 1.6 t and 10.1%), sorghum (RMSE = 1.0 t and 5.9%), winter barley (RMSE = 1.8 t and 11.0%), winter rye (RMSE = 1.9 t and 10.4%). Beet (clay), beet (no clay), and beet (all) simulation results show up with errors of 10.7%, 10.8%, and 11.4%, respectively, and an RMSE of 1.7 t on clay and 2.4 t for the other two. Sunflower and maize results show errors of 12.0% and 11.9% and RMSE values of 1.6 and 2.1 t, respectively. All crops combined in one analysis show up with mid-range error values (RMSE 2.1 t and 12.2%). The percentage error values have been calculated by dividing the RMSE by the mean observed yield (both are in tons per hectare).

Looking at the other statistical measure for model prediction accuracy, the Willmott index of agreement, the

Table 1 Mean for observed and simulated yields, RMSE, percentage error, and WIA for tested crops

	Mean observed	Mean simulated	RMSE	Number	Percentage error	WIA
Maize	17.7	18.2	2.1	31	11.9	0.94
Winter wheat	15.8	16.1	1.6	102	10.1	0.86
Beet (all)	21.0	21.8	2.4	40	11.4	0.77
Beet (clay)	15.9	21.1	1.7	8	10.7	0.94
Beet (no clay)	22.3	21.9	2.4	32	10.8	0.85
Winter barley	16.3	16.1	1.8	6	11.0	0.64
Winter rye	18.3	18.7	1.9	6	10.4	0.73
Sunflower	13.3	13.4	1.6	9	12.0	0.56
Sorghum	17.0	17.3	1.0	5	5.9	0.78
All crops	17.2	17.7	2.1	198	12.2	0.92

WIA, Willmott index of agreement (1 = perfect agreement, 0 = no agreement). Mean observed, mean simulated, and RMSE (root-mean-square error) given in tones dry mass per hectare.

resulting order of the crops is a different one. Now maize and beet (clay) are ranked first, both with a WIA of 0.94 followed by winter wheat (0.86), beet (no clay) (0.85), and sorghum (0.78). The lower ranks are now occupied by beet (all) (0.77), winter rye (0.73), winter barley (0.64), and sunflower (0.56). All crops combined in one analysis have achieved a high WIA of 0.92.

The low WIA values for the winter grains (other than winter wheat) can be explained by an approximately equal over- and underestimation of the observed results (and possibly a low number of samples), whereas the predicted biomass for beet on clay type soils is exclusively overestimated at a similar level (Figure 1). Sorghum yields have been calculated well for the years 2008/2009 but were then overestimated at a high level in 2010 (Figure 2). Sunflower's biomass yield is predicted well in 4 years and then overestimated highly in 2007 (Figure 3). To some extent, this could be the result of a fungus infection (*Sclerotinia sclerotiorum*) which has reportedly [19] damaged the sunflower crops in the extremely rainy summer of 2007.

The overall reaction of the model in response to inter-annual climatic variations is at a good level of accuracy with the curves of the predicted vs. the observed biomass yields following the same pattern (Figure 4). This is particularly true for maize (Figure 5), winter barley and winter rye (Figure 6), winter wheat (Figure 7), and sugar beet (no clay) (Figure 8). The corresponding curves of beet (clay), beet (all) (Figures 1 and 9), sorghum (Figure 2), and sunflower (Figure 3) display some deviations from the inter-annual trend.

For all crops combined in one analysis, a linear regression analysis has been performed. The R^2 value (0.71) for the whole dataset (observed vs. predicted yields) is at a satisfactory level (Figure 10) and has a high correlation (Pearson correlation coefficient of 0.845 at a highly significant level of $\alpha \leq 0.01$).

Evapotranspiration levels

Unlike other crop models, BioSTAR can generate its own crop and phenology-dependent potential transpiration

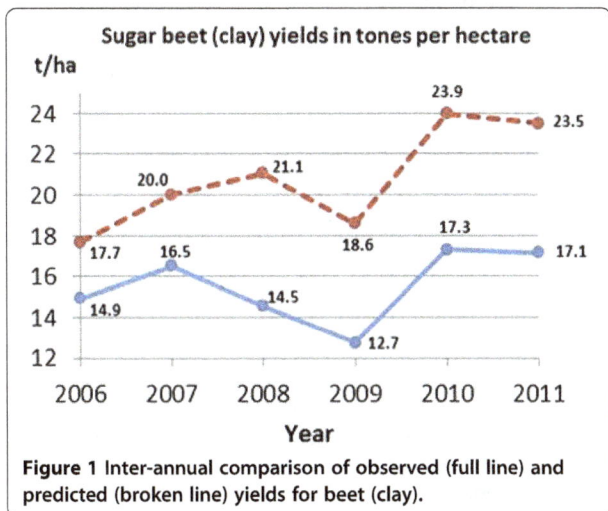

Figure 1 Inter-annual comparison of observed (full line) and predicted (broken line) yields for beet (clay).

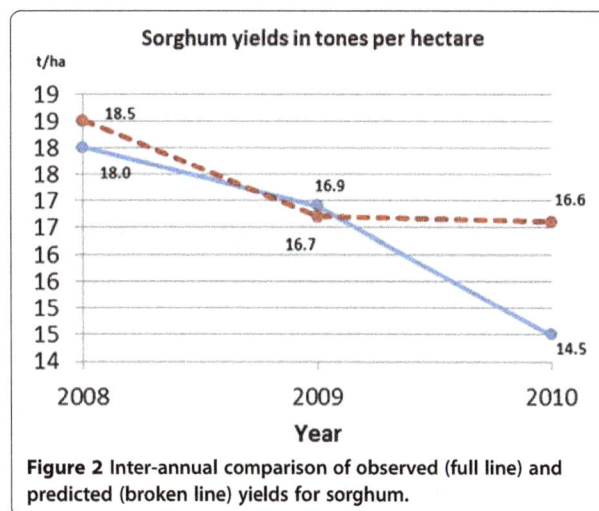

Figure 2 Inter-annual comparison of observed (full line) and predicted (broken line) yields for sorghum.

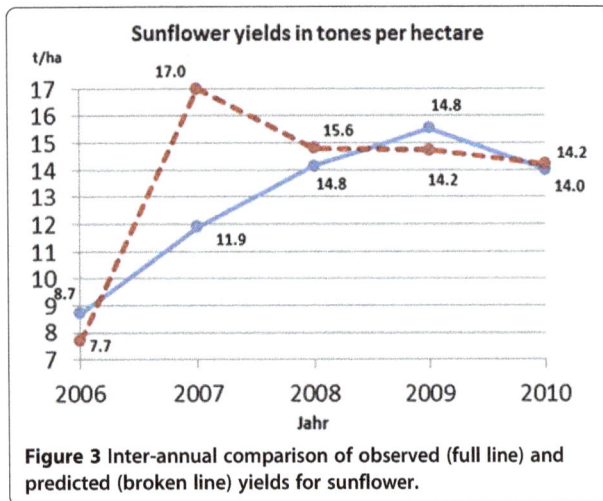

Figure 3 Inter-annual comparison of observed (full line) and predicted (broken line) yields for sunflower.

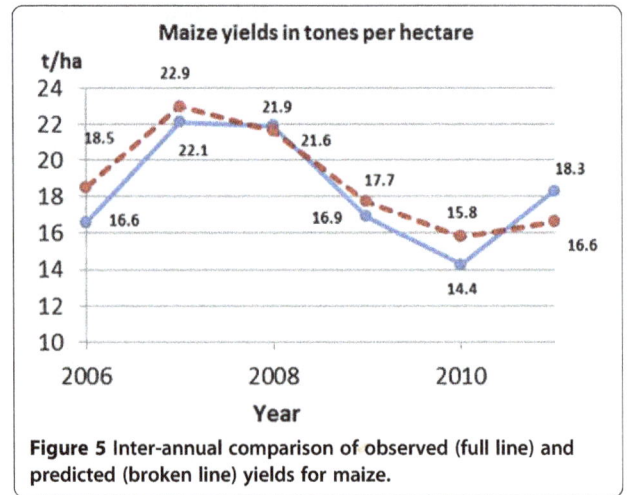

Figure 5 Inter-annual comparison of observed (full line) and predicted (broken line) yields for maize.

rates (see 'Main model processes') to which a leaf area-dependent soil evaporation value is added. In Figure 11, the mean values of all four locations of the simulation and all 7 years of the BioSTAR ET_0 (potential evapotranspiration) and ET_a (actual evapotranspiration) method are displayed along with the corresponding Food and Agriculture Organization (FAO) (Penman-Monteith) values calculated for these years. For both calculations a maize crop with a cropping period from the end of April until the beginning of September was chosen. ET_0 values calculated with the BioSTAR method are considerably lower than their FAO method equivalents. To a lesser extent, this is also true, when the ET_a values of the two methods are compared. The BioSTAR ET_0 and ET_a values range from 543 mm (2008) to 430 mm (2007) and from 423 mm (2007) to 349 mm (2012), respectively. The FAO curves for ET_0 and ET_a follow a similar inter-annual trend but at levels which are approximately 200 mm (ET_0) and 50 mm (ET_a) above the BioSTAR values. The high ET_0 values of

the FAO calculation can be explained by the fact that no crop or phenology parameters have been considered here (grass reference evapotranspiration). Looking at the literature data for ET_a values for northern and eastern Germany, the FAO values appear to be overestimated. Haferkorn [20] and Zenker [21] give ET_a values for various crops measured by lysimeters in eastern Germany, ranging from 280 to 530 mm (April until September), with average values around 350 mm. The DVWK [22] estimates the share of the evapotranspiration from May until September to be about 70% of the year's total precipitation. Since Germany's climate is of a humid character and average annual precipitation values range between 600 and 800 mm, annual evapotranspiration for this climate is not likely to be higher than 600 mm. In fact the DVWK gives an average annual evapotranspiration value (ET_a) of 433 mm for northern Germany. Eulenstein et al. [23] give annual ET_0 values for eastern Germany for the years 1971 to 1998 ranging from 420 to 680 mm (the approximate mean is

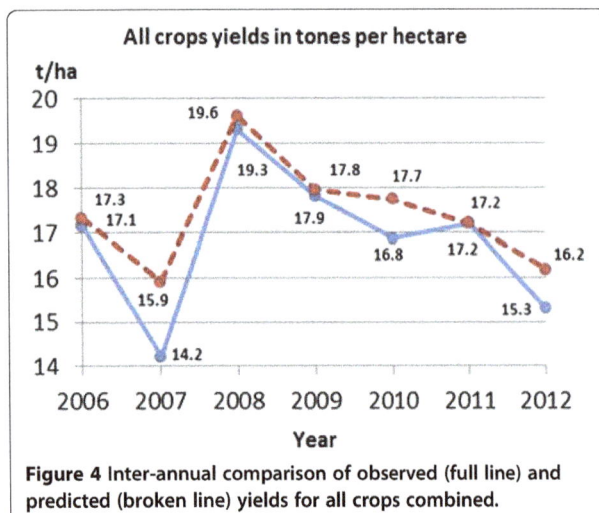

Figure 4 Inter-annual comparison of observed (full line) and predicted (broken line) yields for all crops combined.

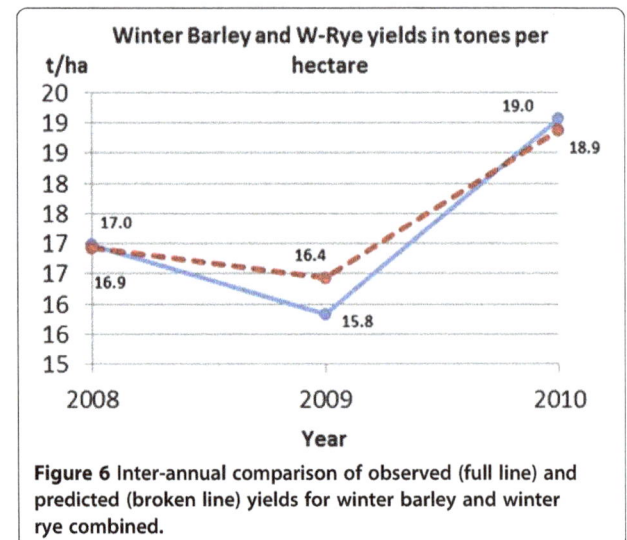

Figure 6 Inter-annual comparison of observed (full line) and predicted (broken line) yields for winter barley and winter rye combined.

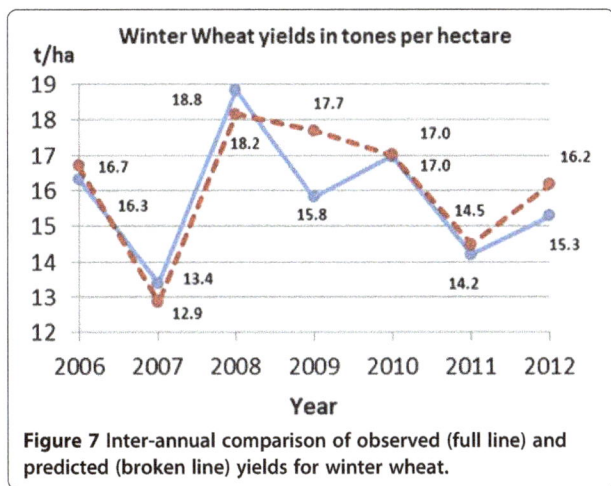

Figure 7 Inter-annual comparison of observed (full line) and predicted (broken line) yields for winter wheat.

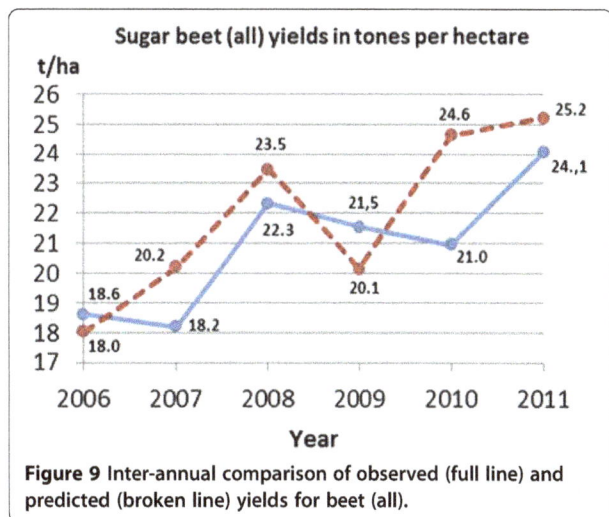

Figure 9 Inter-annual comparison of observed (full line) and predicted (broken line) yields for beet (all).

around 570 mm). Additionally it needs to be mentioned that eastern Germany has a more arid and continental climate than Lower Saxony.

In comparison with this data, the BioSTAR ET_0 and ET_a values seem to be more realistic than the FAO values and underline the relevance of this ET method for the computation of crop biomass potentials.

Conclusions

The performance of the crop model BioSTAR has been tested with datasets from four locations in Lower Saxony, Germany for seven agricultural crops. The model predicts biomass yields for all crops combined at a satisfactory level (mean error of 12.1%). The yields of all crops have been predicted by the model with errors ranging from 8.4% (winter wheat) to 12.1% (maize). The model has proven to be a functioning tool for modeling site-specific biomass potentials at the farm level. Because of its Access® database interface, the model can also

easily be used for the prediction of potential biomass yields of larger areas, like administration districts or states and can therefore serve as a decision support tool when questions of regional and trans-regional crop planning are concerned. Because the model reacts adequately to inter-annual climatic differences, transferability to different climates is probably possible but still needs to be validated. BioSTAR offers its own method for calculating evapotranspiration during the course of crop growth. The model-generated evapotranspiration levels are lower than the ones calculated using the Penman-Monteith approach but seem to be closer to actually measured ET values in northern and eastern Germany.

Out of the seven crops modeled in this study, only limited yield and test site data was available for winter barley, winter rye, sorghum, and sunflower. For further improvement of model performance and model calibration, more trial data and data testing are required here.

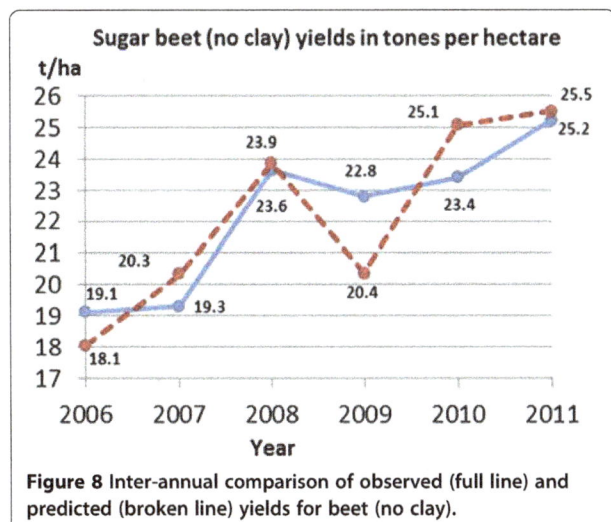

Figure 8 Inter-annual comparison of observed (full line) and predicted (broken line) yields for beet (no clay).

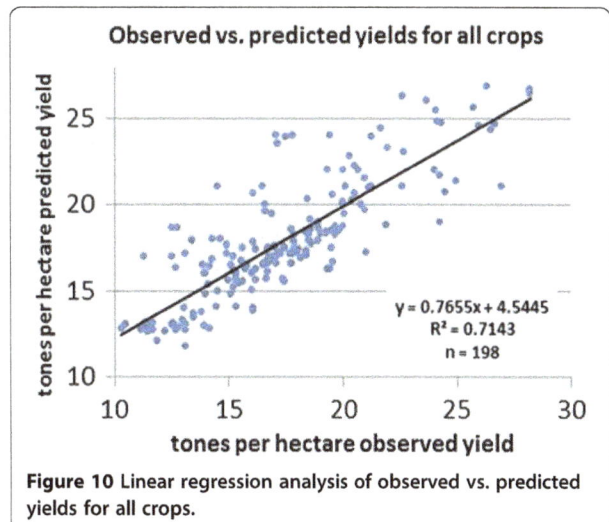

Figure 10 Linear regression analysis of observed vs. predicted yields for all crops.

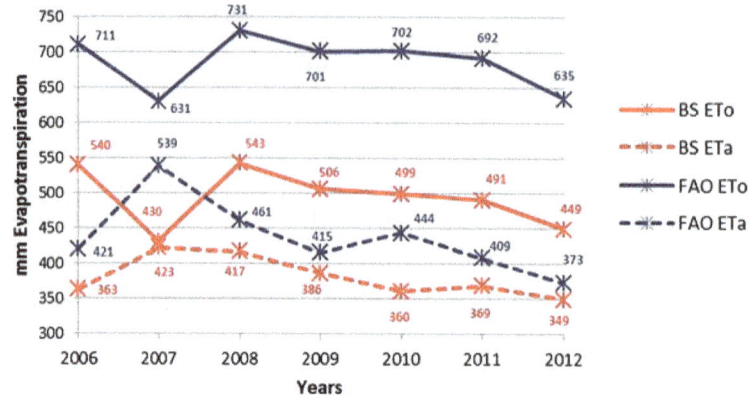

Figure 11 Inter-annual comparison of ET_0 and ET_a values calculated with the BioSTAR and FAO (Penman-Monteith) methods.

The reaction of the sugar beet yield development on clay-type soils still needs to be investigated further and improved in the model. Up to date (September 2013), the model is capable of simulating the general reaction of crops to water and nitrogen stress. To further expand the models' range of application, soil salinity content and related salinity stress reaction of plants should be implemented in the model.

Grasses and perennial cultures like the cup plant (*Silphium perfoliatum*) or short rotation coppices like poplar or willow are potential cultures for bioenergy production in the German agricultural sector. Up to date (September 2013), the model BioSTAR is capable of modeling these cultures, but calibration and validation still have to be performed before the model can be used for yield prediction of these cultures.

Methods
Main model processes
BioSTAR's primary growth engine is carbon-based (see above) and calculates a radiation and temperature-dependent gross CO_2 exchange rate in mmol CO_2 m^{-2} s^{-1} (Equation 1).

Photorespiration (maintenance and growth) and nitrogen-induced photosynthesis inhibition are accounted for in a second step. The remaining fraction of CO_2 (net photosynthesis) is then used to calculate a net photosynthesis-dependent transpiration rate. This is done using the gradients of the water vapor pressure and of the CO_2 concentration inside the leaves to the corresponding pressures of the atmosphere (Equations 2 to 5). Due to this calculation procedure, BioSTAR does not need a separate ET_0 calculation (e.g., Penman, FAO, Turc, or other) to compute crop transpiration (Figure 12):

$$P_G = P_{max} \times 1-\exp^{(-Qe \times PPFDI/P_{max})}, \qquad (1)$$

where P_G is the gross photosynthesis rate (mmol CO_2 m^{-2} s^{-1}), Qe is the initial light use efficiency (mmol CO_2 mol^{-1} light quantum), PPFDI is the intercepted photosynthetic active radiation (mmol m^{-2} s^{-1}), and P_{max} is the maximum photosynthesis rate (mmol CO_2 m^{-2} s^{-1}).

$$H_2O_{grad}: ((VP_{def} \times Vol_{mol})/18) \times 1,000 \qquad (2)$$

$$CO_{2grad}: (CO_{2con}-(CO_{2con} \times C_i/C_a))/1,000 \qquad (3)$$

$$Wat_{use}: (H_2O_{grad}/CO_{2grad}) \times 1.56 \qquad (4)$$

$$Transpot: (P_{rate} \times 3.6 \times L_{day} \times 44 \times 1,000) \times Wat_{use} \qquad (5)$$

$$Evapleaf: \{(1-(Resist_A/250)) \times 0.25\} + 1 \qquad (6)$$

$$P_{reduct}: P_{net} \times (ET_a/ET_0) \times S_{reduct}, \qquad (7)$$

where H_2O_{grad} is the H_2O gradient from leaf to atmosphere (mmol mol^{-1}), VP_{def} is the vapor pressure deficit of the air (g m^{-3}), Vol_{mol} is the volume of 1 mol dry air, CO_{2grad} is the CO_2 gradient from leaf to atmosphere (mmol mol^{-1}), CO_{2con} is the CO_2 concentration of the atmosphere (ppm), C_i/C_a is the internal-external CO_2 ratio dimensionless, range approximately 0.1 to 1.0, Wat_{use} is the H_2O-CO_2 evolution ratio dimensionless, Transpot is the CO_2 assimilation-dependent potential transpiration rate (L day^{-1}), P_{rate} is the CO_2 assimilation rate (mmol CO_2 m^{-2} s^{-1}), L_{day} is the daylight hours, Evapleaf is the aerodynamic resistance-dependent multiplier for leaf evaporation, $Resist_A$ is the aerodynamic resistance (s m^{-1}), P_{reduct} is the stomata conductance-induced photosynthesis reduction (g day^{-1}), P_{net} net photosynthesis (after respiration and nitrogen-induced reduction) (g day^{-1}), and S_{reduct} is the function for water stress-induced photosynthesis reduction.

The transpiration rate calculated by Equation 5 is multiplied by a dimensionless factor (Evapleaf) to account for aerodynamic resistance and leaf evaporation

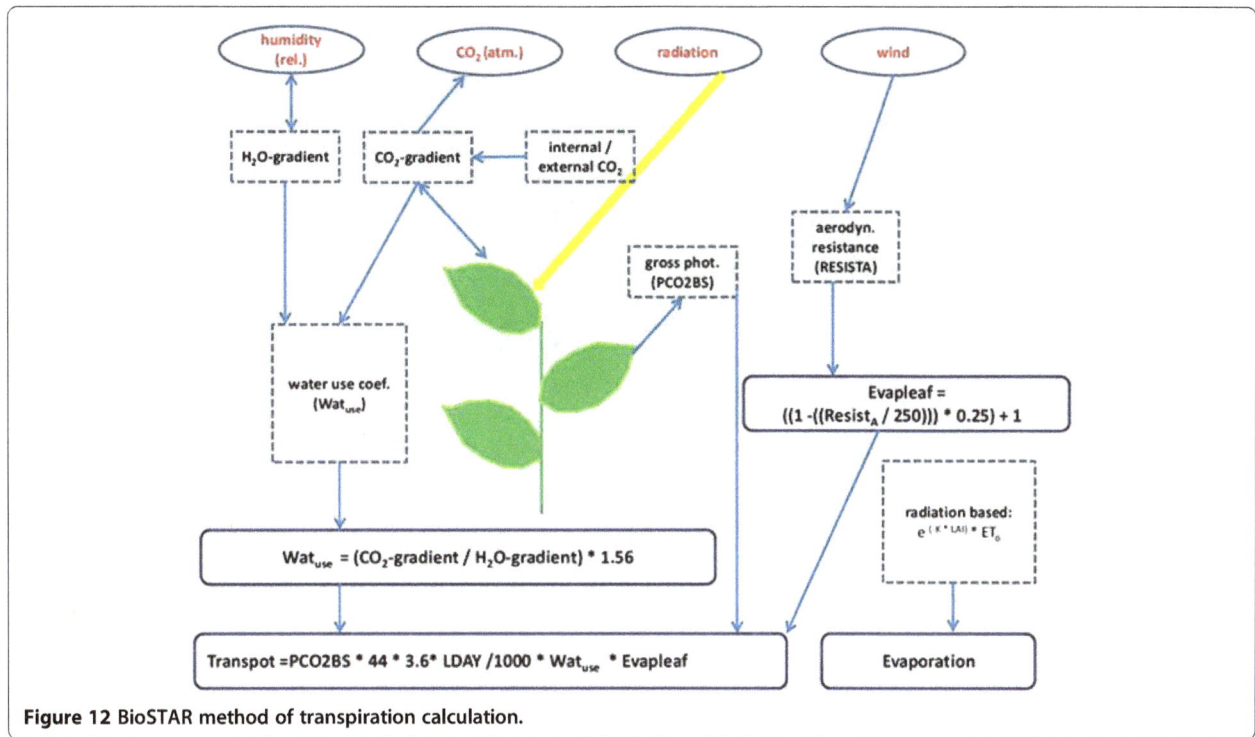

Figure 12 BioSTAR method of transpiration calculation.

(Equation 6) and then added to a leaf area-dependent soil evaporation value. The resulting evapotranspiration value (ET_0) is then used in the soil sub-model to check if enough water for evapotranspiration is available in the rooted layers of the soil profile. Soil water availability is defined by each layer's individual soil water retention curve. If the available soil water, available for evapotranspiration (ET_a), is smaller than the calculated ET_0, biomass accumulation is lowered correspondingly (Equation 7).

Crop development and leaf area index (LAI) development are temperature driven and divided into two main stages: emergence until anthesis (development stages 0 to 1) and anthesis until ripeness (development stages 1 to 2). Maximum LAI is reached at development stage 1, and the curve of LAI development is modeled as a Gaussian integral (normal distribution).

Software architecture and model features

The BioSTAR software is written in Java and uses a connection to Microsoft Access® database tables to read input data and write output data (Figure 13 and Table 2).

Data can easily be imported into these tables from spreadsheets like Excel®, and output data can be exported to a GIS for spatial visualization via the dbf format. One advantage of this software architecture is that all relevant data for running simulations is stored in one Access® database which contains different tables storing location, weather, crop, and soil texture variables. For each simulation run (combination of location and weather data), a new result table is generated in the database. Because all parameters (for crops and soil) are stored in the same database file, editing and comparison of the contents is easily done. Running the model on a PC or laptop requires an installed version of Microsoft Access® (versions 2007 or later) and the installation of Java runtime environment (freeware). The model software itself is contained in an executable JAR file and does not need to be installed on the computer.

Model calibration and input data

The model has been calibrated and tested for different sites and years in Lower Saxony for the winter cereals wheat, rye, triticale, and barley, for maize, sorghum, and sugar beet, and for sunflower. Further cultures which have been implemented in the model are canola (oilseed rape), cup plant (*S. perfoliatum*), and the short rotation coppices poplar and willow, although no validation for these cultures has been performed so far.

For model calibration, harvest and weather data (5 years) and soil data from two locations (Poppenburg and Werlte) in Lower Saxony has been used (Table 1). At these two locations, regular field trials are carried out by the LWK (Chamber of Agriculture Lower Saxony). For further testing of the model, additional harvest and weather data (7 years) from two farms in Lower Saxony (Hedeper and Trögen) were used. Soil qualities at these four locations cover a wide range from deep silt and silt loams to more shallow sands and sand loams and clays. Model testing has been performed for maize, winter wheat, winter barley, winter rye, sugar beet sunflower,

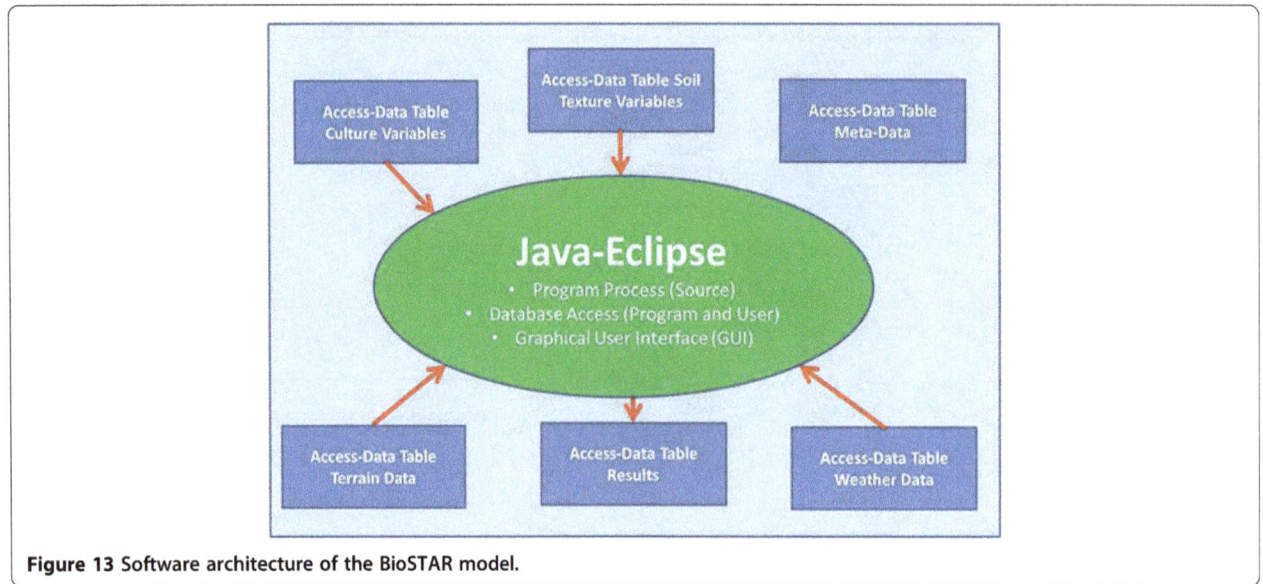

Figure 13 Software architecture of the BioSTAR model.

and sorghum. The growth engine and ET_0 method settings used for these tests are CO_2 and BioSTAR (see above).

Statistical methods used for data analysis

For the interpretation of the model output data (model performance), the *Wilmott index of agreement* [24] (Equation 8) and the RMSE (root-mean-square error) (Equation 9) values have been calculated for all tested crops individually as well as all crops combined in one analysis (Table 1). For all crops combined, a linear regression and the corresponding R^2 value have been calculated (Figure 10). To see how the model reacts to inter-annual variations in climate, an inter-annual comparison of the observed and simulated yields has additionally

been done for all crops combined and individually (Figures 1, 2, 3, 4, 5, 6, 7, 8, 9).

$$d = 1 - \frac{\sum_{i=1}^{n}\left[(P_i - \bar{O}) - (O_i - \bar{O})^2\right]}{\sum_{i=1}^{n}\left[(|P_i - \bar{O}|) + (|O_i - \bar{O}|)^2\right]} \qquad (8)$$

$$\text{RMSE} = \left\lfloor n^{-1}\sum_{i=1}^{n}(P_i - O_i)^2\right\rfloor^{-0.5}, \qquad (9)$$

where P_i is the instance of predicted value, O_i is the instance of observed value, $\bar{O}$ is the mean of observed values, RMSE is the root-mean-square error, and d is the Willmott index of agreement (0 = no agreement, 1 = perfect agreement).

Table 2 Summary of model features

	Features
Software	
Data storage	Fast data reading, writing, and editing due to MS Access® data table interface
Multiple sites	Capability to process either individual sites or large datasets
Program type	Program runs from an executable Java file
Data organization	Soil, weather, crop, and result data are all kept in one database
Crop model	
Growth engines	User can choose between four growth engines and four ET_0 methods
Time step	Daily or monthly climate data can be processed
Minimum data	If data availability is limited, the model can be run with only daily mean temperature
Perennial crops	Modeling of perennial crops like short rotation coppices or cup plant is possible
Soil model	Computation of soil water budget in a one-dimensional 2-m soil profile with decimeter layer increments using van Genuchten soil texture parameters
Crop water stress	Crop water stress simulation enhancement with crop-specific stress phase modeling
Crop development	Crop development tracking with BBCH (EC) stages

Abbreviations

APSIM: agricultural production simulator; AquaCrop: FAO crop model; BBCH/EC scale: growth scale of monocot and dicot plants; BioSTAR: Biomass Simulation Tool for Agricultural Resources; CERES: crop model in the DSSAT family; CROPGRO: crop model in the DSSAT family; CropSyst: cropping systems model; DSSAT: decision support system for agronomy transfer; ET_a: actual evapotranspiration rate; ET_0: potential (reference) evapotranspiration rate; FAO: Food and Agriculture Organization; LINTUL: crop model in the Wageningen family; LWK: Landwirtschaftskammer (Niedersachsen); RMSE: root-mean-square error; RUE: radiation use efficiency; VPD: vapor pressure deficit of the air; WIA: Willmott index of agreement; WOFOST: crop model in the Wageningen family.

Competing interests

The author declares that he has no competing interests.

Acknowledgements

The development of the model BioSTAR has been made possible by funds of the Lower Saxony Ministry of Sciences and Culture (Germany). The research group responsible for developing the model is part of the interdisciplinary research project 'Sustainable use of bioenergy: bridging climate protection, nature conservation and society,' sub-project 2.2 'bioenergy potentials.'

References

1. BMWI-BMU: *Energiekonzept für eine umweltschonende, zuverlässige und bezahlbare Energieversorgung.* Berlin; 2010. http://www.iwo.de/fileadmin/user_upload/Dateien/Standpunkte/Energiekonzept-2010.pdf.

2. FNR: *Fachagentur für nachwachsende Rohstoffe.* Basisdaten Bioenergie; 2011. http://mediathek.fnr.de/media/downloadable/files/samples/f/n/fnr_basisdaten_2012_web_neu.pdf.

3. Karpenstein-Machan M, Weber C: **Energiepflanzenanbau für Biogasanlagen–Veränderungen in der Fruchtfolge und der Bewirtschaftung von Ackerflächen in Niedersachsen.** *Naturschutz und Landschaftsplanung.* 2010, **42:**312–320.

4. Ruppert H: *Wege zum Bioenergiedorf–Leitfaden für eine eigenständige Wärme-und Stromversorgung auf Basis von Biomasse im ländlichen Raum.* 3rd edition. Güstrow: FNR; 2010.

5. *BiS-Project of the University of Goettingen: homepage (English).* http://www.bioenergie.uni-goettingen.de/index.php?id=107.

6. Bouman BAM, van Keulen H, Van Laar HH, Rabbinge R: **The 'School of de Wit' crop growth simulation models: a pedigree and historical overview.** *Agr Syst* 1996, **52:**171–198.

7. De Wit CT: *Photosynthesis of Leaf Canopies,* Agricultural Research Report 663. PUDOC: Wageningen; 1965.

8. Monteith JL: **Climate and the efficiency of crop production in Britain.** *Philos Translat Royal Soc* 1977, **281:**277–294.

9. Steduto P, Hsiao TC, Ferres E: **On the conservative behavior of biomass water productivity.** *Irrig Sci* 2007, **25:**189–207.

10. Tanner CB, Sinclair TR: **Efficient water use in crop production: research or re-search?** In *Limitations to Efficient Water Use in Crop Production.* Madison: American Society of Agronomy; 1983.

11. Boote K, Loomis R: **The prediction of canopy assimilation.** In *Modeling Photosynthesis–from Biochemistry to Canopy,* CSSA Special Publication no.19. Madison: CSSA; 1991.

12. Stöckle C, Donatelli M, Nelson R: **CropSyst, a cropping systems simulation model.** *Eur J Agron* 2003, **18:**289–307.

13. Keating B, Carberry PS, Hammer GL, Probert ME, Robertson MJ, Holzworth D, Huth NI, Hargreaves JNG, Meinke H, Hochman Z, McLean G, Verburg K, Snow V, Dimes JP, Silburn M, Wang E, Brown S, Bristow KL, Asseng S, Chapman S, McCown RL, Freebairn DM, Smith CJ: **An overview of APSIM, a model designed for farming systems.** *Eur J Agron* 2003, **18:**267–288.

14. Jones JG, Porter C, Boote K, Batchelor W, Hunt L, Wilkens P, Singh U, Gijsman AJ, Ritchie JT: **The DSSAT cropping system model.** *Eur J Agron* 2003, **18:**235–265.

15. Todorovic M, Albrizio R, Zivotic L, Abi MT, Stöckle C, Steduto P: **Assesment of aqua crop, CropSyst, and WOFOST models in the simulation of sunflower growth under different water regimes.** *Agron J* 2009, **101:**508–521.

16. Steduto P, Hsiao TC, Ferres E: **AquaCrop—the FAO crop model to simulate yield response to water: I: concepts and underlying principles.** *Agron J* 2009, **101:**426–437.

17. Bauböck R: *GIS-gestützte Modellierung und Analyse von Agrar-Biomassepotenzialen in Niedersachsen – Einführung in das Pflanzenmodell BioSTAR,* PhD thesis. University of Göttingen; 2013. http://hdl.handle.net/11858/00-1735-0000-000E-0ABB-9.

18. Workgroup BioSTAR: *Cartography, GIS and remote sensing section.* University of Goettingen: homepage (English). http://www.uni-goettingen.de/en/431252.html.

19. LWK: *Energiepflanzen in Niedersachsen–Anbauhinweise und Wirtschaftlichkeit.* Landwirtschaftskammer Niedersachsen; 2010. http://www.ml.niedersachsen.de/download/59388.

20. Haferkorn U: *Größen des Wasserhaushaltes verschiedener Böden unter landwirtschaftlicher Nutzung im klimatischen Grenzraum des Mitteldeutschen Trockengebietes–Ergebnisse der Lysimeterstation Brandis,* PhD thesis. University of Göttingen; 2000. http://hdl.handle.net/11858/00-1735-0000-000D-F107-4.

21. Zenker T: *Verdunstungswiderstände und Gras-Referenzverdunstung Lysimeter-untersuchungen Zum Penman-Monteith-Ansatz im Berliner Raum,* Volume PhD thesis. TU Berlin: TU Berlin; 2003. http://nbn-resolving.de/urn/resolver.pl?urn:nbn:de:kobv:83-opus-6956.

22. DVWK: *Merkblätter zur Wasserwirtschaft – Ermittlung der Verdunstung von land- und Wasserflächen.* Bonn: Heft 238/1996.

23. Eulenstein F, Olejnik J, Willms M, Schindler U, Chojnicki B, Meißner R: **Mögliche Auswirkungen der Klimaveränderungen auf den Wasserhaushalt von Agrarlandschaften in Nord-Mitteleuropa.** In *Wasserwirtschaft.* Wiesbaden: Springer-Vieweg; 2006.

24. Willmott CJ: **Some comments on the evaluation of model performance.** *Bull Am Meteorol Soc* 1982, **63:**1309–1313.

2

How to consider engineered nanomaterials in major accident regulations?

Bernd Nowack[*], Nicole C Mueller, Harald F Krug and Peter Wick

Abstract

Major accident regulations aim at protecting the population and the environment from possible accidental releases of chemicals. To achieve this goal, the regulations need to be reassessed in light of the development of new technologies. A currently rapidly growing new technology is nanotechnology, and engineered nanomaterials (ENM) are already produced and used in commercial products. The aim of this work was therefore to evaluate the current knowledge on human and ecotoxicology of ENM and their release and behavior in the environment in the context of major accident prevention. Nano-specific release paths are not to be expected. The established safety standards in the chemical industry are also applicable to ENM, especially the separate storage of flammable solvents and detention reservoirs. The potential of a release to the environment of ENM in powder form is larger than for suspensions; however, it can be minimized by safety measures established for conventional dusts. The considered human toxicology studies show that to date not conclusive enough answers regarding the toxicity of ENM can be made. The effects are dependent not only on the material itself but more on the functionalization, surface reactivity, size, and form. The acute ecotoxicity of ENM seems to be similar to the one of the corresponding microparticles (TiO_2) or the respective dissolved ions (Ag, Zn) with the exception of photocatalytically active nano-TiO_2, which has an increased toxicity. In order to guarantee that all ENM are included in the existing major accident regulations, different classification options are possible and the advantages and disadvantages are discussed. An important step will be the compulsory inclusion of nano-specific data in the Material Safety Data Sheets that serve as the basic medium to transfer information from the manufacturer to downstream users and authorities. We also call for a regular monitoring of the production and uses for 'high production volume ENM' that could have the largest implications for major accident regulations.

Keywords: Nanomaterials; Major accidents; Toxicity; Ecotoxicity

Review

Introduction

The major accident prevention regulations have the goal to protect the general population and the environment from severe damage due to accidents. In Switzerland, a company has to fulfill the requirements of this regulation if they store compounds on their premises in amounts that surpass the thresholds given in the regulation [1]. The determination of these thresholds is based on an approach, which contrasts the properties of a substance to its amount used within the company. Companies that fall under this regulation have to report an estimation of possible damage and formulate scenarios. If major damage of the general population (more than

ten deaths outside the area of the company) or of the environment cannot be excluded, a quantitative risk estimation based on scenarios has to be prepared. The regulation for major accident prevention only considers major damages of people outside the area of the company and is therefore separate from occupational or consumer protection regulations. It only considers acute effects, and chronic effects or exposure is therefore not included in this scenario.

Nanotechnology is a rapidly growing research and application area with increasing importance for economy, research, and society. In line with the precautionary principle, it is therefore important to investigate possible risks and, if necessary, take measures to protect humans and the environment. In the focus of the risk discussion about nanotechnology are the engineered nanomaterials (ENM) because they can show - compared to larger

* Correspondence: nowack@empa.ch
Empa - Swiss Federal Laboratories for Materials Science and Technology, Lerchenfeldstrasse 5, St Gallen 9014, Switzerland

materials - different properties and reactivity. Therefore, they cannot *a priori* be handled in the same way as larger particles. A separate determination of threshold values and criteria for evaluation is therefore necessary, to account for the new characteristics of ENM.

There is to date no international agreement on how to deal with major accident prevention of ENM. Two reports are available about the fire and explosive properties of ENM [2,3]. The European Commission states that ENM with potential for accidents can be categorized within the Seveso II directive [4]. Several countries have identified the issue of nano-accidents as relevant [5-7]. The English Parliament does not consider nano-specific regulation to be necessary [8].

In a document by the British Standards Institution (BSI) about guidelines for safe handling of ENM, also fundamental measures for preventive actions against accidents are listed [9]. BSI asks companies that work with ENM to develop emergency plans for possible accident scenarios and to list the exact sequence of activities and the measures that need to be taken in case of an accident. All persons that would be involved in such an accident should be informed and trained in handling ENM. Furthermore, measures should be taken to prevent the dissipation of ENM in case of an accident. A report of the OECD about major accidents with ENM is in preparation [10].

The basis for the assessment into the major accident regulation is the material safety data sheet (MSDS). However, there is to date no duty to declare the size of particles or indicate if the material contains a nanosized fraction. Downstream users of compounds therefore receive often no information if a product contains ENM. In many cases, the nano-ingredient does not even reach the required threshold value of 1% so that it does not even need to be listed in the MSDS. A clear definition what constitutes an ENM is therefore compulsory so that in the future, appropriate designations can be made on the MSDS.

The aim of this work is the assessment of the human toxicity and ecotoxicity of ENM within a major accident in comparison to conventional chemicals. We try to answer the question if for the determination of threshold amounts in the regulation a mass-based approach can be used for ENM and which major accident scenarios need to be considered for ENM. We performed this assessment using four typical ENM that are covering different classes and that are all produced and used in high quantities: nano-TiO_2, nano-ZnO, carbon nanotube (CNT), and nano-Ag [11]. We use the Swiss regulation as an example but aim to provide general conclusions.

Release into the environment

Potential of ENM for major accidents

A major accident can happen whenever large amounts of a potentially toxic or reactive substance are present.

This can happen during production and manufacturing as well as during storage and transport. The scenarios for major accidents strongly depend on the used ENM and other chemicals on the same site as well as factors specific for the situation or the site:

- Form of ENM: suspended or as powder
- Presence of flammable or explosive substances (e.g., metallic ENM or organic solvents)
- Type of containment
- Risk of accident during transport
- Safety measures.

In general, we can assume that the potential for major accidents is higher when ENM are present as powder, because they are easier dispersible than suspended ENM [12]. This is due to the fact that in the regulation of major accidents, only the first 30 min after the accident are of interest and the acute loss of lives or acute damage of the environment is calculated [1]. As water and soil can be purified and thus major consequences in the long-term are not included in the regulatory measures of major accidents, the velocity of dispersibility of powders compared to presuspended ENM plays no important role in our considerations. The potential is also increased by the storage of easily flammable or explosive compounds in the vicinity of the ENM.

Of 25 major accidents or incidents that were reported in 2008 in Germany (no nano-related cases) [13], 11 occurred during production; 4 during handling; 3 during transport; 2 during maintenance; 1 each during storage, decommissioning, and delivery; and 2 with unknown activity. Also, natural hazards such as lightning, earthquakes, or flooding need to be considered as triggers of major accidents, in addition to man-made disasters such as explosions in neighboring factories or airplane crashes [14,15].

The following release scenarios are considered as realistic causes of major accidents for ENM:

- Major accident scenarios during production
  ENM are produced in very diverse operations [16]. In general, the mechanical-physical top-down approaches can be distinguished from chemical-physical bottom-up approaches. In top-down approaches, ENM are made from larger materials through milling; in bottom-up approaches, ENM are synthesized from atoms, ions, or molecules in a chemical reaction. Milling operations are used for metallic or ceramic ENM with a relatively wide particle size distribution. Chemical-physical approaches have the advantage that the form and size of the particles can be better controlled.

Possible reactions are precipitation and flame-, plasma-, or gas-phase synthesis.

During milling, a major accident with consequences for the general population can practically be excluded, because in this process, only small amounts (around 100 kg) of ENM are used, mainly in aqueous dispersion and of metallic or ceramic ENM; hence, explosions can be excluded. If the process is carried out in organic solvents, then this process needs to be carried out in explosion-proof systems, but due to the small amounts (batch volume of around 100 L with max 10-kg ENM), no hazard for the population is to be expected. In gas-phase processes, a deflagration and, with easily flammable solvents, a fire hazard cannot be excluded. Nevertheless, a chain of different events is necessary so that release of ENM beyond the fabrication site is possible.

A possible scenario is the explosion of a distillation equipment with subsequent fire during which all ENM in the same fire compartment are released. Depending on the safety measures, amounts stored, and situation-specific conditions, larger amounts of ENM could be released into air or wastewater. We also need to consider that carbon-based ENM could be combusted while metallic ENM could be oxidized.

- Accident scenarios during transport

    Transport of toxic substances is connected with a rather high risk because traffic accidents are quite common and because easily flammable substances can be released (e.g., fuel). However, the transported amounts of ENM are currently rather small, compared to the total amount of produced material in any size (compare 5 Mio t of produced TiO_2 of which only 47,000 t are nano-TiO_2 [17], and this is one of the nanomaterials with the highest production amounts). Although there is currently no nano-specific obligation to label, we propose that the containers for transport of ENM should be labeled as a hazardous material and that containers of the highest safety standards be used. An accident during transport could lead to the spilling of dispersions/powder or a fire through which ENM could reach air, soil, wastewater, or natural waters (Figure 1).

- Accident scenarios during manufacturing and storage

    During manufacturing or storage of ENM, fire is a possible hazard, which can result in the release of ENM into air (when stored as powder) as well as into water/wastewater (ENM dispersions). The cause for the accident can be internal (e.g., technical) as well as external factors. Deflagration is possible for metallic ENM, but they are usually stored in dispersion and/or under inert atmosphere. An example is nano-zero-valent iron for which also hydrogen production and the danger of hydrogen explosion need to be considered.

Figure 1 Example of an accident scenario. In 2011, a truck transporting several 750-kg bags of photocatalytic titanium dioxide (TiO_2) has lost part of its cargo. Since TiO_2 is not classified as a hazardous material, workers of the road maintenance department cleaned the road. (Photo from [18]; photographer: Arnaud Viry).

Release and behavior in the environment

Based on the scenarios described above, a primary release of ENM into air, water, and soil is possible. A secondary exposure of soil and surface water through deposition from the atmosphere is likely. Also, the indirect contamination of surface water through the effluent of wastewater treatment plants needs to be considered, especially because a shock load of ENM may destroy the removal capacity of the activated sludge phase of wastewater treatment. The secondary pollution of groundwater can be neglected in the case of accidents because particle mobility in soils is generally rather low [19].

To date, no major accidents with ENM are known, and therefore, we do not have any experience with the dissipation behavior of ENM after an accident. Also, models are not yet available that are able to simulate the distribution of ENM in the environment on a local scale [20]. The only available models predict ENM concentrations on a regional scale [21,22]. In these models, the flows of ENM from production, use, and disposal through technical compartments to the environment are predicted and environmental concentrations are obtained assuming well-mixed environmental compartments. Measurements and modeling have been performed for workplace situations [12,23,24]. One study measured and modeled dispersion of ENM during a (purposely made) failure of ventilation in a laboratory-scale flame synthesis reactor [25]. The authors conclude that coagulation of ENM is negligible inside the room [23,25]. The measurements and calculations are restricted to the production room and can therefore not be transferred to environmental conditions during a major accident.

During a major accident, powdered ENM can be released in larger amounts into air. Particles are not volatile and therefore only be released during an explosion or fire. The first elimination step is primarily agglomeration and subsequently deposition of larger particles. Agglomerates with a diameter of 1 μm have the longest residence time of particles in the atmosphere [26]. Due to agglomeration, it is not likely that single ENM persist in air [27].

The concentration in air is primarily relevant for human exposure after a major accident. The prevailing weather conditions during the accident play a major role (wind speed, wind direction, precipitation). These conditions cannot be considered in a generic scenario. The following generic model parameters are proposed for calculations of local exposure outside the industrial site in the ECHA guidelines [28]:

- The total emissions are distributed radial symmetrically around the source (radius 1 km).
- The height of the source is 10 m.
- The average distance from the source to the boundary of the site is 100 m.
- The concentration decreases linearly from the source to 1 km.

These assumptions are used to calculate a generic local exposure scenario and therefore cannot be used to model a major accident at a specific site. In addition, they are not particle specific.

The input of ENM into surface water can occur in three ways:

1. Release into surface water through storm water collection systems along roads or on the production site or directly into surface water if there is a close proximity of the site of accident and the open water. In both cases, we have to assume that only a fraction of the total emitted ENM amount reaches the water. The probability to reach the water is higher for suspended ENM. However, washing of powdered ENM during rain events or with firefighting water cannot be excluded.
2. Release of ENM into wastewater with direct connection to surface water. Studies have shown that ENM are removed during wastewater treatment with a high efficiency [29]. However, during a shock load of antimicrobial ENM (e.g., silver), the bacterial population may be compromised and therefore the removal efficiency of the water treatment process [30].
3. Deposition from air. In this case, the input is dependent on the surface of the water body. Due to faster dilution, rivers are less at risk than lakes.

The ENM are diluted upon entry into the water bodies. The dilution *de facto* depends on the size of the water body and can have values between 1 and 100,000 [28]. A dilution factor of 10 is proposed if no other data are available. An additional dilution can result from the firefighting water.

Water contamination is even stronger dependent on the local situation than the emission into air. The probability for a contamination of surface waters that has to be classified as a major accident is therefore dependent on the proximity of the site of accident to the next freshwater body. An accident is considered as major when in a water volume of 10^6 m^3 the lethal concentration 50 (LC_{50}) and/or the acute half maximal effective concentration (EC_{50}) for fish or *Daphnia*, respectively, is surpassed [31]. With an $EC_{50(Daphnia)}$ of 1 μg/l [32] for nano-Ag and homogeneous distribution already, a release of 1 kg of nano-Ag could lead to the corresponding concentration. This evaluation neglects relevant elimination processes such as dissolution and precipitation as AgCl; however, they may only play a role at longer time scales that are not relevant for major accident regulations. The contamination of natural waters can result in a secondary exposure of humans through bioaccumulation into fish. However, such long-term effects are not considered in major accident regulations.

A direct release of ENM into soils is only expected during transport accidents. Indirect transfer can occur through wet or dry deposition form air. Although the local soil concentration can be very high after a major accident, a persistent damage of the environment can be practically excluded due to the low particle mobility [33] and the possibility for soil remediation (excavation and landfilling of polluted soil).

Toxicity and ecotoxicity in the framework of major accident prevention

Human toxicity

In the Swiss Ordinance on Major Accidents, the following criteria are used to determine the threshold amounts [1]:

- EU classification
- Acute toxicity; oral, dermal, and inhalation toxicities
- Classification given on MSDS

The comparison of conventional chemicals (e.g., microparticles) with ENM based on these criteria is restricted to the criterion 'acute toxicity' as the ENM have no special EU or MSDS classification. Table 1 summarizes the hazard potential of ENM to humans in comparison with conventional materials.

The published data are very heterogeneous and do not provide a clear effect pattern for each of the materials. The measured endpoint of the experimental data is not the lethal dose concentration (LD_{50}), but more subtoxic events such as inflammation, oxidative stress response, or gene expression profiling over different time scales were assessed, which make a sound comparison of the studies impossible. In addition, the three exposure routes (oral, dermal, and inhalation) provide different symptoms; therefore, the hazard potential of the selected materials is based more on an expert view than on standardized and comparable data sets.

Ecotoxicity

The ordinance of major accidents [1] uses as criteria for the evaluation of the ecotoxicity the EC_{50} for *Daphnia* (swimming disability after 24 h) and the LC_{50} for fish (after 2 to 4 days). The only threshold in the area of ecotoxicity is at 2 t of substance used at a site for an EC_{50} (*Daphnia*) or LC_{50} (fish) <10 mg/l. Tables 2 and 3 contrast EC_{50} and LC_{50} values for the nano- and microparticles (TiO_2) or dissolved metals (Ag, Zn). This comparison evaluates if the overall toxicity of the traditional form and the nano-form is different from each other. It does not consider any difference in the mode of action but simply uses mass as a metric to compare materials.

The only compound that is more critical in its nano-form than as dissolved metal is therefore TiO_2 in its photocatalytically active form. For Ag and ZnO, we cannot see a difference between the toxicity of the nano-form

Table 1 Hazard potential of ENM to humans in comparison with conventional materials (microparticles)

ENM	Hazard potential[a]			Remarks
	Acute toxicity	Chronic effects	Difference (micro/nano)	
Nano-TiO$_2$	Very low	Low	1:10	Although TiO$_2$ was placed in class 3 of the carcinogenic materials, its general toxicity is low as shown by many recent studies. There is a difference in the effects of smaller and larger particles, but this, for a release scenario during a major accident, is only of low relevance [34-39]
Nano-ZnO	Medium	Low	1:1	Application of micro- or nanoscale ZnO particles into the lungs of mice or rats causes a strong but intermediate inflammatory reaction. The strength as well as the course of this reaction is practically identical for micro- or nanoscale ZnO [40,41]. The acute consequence of inhalation of Zn dust is zinc fever; however, for severe effects (deaths), the concentration needs to surpass widely the permissible exposure limit of 5 mg/m^3
CNT	Low	High	n.a.	Dependent on the type of CNT (physicochemical properties, e.g., single-walled, multi-walled, short/long), long-term effects similar to asbestos need to be considered (mesothelioma) [42] if the conditions like for 'WHO fibers' (length >5 μm, diameter <3 μm, and a length/diameter ratio of more than 3:1) are given. So far, most industrially produced CNTs are not considered to be WHO fibers [43-49]
Nano-Ag	Low	Low	1:1	Silver is used since many years in nanoparticulate form (colloidal silver). There is no indication for an acute intoxication of humans with life-threatening degree, unless there is exposure to ultrahigh concentrations. Because Ag particles are added in rather small amounts to products, also the produced and transported amounts are rather small and the risk of a major accident with fatal consequences is therefore limited [50-54]

[a]Acute toxicity: classification on the basis of EU Directive 67-548 Annex VI - medium = R 23/25, low = R 20/22, very low = no labeling; chronic effects: based on actual literature. n.a., not applicable.

Table 2 EC$_{50}$ values (mg/l) for *Daphnia* for different ENM: comparison between nano-form and microform/dissolved metals

	EC$_{50}$ *Daphnia*		
	ENM lowest reported value (mg/l)	Larger particles or dissolved ions[d]	Rounded ratio (nano/micro)
Nano-TiO$_2$	0.03 [55][a]	>100 [56]	>1:100 (for the photocatalytically active form)
Nano-ZnO	0.62 [57][b]	1.86 [58][e]	1:1
CNT	1.8 [57][c]	-	-
Nano-Ag	0.001 [32]	0.0015 [59]	1:1

[a]Other values (in mg/l): 0.8 [60], 3.8 [61]; [b]other values (in mg/l): 1 [56], 3.95 [62]; [c]other value (in mg/l): 50.9 [63]; [d]for Ag and ZnO, the values for the dissolved metals were taken; [e]other value (in mg/l, for micro-ZnO): 1 [56].

compared to the dissolved metals based on the currently available data. For CNT, we cannot make this comparison due to a missing larger particle.

Preventive measures

In the prevention of a major accident, a special emphasis should be placed on safety measures. This is especially important with downstream users. Whereas high safety measures are normally standard at production sites, especially in the chemical industry, the handling of ENM in manufacturing of the final products is much less controlled and can result in a much higher possibility for release [7,72,73]. An important step in this context is the employee training. The safe handling of ENM does not require more action than needed for the handling of powders and conventional chemicals.

Production and manufacturing sites have to comply with existing high safety standards, which are determined by the chemicals (e.g., solvents) that are used during the process. Special emphasis has to be placed on those companies that have no experience in handling conventional chemicals but were founded as pure 'nano'-company. A critical issue is the fact that downstream users of ENM do

Table 3 LC$_{50}$ values (mg/l) for fish for different ENM: comparison between nano-form and microform/dissolved metals

	LC$_{50}$ fish		
	ENM lowest reported value (mg/l)	Larger particles or dissolved ions[e]	Rounded ratio (nano/micro)
Nano-TiO$_2$	2 [55][a]	>100 [64]	>1:10
Nano-ZnO	1.8 [65][b]	1 [58][f]	1:1
CNT	20 [66][c]	-	-
Nano-Ag	0.025 [67][d]	0.003 [59]	10:1

[a]Other values: 125 mg/l [68], 500 mg/l [69]; [b]other value: 4.92 mg/l [68]; [c]LOEC value, no LC$_{50}$ available; [d]other values (in mg/l): 0.028 [32], 1.25 to 1.36/9.4 to 10.6 [70], 7.07 [71]; [e]for Ag and ZnO, the values for the dissolved metals were taken; [f]average 30-day EC$_{50}$ for dissolved Zn (range 0.3 to 1.9 mg/l).

not have other information than those given in the MSDS, because these contain actually no nano-specific descriptions. However, these companies are likely to store only small amounts of ENM on their site - due to the high reactivity of the materials and the normally low concentrations used in final products - thus, the relevance for major accidents is seldom given.

Constructional measures

Constructional measures are indispensable for a safe handling of ENM. However, the established safety procedures used in the chemical industry are deemed to be sufficient. The procedural methods should be distinguished according to the specific form of the ENM. During production, manufacturing, and storage of suspended ENM, a detention basin is needed. The rooms should also not have any direct connection to the sewer system, or the connection needs to be equipped with a possibility for closure during an accident. For ENM in powder form, the ventilation and the configuration of the building envelope are central because they determine if ENM are released through damaged windows/ceilings or through ventilation into the environment. In both cases, fire prevention measures such as fire doors, separate storage rooms for organic solvents, and separate fire compartments are key.

Technical measures

Various technical measures can prevent or restrict a major accident. These include sprinklers in storage rooms, pressure-controlled equipment, and disconnection of ventilation in case of accident. However, these tools are not nano-specific but target the accident prevention of easily flammable compounds, which are stored in the same room. If these conventional measures are adopted consistently, they are also effective for ENM.

Organizational measures

Simple but effective organizational measures are access restrictions and sound employee trainings. All employees working with ENM should get an appropriate training and should be able to have access to personal protective equipment. The plant fire brigade or the local fire brigade should be informed about the presence of ENM and should be trained in suitable firefighting procedures.

First insights of nanomaterials in major accident prevention

The thresholds and criteria in the ordinance on major accident prevention [1] are based on the mass of the compound. Due to the small diameter of ENM, it is questioned if the approach of using a mass basis is adequate for ENM or if the particle number or specific surface area should to be taken instead. However, we consider a mass-based approach for ENM within major accident prevention regulation reasonable because data from (eco)toxicological studies are mainly mass-based. However, indispensable is in any case a clear definition of what an ENM and the proposition for a definition by the EU [74] clearly also influences the major accident regulation as this definition also covers natural and incidental nanomaterials and not only ENM [75,76].

Solubility plays a central role for the assessment of the toxicity of ENM in comparison to microparticles. Nanoparticles of easily soluble materials such as ZnO or metals that can be oxidized and then release ions (e.g., Ag) dissolve, due to their small diameter and the corresponding high surface area, much faster than larger particles. Their toxicity should therefore be compared to that of the corresponding metal ions, e.g., Zn^{2+} and Ag^+. The most important difference is that ENM can enter cells by additional pathways including passive entering into the cells comparable to the Trojan Horse mechanism. However, as shown in our evaluation of ecotoxicological data, the nano-form has overall not a higher toxicity than the dissolved form and thus, within the context of major accident prevention, the total mass of a compound, irrespective of its form, can be used.

In order to guarantee that all ENM are included in the existing major accident regulation, different options are possible (Table 4). In every case, the declaration of ENM on the MSDS is compulsory because without this information downstream users cannot assess if a company has reached a nano-specific threshold. It is also problematic that in certain intermediate products, the nano-content is so low that it does not need to be specified in the list of ingredients.

In our opinion, option 2 is the best option for the following reasons:

- Classification 4 in Table 2 includes all ENM that will be produced in the future. The exemption list given

in the Major Accident Ordinance does not need to be adapted continuously, and there is no danger that new, potentially toxic ENM are regulated by a too generic regulation not strict enough. The precautionary principle is applied.
- The classification is specific for each ENM. An over- or under-regulation is avoided and the specific properties of each ENM are considered. The importance of substance specificity is evident from the following considerations.

Despite a wide breadth of published studies, there are still many significant questions unanswered. The substance-specific factors that affect the toxicity of ENM have not yet been answered in a coherent manner [77]. In addition to the chemical composition, the following properties have been mentioned [78]: primary and secondary particle size; specific surface area; impurities or doting; surface properties (zeta potential, functionalization, coating); redox potential, reactivity; particle form; crystallinity; hydrophobicity/hydrophilicity; solubility; biopersistency; age of particles.

In order to make any conclusions, all these factors need to be characterized and controlled in the studies. Additionally, particle-specific factors such as particle size distribution play a role. These considerations make clear that the terms 'CNT' and 'nano-TiO_2' stand for a whole group of materials which can have very different properties, mainly due to shape, surface functionalization, or doting with other elements. It is, for example, assured that the toxicity of nano-TiO_2 varies tremendously with changing mineral structure or surface coatings [77,79].

For practical reasons, it will be difficult to perform (eco)toxicity tests for each produced or imported ENM. Producers need to be obligated to include the relevant information about toxicity and ecotoxicity of their product in the MSDS.

Based on our evaluation of the ENM, there is no special need for a drastic change in the classification of the

Table 4 Overview on possible classifications for considering ENM in major accident regulations

	Rule	Advantage	Disadvantage
1	All ENM have the same (eco)toxicity as larger particles with the same composition. Some specific ENM are placed in the exemption list	Simple rule; recognizes the higher risk potential of certain ENM	Insufficient data to select exemptions
2	Each manufacturer has to provide specific (eco)toxicological tests for each ENM, independent of the chemical composition of the ENM	The nano-aspect as well as the different properties of various ENM are considered	Problem of definition of ENM; expensive for companies that produce only small amounts or small variations of materials
3	Dispersions are classified like normal chemicals; for powders, a nano-specific regulation is implemented in addition to existing regulations for powders	The different risk of dispersions and powders is accommodated. Simple to define	Possible exemptions need to be formulated (see 1)
4	For ENM in powder form, the mass threshold is reduced by a factor of 10 to accommodate the higher surface area	The precautionary principle and the increased surface area of the nanoparticles are considered	Relatively unspecific
5	ENM are categorized (e.g., soluble/insoluble, metal oxides/metals/organic ENM) and assessed differently	Differences between particle types are considered	Material-specific properties are over-represented; many exemptions

materials, no matter if nano or not. According to the guidelines of the Major Accident Ordinance [80], currently, the following threshold values are valid for potentially nano-scaled compounds:

- 2,000 kg: ZnO, $AgNO_3$
- No threshold: TiO_2, SiO_2
- Not on the list: CNT, CeO_2, carbon black, $CaCO_3$, metallic silver
- Case-specific evaluation needed: iron oxides, pigments

For CNT and all ENM for which their conventional counterpart is not on the list, we suggest that a case-specific classification has to be performed based on the criteria of the regulation [1]. The implementation of option 2 that is favored by us would result in a threshold limit of 2,000 kg for most nano-Ag and nano-ZnO compounds, corresponding to the currently effective limits for conventional Ag and ZnO. For CNT and nano-TiO_2, different threshold limits according to the specific properties of the ENM would come into force. Photocatalytic nano-TiO_2 would get a threshold of 2,000 kg due to the increased toxicity against *Daphnia*. No threshold value would apply for non-photocatalytic nano-TiO_2. For CNT, different thresholds would apply according to the length and stiffness of the fibers. Option 2 thus allows a differentiated regulation of all ENM, under the precondition that a clear definition for ENM and the duty for declaration in the MSDS exists.

Conclusions

As a general conclusion, we can state that ENM are clearly less hazardous than many other chemicals, e.g., solvents or high-reactivity compounds such as certain pharmaceuticals, and can be treated similar to dusts or pigments. However, it is extremely important to note that this conclusion is only valid in relation to major accidents and not for occupational and environmental health or product safety, i.e., the exposure of workers inside the factory premises or of consumers due to products containing ENM. It also does not consider long-term environmental effects due to release during the use of the products. These issues are covered in separate regulations that are not part of our evaluation.

Due to the limited fundamental understanding of ENM fate and effects, our conclusions need to be taken with caution. Until standardized tests for the determination of ENM toxicity are available and the declaration of ENM on the MSDS is standard, it is recommended to perform at regular intervals a monitoring regarding new 'high production volume ENM' and to check if the statements made in this work are still valid. Nevertheless, the multitude of different studies that exist for the ENM considered

in this work and that report relevant results show in general no evidence for a specific need for action. The acute risks of ENM within a major accident are not significantly different from conventional compounds, implying that the current regulation for major accidents is also able to cover ENM.

Competing interests
The authors declare that they have no competing interests.

Authors' contributions
All authors contributed equally to this work. All authors read and approved the final manuscript.

Acknowledgements
This study was financed by the Swiss Federal Office for the Environment. An extended report is available at www.bafu.admin.ch/uw-1301-e. Data from the DaNa website have been used for the evaluation (BMBF; project DaNa No 03X0075A; www.nanoobjects.info).

References
1. CH: *Verordnung über den Schutz vor Störfällen, 27. Februar 1991*. Bern: Swiss Federal Council; 2008.
2. Steinkrauss M, Fierz H, Lerena P, Suter G: *Brand - und Explosionseigenschaften synthetischer Nanomaterialien - Erste Erkenntnisse für die Störfallvorsorge*. Bern: Herausgegeben vom Bundesamt für Umwelt (BAFU); 2010.
3. Holbrow P, Wall M, Sanderson E, Bennett D, Rattigan W, Bettis R, Gregory D: *Fire and Explosion Properties of Nanopowders. RR782 Research Report*. Buxton: Health and Safety Laboratory; 2010.
4. European Commission: **Mitteilung der Kommission an das europäische Parlament, den Rat und den europäischen Wirtschafts- und Sozialausschuss, Regelungsaspekte bei Nanomaterialien.** [http://eur-lex.europa.eu/Notice.do?mode=dbl&lang=en&ihmlang=en&lng1=en,de&lng2=bg,cs,da,de,el,en,es,et,fi,fr,hu,it,lt,lv,mt,nl,pl,pt,ro,sk,sl,sv&val=472573:cs&page=]
5. SRU: *Vorsorgestrategien für Nanomaterialien*. Berlin: Sondergutachten des Sachverständigenrats für Umweltfragen; 2011.
6. Kittel G: *Leitfaden für das Risikomanagement beim Umgang mit Nanomaterialien am Arbeitsplatz, November 2010*. Vienna: Bundesministerium für Arbeit, Soziales und Konsumentenschutz (Wien) im Rahmen des Österreichische Arbeitsschutzstragegie 2007–2012; 2010.
7. Pistner C: **Eco@Work.** [http://www.oeko.de/103/wissen2]
8. UK: **Draft Directive on the control of major accident hazards involving dangerous substances, 14 December 2011.** [http://www.publications.parliament.uk/pa/cm201012/cmselect/428-xliiv/42808.htm]
9. BSI: *Nanotechnologies – Part 2: Guide to Safe Handling and Disposal of Manufactured Nanomaterials*. London: British Standards Institute; 2007.
10. OECD: *Organisation for Economic Co-operation and Development: Environment Directorate Chemicals Committee. Risk of Major Accidents Involving Nanomaterials - Prevention of, Preparedness for and Response to Accidents*. Paris: Organisation for Economic Co-operation and Development; 2011.
11. Piccinno F, Gottschalk F, Seeger S, Nowack B: **Industrial production quantities and uses of ten engineered nanomaterials for Europe and the world.** *J Nanopart Res* 2012, 14:1109.
12. Kuhlbusch TAJ, Asbach C, Fissan H, Gohler D, Stintz M: **Nanoparticle exposure at nanotechnology workplaces: a review.** *Part Fibre Toxicol* 2011, 8:22.
13. Fendler R, Kleiber M, Watorowski J: **Jahresbericht 2008; Zentrale Melde- und Auswertestelle für Störfälle und Störungen in verfahrenstechnischen Anlagen (ZEMA), Umweltbundesamt Dessau.** [http://www.uba.de/uba-info-medien/4130.html]
14. Krausmann E, Cozzani V, Salzano E, Renni E: **Industrial accidents triggered by natural hazards: an emerging risk issue.** *Nat Hazards Earth System Sci* 2011, 11:921–929.
15. Krausmann E, Renni E, Campedel M, Cozzani V: **Industrial accidents triggered by earthquakes, floods and lightning: lessons learned from a database analysis.** *Nat Hazards* 2011, 59:285–300.

16. Raab C, Simkó M, Fiedeler U, Nentwich M, Gazsó A: **Herstellungsverfahren von Nanopartikeln und Nanomaterialien.** *Nano Trust - Dossiers* 2008, **6**. doi:10.1553/ITA-nt-006.

17. Keller A, McFerran S, Lazareva A, Suh S: **Global life cycle releases of engineered nanomaterials.** *J Nanoparticle Res* 2013, **15**:1–17.

18. L'Alsace [http://www.lalsace.fr/actualite/2011/10/10/vieux-thann-des-sacs-d-oxyde-detitane-tombent-d-un-camion-sur- la-rn66]

19. Pan B, Xing BS: **Applications and implications of manufactured nanoparticles in soils: a review.** *Eur J Soil Sci* 2012, **63**:437–456.

20. Johnston JM, Lowry M, Beaulieu S, Bowles E: *State-of-the-Science Report on Predictive Models and Modeling Approaches for Characterizing and Evaluating Exposure to Nanomaterials. EPA/600/R-10/129.* Athens: U.S. Environmental Protection Agency, Office of Research and Development; 2010.

21. Gottschalk F, Sonderer T, Scholz RW, Nowack B: **Modeled environmental concentrations of engineered nanomaterials (TiO₂, ZnO, Ag, CNT, fullerenes) for different regions.** *Environ Sci Technol* 2009, **43**:9216–9222.

22. Mueller NC, Nowack B: **Exposure modeling of engineered nanoparticles in the environment.** *Environ Sci Technol* 2008, **42**:4447–4453.

23. Seipenbusch M, Binder A, Kasper G: **Temporal evolution of nanoparticle aerosols in workplace exposure.** *Ann Occup Hyg* 2008, **52**:707–716.

24. Demou E, Peter P, Hellweg S: **Exposure to manufactured nanostructured particles in an industrial pilot plant.** *Ann Occup Hyg* 2008, **52**:695–706.

25. Walser T, Hellweg S, Juraske R, Luechinger NA, Wang J, Fierz M: **Exposure to engineered nanoparticles: model and measurements for accident situations in laboratories.** *Sci Total Environ* 2012, **420**:119–126.

26. Buseck PR, Adachi K: **Nanoparticles in the atmosphere.** *Elements* 2008, **4**:389–394.

27. OECD: *Joint Meeting of the Chemicals Committee and the Working Party on Chemicals, Pesticides and Biotechnology - Report of the WPMN Expert Meeting on Inhalation Toxicity Testing for Nanomaterials.* Paris: Organisation for Economic Co-operation and Development; 2012.

28. ECHA: *Guidance on Information Requirements and Chemical Safety Assessment.* Helsinki: European Chemicals Agency; 2008.

29. Westerhoff PK, Kiser A, Hristovski K: **Nanomaterial removal and transformation during biological wastewater treatment.** *Environ Eng Sci* 2013, **30**:109–117.

30. Mu H, Zheng X, Chen YG, Chen H, Liu K: **Response of anaerobic granular sludge to a shock load of zinc oxide nanoparticles during biological wastewater treatment.** *Environ Sci Technol* 2012, **46**:5997–6003.

31. BUWAL: *Beurteilungskriterien I zur Störfallverordnung StFV, Richtlinien für Betriebe mit Stoffen, Erzeugnissen oder Sonderabfällen.* Bern: Swiss Federal Office for the Environment; 1996.

32. Kim J, Kim S, Lee S: **Differentiation of the toxicities of silver nanoparticles and silver ions to the Japanese medaka** *(Oryzias latipes)* **and the cladoceran** *Daphnia magna. Nanotoxicology* 2011, **5**:208–214.

33. Grieger KD, Fjordboge A, Hartmann NB, Eriksson E, Bjerg PL, Baun A: **Environmental benefits and risks of zero-valent iron nanoparticles (nZVI) for in situ remediation: risk mitigation or trade-off?** *J Contam Hydrol* 2010, **118**:165–183.

34. NanoCare: *Health Related Aspects of Nanomaterials. Final Scientific Report.* Frankfurt am Main: DECHEMA; 2009.

35. Tang M, Zhang T, Xue Y, Wang S, Huang M, Yang Y, Lu M, Lei H, Kong L, Yuepu P: **Dose dependent in vivo metabolic characteristics of titanium dioxide nanoparticles.** *J Nanosci Nanotechnol* 2010, **10**:8575–8583.

36. Kobayashi N, Naya M, Endoh S, Maru J, Yamamoto K, Nakanishi J: **Comparative pulmonary toxicity study of nano-TiO₂ particles of different sizes and agglomerations in rats: different short- and long-term post-instillation results.** *Toxicology* 2009, **264**:110–118.

37. Park E-J, Yoon J, Choi K, Yi J, Park K: **Induction of chronic inflammation in mice treated with titanium dioxide nanoparticles by intratracheal instillation.** *Toxicology* 2009, **260**:37–46.

38. Li J, Li Q, Xu J, Li J, Cai X, Liu R, Li Y, Ma J, Li W: **Comparative study on the acute pulmonary toxicity induced by 3 and 20 nm TiO₂ primary particles in mice.** *Environ Toxicol Pharmacol* 2007, **24**:239–244.

39. Ma-Hock L, Burkhardt S, Strauss V, Gamer AO, Wiench K, van Ravenzwaay B, Landsiedel R: **Development of a short-term inhalation test in the rat using nano-titanium dioxide as a model substance.** *Inhal Toxicol* 2009, **21**:102–118.

40. Sayes CM, Reed KL, Warheit DB: **Assessing toxicity of fine and nanoparticles: comparing in vitro measurements to in vivo pulmonary toxicity profiles.** *Toxicol Sci* 2007, **97**:163–180.

41. Warheit DB, Sayes CM, Reed KL: **Nanoscale and fine zinc oxide particles: can in vitro assays accurately forecast lung hazards following inhalation exposures?** *Environ Sci Technol* 2009, **43**:7939–7945.

42. Donaldson K, Murphy F, Duffin R, Poland C: **Asbestos, carbon nanotubes and the pleural mesothelium: a review of the hypothesis regarding the role of long fibre retention in the parietal pleura, inflammation and mesothelioma.** *Part Fibre Toxicol* 2010, **7**:5.

43. Murphy F, Schinwald A, Poland C, Donaldson K: **The mechanism of pleural inflammation by long carbon nanotubes: interaction of long fibres with macrophages stimulates them to amplify pro-inflammatory responses in mesothelial cells.** *Part Fibre Toxicol* 2012, **9**:8.

44. Bianco A, Kostarelos K, Prato M: **Applications of carbon nanotubes in drug delivery.** *Curr Opin Chem Biol* 2005, **9**:674–679.

45. Deng X, Jia G, Wang H, Sun H, Wang X, Yang S, Wang T, Liu Y: **Translocation and fate of multi-walled carbon nanotubes in vivo.** *Carbon* 2007, **45**:1419–1424.

46. Pantarotto D, Singh R, McCarthy D, Erhardt M, Briand JP, Prato M, Kostarelos K, Bianco A: **Functionalized carbon nanotubes for plasmid DNA gene delivery.** *Angew Chem Int Ed* 2004, **43**:5242–5246.

47. Qu G, Bai Y, Zhang Y, Jia Q, Zhang W, Yan B: **The effect of multiwalled carbon nanotube agglomeration on their accumulation in and damage to organs in mice.** *Carbon* 2009, **47**:2060–2069.

48. Wang H, Wang J, Deng X, Sun H, Shi Z, Gu Z, Liu Y, Yuliang Z: **Biodistribution of carbon single-wall carbon nanotubes in mice.** *J Nanosci Nanotechnol* 2004, **4**:1019–1024.

49. Yang ST, Guo W, Lin Y, Deng XY, Wang HF, Sun HF, Liu YF, Wang X, Wang W, Chen M, Huang YP, Sun YP: **Biodistribution of pristine single-walled carbon nanotubes in vivo.** *J Phys Chem C* 2007, **111**:17761–17764.

50. Wijnhoven SWP, Peijnenburg WJGM, Herberts CA, Hagens WI, Oomen AG, Heugens EHW, Roszek B, Bisschops J, Gosens I, Van De Meent D, Dekkers S, De Jong WH, van Zijverden M, Sips AJAM, Geertsma RE: **Nano-silver - a review of available data and knowledge gaps in human and environmental risk assessment.** *Nanotoxicology* 2009, **3**:109–138.

51. Kim YS, Kim JS, Cho HS, Rha DS, Kim JM, Park JD, Choi BS, Lim R, Chang HK, Chung YH, Kwon IH, Jeong J, Han BS, Yu IJ: **Twenty-eight-day oral toxicity, genotoxicity, and gender-related tissue distribution of silver nanoparticles in Sprague–Dawley rats.** *Inhal Toxicol* 2008, **20**:575–583.

52. Ji JH, Jung JH, Kim SS, Yoon J-U, Park JD, Choi BS, Chung YH, Kwon IH, Jeong J, Han BS, Shin JH, Sung JH, Song KS, Yu IJ: **Twenty-eight-day inhalation toxicity study of silver nanoparticles in Sprague–Dawley rats.** *Inhal Toxicol* 2007, **19**:857–871.

53. NanoTrust: **Dossier No10: Nanosilber.** [http://nanotrust.ac.at/dossiers.html]

54. Möller M, Eberle U, Hermann A, Moch K, Stratmann B: *Nanotechnologie im Bereich der Lebensmittel.* Bern: TA-SWISS (hrsg) - Zentrum für Technologiefolgen-Abschätzung; 2009.

55. Ma HB, Brennan A, Diamond SA: **Phototoxicity of TiO₂ nanoparticles under solar radiation to two aquatic species:** *Daphnia magna* **and Japanese medaka.** *Environ Toxicol Chem* 2012, **31**:1621–1629.

56. Wiench K, Wohlleben W, Hisgen V, Radke K, Salinas E, Zok S, Landsiedel R: **Acute and chronic effects of nano- and non-nano-scale TiO2 and ZnO particles on mobility and reproduction of the freshwater invertebrate** *Daphnia magna. Chemosphere* 2009, **76**:1356–1365.

57. Zhu XS, Zhu L, Chen YS, Tian SY: **Acute toxicities of six manufactured nanomaterial suspensions to** *Daphnia magna. J Nanopart Res* 2009, **11**:67–75.

58. De Schamphelaere KAC, Lofts S, Janssen CR: **Bioavailability models for predicting acute and chronic toxicity of zinc to algae, daphnids, and fish in natural surface waters.** *Environ Toxicol Chem* 2005, **24**:1190–1197.

59. Ratte HT: **Bioaccumulation and toxicity of silver compounds: a review.** *Environ Toxicol Chem* 1999, **18**:89–108.

60. Amiano I, Olabarrieta J, Vitorica J, Zorita S: **Acute toxicity of nanosized TiO₂ to** *Daphnia magna* **under UVA irradiation.** *Environ Toxicol Chem* 2012, **31**:2564–2566.

61. Dabrunz A, Duester L, Prasse C, Seitz F, Rosenfeldt R, Schilde C, Schaumann GE, Schulz R: **Biological surface coating and molting inhibition as mechanisms of TiO(2) nanoparticle toxicity in** *Daphnia magna. PLoS One* 2011, **6**. doi:10.1371/journal.pone.0020112.

62. Blinova I, Ivask A, Heinlaan M, Mortimer M, Kahru A: **Ecotoxicity of nanoparticles of CuO and ZnO in natural water.** *Environ Pollut* 2010, **158**:41–47.

63. Kennedy AJ, Hull MS, Steevens JA, Dontsova KM, Chappell MA, Gunter JC, Weiss CA: **Factors influencing the partitioning and toxicity of nanotubes in the aquatic environment.** *Environ Toxicol Chem* 2008, **27**:1932–1941.

64. Kronos: **Material safety data sheet for KRONOS titanium dioxide (all types).** [http://kronostio2.com/en/data-sheets-and-literature/safety-datasheets/finish/93/338]

65. Zhu XS, Zhu L, Duan ZH, Qi RQ, Li Y, Lang YP: Comparative toxicity of several metal oxide nanoparticle aqueous suspensions to Zebrafish (Danio rerio) early developmental stage. J Environ Sci Health Part Toxic/ Hazard Subst Environ Eng 2008, 43:278–284.

66. Cheng JP, Flahaut E, Cheng SH: Effect of carbon nanotubes on developing zebrafish (Danio rerio) embryos. Environ Toxicol Chem 2007, 26:708–716.

67. Asharani PV, Wu YL, Gong ZY, Valiyaveettil S: Toxicity of silver nanoparticles in zebrafish models. Nanotechnology 2008, 19:255102.

68. Xiong DW, Fang T, Yu LP, Sima XF, Zhu WT: Effects of nano-scale TiO(2), ZnO and their bulk counterparts on zebrafish: acute toxicity, oxidative stress and oxidative damage. Sci Total Environ 2011, 409:1444–1452.

69. Hall S, Bradley T, Moore JT, Kuykindall T, Minella L: Acute and chronic toxicity of nano-scale TiO(2) particles to freshwater fish, cladocerans, and green algae, and effects of organic and inorganic substrate on TiO(2) toxicity. Nanotoxicology 2009, 3:91–97.

70. Laban G, Nies L, Turco R, Bickham J, Sepúlveda M: The effects of silver nanoparticles on fathead minnow (Pimephales promelas) embryos. Ecotoxicology 2010, 19:185–195.

71. Griffitt RJ, Luo J, Gao J, Bonzongo JC, Barber DS: Effects of particle composition and species on toxicity of metallic nanomaterials in aquatic organisms. Environ Toxicol Chem 2008, 27:1972–1978.

72. Som C, Berges M, Chaudhry Q, Dusinska M, Fernandes TF, Olsen SI, Nowack B: The importance of life cycle concepts for the development of safe nanoproducts. Toxicology 2010, 269:160–169.

73. Kuhlbusch T, Nickel C: Emission von Nanopartikeln aus ausgewählten Produkten in ihrem Lebenszyklus, Umweltbundesamt Dessau. [http://www.uba.de/uba-info-medien/4028.html]

74. EU: Commission Recommendation of 18 October 2011 on the Definition of Nanomaterial (2011/696/EU). O. J. L 275: 38–40. Brussels: European Union; 2011.

75. Lövestam G, Rauscher H, Roebben G, Sokull Klüttgen B, Gibson N, Putaud JP, Stamm H: Considerations on a Definition of Nanomaterial for Regulatory Purposes. Luxembourg: Publications Office of the European Union; 2010. doi:10.2788/98686.

76. Bleeker EAJ, Cassee FR, Geertsma RE, Jong WH, Heugens EHW, Koers Jacquemijns M, Meent D, Oomen AG, Popma J, Rietveld AG, Wijnhoven SWP: Interpretation and Implications of the European Commission Recommendation on the Definition of Nanomaterial, RIVM Letter Report 601358001/2012. Bilthoven: National Institute for Public Health and the Environment; 2012.

77. Menard A, Drobne D, Jemec A: Ecotoxicity of nanosized TiO(2). Review of in vivo data. Environ Pollut 2011, 159:677–684.

78. Ostiguy C, Roberge B, Ménard L, Endo CA: Best practices guide to synthetic nanoparticle risk management; Institut de recherche Robert-Sauvé en santé et en sécurité du travail (IRSST), Commission de la santé et de la sécurité du travail du Québec (CSST) and NanoQuébec; REPORT R-599, Montréal. [http://www.irsst.qc.ca]

79. Auffan M, Pedeutour M, Rose J, Masion A, Ziarelli F, Borschneck D, Chaneac C, Botta C, Chaurand P, Labille J, Bottero JY: Structural degradation at the surface of a TiO$_2$-based nanomaterial used in cosmetics. Environ Sci Technol 2010, 44:2689–2694.

80. BAFU BfU: Mengenschwellen gemäss Störfallverordnung (StFV) - Liste mit Stoffen und Zubereitungen. Bern: Swiss Federal Office for the Environment; 2006.

3

Proposal for a harmonised PBT identification across different regulatory frameworks

Caren Rauert[1]*, Anton Friesen[2], Georgia Hermann[3], Ulrich Jöhncke[4], Anja Kehrer[2], Michael Neumann[4], Ines Prutz[5], Jens Schönfeld[5], Astrid Wiemann[5], Karen Willhaus[2], Janina Wöltjen[3] and Sabine Duquesne[3]

Abstract

European regulatory frameworks for chemicals (i.e. registered under REACH, plant protection products (PPPs), biocides, human and veterinary medicinal products) require that substances undergo an assessment to identify whether they are persistent (P), bioaccumulative (B) and toxic (T), or very persistent (vP) and very bioaccumulative (vB), i.e. to identify them as PBT substances or vPvB substances according to their properties. We screened current practices, evaluated possibilities and made a proposal for a harmonised assessment. Our proposal assumes that it should be possible to identify PBT and vPvB substances on the basis of the data available according to the requirements of the respective legal frameworks. For substances registered as PPPs and mostly also biocides and medicinal products, a 'definitive assessment' is often possible. Within REACH, the registrant has to provide all information necessary for PBT assessment regardless of the yearly tonnage of chemicals. But in cases of limited data availability, we suggest using a weight of evidence approach to account for such differences in data availability and type of data across different frameworks and to make use of valuable additional information. We propose to base the evaluation of persistence on degradation half-lives and to normalise a number of parameters (e.g. type of kinetics used, temperature). But further work is needed, e.g. for deriving $DegT_{50}$ for water and sediment compartments. For the B-criterion, information other than BCF in fish could be considered and more information related to bioaccumulation processes should be gathered (e.g. in species other than fish, different uptake routes). Testing for T identification is focused on standard aquatic species but could also be complemented by e.g. information from other species. Information such as those read-across from structurally related substances and QSAR are often of importance for screening assessments. The aim of PBT and vPvB identification is to reliably target the problematic substances, with as few false negatives or positives as possible, regardless of the regulatory framework. Each aspect was thus considered in the context of the others for a final balanced decision. As the need for conservatism is interpreted differently under the various frameworks, harmonizing this identification is a challenging task.

Keywords: PBT identification; Persistence; Bioaccumulation; Toxicity; REACH; Plant protection products; Biocides; Human medicinal products; Veterinary medicinal products; Harmonisation; Regulatory framework

Background

The different European substance regulations (e.g. REACH [1], plant protection products regulation [2], biocidal products regulation [3], human and veterinary medicinal products regulations [4,5]) have all recognised that substances that are either persistent (P), bioaccumulative (B) and toxic (T) (PBT substances) or very persistent (vP) and very bioaccumulative (vB) (vPvB substances) must be

considered as hazardous for the environment due to their potential for eliciting long-term adverse effects. The goal of preventing exposure of humans and the environment to PBT and vPvB substances is thus shared among all EU-based regulatory frameworks. In a comparison between different European and International regulations, Moermond et al. [6] reported that there are differences in how this goal is achieved, not only regarding technical criteria but also conceptual criteria (e.g. regulatory consequences for PBT and vPvB substances).

In terms of numerical criteria, the identification of PBT and vPvB substances is based on substance properties

* Correspondence: caren.rauert@uba.de

[1]Section International Chemicals Management, Federal Environment Agency (UBA), Wörlitzer Platz 1, 06844 Dessau-Roßlau, Germany

Full list of author information is available at the end of the article

which are addressed by the same trigger values in all European substance regulations (see Table 1).

Therefore, it could be assumed that the identification of a PBT or vPvB substance should be independent of both the use of the substance and the regulatory framework under which it is assessed. However, the decision on whether a substance fulfils the PBT or vPvB criteria not only depends on substance properties, but also on the framework under which the substance is evaluated. For substances whose properties are near the trigger values, this decision may differ due to differences between the different legal frameworks in

(i) The assessment procedures,
(ii) The interpretation of PBT/vPvB criteria,
(iii) Available data within the assessment process (see next subsection), and
(iv) The regulatory consequences

Although based on the same principle (i.e. avoiding emissions of potentially harmful substances to the environment), the mandatory measures imposed by regulation for a substance identified as a PBT or vPvB vary between the different regulatory frameworks:

– For substances registered under REACH, the revised Annex XIII [7] not only gives the criteria but also the information relevant for the screening and assessment of P, vP, B, vB and T properties. Also, ECHA provides extensive guidance on PBT/vPvB assessment [8,9]. Substances of very high concern (SVHC) (e.g. PBT/vPvB) are included in the 'candidate list' for authorisation, based on the outcome of the scientific assessment of intrinsic properties. Substances on the 'candidate list' can then be prioritised for authorisation and finally listed in Annex XIV of the REACH regulation. Some substances may be listed with specific exemptions that do not require authorisation. For each

substance included in Annex XIV, a deadline will be set after which use of that substance in the EU must stop (known as the 'sunset date'), unless authorized by the European Commission.

– The EU Biocidal Products Regulation [3] refers to Annex XIII of REACH [7] and to the ECHA guidance for PBT/vPvB assessment [8,9]. However, for biocides, generally no authorisation will be granted for products containing substances identified as PBT or vPvB. Additionally, biocides that fulfil two of the three criteria are flagged as candidates for substitution and are subjected to a comparative assessment.

– For the assessment of veterinary medicinal products, guidance has been developed by the European Medicines Agency [10], which also refers to ECHA guidance [8,9]. A benefit-risk analysis is conducted in support of the decision on whether to authorise the substance or not [5]. In this context, an identification as a PBT substance is generally regarded as a serious concern.

– For medicinal products for human use (which are usually emitted *via* sewage treatment plants), no restrictions are envisaged as human health is prioritised over environmental issues [4]. ECHA guidance [8,9] is used for the assessment.

– By contrast, the PPP Regulation [2] simply stipulates the PBT/vPvB criteria, oblivious to any other guidance. As for biocides, no authorisation will be granted for those substances identified as PBT or vPvB. Additionally, PPPs that fulfil two of the three criteria are flagged as candidates for substitution and are subjected to a comparative assessment, as is the case for biocides. The draft guidance developed by DG Sanco [11] focuses primarily on the identification of candidates for substitution.

Although the substances differ in their properties, uses, exposure pathways and regulatory consequences,

Table 1 PBT and vPvB criteria across the various European substance regulations

Criterion	PBT identification	vPvB identification
Persistence	• Half-life (degradation) > 60 days in marine water, or	• Half-life (degradation) > 60 days in marine, fresh or estuarine water, or
	• Half-life (degradation) > 40 days in fresh or estuarine water, or	
	• Half-life (degradation) > 180 days in marine sediment, or	• Half-life (degradation) > 180 days in marine, fresh or estuarine sediment, or
	• Half-life (degradation) > 120 days in fresh or estuarine sediment, or	
	• Half-life (degradation) > 120 days in soil	• Half-life (degradation) > 180 days in soil
Bioaccumulation	• BCF > 2,000 L/kg	• BCF > 5,000 L/kg
Toxicity	• NOEC (long-term) < 0.01 mg/L for marine or freshwater organisms, or	• Not applicable
	• Classification as carcinogenic (category 1 or 2), mutagenic (category 1 or 2), or toxic for reproduction (category 1 or 2), or	
	• Other evidence of chronic toxicity, as identified by the classifications STOT RE 1 or STOT RE 2 pursuant to Regulation (EC) No 1272/2008	

the outcome of a PBT/vPvB assessment should not depend on the framework under which they are evaluated, since the protection goals do not differ. This will avoid situations where a substance with different uses is identified as a PBT substance under one regulation, and not under another. This is relevant because, for example, many biocides are also used as PPPs or veterinary medicinal products. Since regulatory authorities should aim at making predictable and consistent decisions, a strategy for harmonised PBT identification across the different frameworks is needed. The focus of this paper is on the assessment procedure; it proposes a reasonable interpretation of the available data while considering not only substance properties but also keeping in mind the regulatory consequences.

Dealing with differences in data requirements and data availability between regulatory frameworks

For substances registered as plant protection products (PPPs), as well as most biocides and medicinal products, requirements for environmental risk assessment usually include sufficient information for PBT/vPvB identification. Therefore, a 'definitive assessment' based on a direct comparison to the trigger values (see Table 1) of the three criteria should be possible in most cases.

For chemicals registered under REACH, data requirements depend on the quantities which are manufactured or imported per year; for high production volume (HPV) chemicals, more data has to be provided, and for substances below 10 t/a, only a basic data set is mandatory. Table 2 provides an overview of test requirements for persistence. The registrant also has to provide all information needed for PBT/vPvB assessment regardless of

the tonnage manufactured or imported annually. However, the necessary information is not always included in the dossiers so a direct comparison of measured data with the criteria is not always possible. Unfortunately, in many cases, registrants submitted dossiers which show deficiencies and do not comply with the requirements of REACH or the Technical Guidance Document, e.g. necessary tests are missing [12,13]. In such cases, only a 'screening assessment' can be performed. In such an assessment screening information such as that listed in Annex XIII of REACH [7] should be considered; this should be supplemented by data from non-standardised tests and from the literature, by information on relevant endpoints based on quantitative structure-activity relationships (QSARs) if adequately documented as well as by read-across from structurally related substances or grouping approaches. Only if the available information is considered unreliable or insufficient and/or does not permit a conclusion on each criterion to be drawn with sufficient confidence, additional testing will be necessary.

Because of the large differences in data requirements across the different frameworks and type of data (e.g. QSARs, non-standardised tests), we recommend considering all available information in a weight of evidence (WoE) approach - as suggested in REACH (Annex XI 1.2 of [1]). The WoE approach means that all available information is assessed through expert judgement for its suitability and reliability, and relative weights are assigned (e.g. quality of the data, consistency of results and/or data, nature and severity of effects, relevance of the information for the given regulatory endpoint). The use of this WoE approach depends on the amount of information needed, the importance of the decision to be taken, and

Table 2 Persistence testing requirements under the various European substance regulations

	Test on ready biodegradability (OECD 301)	Laboratory degradation simulation test in soil (OECD 307)	Laboratory degradation simulation test in a water-sediment system (OECD 308)	Laboratory degradation simulation test in water (OECD 309)	Field degradation test
PPPs	Required	Required in four soils if not readily biodegradable	Required in two test systems if not readily biodegradable	Required if not readily biodegradable	Required if $DegT_{50lab}$ >60 days (20°C, pF2)
Human medicinal products	Required	Required in four soils if not readily biodegradable and $Koc > 10,000$ L/kg	Required in two test systems if not readily biodegradable	Not required	Not required
Veterinary medicinal products	Not required	Required in four soils	Required in two test systems for aquaculture or if risk is identified for aquatic organism	Not required	Not required
Biocides	Required	Required in four soils if not readily biodegradable and/or depending on emission pathway	Required in two test systems if not readily biodegradable depending on substance properties and emission pathway	Required if not readily biodegradable depending on substance properties and emission pathway	Required if $PEC/PNEC_{soils}$ >1 and $DegT_{50lab}$ >60 days (20°C, pF2) or $DegT_{90lab}$ >200 days (20°C, pF2)
Chemicals (REACH)	Required for all substances > 10 t/a	One simulation study in the 'compartment of concern' required for all substances >100 t/a if not readily biodegradable			Not required

the likelihood and consequences of taking a wrong decision [14]. The WoE approach also has the advantage, such that it uses valuable additional information whose submission is not mandatory (e.g. field data, alternative testing strategies). This approach is proposed in REACH, and we suggest that it should also be used to improve PBT assessment under the other frameworks in cases where it could deliver valuable additional information.

Discussion

Persistence, bioaccumulation and toxicity criteria

In this chapter, we introduce the three criteria, i.e. persistence (P), bioaccumulation (B) and toxicity (T). The general scheme is to test sequentially for P, then B and lastly for T, as only substances with high persistence must be considered further for PBT identification. This sequence is further justified by the need to limit testing for animal welfare reasons.

A substance is considered a PBT or vPvB substance if it fulfills the P, B and T or the vP and vB criteria. In the following sections, for each PBT criterion, the criteria for the 'definitive assessment' are presented first. This is followed by the criteria for the 'screening assessment', which is carried out in cases of limited data availability.

Persistence

The persistence of a substance in soil, water or sediment is controlled by its intrinsic properties and also influenced by environmental conditions (e.g. temperature, presence of degraders, pH, moisture content in case of soil, bioavailability of the substance). The assessment of persistence is influenced by the type and quality of available studies and also by the interpretation of the available data.

Persistence is not characterised consistently in the different frameworks, but, as suggested by Boethling et al. [15] and in [9], it is the degradation half-life ($DegT_{50}$) which should be compared to the trigger values in the various compartments, as a mere transfer (dissipation) from one compartment to the other does not affect the persistence of a substance. $DegT_{50}$ values are commonly estimated from the parent substance in extractable residues in degradation simulation studies performed according to e.g. OECD test guidelines 307, 308 or 309 [16-18].

In the context of PBT identification, DT_{50} should not refer to dissipation half-lives ($DissT_{50}$) as these refer to cases where the substance may not be degraded but only redistributed (e.g. from the water phase to sediment or the atmosphere, or leaching, run-off, volatilisation or uptake into plants). This is explicitly mentioned in the revised Annex XIII of the REACH regulation [7], while the PPP Regulation [2] only refers to half-lives in general.

Definitive assessment. If degradation simulation tests have been conducted to meet legal requirements, a sufficient amount of reliable data is usually available to complete a definitive assessment by comparing the $DegT_{50}$ values directly to the criteria as reported in Table 1. However, a number of issues related to these values should be addressed.

Type of kinetics for deriving $DegT_{50}$. We suggest using single first-order (SFO) kinetics to derive $DegT_{50}$ values for the purpose of comparing them to the trigger values. Best fit kinetics such as first-order multi-compartment (FOMC) can be used if they are recalculated by dividing the $DegT_{90}$ values by a factor of 3.32 or by using the degradation rate constant of the slower phase in case of double first-order in parallel model (DFOP) or Hockey stick (HS) kinetics. This proposal is based on the Guidance Document on Estimating Persistence and Degradation Kinetics from Environmental Fate Studies on Pesticides in EU Registration [19]. This document gives detailed guidance on the derivation of $DegT_{50}$ values, which is used for risk assessment of biocides, medicinal products and PPPs and could be applied to other substance groups as well.

Normalisation of the $DegT_{50}$ to specific conditions. Variable conditions such as temperature, humidity, microbial populations and others have a large influence on the rate of degradation of a chemical substance. Simulation studies in the laboratory are conducted at various temperatures: OECD test guidelines 307 and 308 on transformation in soil and in water-sediment systems [16,17] recommend testing at 20°C and 10°C (if the substance is used in colder climates), the US EPA recommends 22°C for soil testing [20] and for many pesticides, tests have been performed both at 20°C and 10°C. For the purpose of comparing different data, half-lives should thus be normalised to a defined temperature since it is a criterion which strongly influences the outcome of a degradation test and, furthermore, is quantifiable unlike other related criteria such as differences in microorganism populations. For PBT identification, normalisation of $DegT_{50}$ values to a temperature of 12°C has been applied to many biocides and other chemicals, as prescribed in the TGD [21], and is included as an option in the REACH guidance on information requirements and chemical safety assessment (Chapter R.7b [8]). This temperature of 12°C is also included in the draft 'Guidance on the assessment of persistent, bioaccumulative and toxic (PBT) or very persistent and very bioaccumulative (vPvB) substances in veterinary medicine' [22]. By contrast, $DegT_{50}$ values for PPPs are normalised to 20°C, both for laboratory and field data for risk assessment; this temperature has also been proposed for PBT identification [11]. For a consistent approach between frameworks, we propose normalising the $DegT_{50}$ values to 12°C because this temperature is established or suggested under the majority of frameworks (i.e. Biocides Regulation, REACH and medicinal products Directives).

Degradation in soil: inclusion of field studies. Laboratory simulation studies can provide information on both route and rate of degradation. This permits the estimation of primary degradation rates ($DegT_{50}$), as well as the measurement of carbon dioxide evolution and of formation of metabolites and bound residues. But the small size of the test systems is a limiting factor for both the test duration and the diversity of the microbial population (which may limit the probability of having competent degraders and which usually also decreases over time).

For PPPs, terrestrial field (dissipation) studies under realistic outdoor conditions are required when the $DegT_{50}$ exceeds 60 days at 20°C in laboratory studies [2]. Such field studies allow DT_{50} values to be derived under conditions going beyond the limitations of laboratory tests. A number of conditions may be more realistic compared to laboratory studies (e.g., prolonged duration, fluctuating temperature and humidity, higher biological activity, larger test system), but may not be as easy to compare or reproduce. A limiting aspect in such testing is that a dissipation half-life, and not a degradation half-life, will be derived. The measured residues result from biotic degradation but also from photolytic transformation and field dissipation processes like volatilisation, leaching and run-off. However, a novel approach to estimate a $DegT_{50}$ in field studies has recently been proposed by EFSA [23] in which dissipation processes on the soil surface are estimated separately from biodegradation in the soil compartment. This $DegT_{50}$ has to be normalised for temperature and humidity for comparability reasons. Although this guidance was developed for the evaluation of persistence of PPPs in field studies, in principle, its rationale and approach should also be applicable to other types of substances. Thus, we suggest including field studies in persistence assessment as additional data, provided that it is possible to derive $DegT_{50}$ values. This is in line with the WoE approach, which considers all available information in the decision process.

Evaluation of $DegT_{50}$ values for the water and sediment compartments. There is no standardised test that measures true degradation only in the sediment compartment. The test on 'aerobic and anaerobic transformation in aquatic sediment systems' (OECD test guideline 308 [17]) is generally used to assess the fate of compounds in water and sediment systems, but reliable separate $DegT_{50}$ values for water and for sediment cannot usually be derived from the study results. Therefore, separate

$DegT_{50}$ values are generally not available for comparison with the trigger values.

Available DT_{50} values for the water phase in many cases only refer to dissipation, as many substances, especially those with high log K_{OC} and low water solubility, are quickly transferred to the sediment. The Aerobic Mineralisation in Surface Water – Simulation Biodegradation Test (OECD 309 [18]) is available to measure degradation in the water phase, but to date it has rarely been used for the assessment of environmental fate.

We suggest comparing the $DegT_{50}$ of the total system to the two trigger values for water and sediment. More specifically, in the case of a substance that is rapidly transferred into the sediment, the $DegT_{50}$ of the total system should be compared with the trigger value for sediment (120 and 180 days for persistence assessment). For substances that mostly remain in the water phase, the $DegT_{50}$ values should be compared with the trigger values for water (40 and 60 days for persistence assessment). There will be many cases where a substance can be found in both the water and sediment phases, which means that clear criteria are necessary to judge which of the trigger values is appropriate.

Selection of an adequate $DegT_{50}$ from multiple studies. As explained above, testing requirements for persistence differ greatly between regulatory frameworks. The information is summarized in Table 2.

If several valid degradation studies for one compartment are available, their quality and reliability are an important consideration. We propose the following approach to processing the information gathered:

– When a small number of studies is available (i.e. up to 4 $DegT_{50}$ values), we suggest selecting the worst-case $DegT_{50}$ rather than performing a weak statistical evaluation.
– When more studies are available (i.e. 5 and more $DegT_{50}$ values), various approaches are possible:
    The most conservative approach would be to use the worst-case $DegT_{50}$;
    A less conservative approach is the geometric mean, which gives a medium degradation rate, but does not provide information on the range of variability of the values and may produce some false negatives, i.e. fail to recognise some persistent substances;
    An intermediate approach is the 90th percentile which may produce some false positives.
– In cases where the studies are not equally reliable, the range and distribution of all available $DegT_{50}$ values should be examined in the WoE approach. This is done by assigning different weights to them according to the quality of the studies, in order to

select an acceptable value (or range of values) to represent the persistence of the substance.

If five or more $DegT_{50}$ values are available, we suggest selecting the geometric mean. This is to be considered together with the decision to normalise the $DegT_{50}$ values to a temperature of 12°C, in order to achieve an overall reasonable conservatism (i.e. target the problematic substances).

This choice is illustrated by the evaluation of a number of substances (64) with 5 and more soil laboratory $DegT_{50}$ values available, whose the degradation data have been recently assessed as part of the regular national PPP registration process in Germany. This exercise indicates how many substances would be identified as P, on the basis of the above evaluation parameters.

- For degradation in soil, only data from laboratory studies was used. Field studies were excluded since no evaluations according to the new EFSA scientific opinion [23] were available yet.
- For water-sediment studies, the total system $DegT_{50}$ was used and compared to the sediment value of 120 days.
- Temperature was normalised both to 12°C and 20°C;
- The maximum $DegT_{50}$ value (worst-case scenario) and the geometric mean of all $DegT_{50}$ were used.

The results indicate that the percentage of substances fulfilling the P criterion varies from about 39% for the least conservative approach (20°C and geometric mean) to 73% for the most conservative approach (12°C and maximum $DegT_{50}$) and is intermediate (i.e. about 53%) when the geometric mean and a temperature of 12°C are used (Table 3).

Screening assessment
For many chemicals, especially those registered under REACH, only a limited data set is available, as explained in Chapter 2. In this case, the assessment of persistence has to be based on screening data. If no further data are presented, the result of the screening assessment (i.e.

screening P) should be used for further regulatory action unless refuted by more definitive data such as valid simulation test data.

In the screening approach, the evaluation of persistence is mainly based on tests on ready biodegradability (OECD 301 A-F [24] and OECD 310 [25]).

Substances which are readily degradable are considered not to be persistent (not P) [26], as the test is fairly stringent. Due to the high rate of mineralisation, an accumulation or formation of relevant metabolites is improbable. On the other hand, substances that are not readily biodegradable are considered to be potentially persistent (potentially P, sometimes also called 'screening P'), until further data are presented. Due to the stringency of the test, this results in a high number of false positives; thus, 'enhanced ready biodegradability' tests (based on ready biodegradability tests but with prolonged test duration, higher concentration of microbial biomass, increased test volumes, but no pre-adaptation) may be used for chemicals registered under REACH to prove their degradability without having to conduct more expensive and time-consuming simulation studies.

Other available data on persistence also have to be examined. This may include data on abiotic degradation, monitoring data, QSAR estimations and read across from structurally related substances.

Main outcomes and open points
Our proposal is to base P identification on degradation, and not on dissipation, as recommended previously [6,17]. Thus, trigger values should be compared to $DegT_{50}$ values. The final identification of substances as persistent should be based on a reasonable conservatism for the various regulatory frameworks and not on a combination of various worst-case choices. Our proposal is summarised in Table 4.

It should be noted that in the P assessment, other information generated in degradation simulation studies such as the formation of non-extractable residues (NER) or the mineralisation rate are not formally considered. Such information can only be taken into account in a

Table 3 Estimation of percentages of substances fulfilling P criterion

Parameter		Persistent in both soil and w/s system	Persistent in soil only	Persistent in w/s system only	Total
Maximum	12°C	45%	17%	11%	73%
		(29 of 64)	(11 of 64)	(7 of 64)	(47 of 64)
	20°C	34%	8%	13%	55%
		(22 of 64)	(5 of 64)	(8 of 64)	(35 of 64)
Geo mean	12°C	30%	3%	20%	53%
		(19 of 64)	(2 of 64)	(13 of 64)	(34 of 64)
	20°C	14%	3%	22%	39%
		(9 of 64)	(2 of 64)	(14 of 64)	(25 of 64)

This evaluation was performed with a dataset of 64 PPP.

Table 4 Main elements of our proposal for a harmonized assessment of persistence across regulatory frameworks

	Persistence	
	Definitive assessment	Screening assessment
Criteria	Half-life	Ready biodegradability
	Water:	Enhanced ready biodegradability
	> 60 days in marine (>60 days)	
	> 40 days in fresh- or estuarine (>60 days)	
	Sediment:	
	> 180 days in marine (>180 days)	
	> 120 days in fresh- or estuarine (>180 days)	
	Soil:	
	> 120 days (>180 days)	
Proposal	Considering DegT$_{50}$ and not DissT$_{50}$	Other information to consider
	Normalising DegT$_{50}$ to 12°C	Abiotic degradation
	Inclusion of field studies if DegT$_{50}$ can be derived	Monitoring data
	Selecting DegT$_{50}$ from multiple studies:	QSAR
	For ≤4 values: worst case	Read-across from structurally related substances
	For ≥5 values: geometric mean, or WoE	
	Water and sediment systems: comparing DegT$_{50}$ to trigger	
	Values of water and sediment; need clear criteria	
	Other information to consider	
	Formation of NER or mineralisation rate	

The P and vP criteria are included for definitive and screening assessments.

WoE approach but should not be ignored. Where NER are the result of degradation and incorporation into the microbial biomass, their formation can be considered as a detoxification step. On the other hand, where the original substance may be remobilised, NER formation should be interpreted as a specific form of compound persistency. The issue of NER is currently being examined in various contexts [27-29].

Bioaccumulation

Substances with a high potential for bioaccumulation are of special concern even if introduced into the environment in low concentrations. They are taken up by biota and their concentrations in the tissues result from the combination of uptake and depuration. Uptake occurs either *via* the surrounding medium - water or soil - (bioconcentration), food (dietary bioaccumulation) or the food chain (biomagnification). These processes are driven by the properties of the substance (mostly hydrophobicity as determined by octanol/water or octanol/air partition coefficient, i.e. K_{OW} or K_{oa}, as well as other characteristics, e.g. protein-binding properties) but they also depend on the environmental matrices (i.e. aquatic, terrestrial) and on the biological, ecological and trophic characteristics of the organisms.

The bioconcentration factor (BCF) is the usual basis for defining the B criterion in PBT assessment although it is not always the most relevant indicator of the environmental bioaccumulation potential of a substance, as stated in [30]. Indeed other available data (e.g. monitoring data, biomagnification, bioaccumulation in other species, literature data) should also be taken into account [30-32]. We propose therefore that all available relevant information on bioaccumulation should be considered in a WoE approach.

Definitive assessment. The definitive assessment of the bioaccumulation potential is based on the bioconcentration factor (BCF) in an aquatic species. The trigger value for the BCF is set at >2,000 L/kg for B assessment and at >5,000 L/kg for vB assessment.

Bioconcentration. The bioconcentration factor (BCF) is defined as the ratio of concentration in fish to the concentration in the surrounding medium. It is generally derived experimentally according to OECD Test Guideline 305 I (bioconcentration flow-through fish test [33]). This test is appropriate for substances with moderate hydrophobicity, so it is necessary to determine the concentration of the test substance in the exposure medium and in fish. The test is mandatory for risk assessment for PPPs, biocides and human medicinal products if log K_{OW} exceeds 3, and for veterinary pharmaceuticals if log K_{OW} exceeds 4. It is also required for REACH substances with a production volume above 100 t/a and a log K_{OW} >3.

The BCF should be normalised to a lipid content of 5% of total body weight (average lipid content of fish commonly used in [33,34]). Potential growth dilution should also be taken into account [26,33], as well as indications of slow or poor depuration.

Bioaccumulation and biomagnification. Bioaccumulation and biomagnification should also be taken into account for species such as fish since the relevant route of exposure may be the food, especially for substances with high log K_{OW} and for substances poorly soluble or non-soluble in water. Indeed for such substances, it may be technically difficult or not feasible (i) to conduct a bioconcentration fish test since aqueous exposure *via* gill uptake is reduced and no longer accounts for the dominant exposure pathway, and/or (ii) to analyse the concentration in water. Instead, a dietary fish test has been developed to determine uptake by ingestion, yielding a biomagnification factor (BMF) [33].

A BMF >1 is a supportive indication of high bioaccumulation and the B criterion is thus considered as fulfilled. Nevertheless, a definitive trigger value for BMF has not yet been established. Hence, a bioaccumulation potential cannot be excluded in cases of BMF < 1. Recalculation of the dietary study data into aquatic BCF values has been suggested [26] but these methods have to be checked carefully for their domain of applicability [35].

Supplementary information. Other reliable data should also be considered, when available:

- High bioaccumulation in organisms other than fish, as mentioned in the Stockholm Convention [36] and the revised Annex XIII criteria in REACH [7]; e.g. bioaccumulation in mussels [37] or oligochaetes [38]. In some cases, lipid normalisation concentrations should be revised since a normalisation to a lipid content of 5% of the body weight - as proposed for fish (see above) - may not be suitable for species with low lipid content (e.g. mussels);
- Indicators other than BCF calculated from environmental data (i.e. measured in mesocosms or in the field), assessing the accumulation of substances from water and diet such as the bioaccumulation factor (BAF), their specific accumulation in food webs expressed as biomagnification factor (BMF) or trophic magnification factor (TMF);
- Bioaccumulation *via* different uptake routes, e.g. terrestrial or benthic oligochaetes [38,39]. As bioaccumulation may differ between water- and air-breathing organisms, aquatic bioaccumulation data should not be transfered to air-breathing organisms [31];
- Accumulation in specific tissues, e.g. substances such as PCBs accumulate extensively in fatty tissues and organs from wildlife samples [40]. In some cases, inefficient or nonexistent detoxification processes may be responsible for high BCF values (e.g. TBT in molluscs) [41,42].
- Non-lipid-based accumulation for substances such as perfluorinated acids, which bioaccumulate in blood plasma proteins [43,44]. Perfluorooctane sulfonate (PFOS) can be considered as bioaccumulative even though the BCF is below the trigger value (<2,000) but it has a long half-life, high toxicity and high biomagnification.

- Toxicokinetic and chronic studies with mammals as well as *in vitro* data on aquatic bioaccumulation as required for human toxicological assessment are also available.

Screening assessment. The screening criterion for assessing bioaccumulative properties is derived from the hydrophobicity of the compound. A substance is considered to potentially fulfil the B criterion when log K_{OW} exceeds a value of 4.5 [21].

Suitable quantitative structure-activity relationship (QSAR) models for estimating the BCF could be used if log K_{OW} is between 4.5 and 6, as the available BCF QSAR models are linear between log K_{OW} 2 to 6 [21,45]. For highly hydrophobic substances (log K_{OW} > 6), the potential for bioaccumulation must be assessed through expert judgement and on a case-by-case basis, taking into account the specific physico-chemical properties of the substance (e.g. molecular size and weight, log K_{OW}) and the available BCF QSAR models (e.g. parabolic equation in [21,46]).

Since hydrophobicity does not drive all bioaccumulation processes, other information has to be considered, e.g. surface activity, structural alerts, high log K_{OA} (octanol/air partition coefficient) as an indicator of a possible bioaccumulation in air-breathing organisms or read-across approaches from structurally related substances.

Main outcomes and open points. Our proposal (see Table 5) is to consider all available endpoints in addition to the BCF for aquatic organisms in the WoE approach, as there are other standardised tests resulting in the derivation of other BAFs (e.g. benthic oligochaetes [38]; terrestrial oligochaetes [39]) and BMFs (fish dietary test guideline [33]). Currently, a comparison between the results of these tests (e.g. BMF, BAF) and a BCF is difficult to make for several reasons such as different uptake mechanisms of different taxa. Also, *in vitro* tests for bioaccumulation (e.g. [47]) are being developed and using information from field studies and TMF (e.g. [42,48]) is being considered. Taking into account these various aspects will improve the assessment of bioaccumulation.

Toxicity

For substances with high persistence and bioaccumulation potential, i.e. fulfilling the P and the B criteria, long-term exposure of organisms can be expected, which may cover the whole life-span of the exposed organisms and even several generations. Therefore, long-term and/or chronic ecotoxicity data, ideally covering the reproductive stages, should be used for assessing the T criterion in the context of a PBT assessment.

Table 5 Main elements of our proposal for a harmonized assessment of bioaccumulation across the regulatory frameworks

	Bioaccumulation	
	Definitive assessment	Screening assessment
Criteria	BCF > 2,000 L/kg (>5,000 L/kg)	Log K_{OW} > 4.5
Proposal	BCF (normalised to 5% lipid content)	QSAR if log K_{OW} is between 4.5 and 6, log K_{OW} > 6: expert judgement, case- by-case
	For high log K_{OW} and poorly soluble or non-soluble: consider BAF, BMF (BMF > 1 indicative)	Other information to consider
	Other information to consider	Surface-activity
	High bioaccumulation in species other than fish	Structural alerts
	Different uptake routes	High log K_{OA} for air-breathing organisms
	Accumulation in specific tissues	Read-across from structurally related substances.
	Non-lipid based accumulation	
	Toxicokinetic and chronic studies with mammals	
	In vitro data on aquatic bioaccumulation	

The B and vB criteria are included for definitive and screening assessments.

Definitive assessment The substance fulfils the T criterion in any of the following situations:

1. Long-term or chronic NOEC or EC_{10} values from ecotoxicological tests with aquatic organisms are below the trigger of 0.01 mg/L, or
2. It is classified as carcinogenic, categories 1 and 2 [49] or 1A and 1B [50], mutagenic, categories 1 and 2 [49] or 1A and 1B [50] or toxic for reproduction, categories 1, 2 and 3 [49] or 1A, 1B and 2 [50], so-called CMR substances, or
3. There is other evidence of chronic toxicity, as identified by the classification of the substance as T, R45, R46, R48, R60 and R61 or Xn, R48, R62, R63, R64 [49], or specific target organ toxicity and repeated exposure (STOT RE category 1 or 2) [50].

If any classification criteria under 2 or 3 are met, there is no need to perform any further aquatic studies for evaluation of toxicity.

Supplementary information. When substances such as PPPs and biocides (i.e. produced to control pests) are tested for toxic effects, it is likely that these will occur both in the target species and in non-target species belonging to the same taxonomic group. Testing for the T criterion refers only to aquatic organisms. However, effects on non-aquatic species could be relevant as well, bearing in mind, however, that the use of vertebrates for testing should generally be minimised.

In this context, we recommend that other reliable data should also be considered where available:

- For PPPs and biocides, results of subchronic, chronic or reproduction studies with birds and

mammals may be available. In this context, a NOEC of ≤30 mg/kg food in a long-term bird study should be considered as an indication that the T criterion is met [21,26].
- It could be relevant to use toxicity data from long-term ecotoxicological tests on terrestrial organisms, e.g. OECD TG 222 [51]. However, in such a case, it has to be clarified whether such an assessment should (i) be based on trigger values comparable to those used in the assessment of toxicity for aquatic organisms (e.g. <0.01 mg/kg dry soil), and (ii) take into account differences in the exposure of the respective organisms.
- For substances with high log K_{ow}, tests performed with sediment-dwelling species may provide better information than tests with pelagic species. Indeed, the results may be more reliable if the substance partitions out of solution. Also, the information gathered may be more useful if it is focused on the compartment in which the substance will likely be found. A way to determine whether a substance has equivalent toxicity in sediment as in the water column should be proposed.

Furthermore, information on aquatic toxicity other than that gathered from standard studies and provided for in relevant test guidelines, i.e. information from non-standard studies and non-standard endpoints, could be used as supporting data. Also, the endocrine disrupting potential of chemical substances is an issue. It is planned to develop a general concept for EDCs which should be consistent and should ensure that endocrine disruptors are dealt with in a consistent and coordinated manner across the different regulatory frameworks [52]. This issue has been discussed by a panel of experts of the German

Federal Environment Agency (UBA) [53]. This could be considered for inclusion of EDC potential in the harmonised assessment of the T criterion.

Screening assessment. The screening criteria for assessing toxicity are based on short-term (acute) toxicity of the compound on aquatic organisms. They are as follows:

- A substance with at least one acute LC_{50}/EC_{50} value below 0.1 mg/L is considered to be potentially toxic (potentially T, sometimes called 'screening T' under REACH). This classification can only be revoked by adequate chronic data, i.e. for determination of definitive criteria for T [26];
- A substance with acute LC_{50}/EC_{50} values below 0.01 mg/L is considered to be toxic (T), as NOEC or EC_{10} values from prolonged or chronic toxicity studies will always be below acute LC_{50} or EC_{50} values for the same taxon [26].

Other possibilities include the use of QSAR and read-across from structurally related substances, if testing of the substance is technically impossible because of its physicochemical properties or in cases of data poor situations.

Main outcomes The criteria for the definitive and the screening assessment of the T criterion are to a large extent consistent between the different regulatory frameworks. However, in situations where data availability or adequacy is too poor to support a definitive or a screening assessment, we propose that other indications of toxicity potential should be taken into account (Table 6). These include, for example, endpoints from toxicity tests with terrestrial organisms, endocrine-disrupting properties, QSAR calculations or harmonised toxicological classifications. Since trigger values have not yet been set for these endpoints, a WoE approach is needed to decide on a case-by-case basis whether a substance fulfils the T criterion as part of PBT assessment.

Considering metabolites and transformation products

Transformation products can be as or even more persistent than their parent compounds and thus must be included in chemical assessment. However, the regulatory frameworks for substances differ with respect to the inclusion of metabolites or transformation products in PBT identification. For PPPs, the consideration of metabolites in the context of PBT identification has been controversially discussed, as PPPs Regulation 1107/2009 [2] does not address this issue precisely and only refers to the relevance of metabolites in a general sense (Art. 3 (32) Annex II point 3.3). However, a current proposal envisages excluding metabolites in the process of PBT identification and in the identification of candidates for

Table 6 Main elements of our proposal for a harmonized assessment of toxicity across regulatory frameworks

		Toxicity	
		Definitive assessment	Screening assessment
Criteria		NOEC (long-term) < 0.01 mg/L, marine or freshwater species	≥1 value for acute LC_{50}/EC_{50} < 0.1 mg/L: potentially toxic
		CMR substances	Acute LC_{50}/EC_{50} values < 0.01 mg/L: toxic
		Other evidence of chronic toxicity	
Proposal		Other information to consider	Other information to consider
		NOEC ≤ 30 mg/kg food, birds and mammals	QSAR
		Long-term tests on terrestrial organisms	Read-across from structurally related substances.
		High log K_{OW}: sediment-dwelling species tests	
		Information from non-standard aquatic studies and endpoints	
		Endocrine-disrupting potential	

The T criteria for definitive and screening assessments is included.

substitution (i.e. substances that meet two of the three criteria) [11]. REACH requires that registration documents for chemicals produced or imported in excess of 100 t/a include information about metabolites and transformation products [54]. They are to be considered in PBT identification if they exceed 0.1% of the substance weight [9,26]. For both human and veterinary medicinal products, transformation products >10% are considered to be 'relevant' and thus included in risk assessment as well as in PBT assessment. In the assessment of biocides, major metabolites (≥10% or ≥5% at two consecutive sampling points or maximum not reached but ≥5% of the active substance at the final time point) are considered in PBT identification and minor metabolites only if data are available or there is any reason for concern. It is questionable whether relevance can be defined in terms of any percentage. Since humans and the environment should be protected from exposure to PBT and vPvB substances, the same criteria should apply to metabolites or transformation products as to the parent compounds. We suggest that in all regulatory frameworks, metabolites and transformation products should be included in the assessment.

Conclusions

The various European regulatory frameworks on chemical substances agree that exposure of humans and the environment to substances identified as PBT or vPvB must be avoided. But the decision-making process is not straightforward and harmonisation between the frameworks is a real challenge. Indeed, despite the same trigger values, the identification of a PBT or vPvB substance will depend on how the criteria are used in the various frameworks to determine.

(i) For persistence, the half-lives in terms of e.g. temperature normalisation, how field studies are considered, appropriate statistics, and others,

(ii) For bioaccumulation, e.g. the endpoints - in addition to the BCF for aquatic organisms - considered in a WoE approach, and

(iii) For toxicity, e.g. indications of toxicity potential considered when availability or adequacy of data is too poor to support a definitive or a screening assessment.

Differences in testing requirements, data availability, data evaluation and interpretation were screened and discussed to identify the best methods. The current proposal is thus characterised by its robustness against those differences and by its reasonable conservatism.

For example, the proposal for the Persistence criterion is based on the outcome of an evaluation performed on a set of selected active substances of PPPs. We recommend (i) normalising the $DegT_{50}$ values to a temperature of 12°C as this is established in biocides legislation and suggested in other regulations (REACH and medicinal products directives), and (ii) selecting the geometric mean if five or more $DegT_{50}$ values are available, or a suitable value or range of $DegT_{50}$ values in a WoE approach if the studies are not equally reliable.

Finally, the differing mandatory measures imposed by the various regulations should be considered together with the present proposal on harmonised PBT/vPvB identification in order to ensure that the truly problematic substances are identified.

Abbreviations
B: Bioaccumulation; BAF: bioaccumulation factor; BCF: bioconcentration factor; BMF: biomagnification factor; CMR: carcinogenic, mutagenic and toxic for reproduction; $DegT_{50}$: degradation half-life; $DissT_{50}$: dissipation half-life; EC_{10}: effect concentration, 10%; EC_{50}: effect concentration, 50%; EDCs: endocrine-disrupting chemicals; EFSA: European Food Safety Agency; EU: European Union; FOMC: first-order multi-compartment; K_{ow}: octanol/water partition coefficient; K_{oa}: octanol/air partition coefficient; LC_{50}: lethal concentration, 50%; NER: non-extractable residues; NOEC: no observed effect concentration; OECD: Organisation for Economic Cooperation and Development; P: persistence; PBT: persistent, bioaccumulative and toxic; PCBs: polychlorinated biphenyls; PFOS: perfluorooctane sulfonate; PPPs: plant protection products; QSAR: quantitative structure-activity relationships; REACH: Registration, Evaluation, Authorisation and Restriction of Chemicals; SETAC: Society for Environmental Toxicology and Chemistry; SFO: single first order; STOT RE: specific target organ toxicity - repeated exposure; SVHC: substances of very high concern; T: toxicity; t/a: tons per year; TBT: tributyltin hydride; TG: test guideline; TGD: Technical Guidance Document; TMF: trophic magnification factor; UBA: Umweltbundesamt (German Federal Environment Agency); US EPA: United States Environmental Protection Agency; vPvB: very persistent and very bioaccumulative; WoE: weight of evidence.

Competing interests
The authors declare that they have no competing interests.

Authors' contributions
CR, AF, GH, UJ, AK, MN, IP, JS, AW, KW, JW and SD contributed to the collection of assessment procedures in PBT assessment for the respective frameworks, and drafted the first manuscript. SD and CR edited the manuscript to include all comments and finalised the manuscript. All authors read and approved the final manuscript.

Acknowledgements
We thank our colleagues from the Federal Environment Agency (Umweltbundesamt, UBA, Germany) and other authorities as well as colleagues from academia and industries for fruitful discussions.

Author details
[1]Section International Chemicals Management, Federal Environment Agency (UBA), Wörlitzer Platz 1, 06844 Dessau-Roßlau, Germany. [2]Section Biocides, Federal Environment Agency (UBA), Wörlitzer Platz 1, 06844 Dessau-Roßlau, Germany. [3]Section Plant Protection Products, Federal Environment Agency (UBA), Wörlitzer Platz 1, 06844 Dessau-Roßlau, Germany. [4]Section Chemicals, Federal Environment Agency (UBA), Wörlitzer Platz 1, 06844 Dessau-Roßlau, Germany. [5]Section Pharmaceuticals, Washing and Cleaning Agents, Federal Environment Agency (UBA), Wörlitzer Platz 1, 06844 Dessau-Roßlau, Germany.

References
1. EC: *Regulation (EC) No 1907/2006 of the European Parliament and of the Council of 18 December 2006 concerning the Registration, Evaluation, Authorisation and Restriction of Chemicals (REACH), establishing a European Chemicals Agency, amending Directive 1999/45/EC and repealing Council Regulation (EEC) No 793/93 and Commission Regulation (EC) No 1488/94 as well as Council Directive 76/769/EEC and Commission Directives 91/155/EEC, 93/67/EEC, 93/105/EC and 2000/21/EC*; 2006. http://eur-lex.europa.eu/legal-content/EN/TXT/PDF/?uri=CELEX:02006R1907-20130701&rid=2 (last accessed 08/04/2014).
2. EC: *Regulation (EC) No 1107/2009 of the European Parliament and of the Council of 21 October 2009 concerning the placing of plant protection products on the market and repealing Council Directives 79/117/EEC and 91/414/EEC*; 2009. http://eur-lex.europa.eu/legal-content/EN/TXT/PDF/?uri=CELEX:32009R1107&rid=2 (last accessed 08/04/2014).
3. EC: *Regulation (EU) No 528/2012 of the European Parliament and of the Council of 22 May 2012 concerning the making available on the market and use of biocidal products*; 2012. http://eur-lex.europa.eu/legal-content/EN/TXT/PDF/?uri=CELEX:32012R0528&rid=2 (last accessed 08/04/2014).
4. EC: *Directive 2001/83/EC of the European Parliament and of the Council of 6 November 2001 on the Community code relating to medicinal products for human use*; 2001. http://eur-lex.europa.eu/legal-content/EN/TXT/PDF/?uri=CELEX:02001L0083-20081230&rid=6 (last accessed 08/04/2014).
5. EC: *Directive 2001/82/EC of the European Parliament and of the Council of 6 November 2001 on the Community code relating to veterinary medicinal products*; 2001. http://eur-lex.europa.eu/legal-content/EN/TXT/PDF/?uri=CELEX:02001L0082-20090306&rid=3 (last accessed 08/04/2014).
6. Moermond CTA, Janssen MPM, de Knecht JA, Montforts MHMM, Peijnenburg WJGM, Zweers PGPC, Sijm DTHM: **PBT Assessment using the revised Annex XIII of REACH – a comparison with other regulatory frameworks.** *IEAM* 2011, 8(2):359–371. http://onlinelibrary.wiley.com/doi/10.1002/ieam.1248/abstract (last accessed 08/04/2014).
7. EC: *Commission regulation (EU) No 253/2011 of 15 March 2011 amending Regulation (EC) No 1907/2006 of the European Parliament and of the Council on the Registration, Evaluation, Authorisation and Restriction of Chemicals (REACH) as regards Annex XIII*; 2011. http://eur-lex.europa.eu/legal-content/EN/TXT/PDF/?uri=CELEX:32011R0253&rid=1 (last accessed 08/04/2014).
8. European Chemicals Agency: *Guidance on information requirements and chemical safety assessment. Chapter R.7b: Endpoint specific guidance*; 2012:1–234. http://echa.europa.eu/documents/10162/13632/information_requirements_r7b_en.pdf.

9. European Chemicals Agency: *Guidance on information requirements and chemical safety assessment. Part C: PBT Assessment*; 2011. http://echa.europa.eu/documents/10162/13643/information_requirements_part_c_en.pdf.

10. European Medicines Agency: *Guidance on the assessment of persistent, bioaccumulative and toxic (PBT) or very persistent and very bioaccumulative (vPvB) substances in veterinary medicine. European Medicines Agency, Committee for Medicinal Products for Veterinary Use (CVMP), EMA/CVMP/ERA/52740/2012*; 2012. http://www.ema.europa.eu/ema/pages/includes/document/open_document.jsp?webContentId=WC500130368 (last accessed 08/04/2014).

11. Directorate General for Health and Consumer Affairs (DG SANCO): *Working Document on "Evidence needed to identify POP, PBT and vPvB Properties for Pesticides", Brussels, 25.09.2012 – rev. 3, European Commission, Health and Consumers Directorate-General, Safety of the Food chain, Chemicals, contaminants, pesticides.* http://ec.europa.eu/food/plant/pesticides/approval_active_substances/docs/wd_evidence_needed_to_identify_pop_pbt_vpvb_properties_rev3_en.pdf (last accessed 08/04/2014).

12. Roberts G: *'Large part' of registration dossiers a concern, says ECHA.* Chemical Watch, 27 February 2013, http://chemicalwatch.com/13957/large-part-of-registration-dossiers-a-concern-says-echa?q=registration%20quality (last accessed on 07/03/2014).

13. European Chemicals Agency: *Facts & Figures. Quality information is required to comply with REACH. EVALUATION REPORT 2012 - MAIN OUTCOMES AND KEY RECOMMENDATIONS FOR INDUSTRY.* February 2013, http://echa.europa.eu/documents/10162/13628/evaluation_report_summary_2012_en.pdf (accessed 07/03/2014).

14. European Chemicals Agency: *Practical guide 2: How to report weight of evidence*; 2010. http://echa.europa.eu/documents/10162/13655/pg_report_weight_of_evidence_en.pdf, (accessed 26/06/2012).

15. Boethling R, Fenner K, Howard P, Meylan W, Klečka G, Madsen T, Snape JR: **Environmental Persistence of Organic Pollutants: Guidance for development and Review of POP Risk Profiles.** *IEAM* 2009, 5(4):539–556. http://onlinelibrary.wiley.com/doi/10.1897/IEAM_2008-090.1/abstract (last accessed on 08/04/2014).

16. OECD: *Aerobic and Anaerobic Transformation in Soil. OECD Guidelines for the Testing of Chemicals Nr. 307. Paris*; 2002. http://www.oecd-ilibrary.org/environment/test-no-307-aerobic-and-anaerobic-transformation-in-soil_9789264070509-en (last accessed on 08/04/2014).

17. OECD: *Aerobic and Anaerobic Transformation in Aquatic Sediment Systems. OECD Guideline for the Testing of Chemicals Nr. 308. Paris*; 2002. http://www.oecd-ilibrary.org/environment/test-no-308-aerobic-and-anaerobic-transformation-in-aquatic-sediment-systems_9789264070523-en (last accessed on 08/04/2014).

18. OECD: *Aerobic Mineralisation in Surface Water – Simulation Biodegradation Test. OECD Guidelines for the Testing of Chemicals Nr. 309. Paris*; 2004. http://www.oecd-ilibrary.org/environment/test-no-309-aerobic-mineralisation-in-surface-water-simulation-biodegradation-test_9789264070547-en (last accessed on 08/04/2014).

19. FOCUS: *Guidance Document on Estimating Persistence and Degradation Kinetics from Environmental Fate Studies on Pesticides in EU Registration. Sanco/10058/2005, version 2.0, June 2006, and Generic Guidance for Estimating Persistence and Degradation Kinetics from Environmental Fate Studies in Pesticides in EU Registration (version 1.0): document based on the official guidance document of FOCUS Degradation Kinetics in the context of 91/414/EEC and Regulation (EC) No 1107/2009*; 2006. http://focus.jrc.ec.europa.eu/dk/doc.html (last accessed 08/04/2014).

20. OPPTS 835.3300 Soil Biodegradation: *Fate, Transport and Transformation Test Guidelines. United States Environmental Protection Agency, Prevention, Pesticides and Toxic Substances (7101). EPA712–C–98–088, January 1998.* http://www.regulations.gov/#!documentDetail;D=EPA-HQ-OPPT-2009-0152-0025 (last accessed 08/04/2014).

21. EC: *Technical Guidance Document on Risk Assessment in support of Commission Directive 93/67/EEC on Risk Assessment for new notified substances, Commission Regulation (EC) No 1488/94 on Risk Assessment for existing substances and Directive 98/8/EC of the European Parliament and of the Council concerning the placing of biocidal products on the market - Part II; Publication No. 20418/EN/2*; 2003. http://ihcp.jrc.ec.europa.eu/our_activities/public-health/risk_assessment_of_Biocides/doc/tgd (last accessed 08/04/2014).

22. European Medicines Agency: *Guidance on the assessment of persistent, bioaccumulative and toxic (PBT) or very persistent and very bioaccumulative (vPvB) substances in veterinary medicine. London, UK: Rapport nr. EMA/CVMP/ERA/52740/2012*; 2012. http://www.ema.europa.eu/docs/en_GB/document_library/Scientific_guideline/2012/07/WC500130368.pdf (last accessed 08/04/2014).

23. European Food Safety Authority: **Guidance for evaluating laboratory and field dissipation studies to obtain DegT50 values of plant protection products in soil. Scientific opinion of the panel on Plant Protection Products and their Residues (PPR).** *EFSA J* 2010, 8(12):1936. http://www.efsa.europa.eu/de/efsajournal/doc/1936.pdf (last accessed 08/04/2014).

24. OECD: *Ready Biodegradability. OECD Guidelines for the Testing of Chemicals Nr. 301. Paris*; 1992a. http://www.oecd-ilibrary.org/environment/test-no-301-ready-biodegradability_9789264070349-en (last accessed 08/04/2014).

25. OECD: *Ready Biodegradability - CO2 in sealed vessels (Headspace Test). OECD Guidelines for the Testing of Chemicals Nr. 310. Paris*; 2006. http://www.oecd-ilibrary.org/environment/test-no-310-ready-biodegradability-co2-in-sealed-vessels-headspace-test_9789264016316-en (last accessed 08/04/2014).

26. European Chemicals Agency: *Guidance on information requirements and chemical safety assessment. Chapter R.11: PBT Assessment.* http://echa.europa.eu/documents/10162/13632/information_requirements_r11_en.pdf.

27. ECETOC: *Understanding the Relationship between Extraction Technique and Bioavailability, Technical Report No. 117, ECETOC, Brussels, May 2013, ISSN-0773-8072-117 (print), ISSN-2079-1526-117 (online)*; 2013. http://www.ecetoc.org/index.php?mact=MCSoap,cntnt01,details,0&cntnt01by_category=22&cntnt01order_by=date%20Desc&cntnt01template=display_list_v2&cntnt01display_template=display_details_v2&cntnt01document_id=7315&cntnt01returnid=59 (last accessed 08/04/2014).

28. ECETOC: *Development of interim guidance for the inclusion of non- extractable residues (NER) in the risk assessment of chemicals, Technical Report No. 118, ECETOC, Brussels, May 2013, ISSN-0773-8072-118 (print), ISSN-2079-1526-118 (online)*; 2013. http://www.ecetoc.org/index.php?mact=MCSoap,cntnt01,details,0&cntnt01by_category=22&cntnt01order_by=date%20Desc&cntnt01template=display_list_v2&cntnt01display_template=display_details_v2cntnt01document&_id=7316&cntnt01returnid=59 (last accessed 08/04/2014).

29. Eschenbach A, Oing K: *Erarbeitung eines gestuften Extraktionsverfahrens zur Bewertung gebundener Rückstände, Gutachten..* Dessau-Roßlau: Umweltbundesamt; 2013. project number 22582.

30. Klecka GM, Muir DCG: *Science-based guidance and framework for the evaluation and identification of PBTs and POPs: summary of a SETAC Pellston workshop. Summary of the SETAC Pellston Workshop on Science-Based Guidance and Framework for the Evaluation and Identification of PBTs and POPs. SETAC Pellston Workshop on Science-Based Guidance and Framework for the Evaluation and Identification of PBTs and POPs; 2008 Jan 28–Feb 1; Pensacola Beach, FL. Pensacola (FL): Society of Environmental Toxicology and Chemistry (SETAC)*; 2008. http://c.ymcdn.com/sites/www.setac.org/resource/resmgr/publications_and_resources/pbtpopsexecutivesummary.pdf (last accessed 08/04/2014).

31. Gobas FAPC, de Wolf W, Burkhard LP, Verbruggen E, Plotzke K: **Revisiting Bioaccumulation Criteria for POPs and PBT Assessments.** *IEAM* 2009, 5(4):624–637.

32. Ehrlich G, Jöhncke U, Drost W, Schulte C: **Problems faced when evaluating the bioaccumulation Potential of substances under REACH.** *IEAM* 2011, 7(4):550–558.

33. OECD: *Bioaccumulation in Fish: Aqueous and Dietary Exposure. OECD Guideline for the Testing of Chemicals Nr. 305. Paris*; 2012. http://www.oecd-ilibrary.org/environment/test-no-305-bioaccumulation-in-fish-aqueous-and-dietary-exposure_9789264185296-en (last accessed 08/04/2014).

34. OECD: *Guidance document on the use of the harmonised system for the classification of chemicals which are hazardous for the aquatic environment. OECD series on testing and assessment Number 27. Paris*; 2001. http://search.oecd.org/officialdocuments/displaydocumentpdf/?cote=env/jm/mono(2001)8&doclanguage=en (last accessed 08/04/2014).

35. Crookes M, Brooke D: *Estimation of fish bioconcentration factor (BCF) from depuration data, Product Code: SCHO0811BUCE-E-E. Bristol, UK: Environment Agency*; 2011. https://www.gov.uk/government/uploads/system/uploads/attachment_data/file/291527/scho0811buce-e-e.pdf.

36. POP-Convention: *Stockholm Convention on Persistent Organic Pollutants. May 22nd*; 2001. http://chm.pops.int/Portals/0/download.aspx?d=UNEP-POPS-COP-CONVTEXT.En.pdf (last accessed 08/04/2014).

37. United States Environmental Protection Agency; Prevention, Pesticides and Toxic Substances (7101): *OPPTS 850.1710. Oyster BCF: Ecological Effects Test Guidelines. EPA712–C–96–127*; 1996. http://www.epa.gov/opptsmnt/pubs/frs/publications/OPPTS_Harmonized/850_Ecological_Effects_Test_Guidelines/Drafts/850-1710.pdf.

38. OECD: *Bioaccumulation in Sediment-dwelling Benthic Oligochaetes. OECD Guideline for the Testing of Chemicals Nr. 315. Paris*; 2008. http://www.oecd-ilibrary.org/environment/test-no-315-bioaccumulation-in-sediment-dwelling-benthic-oligochaetes_9789264067516-en (last accessed 08/04/2014).

39. OECD: *Bioaccumulation in Terrestrial Oligochaetes. OECD Guideline for the Testing of Chemicals Nr. 317. Paris*; 2010. http://www.oecd-ilibrary.org/ environment/test-no-317-bioaccumulation-in-terrestrial-oligochaetes_ 9789264090934-en (last accessed 08/04/2014).

40. Pérez-Fuentetaja A, Lupton S, Vlapsadl M, Samara F, Gatto L, Biniakewitz R, Aga DS: **PCB and PBDE levels in wild common carp (*Cyprinus carpio*) from eastern Lake Erie.** *Chemosphere* 2010, **81**(4):541–547. http://www.sciencedirect. com/science/article/pii/S0045653510007071 (last accessed 08/04/2014).

41. Parkerton TF, Arnot JA, Weisbrod AV, Russom C, Hoke RA, Woodburn K, Traas TP, Bonnell M, Burkhard LP, Lampi MA: **Guidance for evaluating in vivo fish bioaccumulation data.** *IEAM* 2008, **4**(2):139–155. http://onlinelibrary.wiley.com/ doi/10.1897/IEAM_2007-057.1/abstract (last accessed 08/04/2014).

42. Weisbrod AV, Woodburn KB, Koelmann AA, Parkerton TF, McElroy AE, Borgå K: **Evaluation of bioaccumulation using in vivo laboratory and field studies.** *IEAM* 2009, **5**(4):598–623. http://onlinelibrary.wiley.com/doi/10.1897/ IEAM_2009-004.1/full (last accessed 08/04/2014).

43. Martin JW, Mabury SA, Solomon KR, Muir DCG: **Bioconcentration and tissue distribution of perfluorinated acids in rainbow trout (*Oncorhynchus mykiss*).** *Environ Toxicol Chem* 2003, **22**:196–204.

44. Conder JM, Hoke RA, de Wolf W, Russell MH, Buck RC: **Are PFCAs bioaccumulative? A critical review and comparison with regulatory criteria and persistent lipophilic compounds.** *Environ Sci Technol* 2008, **42**:995–1003.

45. Veith GD, Defoe DL, Bergstedt BV: **Measuring and estimating the bioconcentration factor of chemicals in fish.** *J Fish Res Board Canada* 1979, **36**:1040–1048.

46. Bintein S, Devillers J, Karcher W: **Nonlinear dependance of fish biocencentration on n-octanol/water partition coefficient.** *SAR QSAR Environ Res* 1993, **1993**(1):29–39.

47. Nichols JW, Fitzsimmons PN, Burkhard LP: **In vitro-in vivo extrapolation of quantitative hepatic biotransformation data for fish. II. Modeled effects on chemical bioaccumulation.** *Environ Toxicol Chem* 2007, **26**(6):1304–1319.

48. HESI: *HESI Bioaccumulation Project Committee Workshop Summary "Moving Bioaccumulation Assessments to the Next Level: Progress Made and Challenges Ahead.* Alexandria, VA, USA: ILSI Health and Environmental Sciences Institute; 2011. http://www.hesiglobal.org/files/public/Committees/ Bioaccumulation/BioacWkshpSummary070811.pdf (accessed 11/01/2012).

49. EEC: *Council Directive 67/548/EEC of 27 June 1967 on the approximation of laws, regulations and administrative provisions relating to the classification, packaging and labelling of dangerous substances*; 1967. http://eur-lex.europa. eu/legal-content/EN/TXT/PDF/?uri=CELEX:31967L0548&rid=1.

50. EC: *Regulation (EC) No 1272/2008 of the European Parliament and of the Council of16 December 2008 on classification, labelling and packaging of substances and mixtures, amending and repealing Directives 67/548/EEC and 1999/45/EC, and amending Regulation (EC) No 1907/2006*; 2008. http://eur-lex.europa.eu/legal-content/EN/TXT/PDF/?uri=CELEX:02008R1272-20110419&rid=2.

51. OECD: *Earthworm Reproduction Test (Eisenia fetida/Eisenia andrei). OECD Guidelines for the Testing of Chemicals Nr. 222. Paris*; 2004b. http://www. oecd-ilibrary.org/environment/test-no-222-earthworm-reproduction-test-eisenia-fetida-eisenia-andrei_9789264070325-en (last accessed 08/04/2014).

52. European Commission: *4th Report on the implementation of the Community Strategy for Endocrine Disrupters a range of substances suspected of interfering with the hormone systems of humans and wildlife (COM (1999) 706)*; 2011. http://ec.europa.eu/environment/chemicals/endocrine/pdf/sec_2011_1001.pdf.

53. Frische T, Bachmann J, Frein D, Juffernholz T, Kehrer A, Klein A, Maack G, Stock F, Stolzenberg H-C, Thierbach C, Walter-Rohde S: **Identification, assessment and management of "endocrine disruptors" in wildlife in the EU substance legislation—Discussion paper from the German Federal Environment Agency (UBA).** *Toxicol Lett* 2013, **223**(2013):306–309. http://www.sciencedirect. com/science/article/pii/S0378427413001021 (last accessed 07/04/2014).

54. Ng CA, Scheringer M, Fenner K, Hungerbuhler K: **A framework for evaluating the contribution of transformation products to chemical persistence in the environment.** *Environ Sci Technol* 2011, **45**(1):111–117. http://pubs.acs.org/doi/abs/10.1021/es1010237 (last accessed 07/04/2014).

4

Potential water-related environmental risks of hydraulic fracturing employed in exploration and exploitation of unconventional natural gas reservoirs in Germany

Axel Bergmann[1][*], Frank-Andreas Weber[1], H Georg Meiners[2] and Frank Müller[2]

Abstract

Background: The application of hydraulic fracturing during exploration and exploitation of unconventional natural gas reservoirs is currently under intense public discussion. On behalf of the German Federal Environment Agency we have investigated the potential water-related environmental risks for human health and the environment that could be caused by employing hydraulic fracturing in unconventional gas reservoirs in Germany. Here we provide an overview of the present situation and the state of the debate in Germany and summarize main results of the conducted risk assessment.

Results: We propose a concept for a risk assessment considering the site-specific analysis of the geosystem, the relevance of possible impact pathways and the hazard potential of the fracking fluids employed. The foundation of a sound risk analysis is a description of the current system, the relevant impact pathways and their interactions. An evaluation of fracking fluids used in Germany shows that several additives were employed even in newer fluids that exhibit critical properties or for which an assessment of their behaviour and effects in the environment is not possible or limited due to lack of current knowledge. The authors propose an assessment method that allows for the estimation of the hazard potential of specific fracking fluids, formation water, and the flowback based on legal thresholds and guidance values as well as on human- and eco-toxicologically predicted no-effect concentrations. The assessment of a previously employed and a prospectively planed fracking fluids shows that these fluids exhibit a high hazard potential. The flowback containing fracking fluid, formation water, and possibly reaction products can also exhibit serious hazard potentials, requiring environmentally acceptable techniques for its treatment and disposal.

Conclusions: The risk analysis must be conducted always site-specifically and consider regional groundwater flow conditions. The study concludes that currently missing knowledge and data prevent a profound assessment of the risks and their technical controllability in Germany. Missing knowledge and information includes data on the properties of the deep geosystem and of the behaviour and effects of the deployed chemical additives. In this setting the authors propose several recommendations for further action and procedures regarding the application of hydraulic fracturing in unconventional gas reservoirs in Germany.

Keywords: Hydraulic fracturing; Fracking; Unconventional natural gas; Shale gas; Coalbed methane; CBM; Impact pathways; Chemical additives; Flowback; Hazard potential; Risk analysis

* Correspondence: a.bergmann@iww-online.de
[1]IWW Water Centre, Department Water Resources Management,
Moritzstrasse 26, Muelheim 45476, Germany
Full list of author information is available at the end of the article

Background

The application of hydraulic fracturing ("fracking") in the exploration and exploitation of unconventional natural gas reservoirs has been generating intensive public debates in a variety of countries. Major concerns have focused on the potential impacts, hydraulic fracturing may cause on the environment and on human health, especially if fracking fluids contain toxic and environmentally harmful chemical additives.

Unconventional gas reservoirs are proven or presumed to be present in a number of different geological formations. An overview of potential geological host formations of unconventional gas reservoirs in Germany is given in Table 1, differentiating coalbed methane (CBM), shale gas and tight gas reservoirs. According to current estimates [1], the technologically recoverable gas reserves present in shale gas reservoirs in Germany amount to about 1,300 billion m^3 (estimates range from 0.7 to $2.3 \cdot 10^{12}$ m^3), assuming that 10% of the gas in place (GIP) is technologically recoverable. This estimated range of technologically recoverable shale gas reservoirs could, if exploited completely, cover the current annual gas consumption of Germany for 8 to 27 years [2]. The GIP in coalbed methane reservoirs in Germany is estimated to 450 billion m^3 [3], but the technologically recoverable fraction has not yet been analysed. Conventional gas and tight gas reservoirs have been exploited in Germany over

several decades, but current estimates of GIP remaining (100 billion m^3 and 20 billion m^3, respectively [3]) indicate that the remaining reserves are limited.

The mining authorizations that have been issued for the exploration of unconventional gas reservoirs in Germany are shown in Figure 1. Most exploration has yet focused on the recovery and analysis of drilling core material as well as on geophysical methods, but hydraulic fracturing has already been applied in exploration at two sites [4]: at the site Damme 3 in Lower Saxony (3 fracs in the Wealden clay formation in depth of 1,045 – 1,530 m below ground surface using a slickwater fracking fluid in 2008) and at the site Natarp in North Rhine-Westphalia (2 fracs in CBM reservoirs in depth of 1,800 – 1,947 m using a gel fluid in 1995). To our knowledge, no mining authorizations have yet been approved for production-oriented exploitation of shale gas or CBM reservoirs in Germany. In the ongoing exploitation of tight gas and conventional gas reservoirs, however, experience in using hydraulic fracturing has been gained by pumping over 300 fracs over the last decades, mainly in the federal state of Lower Saxony [4]. In general, the exploited tight gas reservoirs in Germany are located in greater depth (often > 3.500 m) than some of the shale gas and CBM reservoirs currently considered for exploration, which vary in depth but are located partly in depth of 1.000 m or less [2,4-6], raising additional

Table 1 Potential unconventional gas reservoirs in Germany

Type of reservoir	Most promising reservoir	Regions
Coal bed methane (source rocks)	Seam-bearing Upper Carboniferous	Northern Ruhr region/Münsterland Basin (NRW)
		Ibbenbühen (NRW)
		Saar Basin (Saarland)
Shale gas (source rocks)	Tertiary clay formations (e.g. Fischschiefer)	Molasse Basin (BW)
	Posidonia Shale (Black Jurassic)*	Northwest German Basin (e.g. Lünne) (NI)
		Molasse Basin (BW)
		Upper Rhine Graben
	Wealden clay formations (e.g. Lower Cretaceous)*	Weser Depression (NRW/NI)
	Permian clay formations (e.g. black shale (stinkschiefe"), copper shale)	Northeast German Basin (NI/SA)
	Carboniferous and Devonian clay formations e.g. alum shale (Lower Carboniferous)*	Northern edge of the Rhenish massif (NRW)
		Northwest German Basin
		Harz Mountain (NI/SA)
	Silurian slates	Northeast German Basin
	Cambro Ordovician clay formations ("alum shale")	(not yet studied in details)
Tight gas (deposit rocks)	Red sandstone	Northwest German basin (NI/SA)
	Permian sandstones (Rotliegend) and carbonates (Zechstein)	Northeast German basin (e.g. Leer) (NI)
	Permian sandstones (Rotliegend) and dolomite (Stassfurt series) sandstones (Triassic)	Thuringian Basin (TH)
	Upper Carboniferous sandstones	Northwest German Basin (e.g. Vechta) (NI)

*indicates most relevant shale gas reservoirs according to [1].

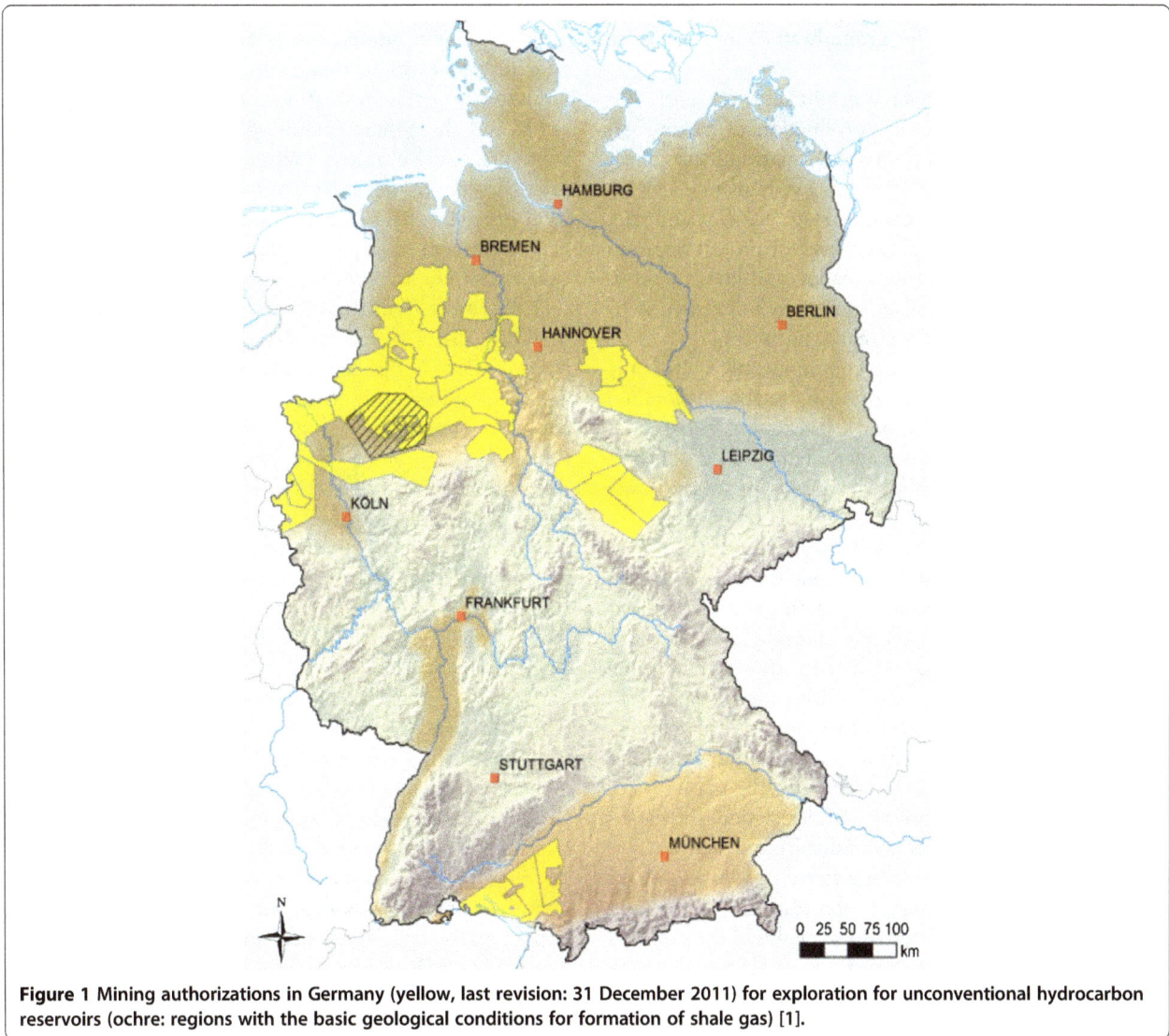

Figure 1 Mining authorizations in Germany (yellow, last revision: 31 December 2011) for exploration for unconventional hydrocarbon reservoirs (ochre: regions with the basic geological conditions for formation of shale gas) [1].

concerns on potential impacts on near-surface groundwater resources.

Driven by reports on the application and risk assessment of hydraulic fracturing in the U.S. [7-11], several risk assessments have recently been conducted on the specific German geological, technical, and legal situation, including an investigation on behalf of the German Federal Environment Agency (UBA) [4], a survey on behalf of the Ministry for Climate Protection, Environment, Agriculture, Nature Conservation and Consumer Protection (MKULNV) of the federal state of North Rhine-Westphalia [5], and an investigation of an independent expert group initiated by ExxonMobil Production Germany GmbH [12]. Given the current state of exploration of shale gas and CBM reservoirs in Germany, most risk assessments were conducted generically (i.e. not site-specific) or focused on some selected geological settings.

Two site-specific investigations on regional situations in northern Hessian and in the river Ruhr watershed have recently been conducted [6,13].

Current state of the debate in Germany

The political debate on hydraulic fracturing in Germany has proceeded as a result of the conducted risk assessments (or independent thereof), and new administrative procedures have been adapted recently.

The State Authority for Mining, Energy and Geology (LBEG) of Lower Saxony has issued minimum requirements for operating plans, criteria, and approval procedure for hydraulic treatments of boreholes in petroleum and natural gas reservoirs [14]. ExxonMobil Production Germany GmbH, a major operator in Germany, has announced that fracking projects in the vicinity of certain mineral spa protection zones are not further pursuit and no further hydraulic fracturing activities are carried out before suitable concepts for groundwater monitoring are implemented [15].

The state of North Rhine-Westphalia is currently not approving any exploration or production of natural gas from unconventional gas reservoirs, if harmful substances are employed [16]. A dialogue process is planned to involve the gas industry and communities, citizens, and relevant institutions in developing criteria for project approval and eliminating deficits of information and knowledge. In this context, borehole investigations, excluding hydraulic fracturing, are discussed for research purposes [17].

According to current press communications [18], the state of Lower Saxony is not approving further exploration and exploitation of shale gas and CBM reservoirs based on the lack of adequate risk assessment, but plans to continue approving exploitation of tight gas reservoirs in sandstone formation in depths > 2.500 m, as long as no environmentally toxic substances are injected underground.

Draft legislations amending the environmental impact assessment (EIA) regulation and of the Water Management Act (WHG) are currently discussed in Germany [19]. The drafts call for a ban of deep drillings involving hydraulic fracturing and the underground disposal of flowback in water protection zones, mineral spa protection zones, and in catchment areas of natural lakes from which raw water is procured directly for the public water supply. Based on the discussed draft legislation, the catchment area of artificial lakes and dams from which water is indirectly obtained for drinking water purposes would not generally be considered an exclusion zone [20].

Two regional investigations have analysed the regional occurrence of shale gas reservoirs in comparison to competing land-use obligations [6,13]. In a study on behalf of the river Ruhr water works consortium (*Arbeitsgemeinschaft der Wasserwerke an der Ruhr e.V.*) and the Ruhr River water board (*Ruhrverband*), we concluded that considering the regional occurrence of the shale gas reservoirs, the exclusion areas proposed by the draft legislations, and adopting criteria for the approval of exploitation involving hydraulic fracturing issued in Lower Saxony, an area of less than 3% of the issued mining authorization is accessible for exploitation of the shale gas reservoirs. Furthermore, a legal expertise commissioned by the Hessian Ministry of Environment, Energy, Agriculture and Consumer Protection (HLUG) has noted [13] that mining authorizations must not be granted if competing obligations among public stakeholders preclude subsequent exploitation of the gas reservoirs in the entire allocated field.

In the so-called "Hannover-Erklärung", the Federal Institute for Geosciences and Natural Resources (BGR), the Helmholtz Centre Potsdam - GFZ German Research Centre for Geosciences and the Helmholtz Centre for Environmental Research (UFZ) have called for the development of environmentally friendly fracking technology and proposed joint demonstration projects involving industry, research institutions, environmental organizations, and the general public [21]. An alliance of water suppliers, the Ruhr River water works consortium, and members of the beverage industry have called for clear legal provisions to protect the safety and purity of water resources from impacts of hydraulic fracturing in the so-called "Gelsenkirchener Erklärung" [22].

Furthermore, the exploitation of shale gas reservoirs is currently discussed controversial from an energy policy point of view. While the Federal Institute for Geosciences and Natural Resources (BGR) concluded that shale gas can contribute to domestic energy security [1], the German Advisory Council on the Environment (SRU) comes to the conclusion that the exploitation of shale gas using hydraulic fracturing is not necessary in Germany from an energy policy point of view and cannot substantially contribute to the transition to renewable energy sources [2].

Objectives

On behalf of the German Federal Environment Agency (UBA), a consortium of IWW Water Centre, ahu AG, [Gaßner, Groth, Siederer & Coll.], and Technical University of Darmstadt, has conducted a comprehensive investigation on potential environmental impacts of hydraulic fracturing related to exploration and exploitation of unconventional natural gas reservoirs, which focused on the framework of a risk assessment, the analysis of potential impact pathways, a method for assessing the hazard potentials of the fracking fluids employed, and on legal regulations and administrative structures [4]. Here we summarize main results of this study and propose recommendations for action and procedures. The study is based mostly on publicly accessible information including the relevant literature available internationally, but also on information provided by German authorities and operating companies.

Results and discussion

For assessing the risks that the application of hydraulic fracturing in unconventional natural gas reservoirs can pose on the water environment, we propose a concept that considers both the possible impact pathways and the potential hazard, any migration of the substances employed or encountered along these impact pathways could cause on exploitable water resources (Figure 2). Only if impact pathways are relevant for substance migration on the time scale considered, the substance-related hazard potentials cause adverse effects on the exploitable water environment. The risk of contamination of exploitable water resources is thus obtained

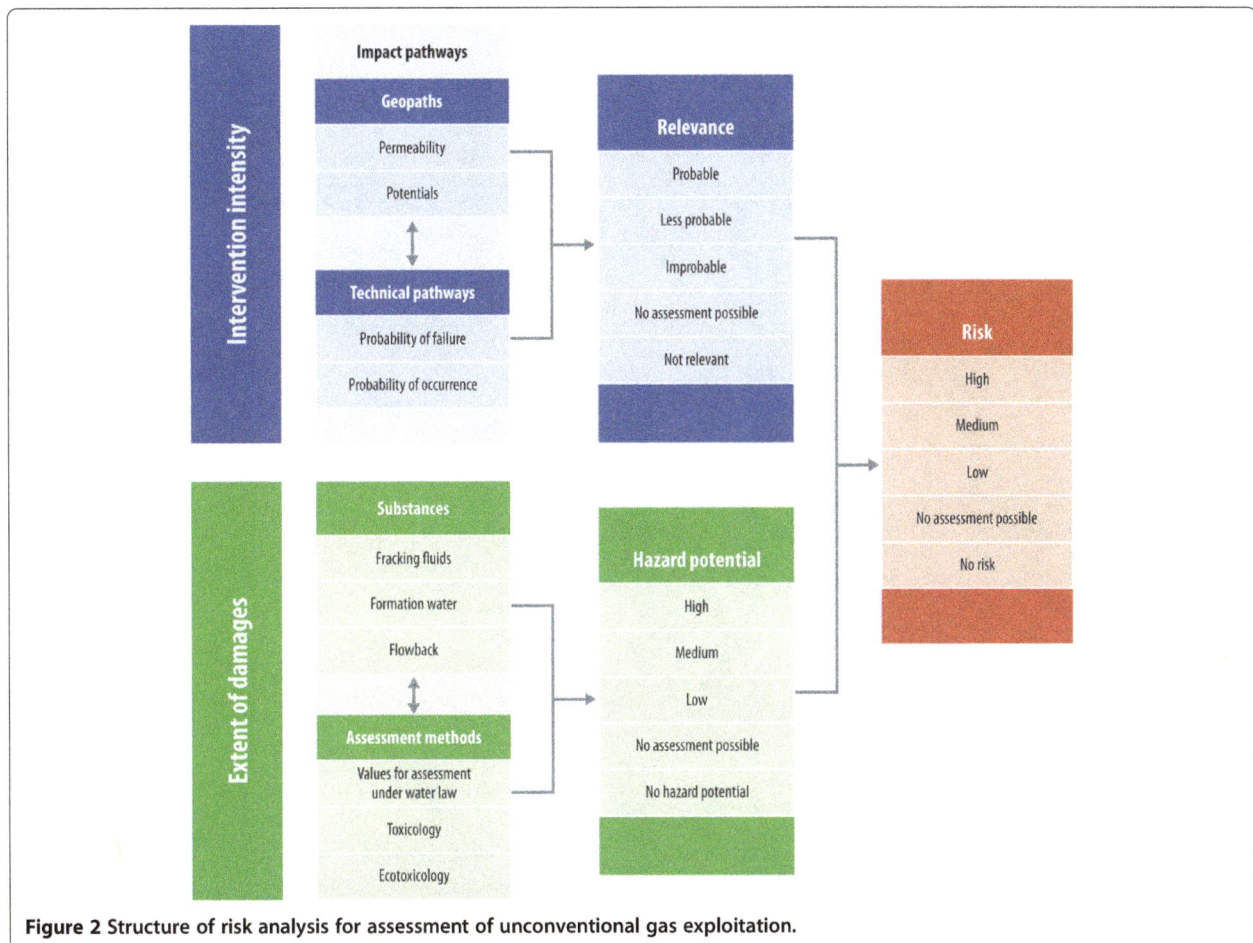

Figure 2 Structure of risk analysis for assessment of unconventional gas exploitation.

by multiplying the relevance of the impact pathway(s) and the hazard potential of the pertinent fluids (fracking fluids and formation water). Since the state of knowledge does currently not allow for numerical calculations, we propose a five-part scale to evaluate the relevance of impact pathways and the hazard potential of the fluids involved (Figure 2).

Impact pathways

Potential water-related impact pathways are shown schematically in Figure 3, considering both technical and geological impact pathways. In most cases, failures of technical systems need to occur (such as failures of the well casing) for activating potential geological impact pathways (such as migration along faults), except in the fracking horizon, where no technical barrier is in place. Technical impact pathways could be quantified by probabilities of occurrence or probabilities of failure if data suitable for the German geological, technical and legal conditions were available. For a geological impact pathway to be relevant for substance migration, both permeability and hydraulic potentials must be considered for each geosystem site-specifically. Without suitable numerical

quantification, however, the relevance of geological impact pathways can be estimated only with great uncertainties, for example using worst-case approaches.

Pathway group 0 refers to (pollutant) discharges that occur directly at the ground surface, and especially in handling of fracking fluids (transport, storage, etc.) or flowback (e.g. via accidents or improper handling).

Pathway group 1 refers to potential (pollutant) discharges and migration along wells, i.e. to artificial underground pathways. With regard to the impact pathways involved, a distinction has to be made between production wells and old wells, such as wells from other explorations and uses.

Pathway group 2 comprises all impact pathways along geological faults. Significantly, the permeability along any given fault can vary, section-wise. Whereas deep-reaching, continuous faults can often be monitored, since the near-surface locations of their outcrops are usually known, faults that affect only parts of the overburden are difficult to monitor.

Pathway group 3 comprises extensive rise, as well as lateral spreading, of gases and fluids through geological strata (for example, via an aquifer), without preferred pathways similar to those described for pathway groups

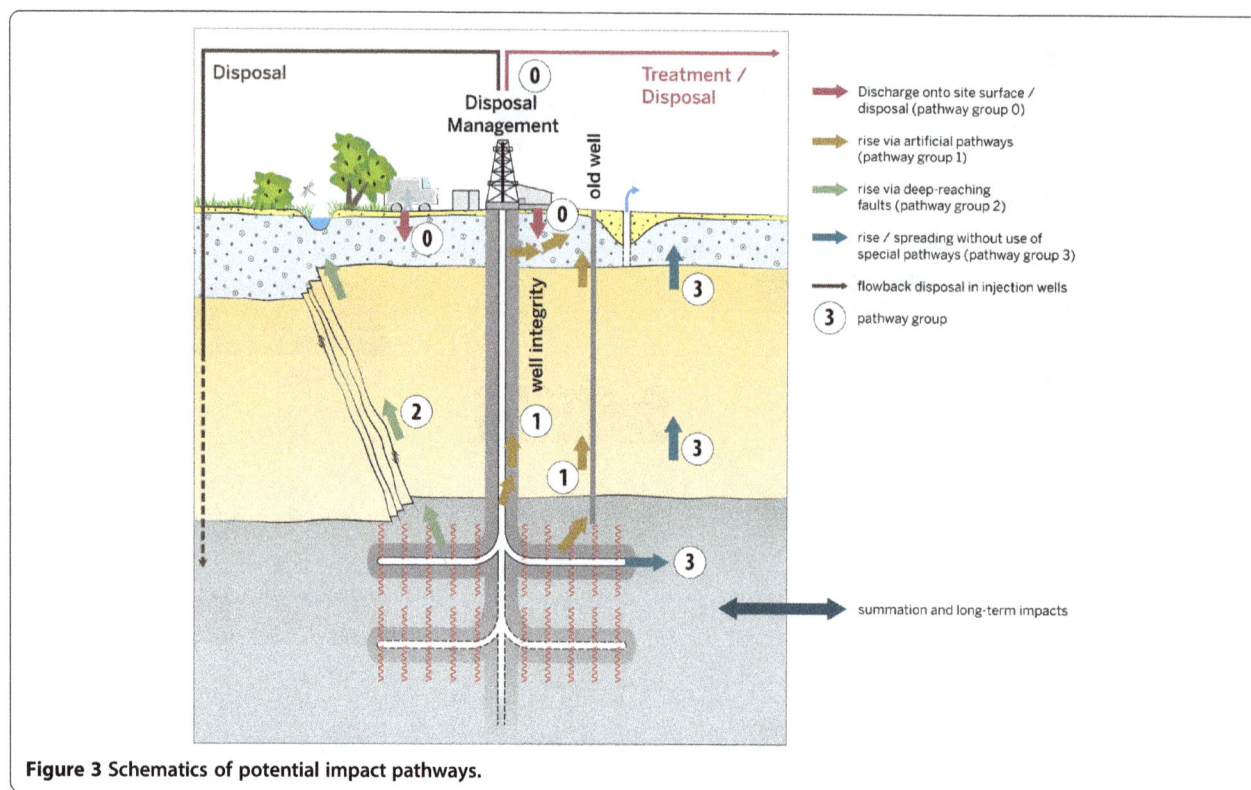

Figure 3 Schematics of potential impact pathways.

1 and 2. Impact pathways in pathway group 3 depend primarily on the prevailing geological and hydrogeological conditions.

Summation and combination effects of the aforementioned impact pathways must be taken into account appropriately. Since many flow processes in the deep underground take place slowly, the relevant long-term impacts need to be considered. Such estimation is possible only on the basis of an extensive understanding of the geological and hydrogeological conditions prevailing in deep underground horizons, although not enough data of the studied geosystems are currently available to support conceptual or even numerical models.

Furthermore, the flowback disposal needs to be assessed as additional impact pathway, especially if flowback disposal is via injection into underground disposal wells.

Fracking fluids
Overview
The fracking fluid is the hydraulic medium used for applying pressure to the rock strata inducing fracturing. With the fracking fluid, proppants (such as quartz sand) are transported into the created fractures in order to keep fractures from closing under the pressure of the surrounding rock and, thus, to ensure that the pathways created remain accessible for gas migration during the production phase. Fracking fluids usually contain a variety of chemical additives, with functions such as facilitating transport of

proppants into fractures, preventing formation of precipitates, microbiological growth, formation of hydrogen sulphide, swelling of clay minerals, corrosion, and reducing fluid friction at high pump rates. Table 2 provides an overview of the functions of certain additives.

In the following we present information on the fracking fluids and additives that have so far been employed in Germany. We then presented a method for assessing the hazard potentials of the fracking fluids employed with regard to groundwater, especially with regard to human use of groundwater as drinking water, and as part of natural cycles. In applying the method we assess selected fracking fluids used in Germany to date and possible new improvements of such fluids.

Fracking fluids used in Germany
We relied primarily on publicly accessible data to obtain information on the fracking fluids used in unconventional reservoirs in Germany [23]; only in some cases information from non-publicly accessible sources were obtainable [24]. The information on the composition of the fracking fluids used is based mainly on analyses of safety data sheets of the commercial products used to prepare fracking fluids. It has been found that these safety data sheets are often the only available source of information on the identity and the concentrations of the additives used. For approval authorities, this situation creates considerable uncertainties and lack of knowledge

Table 2 Functions of additives used in fracking fluids (based on [4,9])

Additive	Function
Proppants	Keeping the fractures created open under the pressure of the surrounding rock and allows gas/fluid to flow to the well bore
Scale inhibitors	Preventing deposits of poorly soluble precipitates, such as carbonates and sulphates
Biocides	Preventing bacterial growth, biofilm formation and formation of hydrogen sulphide by sulphate-reducing bacteria
Iron control	Preventing iron-oxide precipitation
Gelling agents	Improving proppant transport
High-temperature stabilizer (temperature stabilizer)	Preventing gel decomposition at high temperatures within the target horizon
Breakers	Reducing the viscosity of gel-containing fracking fluids for depositing proppants
Corrosion inhibitors	Protecting against equipment corrosion
Solvents	Improving the solubility of additives
pH regulators and buffers (pH control)	Controlling the pH of tracking fluids
Crosslinkers	Increasing viscosity at higher temperatures, to improve proppant transport
Friction reducers	Reducing friction within frac king fluids
Acids	Pretreating perforated sections of the well, and cleaning them of cement and drilling mud; dissolving acid-soluble minerals
Foams	Supporting proppant transport
H2S scavengers	Removing toxic hydrogen sulphide to protect equipment against corrosion
Surfactants	Reducing surface tension of fluids
Clay stabilizers	Reducing swelling and migration of clay minerals

regarding the identity and the quantities of additives actually injected into the borehole.

Quantities used Information on fluid volumes was available for a total of 30 fracking fluids used in various unconventional reservoirs (and in one conventional reservoir) in Germany between 1982 and 2011. Most of the reservoirs in which the fluids were injected were tight gas reservoirs in Lower Saxony. The quantities used varied considerably, depending on the type of fracking fluid and the characteristics of the reservoirs. The quantities of fracking fluids used per frac ranged from less than 100 m^3 to more than 4,000 m^3. With the modern gel fluids used since 2000, an average of about 100 t of proppants and about 7.3 t of additives (of which usually less than 30 kg were biocides) were injected per frac. The quantities used can be quite large especially with multi-frac stimulations and/or use of slickwater fluids: for example, a total of about 12,000 m^3 of water, 588 t of proppants, and 20 t of additives (of which 460 kg were biocides) were injected into the "Damme 3" borehole in three frac operations in 2008.

Commercial hydraulic fracturing products According to the available information, at least 88 different hydraulic fracturing products have been used to prepare fracking fluids in Germany. However, since data are available on only 21 fracking fluids (corresponding to about 21% of the

approximate 300 fracs carried out in Germany), it must be assumed that other products have also been employed. For 80 of the 88 products, we were able to obtain manufacturers' or importers' safety data sheets that were either current or valid at the time the fracs were carried out. Evaluation of the available 80 safety data sheets revealed that:

- 6 products are classified as toxic,
- 6 are classified as dangerous to the environment,
- 25 are classified as harmful,
- 14 are classified as irritant,
- 12 are classified as corrosive, and
- 27 are classified as non-hazardous

according to directives 67/548/EEC or 1999/45/EC, respectively. Several products are classified in more than one hazard class. With respect to the German water hazard classification (Wassergefährdungsklasse WGK), the commercial products were classified as follows according to the information in the safety data sheets:

- 3 preparations are classified as severely hazardous to waters,
- 12 preparations are classified as hazardous to waters,
- 22 preparations are classified as low hazardous to waters,

- 10 preparations are classified as not hazardous for water.

A total of 33 of the safety data sheets available to the study authors provided no information on the water hazard class of the product.

Fracking additives Information on the fracking additives used in the hydraulic fracturing products was available to the study authors for 28 fracking fluids. Those fluids were used in about 25% of 300 fracs carried out in Germany. Evaluation of those 28 fracking fluids showed that, overall, at least 112 substances/substance mixtures have so far been used in Germany. For 76 of the 112 substances/substance mixtures, either unique Chemical Abstracts Service (CAS) numbers were provided or it proved possible to correct or determine the CAS number on the basis of a unique given substance name. A total of 36 substances/substance mixtures could not be uniquely identified, either because their composition was unknown or because the available safety data sheets referred only to unspecific chemical group names (such as aromatic ketones, inorganic salts, etc.).

Hazard potentials of fracking fluids
Comparison of two fracking fluids
Since recipes for fracking fluids are normally tailored to specific reservoirs, the hazard potentials of each fluid need to be assessed site-specifically. Based on the assessment method described in the Methods section, we have assessed the two fluids used to date in shale gas and CBM reservoirs in Germany as two examples. Planned improvements of fracking fluids were taken into account by assessing two fluids mentioned by an operator as potentially being suitable for shale gas reservoirs and,

possibly, CBM reservoirs (improvements of slickwater and gel fluids) [4].

The hazard potentials of the slickwater fluid employed in the shale gas reservoir in 2008 and a planned improved composition are compared in Table 3. The assessment concludes that the slickwater fluid used in 2008 has a high toxicological and ecotoxicological hazard potential. In the improved fracking fluid, three hazardous additives that were still being used in 2008 are replaced by substances with considerably lower hazard potentials. However, also the improved fluid seems to exhibit a high hazard potential, because of employing high concentrations of a formaldehyde-forming biocide, for which little data is available for assessing its behaviour, fate, toxicity, and formation of degradation products. The replacement of the three hazardous additives that were still being used in 2008 by substances with considerably lower hazard potentials must be critically evaluated, since the underlying database for assessing those additives has been available for years, suggesting that service companies, operators, and/or authorities in the past have not always adequately considered the possibilities of substituting hazardous additives.

Current developments aiming at reducing the numbers of additives used, at finding substitutes for substances that are highly toxic, carcinogenic, mutagenic, or toxic for reproduction, and at reducing or replacing biocidal agents, point to potential progress in the development of environmentally compatible fracking fluids. However, the authors can currently not evaluate the feasibility or progress of such efforts.

Flowback
Quantities and composition
After pressure has been applied to the gas-bearing formation, some of the injected fracking fluids are recovered

Table 3 Composition and hazard potential of two slickwater fluids

Function	Fracking fluid used at Damme 3				Planned improvement of a slickwater fluid			
	Additive	Dissolved concentration in fracking fluid	Risk quotient based on toxicological assessment	Risk quotient based on eco-toxicological assessment	Additive	Dissolved concentration in fracking fluid	Risk quotient based on toxicological assessment	Risk quotient based on eco-toxicological assessment
Clay stabilizer	Tetramethyl-ammonium chloride	520 mg/L	1,733,000	Database insufficient (>2,600,000)	Cholinium chloride	750 mg/L	< 43	210
Friction reducer	Hydrotreated light petroleum distillates	220 mg/L	2,200	55,000	Butyl diglycol	350 mg/L	40	6,600
Surfactant	Ethoxylated octylphenol	36 mg/L	120,000	20,000	Polyethylene glycol monohexyl ether	130 mg/L	743	760
Biozide	Isothiazolinone derivative	4 mg/L	7,520	72,000	(Ethylenedioxy)-dimethanol	1,000 mg/L	10,000,000	Database insufficient (139,000)

Assessment of the fracking fluid used 2008 for hydraulic fracturing in a shale gas reservoir at Damme 3 and of a planned improvement based on human- and ecotoxicologically derived risk quotients.

along with formation water and gas extracted from the well. The so-called flowback consists of varying proportions of injected fracking fluids and co-extracted formation water. Initially, fracking fluids account for the larger share of flowback; later, formation water predominates. As a result of various hydrogeochemical processes that can occur within the reservoir horizon (Figure 4), flowback can contain other substances in addition to fracking additives and formation water constituents.

At the high pressures and temperatures prevailing in the target horizon, injected fracking additives may undergo chemical transformation and decomposition reactions in the presence of saline formation water. Microbiological decomposition reactions may occur as soon as the effects of the injected biocides diminish. In the process, metabolites can form that can pose toxicological and ecotoxicological risks potentially even exceeding the hazard posed by the parent substances that were injected.

Because the characteristics of formation water are always reservoir-specific, and because the proportions of extracted fracking additives vary, the characteristics of flowback have to be individually assessed for each site and pertinent time. Little information is available about the constituents of formation water in shale gas and CBM reservoirs in Germany, such as information about primary, secondary, and trace components, dissolved gases, organic substances, and NORM (Naturally Occurring Radioactive Material); regional and depth-oriented data is often missing.

At present, there is a lack of reliable analyses and mass balances that would allow for quantification of the variable mixing fractions, the fraction of the extracted fracking fluid, and possible reaction products. To date, no systematic measurements have been carried out for the purpose of identifying transformation and decomposition products in the flowback. Assessments of flowback from the "Damme 3" borehole carried out by Rosenwinkel et al. [25] concluded that only 8% of injected fracking fluids were being recovered as part of the flowback. Even though that percentage can be expected to increase as production continues, it seems certain that a substantial proportion of the fracking additives injected remains underground.

Disposal of flowback

Possible technical processes for treating flowback have been reviewed by Rosenwinkel et al. [25] concluding that none of those treatment options, at present, qualifies as "best available technology" within the meaning of the German Federal Water Act. In general, the following options are possibly suitable for disposing or recycling of flowback in Germany:

- Underground injection via disposal wells,
- treatment for discharge into surface water,
- treatment for discharge into the sewer system,
- recycle and reuse in future hydraulic fracturing operations.

Operators currently refer to underground disposal of flowback as an important prerequisite for (cost-effective) exploitation of unconventional gas reservoirs. From the perspective of the study authors, flowback disposal via deep-underground injection can entail risks requiring site-specific risk assessment and monitoring.

Conclusions

There is general lack of basic information that would be needed for any well-founded assessment of the pertinent

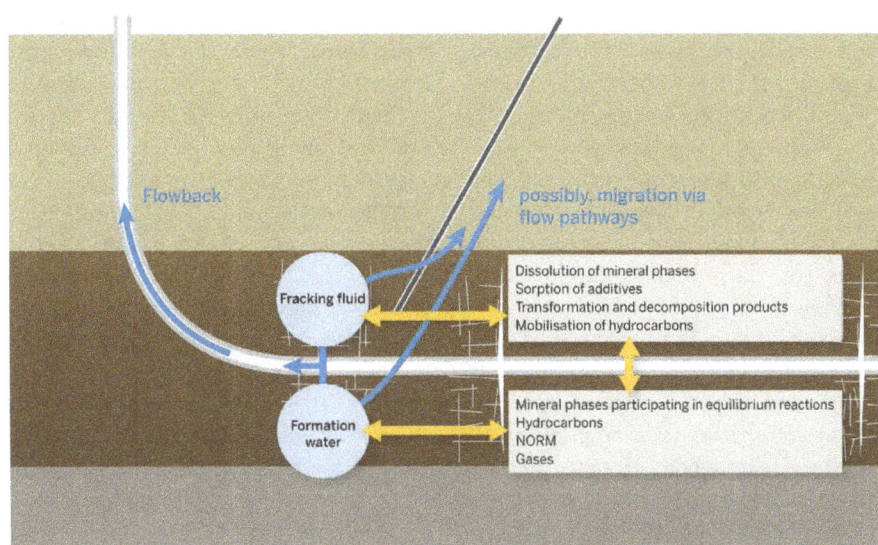

Figure 4 Hydrogeochemical processes affecting flowback formation via mixing of fracking fluids and formation water.

risks and the degree to which they can be controlled by technical means. Examples of such missing data include information regarding the structures and properties of deep geological systems (permeabilities, potential differences), the identities of the fracking additives used, and the chemical and toxicological properties of such additives. There are several reasons for this lack of information and data: (a) the information and data are not (openly) accessible, (b) the information and data have not yet been evaluated, and/or (c) there are gaps in knowledge that can only be closed through additional studies and research.

By studying selected geological systems in which shale gas or CBM reservoirs in Germany are found or assumed [4] we concluded that site-specifically certain impact pathways could be relevant for fluid migration. Little reliable data are currently available that would provide a basis for the reliable exclusion of risks to near-surface water resources. Assessment of selected fracking fluids used in unconventional gas reservoirs in Germany, along with the available information on the characteristics of flowback, have revealed that injected fluids, and fluids requiring disposal, can pose considerable hazard potentials. In summary, the study concludes that currently missing knowledge and data prevent a profound assessment of the risks and their technical controllability in Germany.

Recommendations

In light of the shortcomings of the currently available data, and of the fact that environmental risks cannot be ruled out, we recommend from a standpoint of precautionary water resources management, that above-ground and below-ground activities for unconventional gas exploitation involving fracking should not be approved for exploration or exploitation in water protection areas (classes I through III), water-extraction areas for the public drinking water supply (even if not assigned as water protection areas), mineral spa protection zones, and near mineral water reserves. These areas should be excluded for such activities. This recommendation on denial of approval should be reviewed as more data become available. In areas known to have unfavourable geological and hydrogeological conditions (groundwater potentials and known impact pathways), no exploration and exploitation of unconventional gas (via deep-drilling and hydraulic fracturing) should be allowed.

Site-specific risk assessment should be carried out with regard to any future drilling with fracking, and to drilling and use of underground disposal wells for injection of flowback. Such analyses should take account of all relevant fluids, whether introduced or encountered (fracking additives, formation water and its reaction products, and flowback), and of the relevant geological and technical impact pathways. It is recommended that use of toxicologically and ecotoxicologically hazardous

fluids, and flowback disposal in disposal wells – also in the tight gas reservoirs in Germany that have already been exploited for many years – be reassessed.

Since the potential risks of exploration and exploitation of unconventional gas projects can be reliably assessed only if reliable information on the relevant geological systems (and potential impact pathways) is available, we recommend that any exploration of gas reservoirs provides investigations of the larger regional geological and hydrogeological system.

We further recommended that additional data and experience not yet published or not yet assessed (e.g., cadaster of old wells, cadaster of disposal wells) are evaluated and results are published. We argue however that without new data it will not be possible to answer the question of whether, and where, economically exploitable unconventional gas reserves are present in Germany and which technology (with or without fracking) is suited for exploration. We thus support the idea of carrying out further exploration, including exploration involving deep drilling (but without fracking), and carrying out targeted research in the above-described framework, for the purpose of answering those questions.

We recommend that further actions are taken step-by-step. Clear criteria should be established for deciding whether or not the application of fracking should be allowed at a later time. Such criteria should cover both the hazard potential of fracking additives and the availability of reliable information about the geological and technical impact pathways involved. Clear criteria should be applied for approval of any further exploration and any later production. A catalogue of criteria for approval should be developed step-by-step, applying transparent approaches involving public participation.

We recommend that research and development are intensified in areas such as the long-term integrity of wells, techniques available for forecasting the widths and lengths of fractures generated by fracking, and the development of fracking fluids with lower hazard potential. Practical application of the relevant research findings should be monitored scientifically.

With regard to EIA obligations, we recommend that fracking projects be subject to general federal EIA obligations, and that such obligations include an "opening clause" to allow participation of the German federal states. The public participation required under EIA legislation should be expanded to include a project-monitoring component, since many findings regarding projects' potential environmental impacts cannot be obtained until the projects are actually underway. Careful review of requirements under water law should be assured, via clarification of pertinent requirements, and via a) introduction of an integrated project-approval procedure to be directed by an environmental authority subordinate to the Ministry for the Environment, or b) integration of mining authorities within the environmental administration.

The following two aspects are of central importance with regard to any continuation of exploration and exploitation of unconventional gas in Germany, regardless of the procedures applied: all work processes and results should be fully transparent, and all stakeholders should exercise trust in their dealings with each other. Efforts should include the establishment of a publicly accessible cadaster listing all fracking measures carried out in the past, along with the quantities and the compositions of the fluids used.

In the following sections, we propose special recommendations for further steps towards exploitation of unconventional gas reservoirs in Germany. The focus of the recommendations is on the next phase of pilot exploration, especially, exploration in geological systems for which no information, or very little information, is yet available concerning unconventional gas reservoirs they may contain. The objectives of the recommendations include:

- identifying hydrogeological problematic areas, and possible impact pathways, at an early stage, and proposing measures for ongoing monitoring,
- reducing the hazard potential of the fracking fluid potentially used.

Special recommendations with regard to the area of geological systems and the aquatic environment

The cause-and-effect relationships between deep-reaching and near-surface groundwater flow systems are of particular importance with regard to the water-related environmental impacts of unconventional gas exploitation projects. Such assessments require a detailed understanding of the hydrogeological systems involved, including:

- Conceptual hydrogeological models should be prepared that support reliable risk assessment for all potential impact pathways. The scope of such conceptual models should be large enough to support assessment of the impacts of exploration and exploitation of unconventional gas – via fracking – both for the specific sites and with regard to the large geological systems (system-oriented exploration).
- For areas in which water-related environmental impacts cannot be ruled out, numerical groundwater flow models should be prepared/refined in order to quantify the pertinent risks. This may involve preparing a regional model that can serve as a basis for local numerical models in the exploration area.
- The aforementioned numerical models have to be continually verified and calibrated on the basis of

data and information obtained through monitoring (both prior and during the project). For monitoring to be effective, it must be based on an adequate understanding of the system involved. At the same time, the understanding of the system involved (conceptual or numerical model) can be improved by the monitoring data obtained. Monitoring-based project control requires meaningful indicators and an evaluation system. Ultimately, options must be available for stopping, limiting, or reversing any undesired developments. The models resulting from the aforementioned work steps provide an important basis for authorities' decisions regarding the approval of submitted projects, as well as possible ancillary provisions under water law.

- The necessary regional and local models must be provided by the mining company within the authorization procedure under mining law and water law, based on the requirements imposed by the competent mining and water authorities. A fracking project can be approved only when enough pertinent knowledge has been gained and adequate precaution has been taken to exclude any adverse impact on exploitable water resources.

Special recommendations with regard to the area of substances

Assessment of selected fracking fluids used in unconventional gas reservoirs in Germany, along with the available information on the characteristics of flowback, have revealed that injected fluids, and fluids requiring disposal, can pose considerable hazard potentials. In light of the gaps in knowledge, uncertainties and data deficits identified via research and assessment for the present study, the following recommendations for action are seen as important:

- Complete disclosure of all substances used, with regard to substance identities and quantities.
- Assessment of the toxicological and ecotoxicological hazard potentials of substances used, and provision of all physical-chemical and toxicological substance data required by the mining company. If relevant substance data are lacking, the gaps in the data must be eliminated – if necessary, via suitable laboratory tests or model calculations. In the process, the effects of relevant substance mixtures must be taken into account.
- Substitution of unsafe substances (especially, substances that are highly toxic, carcinogenic, mutagenic, or toxic for reproduction), reduction or substitution of biocides, reduction of the numbers of additives used, lowering of concentrations used.

- Determination and assessment of the characteristics of site-specific formation water, with regard to constituents of relevance to drinking water quality (salts, heavy metals, Naturally Occurring Radioactive Material – NORM, hydrocarbons).
- Determination and assessment of the characteristics of site-specific flowback, with regard to constituents of relevance to drinking water quality (salts, heavy metals, NORM, hydrocarbons), and with regard to additives used (primary substances) and their transformation products (secondary substances); determination and assessment of the proportion of fracking fluids recovered with the flowback.
- Determination of the behaviour and fate of substances in the fracking horizon, via mass balancing of the additives used.
- Modelling of substance transport, for assessment of possible risks to any exploitable groundwater, from any migrating formation water and fracking fluids.
- Technical treatment and "environmentally sound" disposal of flowback, including description of all technically feasible treatment processes and of the possibilities for reusing substances. If injecting flowback into disposal wells, conducting of a site-specific risk analysis is recommended.
- Monitoring and system-oriented examination, including installation of near-surface groundwater observation wells to determine the reference condition with regard to additives and methane; if appropriate, installation of deep groundwater observation wells to determine the characteristics of formation water and the relevant hydraulic potentials.

Methods

Under German water law, the key requirement to be applied in assessing releases of substances into the groundwater is that releases must not adversely affect the water quality (Art. 48 (1) WHG, Federal Water Resources Management Act). An adverse effect on the quality of near-surface groundwater (i.e. of the exploitable groundwater that is integrated in natural cycles) has occurred, if water quality has worsened more than slightly.

An adverse effect on the water quality of groundwater must be assumed if relevant legal and sub-legal limit values, guide values, maximum values, and especially the "Geringfügigkeitsschwellenwerte" (de minimis thresholds, GFS) of the German Federal/State Working Group on Water (LAWA) [26] are exceeded in any exploitable groundwater. These de minimis thresholds are primarily based either on the maximum permitted

concentration specified by the Ordinance on Drinking Water (Trinkwasserverordnung), or, if no maximum permitted concentration has yet been established, on toxicologically and ecotoxicologically derived threshold values. Thus, it is ensured that groundwater remains available as drinking water resource for human consumption, and it remains intact as a habitat and as part of natural cycles.

For the majority of the substances used as fracking additives, no de minimis thresholds or other water-law-based assessment values have yet been established. Therefore, hygienic guidance values for drinking water (GVDW – maximum concentration of a substance in drinking water that can be tolerated for a lifetime without suffering adverse effects on health) or health orientation values (HOV - precautionary value for substances that cannot (or can only partially) be toxicologically assessed [27]) and ecotoxicologically established Predicted No Effect Concentrations (PNEC - maximum concentration of a substance at which no effects on organisms of an aquatic ecosystem are expected [28]) were assessed for such substances, or derived using published methods, following the concept of LAWA [26].

Relevant for the assessment is the concentration at the location where the substance enters exploitable groundwater resources. In case of substances entering groundwater from the surface (pathway group 0, e.g. accidents during transport and preparation of fracking fluids), the relevant substance concentration for the assessment is the concentration at the groundwater surface (see page water). By analogy, in the case of a possible release from the fracking horizon (and related migration via pathway groups 1 through 3), the concentration at the base of the exploitable groundwater aquifer should be used in the assessment.

The relevant substance concentrations can properly assessed only site-specifically. For potential migration and exposure scenarios, suitable models are needed that consider relevant hydraulic and geochemical transport, mixing, decomposition, and reaction processes along the underground flow pathway. No such models are available at present that have the necessary spatial resolution.

As long as suitable models are lacking, we propose to assess hazard potentials on the basis of substance concentrations in (undiluted) fracking fluids and formation water. Based on the current state of knowledge, we consider it not suitable to presume a considerable reduction of their hazard potential due to dilution along the underground flow pathways, because along the flow path dilution occurs mainly by mixing with saline groundwater, which can have considerable hazard potential of its own (see below); thus, mixing with such water would not necessarily reduce the hazard potential of fracking fluids.

The pertinent hazard potentials of the fluids are assessed on the basis of the individual constituents, calculating substance-specific risk quotients of substance concentrations and assessment values (GFS, GVDW, HOV, or PNEC):

$$\mathrm{Risk\,Quotient} = \frac{\text{substance concentration in the fluid}}{\text{assessment value}}$$

When a substance has a risk quotient < 1, no hazard potential is expected, while a risk quotient ≥ 1 represents potentially a toxicological or ecotoxicological hazard (hazard potential). In the present study, a risk quotient > 1,000 is assumed to represent a high hazard potential. This value is given as an example and has not been scientifically established; it needs to be site-specifically reviewed on the basis of exposure scenarios – using numerical models for example.

Competing interests
The authors declare that they have no competing interests.

Authors' contributions
The authors contributed in equal parts to this publication. All authors read and approved the final manuscript.

Acknowledgements
The authors would like to thank the German Federal Environment Agency (UBA) for financing the study and the project partners [Gassner, Groth, Siederer & Coll.] and TU Darmstadt (Prof. Dr. Sass) for their collaboration.

Author details
[1]IWW Water Centre, Department Water Resources Management, Moritzstrasse 26, Muelheim 45476, Germany. [2]ahu AG Wasser Boden Geomatik, Kirberichshofer Weg 6, Aachen 52066, Germany.

References
1. Bundesanstalt für Geowissenschaften und Rohstoffe: **Abschätzung des Erdgaspotenzials aus dichten Tongesteinen (Schiefergas) in Deutschland.** Hannover: 2012. http://www.bgr.bund.de/DE/Themen/Energie/Downloads/BGR_Schiefergaspotenzial_in_Deutschland_2012.pdf?__blob=publicationFile&v=7.
2. Sachverständigenrat für Umweltfragen: Fracking zur Schiefergasgewinnung: **Ein Beitrag zur energie- und umweltpolitischen Bewertung. Aktuelle Stellungnahme Nr. 18.** Berlin: 2013. http://www.umweltrat.de/SharedDocs/Downloads/DE/04_Stellungnahmen/2012_2016/2013_05_AS_18_Fracking.pdf?__blob=publicationFile.
3. Bundesanstalt für Geowissenschaften und Rohstoffe: *Energiestudie 2013. Reserven, Ressourcen und Verfügbarkeit von Energierohstoffen.* Hannover: 2013:112. http://www.bgr.bund.de/DE/Themen/Energie/Downloads/Energiestudie_2013.pdf?__blob=publicationFile&v=5.
4. Umweltbundesamt: **Umweltauswirkungen von Fracking bei der Aufsuchung und Gewinnung von Erdgas aus unkonventionellen Lagerstätten – Risikobewertung, Handlungsempfehlungen und Evaluierung bestehender rechtlicher Regelungen und Verwaltungsstrukturen.** -Gutachten im Auftrag des Umweltbundesamtes. Berlin; 2012. http://www.umweltbundesamt.de/uba-info-medien/4346.html.
5. Ministerium für Klimaschutz, Umwelt, Landwirtschaft, Natur- und Verbraucherschutz des Landes NRW: **Gutachten mit Risikostudie zur Exploration von Erdgas aus unkonventionellen Lagerstätten in Nordrhein-Westfalen und deren Auswirkungen auf den Naturhaushalt, insbesondere die öffentliche Trinkwassergewinnung.** Düsseldorf; 2012. http://www.umwelt.nrw.de/umwelt/wasser/trinkwasser/erdgas_fracking.
6. IWW Rheinisch-Westfälisches Institut für Wasser Beratungs- und Entwicklungsgesellschaft mbH: *Wasserwirtschaftliche Risiken bei Aufsuchung und Gewinnung von Erdgas aus unkonventionellen Lagerstätten im Einzugsgebiet der Ruhr. Gutachten des IWW im Auftrag der Arbeitsgemeinschaft der Wasserwerke an der Ruhr e.V. und des Ruhrverbandes.* Mülheim; 2013. http://www.awwr.de/fileadmin/download/download_2013/studie_fracking_einzugsgebiet_ruhr.pdf, http://www.ruhrverband.de/wissen/forschung-entwicklung/fracking/.
7. U.S. Environmental Protection Agency: **Evaluation of impacts to underground sources of drinking water by hydraulic fracturing of coalbed methane reservoirs.** 2004. http://water.epa.gov/type/groundwater/uic/class2/hydraulicfracturing/wells_coalbedmethanestudy.cfm.
8. U.S. Environmental Protection Agency: **Plan to Study the Potential Impacts of Hydraulic Fracturing on Drinking Water Resources.** Washington; 2011. http://water.epa.gov/type/groundwater/uic/class2/hydraulicfracturing/upload/hf_study_plan_110211_final_508.pdf.
9. Tyndall Centre: **Shale gas: a provisional assessment of climate change and environmental impacts.** Manchester; 2011. http://www.tyndall.ac.uk/shalegasreport.
10. Waxman HA, Markey EJ, Degette D: **Chemicals used in hydraulic fracturing.** In *U.S. House of Representatives Committee on Energy and Commerce Minority Staff.* Washington; 2011. http://democrats.energycommerce.house.gov/sites/default/files/documents/Hydraulic-Fracturing-Chemicals-2011-4-18.pdf.
11. New York State Department of Environmental Conservation: **Revised Draft Supplemental Generic Environmental Impact Statement. Chapter 5: Natural gas development activities & high-volume hydraulic fracturing.** New York; 2011. http://www.dec.ny.gov/docs/materials_minerals_pdf/rdsgeisch50911.pdf.
12. Ewen C, Borchardt D, Richter S, Hammerbacher R: **Risikostudie Fracking – Übersichtsfassung der Studie "Sicherheit und Umweltverträglichkeit der Fracking-Technologie für die Erdgasgewinnung aus unkonventionellen Quellen"** erstellt im Zusammenhang mit dem InfoDialog Fracking. Darmstadt; 2012. http://dialog-erdgasundfrac.de/sites/dialog-erdgasundfrac.de/files/Ex_risikostudiefracking_120518_webprint.pdf.
13. Hessischer Landtag: **60. Sitzung des Ausschusses für Umwelt, Energie, Landwirtschaft und Verbraucherschutz.** Wiesbaden; 2013. http://www.hessischer-landtag.de/icc/med/bb7/bb700690-9433-e31a-628b-31402184e373,11111111-1111-1111-1111-111111111111.pdf.
14. Landesamt für Bergbau, Energie und Geologie Niedersachsen: **Mindestanforderungen an Betriebspläne, Prüfkriterien und Genehmigungsablauf für hydraulische Bohrlochbehandlungen in Erdöl- und Erdgaslagerstätten in Niedersachsen.** Clausthal-Zellerfeld. Rundverfügung vom 31.10.2012. http://www.lbeg.niedersachsen.de/download/72198/Mindestanforderungen_an_Betriebsplaene_Pruefkriterien_und_Genehmigungsablauf_fuer_hydraulische_Bohrlochbehandlungen_in_Erdoel-_und_Erdgaslagerstaetten_in_Niedersachsen.pdf.
15. Hammerbacher Beratung & Projekte: **Statusbericht von ExxonMobil zur Umsetzung der Risikostudie Fracking.** Osnabrück. Protokoll vom 6. November 2012, Osnabrück http://www.erdgassuche-in-deutschland.de/dialog/info_dialog_fracking_status.html.
16. Ministerium für Klimaschutz, Umwelt, Landwirtschaft, Natur- und Verbraucherschutz des Landes Nordrhein-Westfalen, Ministerium für Klimaschutz, Umwelt, Landwirtschaft, Natur- und Verbraucherschutz des Landes Nordrhein-Westfalen: **Pressemitteilung vom 07.09.2012 - Umweltministerium und Wirtschaftsministerium legen Risikogutachten zu Fracking vor.** http://www.umwelt.nrw.de/ministerium/service_kontakt/archiv/presse2012/presse120907_a.php.
17. Remmel J: **Erdgas aus unkonventionellen Lagerstätten.** *gwf Wasser Abwasser* 2012, **11**:1121.
18. Niedersächsisches Ministerium für Wirtschaft, Arbeit und Verkehr: **Gemeinsame Presseinformation von Minister Wenzel und Lies vom 17.03.2014 - Ja zur Erdgasförderung! Nein zu umwelttoxischen Substanzen unter Tage!** http://www.mw.niedersachsen.de/portal/live.php?navigation_id=5459&article_id=123032&_psmand=18.
19. Bundesministerium für Umwelt, Naturschutz und Reaktorsicherheit: **Vorschlag zur Änderung von UVP-V und Wasserhaushaltsgesetz.** http://www.bmu.de/themen/wasser-abfall-boden/binnengewaesser/gesetzesaenderung-zu-fracking.
20. Deutscher Verein des Gas- und Wasserfaches e.V: **Stellungnahme vom 21. März 2013 zum Entwurf eines Gesetzes zur Änderung des Wasserhaushaltsgesetzes vom 7. März 2013 und Entwurf einer Verordnung zur Änderung der Verordnung über die**

Umweltverträglichkeitsprüfung bergbaulicher Vorhaben vom 11. März 2013 in Bezug auf die Umweltverträglichkeitsprüfung bei Bohrungen mit Einsatz der Fracking-Technologie. http://www.dvgw.de/wasser/ressourcen-management/gewaesserschutz/fracking/.

21. BGR, GFZ & UFZ: **Abschlusserklärung zur Konferenz "Umweltverträgliches Fracking?".** Hannover; 2013. am 24./25. Juni 2013 (Hannover-Erklärung). http://www.bgr.bund.de/DE/Gemeinsames/Nachrichten/Veranstaltungen/2013/GZH-Veranst/Fracking/Downloads/Hannover-Erklaerung-Finalfassung.pdf.

22. Gelsenwasser AG, Arbeitsgemeinschaft der Wasserwerke an der Ruhr e.V., Deutscher Brauer-Bund e.V., Verband Deutscher Mineralbrunnen e.V. & Wirtschaftsvereinigung Alkoholfreie Getränke e.V: **Gelsenkirchener Erklärung: Wasserversorger, Bierbrauer, Mineral- und Heilbrunnenbetriebe sowie Erfrischungsgetränkehersteller fordern Schutz vor Fracking.** Gelsenkirchen: 2013. (24.10.2013) http://www.gelsenwasser.de/fileadmin/download/unternehmen/presse/gelsenkirchener_erklaerung.pdf.

23. ExxonMobil Central Europe Holding GmbH: *Frack-Flüssigkeiten*; http://www.erdgassuche-in-deutschland.de/erkundung_foerderung/frac_fluessigkeiten/index.html.

24. Bezirksregierung Arnsberg: *Gewinnung von Erdgas aus unkonventionellen Lagerstätten – Erkundungsmaßnahmen der CONOCO Mineralöl GmbH in den Jahren 1994 – 1997.* Arnsberg; 2011. 61.01.25-2010-9.

25. Rosenwinkel KH, Weichgrebe D, Olsson O: **Gutachten Stand der Technik und fortschrittliche Ansätze in der Entsorgung des Flowback des Instituts für Siedlungswasserwirtschaft und Abfall (ISAH) der Leibniz-Universität Hannover zum Informations- und Dialogprozess über die Sicherheit und Umweltverträglichkeit der Fracking-Technologie für die Erdgasgewinnung.** Hannover: 2012. http://dialog-erdgasundfrac.de/sites/dialog-erdgasundfrac.de/files/Gutachten%20zur%20Abwasserentsorgung%20und%20Stoffstrombilanz%20ISAH%20Mai%202012.pdf.

26. LAWA – Bund/Länder-Arbeitsgemeinschaft Wasser: **Ableitung von Geringfügigkeitsschwellen für das Grundwasser.** Düsseldorf: 2004. http://www.lawa.de/documents/GFS-Bericht-DE_a8c.pdf.

27. Umweltbundesamt: **Bewertung der Anwesenheit teil- oder nicht bewertbarer Stoffe im Trinkwasser aus gesundheitlicher Sicht. Empfehlung des Umweltbundesamtes nach Anhörung der Trinkwasserkommission des Bundesministeriums für Gesundheit.** *Bundesgesundheitsbl Gesundheitsforsch Gesundheitsschutz* 2003, **46**:249–251.

28. European Commission: **Technical Guidance Document in support of Commission Directive 93/67/EEC on Risk Assessment for new notified substances, Commission Regulation (EC) No 1488/94 on Risk Assessment for existing substances and Directive 98/9/EC of the European Parliament and of the Council concerning the placing of biocidal products on the market, Part II.** Ispra; 2003. http://ihcp.jrc.ec.europa.eu/our_activities/public-health/risk_assessment_of_Biocides/doc/tgd.

5

Possibilities of using the German Federal States' permanent soil monitoring program for the monitoring of potential effects of genetically modified organisms (GMO)

Andreas Toschki[1*], Stephan Jänsch[2], Martina Roß-Nickoll[3], Jörg Römbke[2] and Wiebke Züghart[4]

Abstract

Background: In the Directive 2001/18/EC on the deliberate release of genetically modified organisms (GMO) into the environment, a monitoring of potential risks is prescribed after their deliberate release or placing on the market. Experience and data of already existing monitoring networks should be included. The present paper summarizes the major findings of a project funded by the Federal Agency for Nature Conservation (Nutzungsmöglichkeiten der Boden—Dauerbeobachtung der Länder für das Monitoring der Umweltwirkungen gentechnisch veränderter Pflanzen. BfN Skripten, Bonn-Bad Godesberg 369, 2014). The full report in german language can be accessed on http://www.bfn.de and is available as Additional file 1. The aim of the project was to check if it is possible to use the German permanent soil monitoring program (PSM) for the monitoring of GMO. Soil organism communities are highly diverse and relevant with respect to the sustainability of soil functions. They are exposed to GMO material directly by feeding or indirectly through food chain interactions. Other impacts are possible due to their close association to soil particles.

Results: The PSM program can be considered as representative with regard to different soil types and ecoregions in Germany, but not for all habitat types relevant for soil organisms. Nevertheless, it is suitable as a basic grid for monitoring the potential effects of GMO on soil invertebrates.

Conclusions: PSM sites should be used to derive reference values, i.e. range of abundance and presence of different relevant species of soil organisms. Based on these references, it is possible to derive threshold values to define the limit of acceptable change or impact. Therefore, a minimum set of sites and minimum set of standardized methods are needed, i.e. characterization of each site, sampling of selected soil organism groups, adequate adaptation of methods for the purpose of monitoring of potential effects of GMO. Finally, and probably most demanding, it is needed to develop a harmonized evaluation concept.

Keywords: GMO, Permanent soil monitoring, PSM sites, Soil organisms

Background

In the Directive 2001/18/EC on the deliberate release of genetically modified organisms (GMO) into the environment, a monitoring of potential adverse effects, including cumulative long-term effects is prescribed after placing on the market of GMO. The aim of this monitoring is to trace and identify any direct or indirect, immediate, delayed or unforeseen effects on human health or the environment to enable fast action if necessary. Therefore, the Directive divides the monitoring into two principal components (1) case-specific monitoring and (2) general surveillance [2]. The former is focused on a constant check whether the outcome of the risk assessment performed when notifying a GMO is correct. In the latter, the emphasis is to survey those effects in the environment that are unexpected. According to Directive 2001/18/EC,

*Correspondence: toschki@gaiac.rwth-aachen.de
[1] gaiac Research Institute for Ecosystem Analysis and Assessment, Kackerstr. 10, 52072 Aachen, Germany
Full list of author information is available at the end of the article

the notifier of the GMO in question is responsible for the execution of the monitoring. Where existing monitoring networks are suitable, experience and data should be included in the monitoring and interpretation process [3]. The natural function of soil as a habitat for soil organisms and thus soil biodiversity are one of the protection goals to be considered in this context. Against this background, the question needed to be answered whether the existing network of permanent soil monitoring (PSM) sites (in German: BDF = Bodendauerbeobachtungsflächen) is suitable for the purpose of monitoring of GMO. The overall aim of this contribution was to assess whether it is possible and sensible to use the PSM program of the German Federal States as part of the monitoring program required by Directive 2001/18/EC on the deliberate release into the environment of GMO [3, 4].

To this end, we discussed the following issues:

1. Relevance of PSM measurement parameters for GMO monitoring: Are the currently investigated site and soil parameters of the German PSM program relevant for the monitoring of potential effects of GMO cultivation?

2. Representativeness of PSM sites: Are the sites of the PSM program and their properties representative regarding the main ecological regions of Germany?

3. Exposure of PSM towards GMO: Have the sites of the PSM program already been exposed to GMO in the past?

4. Practicability: Which basic conditions concerning the monitoring sites have to be considered?

In conclusion, possible adaptions of the federal permanent soil monitoring and/or complementary monitoring modules for the GMO monitoring were formulated.

Theoretical background
The German permanent soil monitoring program
The aim of the German permanent soil monitoring program is, amongst others, to assess the current condition of soils and their monitoring in the long term to detect harmful soil changes and to project the function of soils [5]. To this end, there are 794 German PSM sites representing the regional soil forms, main land use and special pollution scenarios (Fig. 1). 344 of these are field sites, 146 grassland, 247 forests and the rest are special sites. Bremen, Berlin and Rhineland-Palatinate have no recorded PSM

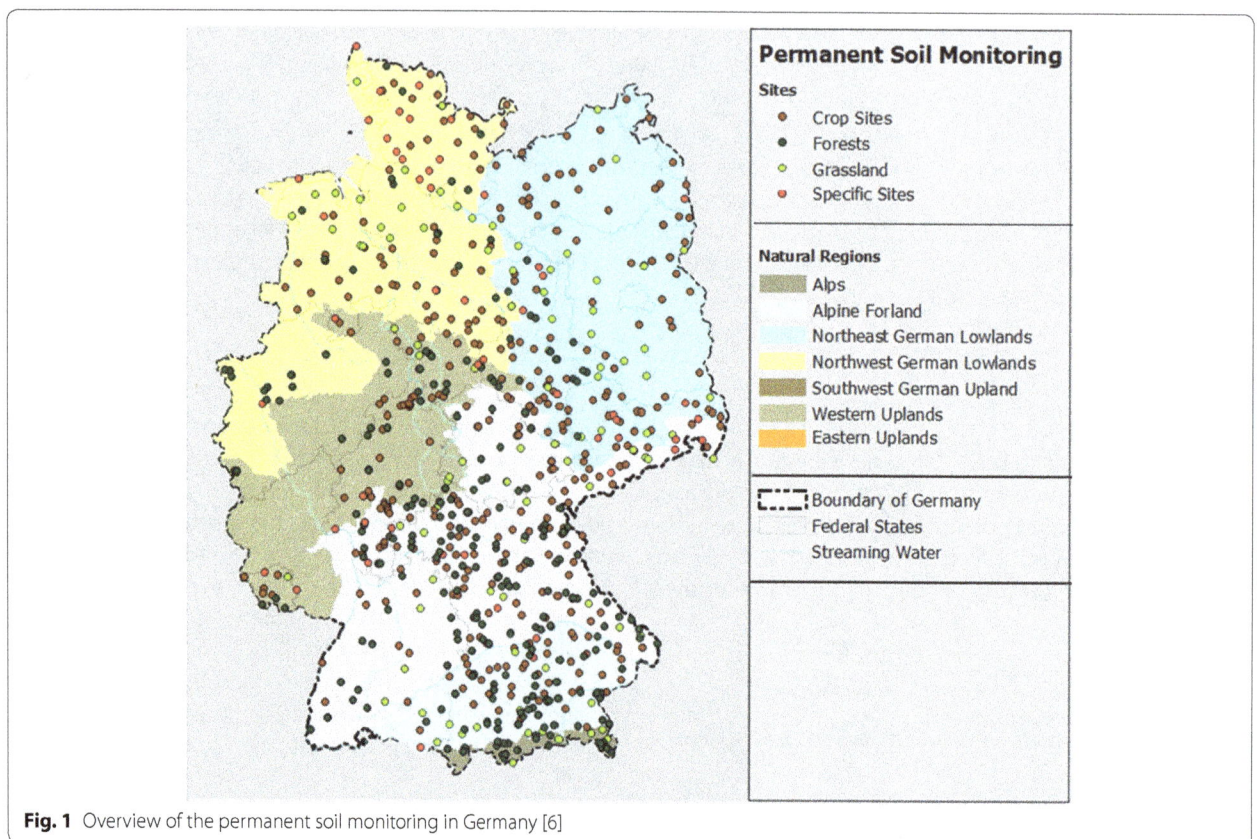

Fig. 1 Overview of the permanent soil monitoring in Germany [6]

Possibilities of using the German Federal States' permanent soil monitoring...

49

sites at the time. The largest federal state Bavaria harbours the highest share of all federal states (289 sites). Since 1990, the establishment and maintenance of PSM sites are standardized according to ISO guideline 16133 (ISO 2004). Sampling takes place on about 2500 m² large plots that are divided into a core area (100 m²) and a surrounding border area for special investigations and containing a soil profile pit. There are mostly "basic" and a few "intensive" PSM sites. On the basic PSM sites, the long-term survey of changes of the soil condition is performed through comprehensive measurement of biological, chemical and physical soil parameters and additional information such as management practices. Depending on the parameter, the measurements are repeated in intervals ranging from 1 to 10 years. On the intensive PSM sites, there are additionally numerous continuous measurement devices, also recording atmospheric deposition and the chemical composition of liquid soil phase to investigate dynamic soil processes whose changes may occur in the short term. Measurement results are consolidated in an IT system (database "BDF— Bodendauerbeobachtungsflächen") at the German Federal Environmental Agency within the scope of the administrative agreement of federal and state agencies regarding data exchange to enable cross-national evaluations.

Structure and function of soil organism communities

Soil biota are thought to harbour a large part of the world's biodiversity and to govern processes that are regarded as globally important components in the cycling of organic matter, energy and nutrients [7]. Rough estimates of the soil biodiversity indicate several thousands of invertebrate species apart from the largely unknown microbial and protozoan diversity. By far the most dominant groups of soil organisms, in terms of numbers and biomass, are the microbial organisms, i.e. bacteria and fungi [8]. Besides these organisms, soil ecosystems generally contain a large variety of animals, such as protozoa (bacterivores, omnivores, predators), nematodes (bacterivores, fungivores, omnivores, herbivores and predators), micro-arthropods such as mites (bacterivores, fungivores, predators) and collembolans (fungivores and predators), enchytraeids and earthworms (both mainly saprophagous). In addition, a high number of macrofauna species (mainly arthropods such as beetles, spiders, diplopods and chilopods or snails) are living in the uppermost soil layers, the soil surface and the litter layer. Anthropogenic activities clearly influence soil biota; most strongly in industrial or urban areas where only very few species can survive. Modern agricultural practices characterized by high levels of inorganic fertilizer additions, the use of pesticides and soil tillage are known to affect the diversity of the soil community, leading to the local loss or extinction of various groups of organisms [9].

With its large diversity and complexity, the soil community has a strong impact on soil processes, and the way in which these processes may vary in time and space. Most noteworthy are:

- decomposition of organic matter, thus regulating the cycling of nutrients;
- fixation of nitrogen from the air, making it available for plants;
- degradation of anthropogenic compounds such as pesticides;
- stabilization of soil aggregates, specifically by building clay-humus complexes;
- improval of soil porosity due to burrowing activities;
- influencing soil pH by nitrification and denitrification;
- being prey for many aboveground organisms.

As these processes also determine nutrient availability for take-up by plants, the belowground decomposer food web interactions also influence aboveground primary productivity and carbon sequestration [10]. In fact, plant productivity appears to increase by a reduced turnover of the microbial biomass due to stabilized carbon content and soil pH. The soil biomass is known to process over 100,000 kg of fresh organic material each year per hectare (25 cm top soil layer) in many agricultural systems. This processing includes the decomposition of dead organic matter by microbes as well as the consumption and production rates in the soil community food web [11–13]. The soil food web is defined as the structure and interactions across and between the communities of soil-living organisms and which are linked by conversions of energy and nutrients as one organism eats another. Therefore, most food web models merely provide a way to connect the dynamics of populations to the dynamics in ecological pathways within the cycling of matter, energy and nutrients [14].

Exposure of soil organisms towards GMO

Information on the exposure of soil organism towards GMO can be found in literature [15–19]. Furthermore, the exposure of soil organisms has been intensively studied with organic chemicals [20–23]. These sources have been used to compile an overview on the exposure of soil organisms towards GMO.

Exposure pathways of GMO for soil organisms

The definition and assessment of exposure pathways serve the purpose to estimate, whether or to what extent soil organisms (individual populations of a single species or whole communities) are confronted with any of the following impacts during GMO cultivation [20, 21]:

- GMO-specific active substances (e.g. *Bacillus thuringiensis* Cry proteins) or their metabolites;
- physiologically altered GM plant components (e.g. a modified starch content);
- agricultural management practices that would not be performed to the same extent without GMO cultivation (e.g. an increased application of herbicides).

Thus, the possible exposure pathways can be classified in various ways:

- organism-related approach [22]: direct and indirect exposure towards living or dead GMO materials or by uptake through the food chain; this classification is mainly based on experience with GM maize and the *Bt* toxin.
- GMO-related approach [24]: exposure through transport of GMO materials via pollen dispersal.

Factors influencing the exposure of soil organisms to GMO

To assess the potential risks of GMO, the genetically modified plant as a whole and not only the isolated genetically modification must be taken into consideration. An affected organism can only react on that part of the GMO (toxin, dead plant material, pollen, etc.) which is bioavailable [20] or bioaccessible [23, 25]. The fraction of potentially harmful GMO material that is reaching the body tissues especially blood or lymph is relevant independently from the respective uptake pathway (e.g. water or food). The pathways of uptake depend on the individual species, while the bioavailability depends on site and soil characteristics. In addition, processes such as the biological degradation of the GMO material have to be taken into account, which are a function of time.

There are additionally also morphological, physiological and behavioural factors which strongly influence the exposure of soil organisms towards GMO [21]. Based on their biology, two groups of soil organisms can be identified which differ strongly in their way to take up chemicals: soft-bodied organisms, i.e. nematodes, enchytraeids, earthworms and hard-bodied organisms, i.e. mainly arthropods such as spiders, mites, collembolans, diplopods, isopods, or chilopods. Arthropods have special organs for water and oxygen uptake, while soft-bodied organisms use the body surface for these purposes. In addition, both groups can take up harmful substances by food. Different feeding types can be differentiated in soil organisms [26–28]: saprophages (feeding on dead organic material), microphages (feeding on bacteria), fungiphages (feeding on fungi), phytophages (feeding on living plants) and zoophages (predators) are the most common. However, evidence increases that many soil organisms are able to use different food resources [29–31]. Based on

their mobility or preferences for different soil layers, most soil organism species can be classified into a limited number of ecological (trait) groups, e.g. epigeic, endogeic and anecic groups of earthworms [32].

Principles of biodiversity monitoring

To facilitate the use of biocoenotical data at the landscape level, a standard frame of reference is needed. A reference system can be described as certain values for the biocoenosis at certain habitat types by evaluating biocoenosis-site-relationships and will ultimately lead to the identification of threshold values with which a significant change of the biocoenosis can be indicated (Fig. 2).

Thus, a reference system for the site-specific diversity of soil organisms consists of:

- reference values: lists of species expected to occur at a certain site with its specific conditions (e.g. climate, soil factors, region);
- a quantification of deviations from these reference values that indicate impacted habitat function.

To develop reference values that link soil and site parameters with the occurrence of soil organisms, the landscape had to be classified into a limited number of "site categories". The use of the habitat classification concept compiled in the German Red Data Book on endangered habitats [34] ensures the compatibility with other monitoring approaches, nature conservation management, and prospectively also pesticide registration. When analysing different organism groups, correlations between the occurrence of species and the corresponding hierarchical level within the system of habitat types became apparent [6]. Further analysis demonstrated that the composition of communities depends on site properties. A comprehensive ecological assessment of sites requires the integration of different relevant organism groups on the species level thus at the same time covering their function (e.g. organic matter decomposition).

Methods

Relevance of PSM measurement parameters for GMO monitoring

Information on the structure and function of soil organism communities in Germany, especially at agricultural sites, was taken from the data base Bo-Info [6]; today part of the database Edaphobase [35, 36]. It contains both site-specific abiotic (e.g. soil properties) as well as biological as, for instance, species lists or data on the abundance of a certain organism group. 1744 sites (including 60 PSM sites) were covered, yielding about 42,473 datasets, 2000 of which are from PSM sites. The latter were contributed by five federal states: Brandenburg, Hamburg, North

Rhine-Westphalia, Schleswig–Holstein and Thuringia (Fig. 1).

In parallel, information was compiled when developing a guideline for the monitoring of effects of GMO on soil organisms by the Association of German Engineers (VDI; [27]). This guideline and an explanatory paper on the same subject [28] represent suitable complements to the work described in this paper.

Representativeness of PSM sites

To evaluate the question whether selected PSM sites can represent German landscapes, soil and site parameters and the occurrence of soil organisms have to be linked in a way that the whole landscape is classified into a limited number of "site categories". For this, the habitat classification concept, compiled in the "German Red Data Book on endangered habitats", was used [34, 37] (Table 1). It comprises 44 basic (first level) types with approximately 1000 hierarchically derived subtypes. This concept is already accepted by the German authorities and has been used in the areas of the European Habitats Directive [53], nature conservation management, GMO authorization and prospectively also pesticide registration. Habitat types harbour synecologically specific soil animal communities and additionally integrate the factors (in particular soil type, moisture and nutrient supply) relevant for a differentiation of communities [38, 39].

Exposure of PSM towards GMO

PSM sites are potentially exposed towards GMO, possibly influencing their usability within a monitoring program. The following exposure scenarios are generally possible:

1. GMO cultivation on the PSM site;
2. the PSM site is located outside the area of GMO influence;
3. no GMO cultivation on the PSM site itself but PSM site lies within the area of direct influence of the GMO.

Generally, it is permitted to cultivate GMO on PSM sites. However, until today GMO cultivation on PSM sites was not recorded. Hence, in the following, only the impact of GMO from "outside" influencing the PSM sites was assessed. Only pollen dispersal was considered as a pathway with a profound impact potential, while other pathways were regarded as being less relevant due to spatially and temporarily limited GMO release and the strong alteration (e.g. decomposition) of GMO material in dung, manure or sewage sludge. Literature data regarding the dispersal of pollen vary from a few metres to several kilometres [40, 41]. To ensure a sustainable use of the PSM program for the monitoring of GMO, the question had to be answered to what extent existing PSM sites had already been exposed towards GMO. The insect resistant maize variety MON810 was chosen as an example of GMO, the only GMO cultured on a broader scale in Germany so far until its cultivation was prohibited in 2009 (Fig. 3). Pollen dispersal radii of 50, 150 and 1000 m were assumed in analogy to different European buffer zone regulations towards conventional fields (e.g. Spain: 50 m, Germany: 150 m; [42]) and recommendations from field trials on maize pollen dispersal [43, 44].

These scenarios were exemplarily investigated for the federal states of Brandenburg, Hesse, Lower Saxony and Schleswig–Holstein. For this purpose, the Integrated Administration and Control System (IACS) data for these

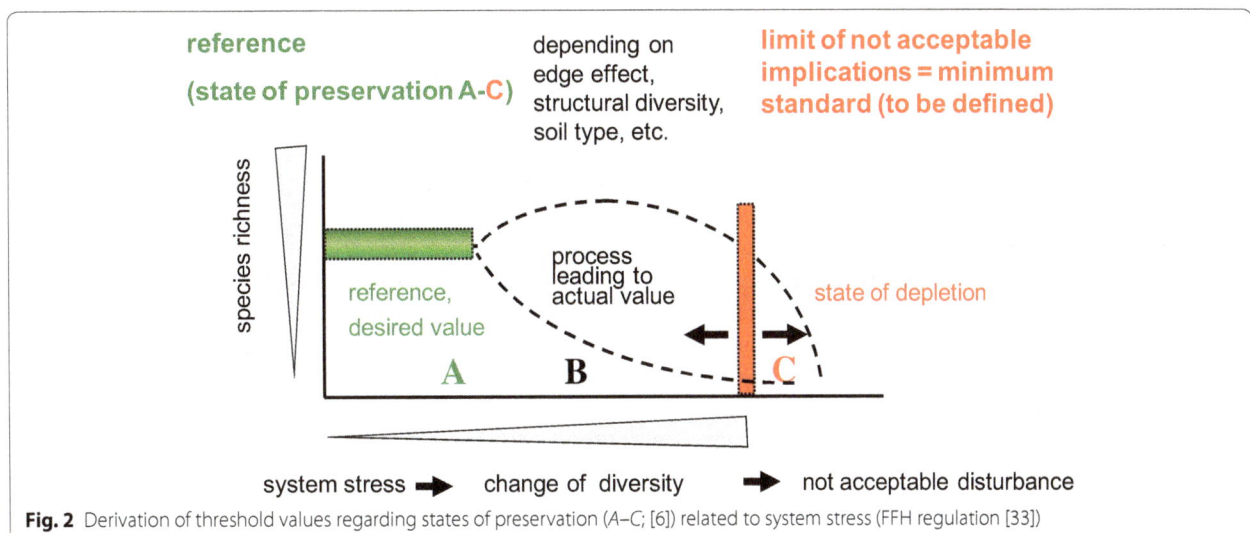

Fig. 2 Derivation of threshold values regarding states of preservation (A–C; [6]) related to system stress (FFH regulation [33])

Table 1 Habitat types, derived from the German Red Data Book on endangered habitats [33, 36], used in this study for the establishment of a reference system to evaluate the biological state of the soil

Habitat type number	Description
33.	Arable and fallow land (in the following abbreviated 'arable land')
33.01	Farmed and fallow land on shallow skeletic calcareous soil
33.02	Farmed and fallow land on shallow skeletic silicaceous residual soil
33.03	Farmed and fallow land on sandy soil
33.04	Farmed and fallow land on loess, loam or clay soil
33.05	Farmed and fallow land on peaty or half-bog soil
34.	Natural dry grasslands and grasslands of dry to humid sites (in the following abbreviated 'grassland')
34.01	Xeric grassland
34.02	Semi-dry grassland
34.03	Steppic grassland (subcontinental, on deep soil)
34.04	Dry sandy grassland
34.05	Heavy-metal grassland
34.06	Mat-grass swards
34.07	Species-rich grassland on moist sites
34.08	Species-poor intensive grassland on moist sites
34.09	Trampled grass and park lawns
43.	Deciduous and mixed woodlands and forest plantations (deciduous share >50 %) (in the following abbreviated 'deciduous forest')
43.01	Birch bog woodland
43.02	Carr woodland
43.03	Swamp forest (on minerogenic soil)
43.04	Alluvial forest
43.05	Tidal alluvial forest
43.06	Ravine, boulder-field and scree forests
43.07	Deciduous and mixed forest on damp to moist sites
43.08	Deciduous (mixed) forest on dry or warm dry sites
43.09	Deciduous (mixed) plantations with native tree species
43.10	Deciduous (mixed) plantations with introduced tree species (including subspontaneous colonisations)
44.	Coniferous (mixed) woodlands and forest plantations (in the following abbreviated 'coniferous forest')
44.01	Bog woodland (coniferous)
44.02	Natural and near-natural dry to intermittently damp pine forest
44.03	Spruce/fir (mixed) forest and spruce (mixed) forest
44.04	Coniferous (mixed) plantations with native tree species
44.05	Coniferous (mixed) plantations with introduced tree species (including subspontaneous colonisations)

Italic types at first hierarchical level. Normal types at second hierarchical level

federal states that identify agriculturally managed parcels of land were combined with the coordinates of the PSM sites and of the MON810 field sites notified to the Public Location Register of the German Federal Office of Consumer Protection and Food Safety (BVL) in a geographic information system (GIS). The pollen dispersal radii were projected as buffer zones around the MON810 field sites. The resulting maps have a residual uncertainty regarding the exact position of the MON810 field sites but represent a sufficient approximation for this exemplary exercise.

Practicability

Issues of practicability in the context of GMO monitoring can be divided into two areas: performance of biological soil monitoring in general and GMO-specific issues to be considered in such programs. The first issue will be discussed on the basis of own experiences when assessing biological soil quality at a high number of sites, both within the German PSM program but also European projects. The second issue was compiled after discussing how GMO crops are cultivated in general with farmers and representatives of the responsible agencies.

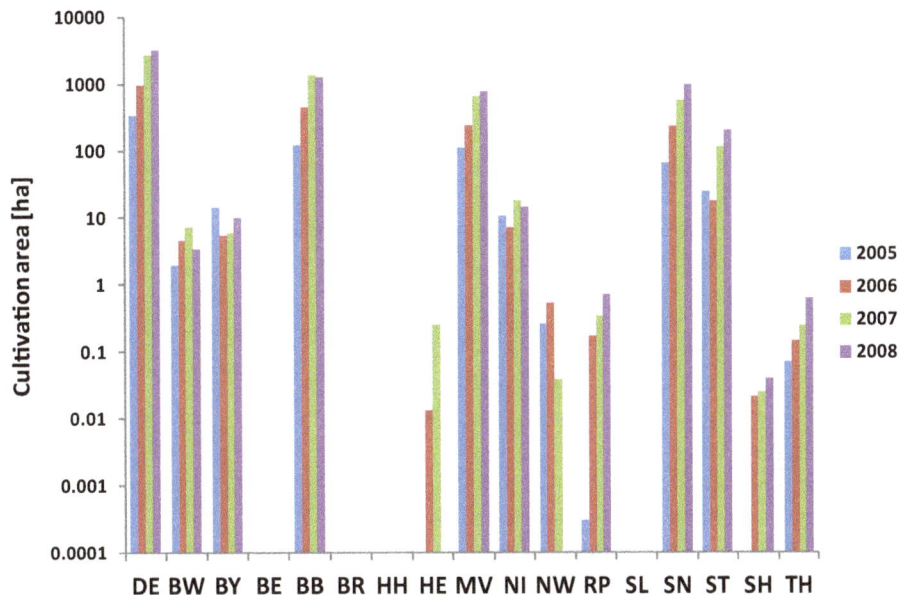

Fig. 3 GM maize variety MON810 in Germany and the federal states for the years 2005–2008. Area under cultivation, *DE* Germany, *BW* Baden-Wuerttemberg, *BY* Bavaria, *BE* Berlin, *BB* Brandenburg, *BR* Bremen, *HH* Hamburg, *HE* Hesse, *MV* Mecklenburg-Western Pommerania, *NI* Lower Saxony, *NW* North Rhine-Westphalia, *RP* Rhineland-Palatinate, *SL* Saarland, *SN* Saxony, *ST* Saxony-Anhalt, *SH* Schleswig–Holstein, *TH* Thuringia. Data source: public location register of the German Federal Office of Consumer Protection and Food Safety (BVL)

Results and discussion

Relevance of PSM measurement parameters for GMO monitoring

So far the German permanent soil monitoring program is focusing on environmental and soil parameters alone, despite the fact that several biological parameters (including soil organisms) were recommended as part of the standard sampling program [45]. Data on vegetation are only recorded in the federal state of Schleswig–Holstein. For seven federal states, data on soil microbial respiration are regularly available. In five federal states (Brandenburg, Hamburg, North Rhine-Westphalia Schleswig–Holstein and Thuringia), zoological data have been gathered (Table 2). However, even in those states only earthworms and enchytraeids have been sampled so far and usually only 2–3 times over the last 15 years. A uniform classification of all sites according to habitat type, vegetation units, etc. which is essential for ecological comparison purposes, is not applied at all.

The data sets originating from PSM sites contain little soil biological data and the data inventory fundamentally lacks comprehensive data for most organism groups:

– Lumbricidae (best data packet): data from 97 PSM sites (of 795), representative for grassland, agricultural sites and forests but allocation of PSM sites to further levels of habitat types only rudimentarily possible, gaps for some federal states;

Table 2 Number of datasets on zoology, vegetation and microbiology within the permanent soil monitoring program of the federal states of Germany

Federal state	No. of sites	Sum of data sets		
		Zoology	Microbiology	Vegetation
Baden-Württemberg	33	0	0	0
Bavaria	289	0	124	0
Berlin	0	0	0	0
Brandenburg	32	3241	178	0
Bremen	0	0	0	0
Hamburg	3	137	0	0
Hesse	68	0	0	0
Lower Saxony	46	0	0	0
Western Pomerania	90	0	635	0
North Rhine-Westfalia	21	743	60	0
Rhine land-Palatinate	0	0	0	0
Saarland	11	0	0	0
Saxony	55	0	0	0
Saxony-Anhalt	78	0	727	0
Schleswig-Holstein	37	1599	1576	2478
Thuringia	32	77	120	0
Sum	795	5797	3420	2478

– Enchytraeidae: data from 60 PSM sites; no regional representativeness (e.g. little data from Eastern Germany, Bavaria or Rhineland-Palatinate);

- Collembola, Oribatida, and other soil invertebrates in general: no data sets from PSM sites;
- Microbes: thus far no suitable data for biodiversity assessment.

In summary, the results indicate that the data basis on the occurrence of the most important soil organism groups is at date not sufficient to be used for a monitoring of GMO.

Additional sampling on representative PSM sites is recommended while all PSM sites need to be classified at least according to a standardized list of habitat types beforehand.

Up to now, there is little experience regarding the assessment of GMO-related effects in the field [39, 46–50]; see also the overview compiled by Theißen and Russel [51]. Based on the experience and considerations presented so far a technical committee of the Association of German Engineers VDI developed a guideline with proposals for identifying organism groups within the scope of GMO monitoring [27, 28]. To assess the biological soil quality, taxa should be selected according to the following criteria:

- important ecological function within the ecosystem, representativeness for a trophic level;
- close association with the mineral soil or the litter layer;
- sufficient species diversity to differentiate between sites;
- good taxonomical and ecological knowledge;
- wide distribution in Central Europe;
- existing standardized sampling methods;
- potential for routine use, e.g. regarding simplified determination methods;
- availability of data from existing monitoring programs.

At least four different taxa should be used that facilitate the inclusion of different trophic (epigeic–endogeic) as well as functional (feeding type) levels. Which taxa are most appropriate depends on the region and the land use type of the site to be monitored. Details are given in [27, 28].

Representativeness of PSM sites
The establishment of a biodiversity monitoring of GMO requires generating a data basis representative for Germany, to enable the derivation of reference values for relevant habitat types. To assess the representativeness of PSM sites/data for a monitoring of effects of GMOs on the soil biocoenosis, the classification of habitat types according to Riecken et al. [34, 37] was used. The PSM sites are representative regarding different basic soil types [45] and also according to different natural regions and ecoregions in Germany. Considering all ecoregions of Germany only three types, Swabian cuesta, Swabian alps and the Frisian marshlands are not well represented. In these three ecoregions, cultivation of genetically modified plants is highly unlikely. According to Riecken et al. [33, 36], 21 of 44 basic habitat types can be assumed as relevant for soil organism diversity [6]. Almost all PSM sites belong to three following listed relevant habitat types:

Habitat code	No. of PSM sites	Abbreviated name
33	351	Arable land
34	102	Grassland
43	242	Deciduous forest

There is a generally good representativeness of German PSM sites for arable land. However, this statement is limited to the basic habitat type "arable and fallow land" [34]. Due to their life form, for soil organisms, additional site-specific parameters (soil properties, nutrient supply, moisture, etc.) are relevant for their distribution. These parameters are reflected in the further subdivision of habitat types (2nd and 3rd level). For an allocation of PSM sites to certain habitat types, often detailed data are missing, e.g. regarding bedrock (lime, silicate, sand, etc.) or general nutrient availability (extensive, species rich, intensive, nutrient rich, species poor, etc.). A standardized and detailed data collection should be pursued since the distribution of soil animals shows the strongest correlation at lower levels of site classification [38, 39].

Exposure of PSM towards GMO
In Fig. 4, the distribution of PSM sites and MON810 field sites on the four exemplary federal states of Brandenburg, Hesse, Lower Saxony and Schleswig–Holstein are displayed.

Brandenburg is the federal state with the most intensive cultivation of MON810 in Germany in the past (ca. 1350 ha in 2007). In particular, four PSM sites within the administrative districts of Maerkisch-Oberland, Oberhavel and Uckermark were probably already exposed to the adjacent GMO cultivation in the past (Fig. 5) when assuming a pollen dispersal radius of 1000 m. Referring to the total number of 109 PSM sites located in the four considered federal states, this is a share of 3.7 %. Referring to the total number of PSM sites located within Brandenburg (20), this is a percentage of 20 %. In the remaining three exemplary federal states with a much lower MON810 cultivation in the past, none of the PSM sites appear to have already been exposed to MON810

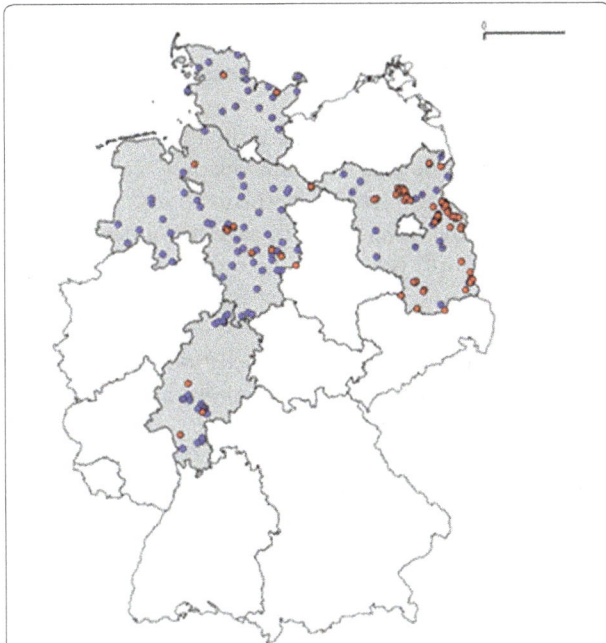

Fig. 4 Distribution of PSM sites (*blue dots*) and MON810 field sites (*red dots*) on exemplary federal states (Brandenburg, Hesse, Lower Saxony and Schleswig–Holstein). Data sources: environmental agencies of the federal states for the PSM sites; public location register of the German Federal Office of Consumer Protection and Food Safety (BVL) for the MON810 field sites

pollen in the past. The cultivation of MON810 on PSM sites could not be observed albeit it is generally permitted. From this example, it can be assumed that with a future intensification of GMO cultivation in Germany, more PSM sites on arable land would likely become exposed towards GMO.

Practicability

To run a PSM site system in a sustainable way and on a long term, several issues of practicability have to be taken into account and clarified in general.

To be able to use PSM sites as non-influenced reference sites in the future, measures would be needed to prevent these sites from exposure towards GMO. These would need to include prohibition of GMO cultivation on PSM sites and the definition of GMO-free buffer zones around these sites as well as a minimum time lag since the last possible exposure towards GMO. Such spatial and temporal distances would have to be fixed on a scientific basis as much as possible but may vary, e.g. according to GMO crop type and expected exposure pathways. If GMO crops have been previously cultured at this site, it cannot be used as a reference site in the monitoring program, i.e. a historically GMO use has to be excluded definitely. Of course, a site at which GMO cultivation has been practised for several years

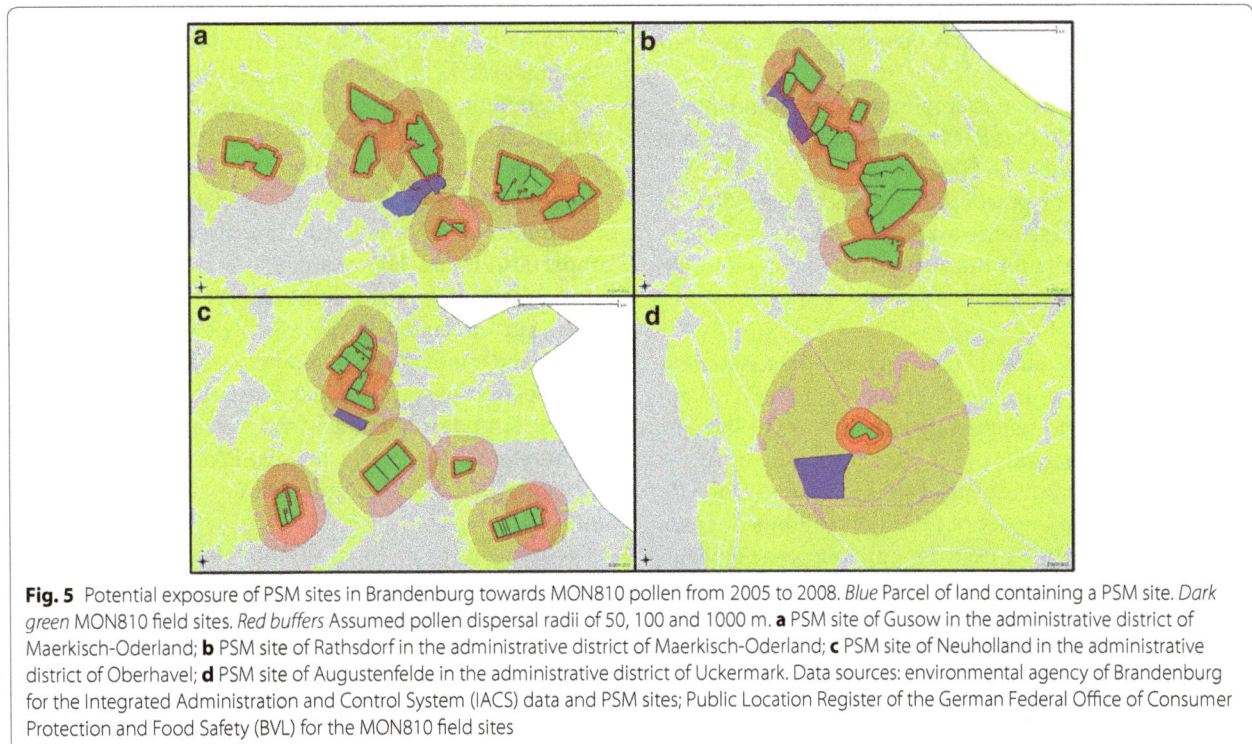

Fig. 5 Potential exposure of PSM sites in Brandenburg towards MON810 pollen from 2005 to 2008. *Blue* Parcel of land containing a PSM site. *Dark green* MON810 field sites. *Red buffers* Assumed pollen dispersal radii of 50, 100 and 1000 m. **a** PSM site of Gusow in the administrative district of Maerkisch-Oderland; **b** PSM site of Rathsdorf in the administrative district of Maerkisch-Oderland; **c** PSM site of Neuholland in the administrative district of Oberhavel; **d** PSM site of Augustenfelde in the administrative district of Uckermark. Data sources: environmental agency of Brandenburg for the Integrated Administration and Control System (IACS) data and PSM sites; Public Location Register of the German Federal Office of Consumer Protection and Food Safety (BVL) for the MON810 field sites

could serve as a "positive" GMO reference site—but just for one specific GMO. In any case, the sites should have a sufficient size (minimum: 1 ha), and should be easily accessible, which means that the owners have to be integrated in the monitoring program in a contractually embedded long-term approach. Also, the difficulty to ensure a stable land use on each reference PSM site over time has to be realized.

Conclusion regarding the current status of the German PSM program

The information compiled in this chapter can be summarized as follows: The German PSM is suitable for monitoring potential effects of GMO on soil organism communities, because:

- There is already an existing network of 795 PSM sites in Germany.
- Soil organism communities are an important protection goal and could be exposed to GMO via different pathways.
- A theoretical concept to utilize soil biodiversity data in monitoring concepts is available.
- Basic parameters needed for such a monitoring program (e.g. the characterization of site and soil properties as well as the history of the sites) are already measured at the PSM.
- The PSM is representative for the different biogeographic regions in Germany as well as for the distribution of agricultural land potentially to be used for GMO crops.
- As exemplified by an example site from Brandenburg, it is highly likely that PSM is located in the same area as sites cultivated with GMO crops.
- Finally, issues of practicability do not contradict the use of PSM as reference sites or (less likely) as a positive reference site for the assessment of potential side-effects of GMO crops.

This approach is already proposed for the assessment of soil quality, e.g. in the Netherlands (BISQ), where both structural and functional endpoints are utilized for various organism groups [52]. The development of references on GMO uninfluenced sites is necessary for the assessment of the impact of GMO on soil biocoenoses. If PSM sites are to be used for this purpose, it would be necessary to protect them from exposure towards GMO.

However, when discussing these issues, several shortcomings of the use of PSM have been identified. To overcome these, several modifications of the PSM program have to be performed. They will be listed and discussed in the following chapter.

Outlook: formulation of possibilities to expand or adapt the federal permanent soil monitoring program and/or complementary monitoring modules for the GMO monitoring

Based on the experiences made in the course of this research, the following recommendations can be given how to adapt the German permanent soil monitoring program for the assessment of potential side-effects of GMO. Therefore, the following recommendations will also be useful for biological soil monitoring in general. For a minimum set of sites, it is recommended to use a grid, based on the distribution of existing PSM sites. The sites should be evenly distributed among all federal states and should be nationally coordinated to ensure a harmonized approach. The major habitat types (arable land, grassland, deciduous and coniferous forests), integrating several subtypes [6, 34, 37], with ten sites each (i.e. roughly 160–200 sites), should be covered and sampled within the course of 5 years. Standardization regarding both, point in time and method of sampling should strongly reduce variability and strengthen data comparability. The sites should be representative regarding the soil factors in those ranges relevant for Germany: pH value, soil texture, surface soil conditions (humus form, litter layer/mineral soil), and geographical regions. Finally, site selection should allow integration into European monitoring programs.

Recommendations of parameters for a minimum soil characterization (all measurements should be performed according to available ISO guidelines or other comparable standards); [14, 45, 53, 54]: pH value ($CaCl_2$, KCl), SOM content, cation exchange capacity, soil dry mass, texture, soil density. With respect to the biological monitoring focus, nitrogen content, C/N ratio, water holding capacity and humus form (especially for forest sites) should also be recorded. Additionally, the following site properties should be recorded: site history (land use, prior samplings), exact geographical location (coordinates), current land use type, climate data (at least: mean annual and monthly air temperature and precipitation; annual course of surface soil temperature), ground-water level, anthropogenic impact (concentrations of common contaminants, e.g. heavy metals, PAH); physical stress (management practice, compaction, fertilization, erosion, etc.).

Recommendations for a methodological standard for biological monitoring comprise the organism groups of Oribatida, Collembola, Lumbricidae, Enchytraeidae and the diversity of microorganisms. This list is in fact an expansion of a list developed for general soil biological monitoring in the EU-project 'ENVASSO' [55]. Sampling should be seasonally matched (spring/autumn) and performed according to available ISO guidelines. Vertical distribution between litter and mineral soil layer should be addressed where appropriate, and sampling should be repeated with a frequency of 3–5 years to get a chronological update of possible changes.

The data raised in such an improved biological soil monitoring program can thus be utilized to fill existing data gaps regarding the occurrence of soil organism taxa at different habitat types. Subsequently, the biological soil-quality assessment approach presented above can be subjected to a validation step and then be implemented for routine practical application.

Any deviation from the reference values determined on PSM sites needs to be evaluated according to previously determined threshold values (Fig. 5). So far there is no regulation regarding effects on the soil biocoenosis in the German Federal Soil Protection Act or other laws, but only precautionary, trigger and action values for substances in the German Federal Soil Protection Ordinance (i.e. certain concentrations of single chemicals must not be exceeded in soils with a certain land use [56]).

The usability of data for a nationwide monitoring of GMO requires a standardized collection and management of data among all federal states to facilitate a central evaluation of the results (e.g. landscape- or culture-based [39, 57, 58]). In this context, it has to be noted that:

1. The various federal state PSM programs currently differ in their structure regarding data collection, management and evaluation;

2. The data will become more valuable through long-term, comparable measurements. Hence, continuity regarding data management should be established through a centralized coordination.

3. Uniform and comparable data necessary for the evaluation within a GMO monitoring need to be analysed and discussed;

4. Minimum standards for data flow and management need to be provided.

5. Qualified and independent committees need to be nominated for performing nationwide data evaluation. The competent authority could resort to the already existing working group for the evaluation of PSM data, consisting of representatives from both federal agencies and federal state authorities.

6. The question of the financial contributions (e.g. GMO commercializing companies or agencies) for the preparation and use of nationwide monitoring data must be practically and adequately solved.

Authors' contributions
AT, SJ, MRN and JR have designed the study, preformed it, assembled and interpreted the data material as well as written the manuscript. WZ has been involved during the interpretation of the results as well in writing the manuscript. All authors read and approved the final manuscript.

Author details
[1] gaiac Research Institute for Ecosystem Analysis and Assessment, Kackerstr. 10, 52072 Aachen, Germany. [2] ECT Oekotoxikologie GmbH, Böttgerstr. 2-14, 65439 Flörsheim, Germany. [3] RWTH Aachen University, Institute for Environmental Research, Worringer Weg 1, 52074 Aachen, Germany. [4] Federal Agency for Nature Conservation (BfN), Konstantinstr. 110, 53179 Bonn, Germany.

Acknowledgements
We want to thank the UBA especially Dr. Frank Glante for his collaboration within this project. We want to thank the different federal state agencies for providing their data and the reviewers for providing their help in the review process. This study was funded by the German Federal Ministry for the Environment, Nature Conservation, Building and Nuclear Safety on behalf of the German Federal Agency for Nature Conservation.

Competing interests
The authors declare that they have no competing interests.

References
1. Römbke J, Jänsch S, Roß-Nickoll M, Toschki A (2014) Nutzungsmöglichkeiten der Boden—Dauerbeobachtung der Länder für das Monitoring der Umweltwirkungen gentechnisch veränderter Pflanzen. BfN Skripten, Bonn-Bad Godesberg 369. [http://www.bfn.de]
2. EU (2001) Directive 2001/18/EC of the European parliament and of the Council of 12 March 2001 on the deliberate release into the environment of genetically modified organisms and repealing Council Directive 90/220/EC. Commission Declaration :1–39
3. Züghart W, Raps A, Wust-Saucy AG, Dolezel M, Eckerstorfer M (2011) Monitoring of genetically modified organisms. A policy paper representing the view of the National Environment Agencies in Austria and Switzerland and the Federal Agency for Nature Conservation in Germany. Umweltbundesamt GmbH
4. Middelhoff U (2006) Hildebandt J. Die Ökologische Flächenstichprobe als Instrument eines GVO-Monitoring. BfN Skripten, Breckling B, p 172
5. UBA (2011) Bodendauerbeobachtung in Deutschland. Umweltbundesamt
6. Römbke J, Jänsch S, Roß-Nickoll M, Toschki A, Höfer H, Horak F, Russell D, Burkhardt U, Schmitt H (2012) Erfassung und Analyse des Bodenzustands im Hinblick auf die Umsetzung und Weiterentwicklung der Nationalen Biodiversitätsstrategie. Umweltbundesamt
7. Faber JH, Creamer RE, Mulder C, Römbke J, Rutgers M, Sousa JP, Stone D, Griffiths BS (2013) The practicalities and pitfalls of establishing a policy-relevant and cost-effective soil biological monitoring scheme. Integr Environ Assess Manag (IEAM) 9:276–284
8. Curtis TP, Sloan WT, Scannell JW (2002) Estimating prokaryotic diversity and its limits. Proc Natl Acad Sci USA 99:10494–10499
9. Hedlund K (2012) SOILSERVICE—Conflicting demands of land use, soil biodiversity and the sustainable delivery of ecosystem goods and services in Europe. Report for the European Union (FP7), University of Lund
10. Wardle DA (2002) Communities and ecosystems: Linking the aboveground and belowground components. Princeton University press
11. De Ruiter PC, Moore JC, Zwart KB, Bouwman LA, Hassink J, Bloem J, de Vos JA, Marinissen JCY, Didden WAM, Lebbink G, Brussaard L (1993) Simulation of nitrogen mineralization in the belowground food webs of two winter wheat fields. J Appl Ecol 30:95–106
12. Scheu SF, Falca M (2000) The soil food web of two beech forests (Fagus sylvatica) of contrasting humus type: stable isotope analysis of a macro- and a mesofauna-dominated community. Oecologia 123:285–296

13. Bardgett RD, Usher MB, Hopkins DW (2005) Biological diversity and function in soils. Cambridge University Press, Cambridge

14. Turbé A, De Toni A, Benito P, Lavelle P, Ruiz N, Van der Putten W, Labouze E, Mudgal S (2010) Soil biodiversity: functions, threats, and tools for policy makers. BioIntelligence Service, IRD, and NIOO, Report for European Commission (DG Environment)

15. Priesnitz K (2011) Potential impact of Diabrotica resistant Bt-maize expressing Cry3Bb1 on ground beetles (Coleoptera: Carabidae). Dissertation, RWTH University Aachen

16. Marquard E, Durka W (2005) Auswirkungen des Anbaus gentechnisch veränderter Pflanzen auf Umwelt und Gesundheit: Potenzielle Schäden und Monitoring. UFZ-Umweltforschungszentrum, Leipzig-Halle

17. Dolezel M, Heissenberger A, Gaugitsch H (2005) Ökologische Effekte von gentechnisch verändertem Mais mit Insektizidresistenz und/oder Herbizidresistenz. Umweltbundesamt, Wien

18. Saxena D, Stotzky G (2001) Bacillus thuringiensis (Bt) toxin released from root exudates and biomass of Bt corn has no apparent effect on earthworms, nematodes, protozoa, bacteria and fungi in soil. Soil Biol Biochem 33:1225–1230

19. Vercesi ML, Krogh PH, Holmstrup M (2006) Can Bacillus thuringiensis (Bt) corn residues and Bt-corn plants affect life-history traits in the earthworm Aporrectodea caliginosa? Appl Soil Ecol 32:180–187

20. ISO (2008) ISO 17402, Soil quality—guidance for the selection and application of methods for the assessment of bioavailability of contaminants in soil and soil materials. ISO (International Organization for Standardization)

21. Peijnenburg W, Jensen J, Kula C, Liess M, Capri E, Luttik R, Montforts M, Nienstedt K, Römbke J, Sousa JP (2012) Evaluation of exposure metrics for effect assessment of soil invertebrates. Crit Rev Environ Sci Technol 42:1862–1893

22. Hilbeck A, Jänsch S, Meier M, Römbke, J (2008) Analysis and validation of present ecotoxicological test methods and strategies for the risk assessment of genetically modified plants. BfN Skripten 236

23. Naidu R, Semple KT, Megharaj M, Juhasz AL, Bolan NS, Gupta S, Clothier B, Schulin R, Chaney R (2008) Bioavailability, definition, assessment and implications for risk assessment. In: Naidu et al. (ed) Chemical bioavailability in terrestrial environment. Elsevier, Amsterdam, 39–52

24. Eschenbach C, Windhorst W (2009) Indikatoren für die nationale Strategie zur biologischen Vielfalt: Gentechnik in der Landwirtschaft. Bericht für das Bundesamt für Naturschutz, Bonn

25. Semple KT, Doick KJ, Jones KC, Burauel P, Craven A, Harms H (2004) Defining bioavailability and bioaccessibility of contaminated soil and sediment is complicated. Environ Sci Technol 38:228A–231A

26. Beck L (1993) Zur Bedeutung der Bodentiere für den Stoffkreislauf in Wäldern. Zur Bedeutung der Bodentiere für den Stoffkreislauf in Wäldern 23:286–294

27. VDI (2014) VDI 4331, monitoring the effects of genetically modified organisms—Effects on soil organisms. Verein Deutscher Ingenieure; Berlin

28. Ruf A, Seitz H, Römbke J, Roß-Nickoll M, Theißen B, Toschki A, Züghart W, Blick T, Russell DJ, Beylich A, Rueß L, Höss S, Büchs W, Glante F (2013) Soil organisms as an essential element of a monitoring plan to identify the effects of GMO cultivation. requirements—methodology—standardisation. BioRisk 8:73–87

29. Chamberlain PM, Bull ID, Black HJ, Ineson P, Evershed RP (2006) Collembolan trophic preferences determined using fatty acid distributions and compound-specific stable carbon isotope values. Soil Biol Biochem 38:1275–1281

30. Chahartaghi M, Langel R, Scheu S, Ruess L (2005) Feeding guilds in Collembola based on nitrogen stable isotope ratios. Soil Biol Biochem 37:1718–1725

31. Erdmann G, Otte V, Langel R, Scheu S, Maraun M (2007) The trophic structure of bark-living oribatid mite communities analysed with stable isotopes (N-15, C-13) indicates strong niche differentiation. Pedobiologia 41:1–10

32. Bouché MB (1977) Strategies lombriciennes. Ecol Bull 25:122–132

33. EU (1992) Council Directive 92/43/EEC 1992 on the conservation of natural habitats and of wild fauna and flora. European Parliament

34. Riecken U, Finck P, Raths U, Schröder E, Ssymnak A German red data book on endangered habitats (short version). July 2009 [http://www.bfn.de/0322_biotope+M52087573ab0.html]

35. Burkhardt U, Russell DJ, Decker P, Döhler M, Höfer H, Römbke J, Trog C, Vorwald J, Wurst E, Xylander WE (2014) The Edaphobase project of GBIF-Germany—a new online soil-organism zoological data warehouse. Applied Soil Ecology 83:3–12

36. Edaphobase Portal. GBIF Datenbank Bodenzoologie; Informationssystem für Taxonomie, Literatur und Ökologie. http://portal.edaphobase.org/

37. Riecken U, Finck P, Raths U, Schröder E, Ssymank A (2003) Standard-Biotoptypenliste für Deutschland. Schriftenreihe für Landschaftspflege u. Naturschutz, 2. Fassung

38. Roß-Nickoll M, Lennartz F, Fürste A, Mause R, Ottermanns R, Schäfer S, Smolis M, Theißen B, Toschki A, Ratte HT (2004) Die Arthropodenfauna von grasigen Feldrainen (off crop) und die Konsequenzen für die Bewertung der Auswirkungen von Pflanzenschutzmitteln auf den terrestrischen Bereich des Naturhaushaltes. Umweltbundesamt, Dessau-Roßlau

39. Toschki A (2008) Eignung unterschiedlicher Monitoring-Methoden als Grundlage zum Risk-Assessment für Agrarsysteme—Am Beispiel einer biozönologischen Reihenuntersuchung und einer Einzelfallstudie—Dissertation, RWTH-Aachen

40. Biosicherheit (2006) Coexistence of genetically modified and nongenetically modified maize: making the point on scientific evidence and commercial experience. http://www.pgeconomics.co.uk/pdf/Co-existence_maize_10october2006.pdf. Accessed 20 Oct 2015

41. Hofmann F, Janicke U, Janicke L, Wachter R, Kuhn U (2008) Modellrechnungen zur Ausbreitung von Maispollen unter Worst-Case-Annahmen mit Vergleich von Freilandmeßdaten. http://www.bfn.de/fileadmin/MDB/documents/service/Hofmann_et_al_2009_Maispollen_WorstCase_Modell.pdf. Accessed 20 Oct 2015

42. GenTPflEV (2008) Verordnung über die gute fachliche Praxis bei der Erzeugung gentechnisch veränderter Pflanzen. Gentechnik-Pflanzenerzeugungsverordnung. Gentechnik-Pflanzenerzeugungsverordnung vom 7. April 2008 BGBl

43. Felke M, Langenbruch G-A (2005) Auswirkungen des Pollens von transgenem Bt-Mais auf ausgewählte Schmetterlingslarven. Bonn-Bad Godesberg, BfN Skripten, p 157

44. Hofmann F, Epp R, Kalchschmid A, Kruse L, Kuhn U, Maisch B, Müller E, Ober S, Radtke J, Schlechtriemen U, Schmidt G, Schröder W, von der Ohe W, Vögel R, Wedl N, Wosniok W (2008) GVO-Pollenmonitoring zum Bt-Maisanbau im Bereich des NSG/FFH-Schutzgebietes Ruhlsdorfer Bruch. Umweltwiss Schadst Forsch 20:275–289

45. Barth N, Brandtner W, Cordsen E, Dann T, Emmerich KH, Feldhaus D, Kleefisch B, Schilling S, Utermann J (2000) Boden-Dauerbeobachtung. Einrichtung und Betrieb von Boden-Dauerbeobachtungsflächen. In: Rosenkranz D, Bachmann G, König W, Einsele G (eds) Bodenschutz., Volume 32. XI/00Erich Schmidt Verlag, Berlin

46. Candolfi MP, Brown K, Grimm C, Reber B, Schmidli H (2004) A faunistic approach to assess potential side effects of genetically modified Bt-corn on non-target arthropods under field conditions. Biocontrol Sci Tech 14:129–170

47. Cortet J, Andersen MN, Caul S, Griffiths B, Joffre R, Lacroix B, Sausse C, Thompson J, Krogh PH (2006) Decomposition processes under Bt (Bacillus thuringiensis) maize: results of a multi-site experiment. Glob Change Biol 38:195–199

48. Ludy C, Lang A (2006) Bt maize pollen exposure and impact on the garden spider. Araneus diadematus. Entomologia Experimentalis et Applicata 118:145–156

49. Toschki A, Hothorn LA, Ross-Nickoll M (2007) Effects of cultivation of genetically modified Bt maize on epigeic arthropods (Araneae; Carabidae). Environ Entomol 36:967–981

50. Priestley AL, Brownbridge M (2009) Field trials to evaluate effects of Bt-transgenic silage corn expressing the Cry1Ab insecticidal toxin on non-target soil arthropods in northern New England USA. Transgenic Res 18:425–443

51. Theißen B, Russell D (2009) Zur Bedeutung von Collembolen im GVO-Monitoring. Gefahrstoffe Reinhaltung der Luft 69:391–394

52. Rutgers M, Mulder C, Schouten AJ, Bloem J, Bogte JJ, Breure AM, Brussaard L, De Goede RGM, Faber JH, Jagers op Akkerhuis GAJM, Keidel H, Korthals GW, Smeding FW, Ter Berg C, Van Eekeren N (2008) Soil ecosystem profiling in the Netherlands with ten references for biological soil quality. RIVMReport 607604009

53. Römbke J, Labes G, Woiwode J (2002) Ansätze für Strategien zur Bewertung des Bodens als Lebensraum für Bodenorganismen. Bodenschutz 2(02):62–69

54. ISO (2008) ISO 16133, Soil quality—guidance on the establishment and maintenance of monitoring programmes. ISO (International Organization for Standardization)

55. Bispo A, Cluzeau D, Creamer R, Dombos M, Graefe U, Krogh PH, Sousa JP, Peres G, Rutgers M, Winding A, Römbke J (2009) Indicators for Monitoring Soil Biodiversity. Integr Environ Assess Manag 5:717–719

56. BBodSchV (1999) Bundes-Bodenschutz—und Altlastenverordnung. Verordnung zur Durchführung des Bundes-Bodenschutzgesetzes. BGBl 36:1554–1582

57. Plachter H, Bernotat D, Müssner R, Riecken U (2002) Entwicklung und Festlegung von Methodenstandards im Naturschutz. Schr.R. f. Landschaftspl. U. Naturschutz 70

58. Züghart W, Benzler A, Berhorn F, Sukopp U, Graef F (2008) Determining indicators, methods and sites for monitoring potential adverse effects of genetically modified plants to the environment: the legal and conceptional framework for implementation. Euphytica 164:845–852

6

Can perfluoroalkyl acids biodegrade in the rumen simulation technique (RUSITEC)?

J. Kowalczyk[1*], S. Riede[2], H. Schafft[1], G. Breves[2] and M. Lahrssen-Wiederholt[1]

Abstract

Background: The behaviour of perfluoroalkyl acids (PFAAs) in tissues of ruminants has been shown to differ from that of monogastrics (J Agric Food Chem 61(12):2903–2912 doi:10.1021/jf304680j, 2013; J Agric Food Chem 62(28):6861–6870, 2014). This may be a consequence of the complex microbial ecosystem in the rumen. To evaluate this hypothesis, the recovery of PFAAs was studied using the rumen simulation technique as an indication for biodegradation in rumen. The PFAA-recovery from a microbial fermentation of feed containing PFAAs was compared to the same feed in the absence of ruminal microorganisms (MOs).

Results: Release of PFAAs from feed into fermentation fluid was found to be faster for perfluorobutane sulfonic acid (PFBS) than for perfluorooctane sulfonic acid (PFOS). Differences between perfluoroalkyl carboxylic acids (PFCAs) could not be observed. Proportions of PFAAs recovered in the fermentation fluids decreased by increasing chain lengths for the perfluoroalkyl sulfonic acids (PFSAs) (31 % PFBS, 28 % perfluorohexane sulfonic acid [PFHxS], 20 % perfluoroheptane sulfonic acid [PFHpS], 11 % PFOS) and PFCAs (33 % perfluorohexane carboxylic acid [PFHxA], 32 % perfluoroheptane carboxylic acid [PFHpA], 24 % perfluorooctanoic acid [PFOA]). In contrast, levels in feed increased with increasing chain length for both PFSAs and PFCAs.

Conclusion: The attachment of MOs to feed particles was assumed to account for higher PFAA levels in fermented feeds and for lower levels in the fermentation fluids. Total recovery of PFAAs was significantly lower in presence of ruminal MOs compared to experimental procedure under sterile conditions. Although, there are optimal reductive conditions for MOs in rumen, our results do not univocally indicate whether PFAAs were degraded by ruminal fermentation.

Keywords: Perfluoroalkyl acids, PFOS, PFOA, Biodegradation, Rumen microorganisms, RUSITEC

Background

Perfluoroalkyl acids (PFAAs) belong to the chemical group of per- and polyfluoroalkyl substances (PFAS) [3]. PFAS are used in numerous industrial applications (e.g. electroplating, fluoropolymer production, photography), fire extinguishing agents and as processing aids in impregnation agents for a large number of consumer products (e.g. textile, carpet and leather protectors and food contact papers) since the 1950s [4–6]. Because of their various applications, large amounts of PFAS have been released into the environment, leading to their presence and detection in wildlife and humans [7–12]. The most widely studied PFAAs are perfluoroalkyl carboxylic acids (PFCAs) and perfluoroalkyl sulfonic acids (PFSAs) with the most frequently detected eight-carbon homologues perfluorooctanoic acid (PFOA) and perfluorooctane sulfonic acid (PFOS). These substances have received most attention because of their persistence [3, 13], toxic effects to mammals [14] and their strong tendency for bioaccumulation [15] and biomagnification in food chain [16, 17]. The chemical and biological persistence of PFAAs is generally believed to be a result of the carbon–fluorine bond, which is one of the strongest covalent bonds in nature [18]. Concerns about the potential of toxicity, environmental persistence and bioaccumulation have led to restricting use and inclusion of

*Correspondence: Janine.Kowalczyk@bfr.bund.de
[1] Federal Institute for Risk Assessment, Max-Dohrn-Str. 8-10, 10589 Berlin, Germany
Full list of author information is available at the end of the article

PFOS and its related substances in the stockholm convention on persistent organic pollutions (POPs) in 2009 [19]. Also for PFOA, a stewardship agreement between the US environmental protection agency (US EPA) and the leading global manufacturers intends to eliminate the environmental emission until 2015 [20]. In the European union, the phase-out of long-chain PFCAs (C9, C11–C14), PFOA and its ammonium salt (APFO) is controlled using the European chemical regulation (REACH EC No. 1907/2006). These substances have been identified as substances of very high concern (SVHC) and are included into the candidate list as substances proposed for authorization. Moreover, PFOA is discussed for inclusion into the restriction proceedings regulated in REACH Annex XVII (Restriction, Art. 67) [21, 22].

Since PFAAs raised concern due to their high persistence, some studies have been performed to investigate methods for their complete decomposition [23–26]. Although various methods have shown to be more or less successful, they often are very energy-consuming, require extreme conditions or lead to by-product formation with unknown adverse effects [27].

The degradation of PFAAs in sewage sludge has been examined for wastewater treatment plants. The study [28] recognized that PFOA and PFOS disappeared quite rapidly under anaerobic reductive conditions (redox potential of −380 mV). A possible microbially-mediated transformation of PFOA was observed [29] during the reductive dechlorination of trichloroethene when sewage treatment plant samples were used as inoculum for in vitro incubation. However, transformation products could not be clearly identified, thus, PFOA biodegradation was not proven successful. Only one study was found that reports the degradation of PFOS under ambient conditions [30]. Ochoa-Herrera and coworkers demonstrated that PFOS defluorination was catalyzed by cobalamin and simultaneous presence of titanium(III)citrate as reducing agent to achieve corresponding reductive conditions [30]. The complete degradation of PFOS (mineralization) was confirmed by measurement of fluoride ion release during the reductive dehalogenation process. In this study cobalamin was shown to be an important biomolecule for the reductive dehalogenation of PFOS. In nature, cobalamin is frequently involved in reductive dehalogenase catalysis [31]. This essential biomolecule is required by animals and humans, but can only be synthesized by bacteria and archaea [32].

In cows, the rumen contains a variety of bacterial species able to produce cobalamin and its analogues [33]. In combination with the reductive condition of about −300 to −400 mV, the microbial fermentation in the rumen might provide good conditions for the degradation of PFAAs and could affect the PFAA-adsorption downstream.

Interestingly, transfer studies on dairy cows and fattening pigs showed a different behaviour of PFAAs regarding partitioning to blood plasma, liver and edible tissues [1, 2]. Both species were exposed to PFAAs of different chain-length through the same naturally contaminated hay. Although there was an ad libitum feeding of PFAA-contaminated hay for dairy cows (app. 14 % of diet) and a restrictive feeding of a PFAA-hay containing pelleted feed for fattening pigs (app. 17 % of diet), the results of the percent of PFAAs accumulation in the estimated mass of tissues are well comparable and indicate different toxicokinetics in both species. The results showed that PFBS and PFOA had no tendency of accumulation in blood plasma of dairy cows compared to PFHxS and PFOS, while the affinity for blood plasma were evenly high for all PFAAs in fattening pigs. Considering the PFAA accumulation in edible tissue, similar amounts of all examined PFAAs (40–49 %) were found in the meat of fattening pigs. This was in particular contrast to the different amounts of PFOS (43 %), PFOA (<1 %) and PFBS (<0.001 %) in the muscle tissues of the dairy cows [1, 2]. Furthermore, excretion of PFBS and PFOA via milk throughout the feeding study was very low (<1 %) and cannot explain the low affinity for muscle in dairy cows. Because of the different findings, it was assumed that the difference of toxikokinetics in ruminants and monogastrics might be the consequence of their species-specific digestive system. The hypothesis was put forth that the microbial fermentation in rumen of bovine species might have an effect on PFAA recovery. It was posited that ruminal MOs may have effects on PFAA degradation because numerous studies in ruminants with other undesirable substances in feed, such as secondary plant compounds, mycotoxins, but also antibiotics have demonstrated the capacity of ruminal microorganisms to degrade potentially toxic compounds and thus protecting the host from detrimental effects [34–36]. Overall, MOs in the rumen are presumed to have an unexpectedly high capacity to degrade or transform fluorinated compounds.

The aim of this work was to evaluate the PFAA recovery as an indication of microbial degradation of PFAAs in the rumen in standardized conditions by using the in vitro rumen simulation technique (RUSITEC). The in vitro technique was chosen due to the ability to analyze the PFAA concentration in the fermented feed and liquid fractions of rumen at different points in time during the ruminal fermentation. Besides PFAA recovery, the effects of PFAAs on the fermentation characteristics (e.g. pH value, redox potential) and on microbial digestion of feed nutrients were investigated.

Result and discussion

Fermentation characteristics

The chemical compositions of the PFAA and PFAA-free hay are shown in Table 1. The PFAA hay had higher levels of crude fiber (CF, 34.6 %) than the PFAA-free hay (30.8 %), indicating that the grass grown on a PFAA contaminated farmland was cut later than the non-contaminated grass. Differences in crude protein (CP) level between PFAA hay (7.07 %) and PFAA-free hay (18.78 %) were also observed.

During the control period, the pH value in the fermentation vessels remained constant (Additional file 1: Figure S1A). The pH value decreased significantly when the PFAA-free hay was exchanged with the PFAA hay. The decrease in pH value was more pronounced for the vessels with high PFAA levels than for vessels with low PFAA levels (Table 2). The differences in pH value might be related to the different production level of volatile fatty acids (VFA).

The average total VFA concentration (22.55 mmol/day) was affected after exchanging the PFAA-free hay with the low (25.90 mmol/day) and high (28.38 mmol/day) PFAA hay (Table 2). It was assumed that the higher levels of CF and nitrogen-free extracts (NfE) in the PFAA hay contribute to the increase in total VFA production. PFAA hay depressed production of acetate and butyrate and significantly increased propionate production. Higher proportions of propionate in rumen are generally observed by supplementation of readily fermentable starch-rich diets [37]. The higher level of NfE (readily fermentable carbohydrates such as starch and sugar) in the PFAA hay likely enhanced the growth of propionate producers.

During control period, the average ammonia concentration was 11.79 mmol/l, but significantly decreased in the fermentation vessels after receiving the PFAA hay (Additional file 1: Figure S1B), whereas the effect was less pronounced for the low PFAA dose (9.82 mmol/l)

Table 1 Ingredients of the fermentation vessels, chemical composition and PFAA concentration of the PFAA hay and the PFAA-free hay

Ingredients, g/vessel	PFAA feed			PFAA-free feed[c]
	Low dosage[a]		High dosage[b]	
PFAA hay	7.998 ± 0.002		16.001 ± 0.004	
PFAA-free hay		8.002 ± 0.002		15.999 ± 0.001
Concentrate*	2.000 ± 0.002	1.997 ± 0.003	3.997 ± 0.003	3.997 ± 0.002
	PFAA hay			**PFAA-free hay**
Chemical composition, % DM				
Dry matter (DM, %)	92.37			88.90
Crude ash (CA)	7.85			7.08
Crude protein (CP)	7.07			18.78
Crude lipids (CL)	1.11			1.69
Crude fiber (CF)	34.62			30.82
Nitrogen free extracts (NfE)	53.40			41.62
PFAA concentration, µg/kg				
PFBS	779.0 ± 27.9			<LOD[d]
PFHxA	235.6 ± 4.9			<LOD[d]
PFHpA	37.7 ± 0.7			<LOD[e]
PFHxS	489.0 ± 20.5			<LOD[e]
PFOA	91.6 ± 1.4			<LOD[e]
PFHpS	20.8 ± 0.3			<LOD[e]
PFNA	0.6 ± 0.0			<LOD[e]
PFOS	611.9 ± 9.1			<LOD[e]

[a] Three vessels with one nylon bag containing PFAA hay and a second nylon bag filled with PFAA-free hay

[b] Three vessels with two nylon bags with PFAA hay

[c] Six vessels contained two nylon bags filled with PFAA-free hay

[d] Limit of detection: 0.5 µg/kg

[e] Limit of detection: 0.2 µg/kg

* Ingredients: amount (g/kg dry matter), percentage, or international units (IU/kg) ash 69.1, crude protein 196.2, crude fat 42.4, crude fibre 94.8, acid detergent fibre 149.4, neutral detergent fibre 305.5, acid detergent lignin 55.7, organic matter 93.09, calcium 1.5 %, phosphorous 0.55 %, sodium 0.25 %, vitamin A 20,000 IU, vitamin D3 1600 IU

Table 2 Fermentation characteristics of PFAA-free hay and PFAA hay with low and high PFAA levels

Item	PFAA level	PFAA-free hay[c] (day 7–12)	PFAA hay (day 13)	SEM	P value[d]
pH	Low[a]	6.87 ± 0.03	6.77 ± 0.08	1.60	≤0.05
	High[b]	6.86 ± 0.02	6.69 ± 0.09	2.16	≤0.05
Redox potential, mV	Low[a]	−271 ± 19	−309 ± 30	29.7	≤0.05
	High[b]	−267 ± 24	−298 ± 27	31.9	≤0.05
Ammonia, mmol/l	Low[a]	11.77 ± 0.50	9.82 ± 1.12	1.87	0.09
	High[b]	11.81 ± 0.61	6.85 ± 0.42	0.91	≤0.05
Volatile fatty acids (VFA), mmol/day					
Total VFA	Low[a]	22.32 ± 1.00	25.90 ± 0.78	1.60	≤0.05
	High[b]	22.78 ± 0.85	28.38 ± 0.52	2.16	≤0.05
Molar proportion (%)					
Acetate	Low[a]	66.46 ± 0.50	64.03 ± 0.20	0.99	≤0.05
	High[b]	65.55 ± 1.92	61.80 ± 1.18	2.25	≤0.05
Propionate	Low[a]	25.39 ± 0.49	28.65 ± 0.26	1.25	≤0.05
	High[b]	26.46 ± 2.02	31.41 ± 1.22	2.60	≤0.05
Butyrate	Low[a]	8.15 ± 0.27	7.32 ± 0.14	0.39	≤0.05
	High[b]	8.00 ± 0.38	6.79 ± 0.10	0.56	≤0.05

[a] Three vessels with one nylon bag containing PFAA hay and a second filled with PFAA-free hay

[b] Three vessels with two nylon bags with PFAA hay

[c] Six vessels contained two nylon bags filled with PFAA-free hay

[d] Comparison of the mean was performed using t test

than for the high PFAA dose (6.85 mmol/l) (Table 2). The results indicated that ammonia was produced in a dose-related manner depending on the different protein levels of the hay.

Two electrodes used for the measurement of the redox potential showed an average result of −269 mV during the control period (Additional file 1: Figure S2). The redox potentials in the fermentation vessels decreased to −309 and −298 mV after receiving the low and high doses of PFAA hay, respectively (Table 2). It was assumed that this was associated with the increased production of propionate due to its higher demand for hydrogen.

The authors concluded that changes in fermentation characteristics were affected by the different nutrient composition of the hay, and most probably not because of the contamination with PFAAs. The constantly stable fermentation process during the whole study further indicates there were no detectable detrimental effects of PFAAs on the microbial fermentation in rumen.

Levels in the fermentation liquids

In the fermentation liquids, highest levels were found at different time points for PFBS (2 h) and PFOS (8 h), which indicates that the release of short-chain PFSAs from feed to fermentation liquid was faster (Fig. 1). No significant differences in dilution rate could be observed for PFCAs (Additional file 1: Figure S3). Different dilution rates were found between the analytical and predicted values, whereas PFOS and PFOA showed highest average differences. The effect of PFAA adhesion to the equipment surfaces were excluded by the performing an experiment B (same experimental procedure as experiment A, but under sterile conditions). In experiment B, less than 1 % of PFAAs were recovered in leachates after washing the equipment with methanol. Apart from the adhesion of PFAAs to equipment surfaces, differences between the analytical and predicted values are posited to be an effect of PFAA inclusion in biofilms formed on the RUSITEC surfaces and possibly affected the release of PFAAs into the fermenter liquid. However, this conclusion is only hypothetical because biofilms were not examined in experiment A (+MOs).

Partitioning between liquids and feed

To calculate the proportion of PFAAs in the liquid and solid phase after fermentation, the mean concentration in each sample was referred to its initial PFAA concentration in feed (0 h) set as 100 %. The PFSA proportion in the fermentation liquids was seen to increase with decreasing chain length of the substance (Fig. 2). PFBS was higher (31 %) than the respective proportions that have been observed for PFHxS (28 %), PFHpS (20 %),

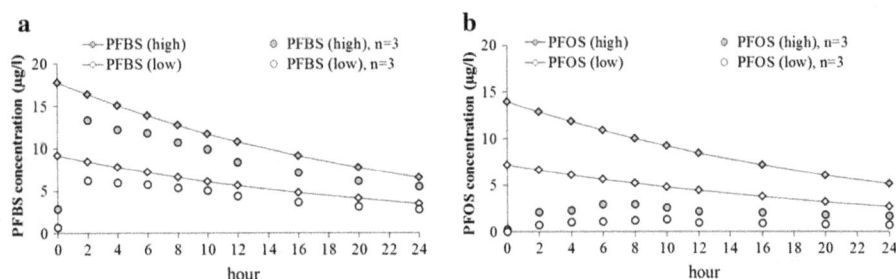

Fig. 1 Comparison of predicted (*diamond lined*) and analyzed (*point*) concentration of PFBS (**a**) and PFOS (**b**) during 24 h fermentation

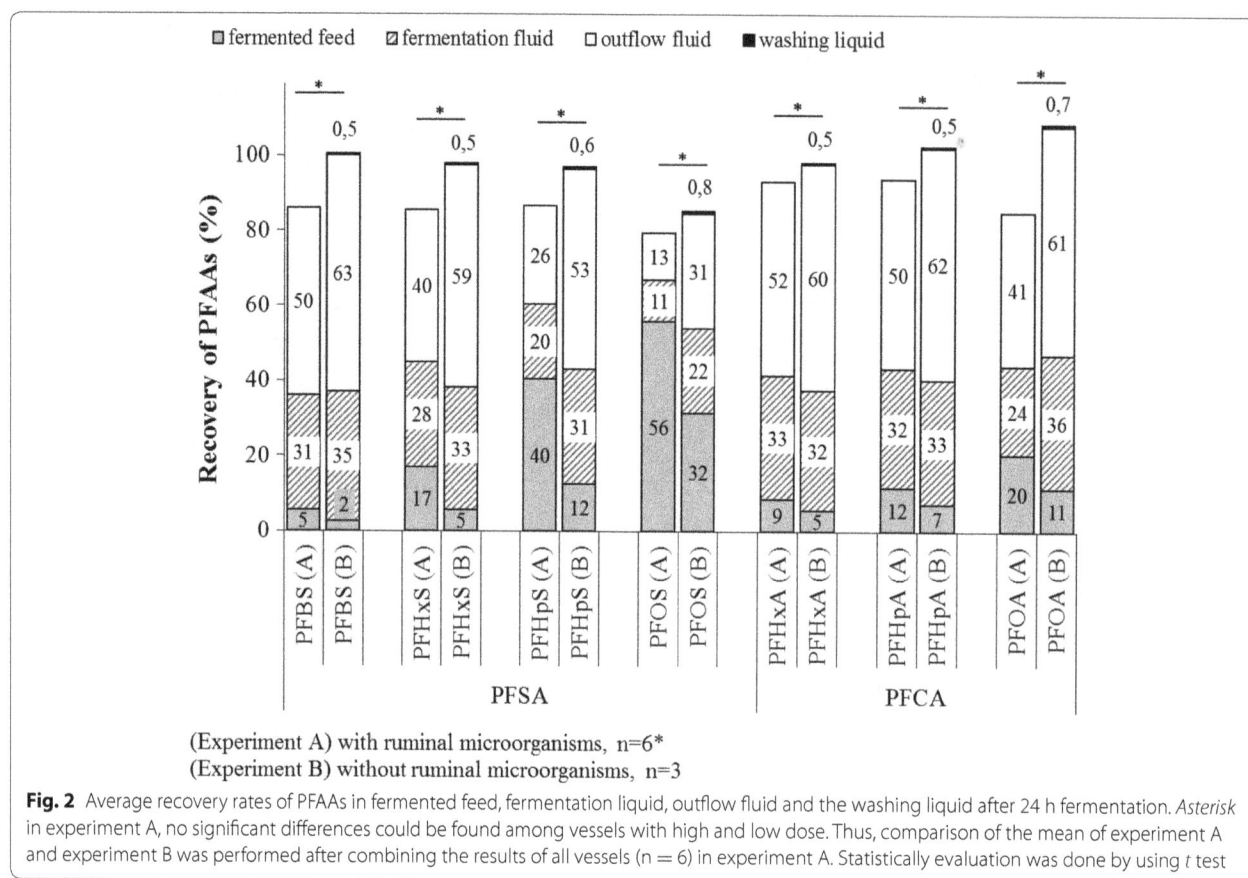

Fig. 2 Average recovery rates of PFAAs in fermented feed, fermentation liquid, outflow fluid and the washing liquid after 24 h fermentation. *Asterisk* in experiment A, no significant differences could be found among vessels with high and low dose. Thus, comparison of the mean of experiment A and experiment B was performed after combining the results of all vessels (n = 6) in experiment A. Statistically evaluation was done by using *t* test

and PFOS (11 %). A similar trend was observed for the PFCAs, whereas proportions of PFHxA and PFHpA did not differ significantly. The same distribution pattern but to a slightly higher extent was observed for PFSAs and PFCAs in the outflow liquids. Inverse proportions of PFAAs related to chain length were observed in the fermented feed. Regardless of PFAA dose, lowest levels in fermented feed were found for PFBS (5 %), which was three-times lower than for PFHxS (17 %) and tenfold lower than for PFOS (56 %). Levels in fermented feed increased with increasing chain length also for PFCAs, but to a lower extent (Fig. 2).

The partitioning behaviour of PFAAs is a function of their physicochemical properties [38–40]. The partitioning behaviour of PFAAs between liquid and solid phase in this study is similar to the partitioning between water and sediment in other studies [41, 42]. Ahrens and co-workers showed that short-chain PFCAs ($C \leq 7$) could be found solely in pore water, while the longer PFCAs were present in sediment [41]. The hydrophobic character of perfluoroalkyl acids is stronger with increasing chain length [43]. Thus, longer-chain PFAAs have stronger hydrophobic interactions with the organic matter [39, 40]. Comparison of PFSAs and PFCAs with the

same chain length indicated that partitioning between liquids and feed is also influenced by functional groups. The stronger absorption of the PFSAs to organic matter indicates that the sulfonate group exhibits lower water solubility and stronger hydrophobicity than the carboxylate group [40].

Experiment B (−MOs) showed significant differences of PFAAs proportion in the fermentation liquids and feed compared to experiment A (+MOs) (Fig. 2). This effect can be explained with the fact that 70–80 % of the total microbial population in the rumen is attached to feed particles [44]. Because of the attachment of MOs to feed particles in experiment A (+MOs), less PFAAs could dissolve in the fermentation liquids, thus accounting for the higher levels in fermented feeds compared to experiment B (−MOs).

Total recovery

To calculate the recovery of PFAAs, the mean concentration in each sample was referred to its initial PFAA concentration in feed (0 h) and set as 100 %. In experiment A (+MOs), the statistical evaluation showed no differences in PFAA recovery among the vessels with high and low PFAA doses (data not shown). For this reason, the results of PFAA recovery with low and high doses of experiment A (+MOs) were combined and statistically compared with the results of PFAA recovery of experiment B (−MOs). Figure 2 shows the percentages of PFAAs recovered in the solid phase (fermented feed) and the liquid phases (fermentation liquid, outflow liquid) for the samples of experiment A and experiment B.

In experiment A (+MOs), total recovery was almost constant for PFCAs and PFSAs up to carbon chain-lengths of ≤C7 of 93–94 % and 85–87 %, respectively. Lowest total recovery was examined for PFOA (85 %) and PFOS (80 %). Without the presence of MOs, total recovery was found to be 98 % for PFHxA, 103 % for PFHpA and 109 % for PFOA which was significantly higher compared to the total recovery of experiment A (Additional file 1: Table S1). Significantly higher levels were also found for PFSAs with lowest total recovery for PFOS (PFBS 101 %, PFHxS 98 %, PFHpS 97 %, PFOS 85 %). However in experiment A (+MOs), the missing percentages of PFAAs to total recovery of 100 % could not be explained by the standard deviation. The experimentally calculated standard deviation for recovery replicates was very low (5.6 % for PFHpS and ≤2.8 % for all other substances). The interpretation of the results is markedly complicated for PFOS because the total recovery resulted in a PFOS disappearance of 15 % in the absence of MOs. Overall, by using RUSITEC, the mechanism of PFAA disappearance in the presence of MOs remains unclear.

In studies on biochemical degradation of perfluorinated compounds in sewage sludge, disappearance of PFOS and PFOA could only be observed under anaerobic conditions with redox potentials below −300 mV [28]. However, since neither fluorine ions nor transformation products could be detected, it remained unclear if disappearance of PFOA and PFOS was a microbially-mediated degradation leading to mineralization. Ochoa-Herrera and co-workers studied the PFOS degradation under reductive conditions without presence of MOs and by the use of cobalamin and titanium(III)citrate and observed significant reduction of PFOS [30]. They suggested that both molecules play a key role in the kinetics of PFOS degradation. Based on chromatography analysis, the researchers noticed that branched PFOS isomers, but not linear PFOS isomers disappeared during the reductive defluorination. However, in the current study, only the linear PFOS isomers were quantified, so the authors can rule out that disappearance of PFOS of 15 % was affected by the degradation of branched chain PFOS isomers.

In the literature, researchers also discuss whether the reductive defluorination of PFOA might be a cometabolic reaction carried out by mixed microbial communities [29]. A cometabolic reaction is defined as a breakdown of a contaminant by an enzyme with a cofactor that is produced during microbial metabolism of another compound under anaerobic conditions [45]. In fact, the rumen includes diverse MOs producing a complex mixture of enzymes and metabolites. In anaerobic MOs, some dehalogenating enzymes have been characterized. For the catalytic activity of the dehalogenating enzymes, the involvement of cobalamin has also been shown in MOs [29]. Fermentation in rumen is known to be a primary source of cobalamin for ruminants. Several organisms were identified to be responsible for cobalamin production in the rumen, e.g. *Selenomonas ruminantium* and *Peptostreptococcus elsdenii* [33]. The total mass of MOs in the rumen consists of bacteria (10^{11}/g), protozoa (10^4/g) and fungi (10^5/g) [46] which are mutually involved in complex metabolic processes, and theoretically provide a large gene pool, e.g. for recruiting specific enzymes that can carry out a cometabolic reaction as a critical step for biodegradation [47]. Furthermore, there are optimal reductive conditions for MOs in rumen (−300 mV) by VFA production through fermentation of structural carbohydrates. Taking into account the commonly available conditions in the rumen, such a cometabolic reaction might explain the indirect microbiologically driven disappearance of PFAAs in rumen. Unfortunately, the ultimate proof for the degradation of PFAAs by MOs was not provided because an analysis of fluoride ions or transformation products was not performed.

Available literature data about PFAAs biodegradation generally failed to provide results of fluoride ion loss as indication for PFAAs mineralization [28, 29, 48, 49]. Although the reductive defluorination is supposed to be energetically favourable for bacteria under anaerobic conditions, it may be hindered by the high strength of the C-F bond [18, 31, 50]. This is the reason why PFAAs are generally considered to be microbiologically resistant to degradation.

Conclusion

This study shows that the partitioning of PFAAs between the solid and liquid phase in the rumen is influenced by their physicochemical properties, but also by the presence of ruminal MOs having an effect on the dilution rate of PFAAs from the feed into the fermentation liquid. This conclusion is confirmed by the lower recovery rates of PFAAs in fermented feed in the presence of ruminal MOs.

For PFAA disappearance in experiment A (+MOs), the potential effects of a cometabolic reaction as an indirect microbiologically-mediated PFAA degradation mechanism was discussed on the one hand and on the other the formation of biofilms as a sink for PFAAs in a microbial system. The mechanisms for PFAA disappearance in RUSITEC still remain inconclusive, since analysis of fluoride ions or transformation products was not performed.

PFAAs are accumulated more to a larger extent in muscle tissues of monogastrics compared to ruminants [1, 2]. In the context of consumer risk assessment, the hypothesis was formulated that this may be due to a disappearance (low recovery) as a consequence of the complex microbial ecosystem in the rumen. Using the RUSITEC we found indications for a small but statistically significant disappearance of PFOS (15 %). This is however not enough to explain the said difference between monogastric and ruminants (40–49 % vs <0.001–43 % accumulation of ingested PFAAs in muscle tissue). Beyond this, an interesting question is whether biofilms are an important sink for PFAAs in a microbial system.

Methods
Hay cultivation and preparation
PFAA contaminated hay was cultivated and harvested from a PFAA-contaminated agricultural land in Lower-Saxony, Germany. The agricultural land was contaminated by the use of a soil improver containing industrial waste with high concentrations of PFAAs. The soil improver provided by a recycling company was spread by the farmer on the agricultural land for several years before the environmental contamination became evident [51].

The second-cut PFAA hay was stored in square bales at the Federal Institute for Risk Assessment (BfR) and analyzed for PFAAs concentration at the Chemical and Veterinary Analytical Institute in Münster (Table 1). The PFAA-free hay (1st cut) was cultivated and harvested from a PFAAs unpolluted farmland in Lower-Saxony, Germany, provided by the Department Institute of Physiology of the University of Veterinary Medicine Hannover, Foundation. For sample preparation, PFAA-free hay and PFAA hay were cut with a sheep shearing machine into small pieces of 0.5–1 cm. Thereafter, the hay were divided homogeneously into 10 g samples by a rotary sample divider.

Experimental procedure and sampling
Experiment A (with ruminal MOs)
The RUSITEC experiment with six fermentation vessels was carried out similar as described previously [52]. Rumen inoculum was collected in the morning (3 h after feeding) from a ruminal fistulated Hinterwälder cow (13 years old, 650 kg body weight, non-lactating). Donor animal received hay and concentrate of 9 kg and 200 g per day, respectively. Concentrate is usually used as part of feed for ruminants. As this in vitro study should represent normal feeding conditions it was necessary to add concentrate provided by DEUKA Schaffutter, Erfurt,

Germany. At the beginning of the experiment, each fermentation vessel was inoculated with 80 g of solid rumen content and 700 ml rumen liquid.

The experiment consisted of an equilibration period of 6 days and a consecutive control period of 6 days followed by an experimental period of 24 h. During equilibration and control period nylon bags (11.5 × 6.5 cm, pore size 150 μm) comprised 8 g of PFAA-free hay and 2 g concentrate. After 24 h of equilibrium period, one nylon bag in each vessel was replaced with a nylon bag containing PFAA-free hay and concentrate. Throughout the overall experimental period the retention time of each nylon bag was adjusted to 48 h, i.e. the nylon bags were exchanged alternately at 24 h intervals. At the beginning of the experimental period, three fermentation vessels received two nylon bags with 8 g PFAA hay and 2 g concentrate (high dosage), whereas the remaining three fermentation vessels received one nylon bag containing 8 g PFAA hay and 2 g concentrate and a second nylon bag filled with 8 g PFAA-free hay and 2 g concentrate (low dosage).

For simulation of saliva secretion each fermentation vessel was connected to a fermenter pump about which buffer solution was constantly infused (0.7 l/day). The dilution rate of PFAAs in the fermentation liquids was determined during experimental period by relating the initial PFAA concentration in feed to the turn over rate of the fermentation fluid in each vessel. The time course of the diluted PFAA in the fermentation liquid (C_d) was calculated using equation:

$$C_d = a_0 \cdot e^{-kt}$$

where a_0 is the initial concentration of the PFAA (μg/l), $-k$ is the dilution value estimated by division of fermentation vessel volume (l) and turn over rate of fermenter pump (l/day), and t is the duration of the experiment expressed in days. The buffer solution (28 mmol/l NaCl; 7.69 mmol/l KCl; 0.5 mmol/l 1 N HCl; 0.22 mmol/l $CaCl_2 \cdot 2H_2O$; 0.63 mmol/l $MgCl_2 \cdot 6H_2O$; 5 mmol/l NH_4Cl; 10 mmol/l $Na_2HPO_4 \cdot 12H_2O$; 10 mmol/l $NaH_2PO_4 \cdot H_2O$; 97.90 mmol/l $NaHCO_3$) was infused continuously to reach a liquid turnover of once a day. Effluents of vessels were collected in air tight glass flasks which were stored on ice to inhibit microbial fermentation. The fermentation gases were simultaneously collected in gas-tight bags (Plastigas, Linde AG, München, Germany). PFAA-free hay, PFAA hay, concentrate, distilled water used for buffer preparation and nylon bags were analyzed for contents of PFAAs. In addition, chemical composition of PFAA-free and PFAA hay was determined (Table 1).

Determination of pH and redox potentials were done using the Digital-pH-Meter 646 (Knick GmbH & Co. KG, D-14163 Berlin) and the redox-electrode InLab® RedoxPro (Mettler-Toledo GmbH, D-35353

Gießen), respectively. The measurements were carried out throughout the experiment to monitor the anaerobic status of the system (redox potential normally varies between −100 and −300 mV) [53, 54] and to ensure adequate environmental conditions (pH between 5.5 and 7.0) for microbial survival.

Concentration of VFA in effluents was analyzed daily as described previously [55]. VFA production was determined by multiplying concentrations and volumes of the effluent.

For analyzes of ammonia concentrations 1 ml of the effluents was centrifuged at 4600 g for 10 min followed by mixing 50 µl of the supernatant with 5 ml phenol solution (106 mM phenol, 0.17 mM sodium nitroprusside dihydrate) and 5 ml sodium hypochlorite solution [1 % (v/v) sodium hypochlorite; 125 mM NaOH]. After an incubation step at 60 °C in a water bath for 10 min, concentrations of ammonia were determined photometrically at 546 nm in a spectrometer (DU 640, Beckman Coulter GmbH, Krefeld, Germany) using a NH_4Cl standard solution (5 mM).

During the experimental period, 40 ml of fermentation liquid were retrieved from each vessel for determination of PFAA concentration at following points after exchanging feed: 0, 2, 4, 6, 8, 10, 12, 16, 20, 24 h. In addition, the infused buffer volume was recorded at each point. Counteracting the continuous loss of fermentation liquid by sampling, the deviating volume was refilled from further six fermentation vessels which were treated equally.

At the end of the experiment both nylon bags were removed, dried at 65 °C and analyzed for concentration of PFAAs. Finally, volume of fermentation liquid was measured.

Experiment B (without ruminal MOs)
A second experiment with three fermentation vessels was conducted under sterile conditions. Therefore, the nylon bags and the feed were sterilized in an autoclave by subjecting them to high pressure saturated steam at 103 and 121 °C, respectively. At the beginning of the experiment, each fermentation vessel was filled with two nylon bags containing 8 g PFAA hay and 2 g concentrate and 750 ml buffer solution (for composition see "Experiment A"). For PFAA analysis, 50 ml of liquid was retrieved from each fermentation vessel. The buffer solution was infused continuously to reach a liquid turnover of once a day.

After 24 h, samples of 50 ml were taken from the liquid of each fermentation vessel and from the effluents and were analyzed for PFAA concentrations. In addition, total liquid volume in fermentation vessels and volume of effluents were measured. Both nylon bags were dried at 65 °C and analyzed for PFAA concentrations.

Subsequently, each fermentation vessel was washed with 100 ml of methanol solution (25 %) to dissolve adsorbed PFAAs from vessel surfaces. The washing solution was collected and analyzed for PFAA concentrations.

PFAA analysis
Sample preparation
Because of possible interferences between PFAA and proteins, matrix specific sample preparations were performed. For sample storage and sample preparation, only vessels made of polypropylene (PP) were used and served as blank in each series of PFAA analyzes. 1–5 g feed samples were extracted with methanol and an aliquot of this solution was diluted with water (VDLUFA-Method) [56]. About 5 ml (10 ml) methanol were added to 10 ml (20 ml) RUSITEC sample and afterwards the pH-value was adjusted to 4.5–5.5. All sample solutions were purified and concentrated using solid phase extraction on an OasisWAX column (150 mg/6 ml) [57]. The final extract was reconstituted in 100–1000 µl methanol/water (50/50, v/v), depending on the expected concentration.

Measurement
The purified solutions were measured using high-performance liquid chromatography (HPLC) with tandem mass spectrometry (MS/MS) in negative ion multiple reaction monitoring (MRM) mode. The separation was performed on an Agilent 1200 SL HPLC-System. A mixture of 2 mM ammonium acetate/acetonitrile (95/5, v/v) and a mixture of methanol/acetonitrile (40/60, v/v) were used as solvents in a gradient elution. MS/MS-detection was performed with an Agilent 6460 triple quadrupole mass spectrometer equipped with an electrospray interface (ESI) operating in the negative ion mode. The MRM-settings are published elsewhere [58]. In each sample sequence a blank-sample and a none-contaminated sample, which was spiked with isotope labeled PFHxS, PFOA and PFOS, was measured. The recovery was between 90 and 100 %. The analytical results were corrected for recovery rates. The a priori relative standard deviations were about 20 % for concentrations near the limit of quantification and 10–15 % for higher concentrations.

Quantification
Quantification was performed with isotope labeled standards and a seven-point calibration curve. ^{18}O-PFHxS and later ^{13}C-PFHxS were used as internal standards for PFBS and PFHxS. ^{13}C-PFOA was used as internal standard for PFOA and ^{13}C-PFOS for PFOS. The internal standards were added at the beginning of the sample preparation. The limit of detection was defined as the signal to noise ratio of 3:1 of the qualifier ion. The limit of quantification is defined as the concentration on

which a substance is identified unequivocally and quantified with a relative standard deviation of 20 % or lower. The measurement range was between 0.2 and 100 µg/l. In the case of PFOS, it could be ensured that only the linear substance was quantified. Due to the fact that standard substances of branched PFHpS and PFHxS were not available, it was not possible to ascertain that all the branched substances are separated completely from the linear ones. Only peaks with the same retention time as the linear substances were integrated. The ratios of the two most intensive MRM-transitions of the integrated PFHpS and PFHxS are the same as in the linear standard substance. Due to the finding that the transitions of the branched isomers differ from the transitions of the linear substances in the case of PFOS, this is an indication that the uncertainty caused by the branched isomers is small for PFHpS and PFHxS. The PFCA concentrations in the analyzed matrices were very low, so that a differentiation between the linear and the branched substances was not possible. It is described in literature that 78 % of the PFOA manufactured by the 3 M Company is linear [59, 60]. The ratios of the two most intensive MRM-transitions of the integrated PFCA were the same as in the linear standard substances. So it can be assumed that also in the case of PFCA the potential uncertainty caused by the branched isomers is small.

A short HPLC-column was placed as a pre-column between purge valve and autosampler to separate background PFCA and PFAS from the analytes of the samples. An injector program was used to minimize potential cross-contamination from heavily contaminated samples as far as possible. Interferences of PFOS with taurodeoxycholic acid are excluded, because both substances are separated chromatographically and furthermore the relation of the two most intensive transitions of PFOS in comparison to a standard solution was used to check possible interferences. Taurodeoxycholic acid does not show the m/z transition 499 to 99, which is specific for PFOS. The analytical method is described in detail by Ehlers [61].

Statistical evaluations

Data were processed using the statistic program SPSS1201 version 7.0.1.4. Effects on fermentation parameters were determined using t test if normal distribution of data was given. If not, Kruskal–Wallis test was used. For testing normal distribution Kolmogorov–Smirnov test was used. The homogeneity of variances was tested using the Levene's test. The significant differences of redox potential between each time point were determined by one-way analysis of variance. Here, the Dunett-T3 test was used when homogeneity of variance was not given. Comparison of the mean PFAA levels between

samples of experiment A and experiment B of the same matrix was performed using t test, whereas normal distribution was assumed. The data are reported as the mean ± standard deviation. All differences were considered to be significant when $P \leq 0.05$.

Abbreviations

APFO: ammonium perfluorooctanoate; CF: crude fiber; CP: crude protein; DM: dry matter; ESI: electrospray interface; HPLC: high-performance liquid chromatography; MOs: ruminal microorganisms; MRM: multiple reaction monitoring; mV: millivolt; NfE: nitrogen free extracts; PFAAs: perfluoroalkyl acids; PFAS: per- and polyfluoroalkyl substances; PFBS: perfluorobutane sulfonic acid; PFCAs: perfluoroalkyl carboxylic acids; PFHpA: perfluoroheptane carboxylic acid; PFHpS: perfluoroheptane sulfonic acid; PFHxA: perfluorohexane carboxylic acid; PFHxS: perfluorohexane sulfonic acid; PFOA: perfluorooctanoic acid; PFOS: perfluorooctane sulfonic acid; PFSAs: perfluoroalkyl sulfonic acids; POP: persistent organic pollutant; PP: polypropylene; REACH: European chemicals regulation, EC No. 1907/2006; RUSITEC: rumen simulation technique; SVHC: substances of very high concern; SEM: standard error of the mean; US EPA: United States environmental protection agency; VFA: volatile fatty acids.

Authors' contributions

GB and SR carried out the RUSITEC experiment. ML-W and HS conceived of the study, and participated in its design and coordination. All authors read and approved the final manuscript.

Author details

[1] Federal Institute for Risk Assessment, Max-Dohrn-Str. 8-10, 10589 Berlin, Germany. [2] Department Institute of Physiology, University of Veterinary Medicine Hannover, Foundation, Bischofsholer, Damm 15, 30173 Hannover, Germany.

Acknowledgements

The authors gratefully acknowledge P. Fürst and S. Ehlers of the Chemical and Veterinary Analytical Institute in Münster for performing the PFAA analyzes. Jorge Numata is acknowledged for his support in preparing the manuscript.

Competing interests

The authors declare that they have no competing interests.

References

1. Kowalczyk J, Ehlers S, Oberhausen A, Tischer M, Furst P, Schafft H et al (2013) Absorption, distribution, and milk secretion of the perfluoroalkyl acids PFBS, PFHxS, PFOS, and PFOA by dairy cows fed naturally contaminated feed. J Agric Food Chem 61(12):2903–2912. doi:10.1021/jf304680j
2. Numata J, Kowalczyk J, Adolphs J, Ehlers S, Schafft H, Fuerst P et al (2014) Toxicokinetics of seven perfluoroalkyl sulfonic and carboxylic acids in pigs fed a contaminated diet. J Agric Food Chem 62(28):6861–6870
3. Buck RC, Franklin J, Berger U, Conder JM, Cousins IT, de Voogt P et al (2011) Perfluoroalkyl and polyfluoroalkyl substances in the environment: terminology, classification, and origins. Integr Environ Assess Manag 7(4):513–541. doi:10.1002/ieam.258
4. Footitt A, Nwaogu TA, Brooke D (2004) Risk reduction strategy and analysis of advantages and drawbacks for perfluorooctane sulphonate (PFOS). Final report. Department for environment, food and rural affairs & the environmental agency for England and Wales. https://www.gov.uk/

government/uploads/system/uploads/attachment_data/file/183154/pfos-riskstrategy.pdf. Accessed 24 Sept 2015

5. Prevedouros K, Cousins IT, Buck RC, Korzeniowski SH (2006) Sources, fate and transport of perfluorocarboxylates. Environ Sci Technol 40(1):32–44. doi:10.1021/es0512475

6. 3 M. Fluorochemical use, distribution and release overview. http://www.chemicalindustryarchives.org/dirtysecrets/scotchgard/pdfs/226-0550.pdf1999. Accessed on May 26, 1999

7. Giesy JP, Kannan K (2001) Global distribution of perfluorooctane sulfonate in wildlife. Environ Sci Technol 35(7):4

8. Houde M, Martin JW, Letcher RJ, Solomon KR, Muir DC (2006) Biological monitoring of polyfluoroalkyl substances: a review. Environ Sci Technol 40(11):3463–3473

9. Sedlak MD, Greig DJ (2012) Perfluoroalkyl compounds (PFCs) in wildlife from an urban estuary. J Environ Monit 14(1):146–154. doi:10.1039/C1EM10609K

10. Rigét F, Bossi R, Sonne C, Vorkamp K, Dietz R (2013) Trends of perfluorochemicals in Greenland ringed seals and polar bears: indications of shifts to decreasing trends. Chemosphere 93(8):1607–1614. doi:10.1016/j.chemosphere.2013.08.015

11. Bossi R, Riget FF, Dietz R, Sonne C, Fauser P, Dam M et al (2005) Preliminary screening of perfluorooctane sulfonate (PFOS) and other fluorochemicals in fish, birds and marine mammals from Greenland and the Faroe Islands. Environ Pollut 136(2):323–329. doi:10.1016/j.envpol.2004.12.020

12. Pérez F, Nadal M, Navarro-Ortega A, Fàbrega F, Domingo JL, Barceló D et al (2013) Accumulation of perfluoroalkyl substances in human tissues. Environ Int 59:354–362. doi:10.1016/j.envint.2013.06.004

13. OECD. Hazard assessment of perfluorooctane sulfonate (PFOS) and its salts. http://www.oecd.org/env/ehs/risk-management/perfluorooctanesulfonatepfosandrelatedchemicalproducts.htm2002 21-Nov-2002. Report No.: ENV/JM/RD(2002)17/FINAL

14. Stahl T, Mattern D, Brunn H (2011) Toxicology of perfluorinated compounds. Environ Sci Eur 23(38):108

15. Fromme H, Mosch C, Morovitz M, Alba-Alejandre I, Boehmer S, Kiranoglu M et al (2010) Pre- and postnatal exposure to perfluorinated compounds (PFCs). Environ Sci Technol 44(18):7123–7129. doi:10.1021/es101184f

16. van Asselt ED, Rietra RPJJ, Römkens PFAM, van der Fels-Klerx HJ (2011) Perfluorooctane sulphonate (PFOS) throughout the food production chain. Food Chem 128(1):1–6. doi:10.1016/j.foodchem.2011.03.032

17. EFSA (2012) Perfluoroalkylated substances in food: occurrence and dietary exposure. European Food Safety Authority Report No.: EFSA J 10(6):2743. http://www.efsa.europa.eu/sites/default/files/scientific_output/files/main_documents/2743.pdf

18. Parsons JR, Saez M, Dolfing J, de Voogt P (2008) Biodegradation of perfluorinated compounds. Rev Environ Contam Toxicol 196:53–71

19. UNEP (2009) Governments unite to step-up reduction on global DDT reliance and add nine new chemicals under international treaty. United Nations Environment Programme. http://www.unep.org/Documents.Multilingual/Default.Print.asp?DocumentID=585&ArticleID=6158&

20. USEPA (2015) 2010/2015 PFOA Stewardship Program. U.S. Environmental Protection Agency. http://www.epa.gov/oppt/pfoa/pubs/stewardship/. Accessed on September.24, 2015

21. Vierke L, Staude C, Biegel-Engler A, Drost W, Schulte C (2012) Perfluorooctanoic acid (PFOA)—main concerns and regulatory developments in Europe from an environmental point of view. Environ Sci Eur 24(1):1–11. doi:10.1186/2190-4715-24-16

22. ECHA (2015) Risk management option analysis conclusion document for substance name: Perfluorononan-1-oic acid (2,2,3,3,4,4,5,5,6,6,7,7,8,8,9,9,9-heptadecafluorononanoic acid and its sodium and ammonium salts. European Chemicals Agency. http://echa.europa.eu/documents/10162/233c4b5c-2dcf-41a7-ab9a-1c6892978956. Accessed on September 24, 2015

23. Krusic PJ, Roe DC (2004) Gas-phase NMR technique for studying the thermolysis of materials: thermal decomposition of ammonium perfluorooctanoate. Anal Chem 76(13):3800–3803. doi:10.1021/ac049667k

24. Vecitis CD, Park H, Cheng J, Mader BT, Hoffmann MR (2008) Kinetics and mechanism of the sonolytic conversion of the aqueous perfluorinated surfactants, perfluorooctanoate (PFOA), and perfluorooctane sulfonate (PFOS) into inorganic products. J Phys Chem A 112(18):4261–4270. doi:10.1021/jp801081y

25. Park H, Vecitis CD, Cheng J, Choi W, Mader BT, Hoffmann MR (2009) Reductive defluorination of aqueous perfluorinated alkyl surfactants: effects of ionic headgroup and chain length. J Phys Chem A 113(4):690–696. doi:10.1021/jp807116q

26. Zhang K, Huang J, Yu G, Zhang Q, Deng S, Wang B (2013) Destruction of perfluorooctane sulfonate (PFOS) and perfluorooctanoic acid (PFOA) by ball milling. Environ Sci Technol 47(12):6471–6477. doi:10.1021/es400346n

27. Lutze H, Panglisch S, Bergmann A, Schmidt T (2012) Treatment options for the removal and degradation of polyfluorinated chemicals. In: Knepper TP, Lange FT (eds) Polyfluorinated chemicals and transformation products. The handbook of environmental chemistry. Springer, Berlin, pp 103–125

28. Schröder HF (2003) Determination of fluorinated surfactants and their metabolites in sewage sludge samples by liquid chromatography with mass spectrometry and tandem mass spectrometry after pressurised liquid extraction and separation on fluorine-modified reversed-phase sorbents. J Chromatogr 1020(1):131–151. doi:10.1016/S0021-9673(03)00936-1

29. Liou JS, Szostek B, DeRito CM, Madsen EL (2010) Investigating the biodegradability of perfluorooctanoic acid. Chemosphere 80(2):176–183. doi:10.1016/j.chemosphere.2010.03.009

30. Ochoa-Herrera V, Sierra-Alvarez R, Somogyi A, Jacobsen NE, Wysocki VH, Field JA (2008) Reductive defluorination of perfluorooctane sulfonate. Environ Sci Technol 42(9):3260–3264. doi:10.1021/es702842q

31. Smidt H, de Vos WM (2004) Anaerobic microbial dehalogenation. Annu Rev Microbiol 58:43–73. doi:10.1146/annurev.micro.58.030603.123600

32. Martens JH, Barg H, Warren M, Jahn D (2002) Microbial production of vitamin B12. Appl Microbiol Biotechnol 58(3):275–285. doi:10.1007/s00253-001-0902-7

33. Dryden LP, Hartman AM, Bryant MP, Robinson IM, Moore LA (1962) Production of vitamin B12 and vitamin B12 analogues by pure cultures of ruminal bacteria. Nature 195:201–202

34. Atef M, Salem AA, Al-Samarrae SA, Zafer SA (1979) Ruminal and salivary concentration of some sulphonamides in cows and their effect on rumen flora. Res Vet Sci 27(1):9–14

35. Craig AM, Latham CJ, Blythe LL, Schmotzer WB, O'Connor OA (1992) Metabolism of toxic pyrrolizidine alkaloids from tansy ragwort (*Senecio jacobaea*) in ovine ruminal fluid under anaerobic conditions. Appl Environ Microbiol 58(9):2730–2736

36. Zinedine A, Soriano JM, Moltó JC, Mañes J (2007) Review on the toxicity, occurrence, metabolism, detoxification, regulations and intake of zearalenone: an oestrogenic mycotoxin. Food Chem Toxicol 45(1):1–18. doi:10.1016/j.fct.2006.07.030

37. Martínez ME, Ranilla MJ, Tejido ML, Ramos S, Carro MD (2010) Comparison of fermentation of diets of variable composition and microbial populations in the rumen of sheep and Rusitec fermenters. I. Digestibility, fermentation parameters, and microbial growth. J Dairy Sci 93(8):3684–3698. doi:10.3168/jds.2009-2933

38. Higgins CP, Luthy RG (2007) Modeling sorption of anionic surfactants onto sediment materials: an a priori approach for perfluoroalkyl surfactants and linear alkylbenzene sulfonates. Environ Sci Technol 41(9):3254–3261. doi:10.1021/es062449j

39. Ahrens L, Taniyasu S, Yeung LW, Yamashita N, Lam PK, Ebinghaus R (2010) Distribution of polyfluoroalkyl compounds in water, suspended particulate matter and sediment from Tokyo Bay, Japan. Chemosphere 79(3):266–272. doi:10.1016/j.chemosphere.2010.01.045

40. Zhao L, Zhu L, Yang L, Liu Z, Zhang Y (2012) Distribution and desorption of perfluorinated compounds in fractionated sediments. Chemosphere 88(11):1390–1397. doi:10.1016/j.chemosphere.2012.05.062

41. Ahrens L, Yamashita N, Yeung LWY, Taniyasu S, Horii Y, Lam PKS et al (2009) Partitioning behavior of per- and polyfluoroalkyl compounds between pore water and sediment in two sediment cores from Tokyo Bay, Japan. Environ Sci Technol 43(18):6969–6975. doi:10.1021/es901213s

42. Ahrens L, Yeung LW, Taniyasu S, Lam PK, Yamashita N (2011) Partitioning of perfluorooctanoate (PFOA), perfluorooctane sulfonate (PFOS) and perfluorooctane sulfonamide (PFOSA) between water and sediment. Chemosphere 85(5):731–737. doi:10.1016/j.chemosphere.2011.06.046

43. Krafft MP, Riess JG (2009) Chemistry, physical chemistry, and uses of molecular fluorocarbon—hydrocarbon diblocks, triblocks, and related compounds—unique "apolar" components for self-assembled colloid

and interface engineering. Chem Rev 109(5):1714–1792. doi:10.1021/cr800260k

44. McAllister TA, Bae HD, Jones GA, Cheng KJ (1994) Microbial attachment and feed digestion in the rumen. J Anim Sci 72(11):3004–3018

45. EPA U. Engineered Approaches to in situ bioremediation of chlorinated solvents: fundamentals and field applications. EPA 542-R-00-0082000

46. Lee SS, Ha JK, Cheng KJ (2000) Relative contributions of bacteria, protozoa, and fungi to in vitro degradation of orchard grass cell walls and their interactions. Appl Environ Microbiol 66(9):3807–3813

47. Janssen DB, Dinkla IJ, Poelarends GJ, Terpstra P (2005) Bacterial degradation of xenobiotic compounds: evolution and distribution of novel enzyme activities. Environ Microbiol 7(12):1868–1882. doi:10.1111/j.1462-2920.2005.00966.x

48. Meesters RJ, Schroder HF (2004) Perfluorooctane sulfonate—a quite mobile anionic anthropogenic surfactant, ubiquitously found in the environment. Water Sci Technol 50(5):235–242

49. Schroder HF, Jose HJ, Gebhardt W, Moreira RF, Pinnekamp J (2010) Biological wastewater treatment followed by physicochemical treatment for the removal of fluorinated surfactants. Water Sci Technol 61(12):3208–3215. doi:10.2166/wst.2010.917

50. Natarajan R, Azerad R, Badet B, Copin E (2005) Microbial cleavage of CF bond. J Fluor Chem 126(4):424–435. doi:10.1016/j.jfluchem.2004.12.001

51. LANUV (2011) Verbreitung von PFT in der Umwelt Ursachen—Untersuchungsstrategie—Ergebnisse—Maßnahmen: Landesamt für Natur, Umwelt und Verbraucherschutz Nordrhein-Westfalen

52. Czerkawski JW, Breckenridge G (1977) Design and development of a long-term rumen simulation technique (RUSITEC). Br J Nutr 38(3):371–384

53. Barry TN, Thompson A, Armstrong DG (1977) Rumen fermentation studies on two contrasting diets. 1. Some characteristics of the in vivo fermentation, with special reference to the composition of the gas phase, oxidation/reduction state and volatile fatty acid proportions. J Agric Sci 89(01):183–195. doi:10.1017/S0021859600027362

54. Marounek M, Brezina P, Simunek J, Bartos S (1991) Influence of redox potential on metabolism of glucose in mixed cultures of rumen microorganisms. Arch Tierernahr 41(1):63–69

55. Koch M, Strobel E, Tebbe CC, Heritage J, Breves G, Huber K (2006) Transgenic maize in the presence of ampicillin modifies the metabolic profile and microbial population structure of bovine rumen fluid in vitro. Br J Nutr 96(5):820–829

56. VDLUFA (2006) Handbuch der Landwirtschaftlichen Versuchs- und Untersuchungsmethodik. Band III. Die chemische Untersuchung von Futtermitteln. VDLUFA-Verlag, Darmstadt, Deutschland.

57. Taniyasu S, Kannan K, So MK, Gulkowska A, Sinclair E, Okazawa T et al (2005) Analysis of fluorotelomer alcohols, fluorotelomer acids, and short- and long-chain perfluorinated acids in water and biota. J Chromatogr 1093(1–2):89–97. doi:10.1016/j.chroma.2005.07.053

58. Bernsmann T, Fürst P (2008) Determination of perfluorinated compounds in human milk. Organohalog Compd 70:718–721

59. De Silva AO, Benskin JP, Martin LJ, Arsenault G, McCrindle R, Riddell N et al (2009) Disposition of perfluorinated acid isomers in Sprague-Dawley rats; part 2: subchronic dose. Environ Toxicol Chem 28(3):555–567. doi:10.1897/08-254.1

60. Stevenson L (2002) Comparative analysis of fluorochemicals in human serum samples obtained commercially. Public Docket AR-2261150: 3M Environmental Laboratory

61. Ehlers S (2012) Analytik von Perfluoralkylsäuren in verschiedenen Matrices zur Klärung der Toxikokinetik in Tierarten, die der Lebensmittelgewinnung dienen.: Westfälische Wilhelms-Universität Münster

7

Computing the biomass potentials for maize and two alternative energy crops, triticale and cup plant (*Silphium perfoliatum* L.), with the crop model BioSTAR in the region of Hannover (Germany)

Roland Bauböck[1*], Marianne Karpenstein-Machan[2] and Martin Kappas[1]

Abstract

Background: Lower Saxony (Germany) has the highest installed electric capacity from biogas in Germany. Most of this electricity is generated with maize. Reasons for this are the high yields and the economic incentive. In parts of Lower Saxony, an expansion of maize cultivation has led to ecological problems and a negative image of bioenergy as such. Winter triticale and cup plant have both shown their suitability as alternative energy crops for biogas production and could help to reduce maize cultivation.

Results: The model Biomass Simulation Tool for Agricultural Resources (BioSTAR) has been validated with observed yield data from the region of Hannover for the cultures maize and winter wheat. Predicted yields for the cultures show satisfactory error values of 9.36% (maize) and 11.5% (winter wheat). Correlations with observed data are significant ($P < 0.01$) with $R = 0.75$ for maize and 0.6 for winter wheat. Biomass potential calculations for triticale and cup plant have shown both crops to be high yielding and a promising alternative to maize in the region of Hanover and other places in Lower Saxony.

Conclusions: The model BioSTAR simulated yields for maize and winter wheat in the region of Hannover at a good overall level of accuracy (combined error 10.4%). Due to input data aggregation, individual years show high errors though (up to 30%). Nevertheless, the BioSTAR crop model has proven to be a functioning tool for the prediction of agricultural biomass potentials under varying environmental and crop management frame conditions.

Keywords: Alternative energy crops; BioSTAR; Biomass potentials; Integrative concepts

Background

Current state biogas production

Bioenergy from agricultural substrates (plant material and manure) has become an important input to the renewable energy mix in Germany [1] particularly in Lower Saxony [2]. Lower Saxony is the most important agricultural state in Germany [3], with the highest installed electric capacity from biogas [4]. The dominant plant substrate used in these facilities is maize grown for silage [5]. Silage maize is well established in Germany as a feed culture in cattle farms. Through years of successful breeding, the thermophile C_4 crop has been adapted to the central European climate, and with ample water and temperature, high biomass yields are possible. This aspect makes maize the most widely used energy crop for biogas production in Lower Saxony and Germany in general. The marginal returns are high; especially farmers in regions with a high cattle stocking are familiar with this crop, and risks associated with the production of silage maize are low and known to the farmers. With a boom of bioenergy facilities stimulated by the German renewable energies act (first version 2000), a strong expansion of maize

* Correspondence: rbauboe1@gwdg.de
[1]Department of Cartography, GIS and Remote Sensing, Research Project 'BIS', University of Göttingen, Goldschmidtstraße 5, Göttingen 37077, Germany
Full list of author information is available at the end of the article

production in Germany and in Lower Saxony has been triggered [6]. From 2001 to 2012, the number of biogas-producing facilities has increased from 148 to 1,480 [7]. As a result of this strong expansion of the maize share in the crop mix on agricultural land (some of the expansion was generated by turning pasture into cropland or by using low-yielding lands which had been used as fallow and thus to some degree for nature conservation), criticism against a further expansion of bioenergy crop production and against existing biogas (methane) facilities arose. The main points of concern with regard to the strong expansion of the maize production are as follows [8]:

1. A reduction of biodiversity (in species related to crops and crop rotations)
2. A negative impairment of the characteristic landscape (dominated by cereals for centuries)
3. Ecologically adverse effects like nitrate leaching, soil erosion and a reduction of humus content in the soils

According to the German environmental protection law, farmers need to follow certain agricultural guidelines referred to as 'good professional practice' [9]. According to paragraph 1 of the German environmental protection law, the diversity, character and beauty of the landscape have to be protected; their state is not to be worsened and has to be restored where damage or impairment has occurred. According to paragraph 5 of this law and according to the guidelines of a good professional practice, the long-term maintenance of soil fertility and the practice of maintaining a crop rotation with three different crops have to be established. Furthermore, it is stipulated that the natural fitting of soil, water, flora and fauna should not be impaired beyond the measure of a renewable yield of an agricultural site [10].

Nevertheless, large-scale cultivation of maize for silage (as is the case in some parts of Lower Saxony) leads to monocultures and can cause the environmental problems mentioned above.

One possibility to counteract the negative impacts associated with the large-scale production of silage maize would be the introduction of new bioenergy crops into the existing crop rotations and, beyond this, to create new 'integrative concepts' [11,12].

Integrative concepts with new energy crops

The term 'integrative concept' is used for the scientific approach which combines different land use options to produce food, fodder and energy while promoting biodiversity and ecosystem services in agricultural land-scapes [13-16]. Integrative cultivation concepts should harmonize utilization and protection of a landscape. Due to external costs of non-sustainable systems, only sustainable concepts are economically sound in the long run. The vision of integrative concepts is to contribute to a more diverse agricultural landscape, keep nature in balance and conserve ecosystems. The integration of new annual energy crops into the crop rotation, as well as the cultivation of perennial crops on problematic soils, offers great opportunities to mitigate negative impacts of agriculture on biodiversity and ecosystem services. In this article, we will focus on two new energy crops (triticale and *Silphium perfoliatum* L.) which have the potential to increase crop and landscape diversity.

Triticosecale wittmack (triticale)

As an energy crop, winter triticale is suitable for locations with cool and moderate climates and locations that lack a high summer precipitation. Triticale utilizes winter soil moisture to produce biomass in the spring. It already reaches the maximum biomass yield in the first half of the year. Therefore, it is hardly affected by summer dryness. Dry matter yields range between 12 and 16 t/ha. Triticale is grown on good soil quality locations in southern Lower Saxony, and it is the most productive winter crop for biogas production. Even on poorer soils, triticale produces high biomass yields.

In combination with field grass (double cropping system), triticale can broaden a narrow maize rotation, increase biodiversity and improve the humus balance [17].

Silphium perfoliatum L.

A very long useful life is anticipated for Silphie (*S. perfoliatum* L.), also known as 'cup plant'. With its cupped leaves, Silphie can collect air moisture; it is therefore relatively resistant to dry conditions. It is adapted to the moderate climate conditions of eastern North America and can be cultivated 400 m above sea level. Silphie has been cultivated as fodder for cattle in North America and in the former German Democratic Republic (GDR). It was tested as an alternative biogas crop in field trials in Germany from 2005 onward. In 2010, farmers cultivated Silphie on about 20 ha of farmland [18]. The best results have been obtained when the seeds are sown and nursed in greenhouses and transplanted as young plants with three or four leaves into the fields in May or June. In the first year, the crop should establish itself in the soil and the plants should only build a leaf rosette before winter. In the following spring, the plants grow very quickly and can deliver their first harvest in early autumn [19]. The first results show that Silphie has a very high yield which is similar to that of maize [20,21]. One big advantage of this crop is the fact that after the first year, it needs no further weed control and no additional pesticides. Further advantages are given by the absence of a yearly tillage, which induces CO_2 storage in the soil and by the long flowering period of Silphie which is used by honeybees to collect pollen and nectar (bee bread) to survive and reproduce. However, the seed quality of this crop still must be improved to help broaden Silphie's use as a commonly used energy crop.

Results and discussion

Model validation

To verify the biomass results of the model Biomass Simulation Tool for Agricultural Resources (BioSTAR)

[22], winter wheat and maize yields for the years 1981 to 2007 for the region of Hannover have been calculated and then compared with observed yield data from the same years and region and then statistically analysed.

In Table 1, the observed (obs.), the predicted (pred.), the percent error of the predicted (e%), the adjusted value of the prediction (adjust) and the percent error of the adjusted value of prediction (ae%) are displayed for both maize and winter wheat for the years 1981 to 2007.

Table 1 Comparison of observed (statistical data) and predicted (modelled) yields for maize and winter wheat, 1981–2007

Year	Maize					Winter wheat				
	obs.	pred.	e%	adjust	ae%	obs.	pred.	e%	adjust	ae%
1981	482.9	753.1	55.9	542.2	12.3	55.1	90.7	64.7	72.6	31.7
1982	491.6	650.7	32.4	468.5	−4.7	64.5	78.7	22.0	62.9	−2.4
1983	400.4	533.2	33.2	383.9	−4.1	61.3	85.8	39.9	68.6	11.9
1984	413.2	642.9	55.6	462.9	12.0	63.3	96.5	52.5	77.2	22.0
1985	468.3	794.2	69.6	571.8	22.1	61.8	87.5	41.5	70.0	13.2
1986	460.5	715.7	55.4	515.3	11.9	78.3	82.0	4.7	65.6	−16.3
1987	427.6	703.0	64.4	506.2	18.4	75.4	83.5	10.8	66.8	−11.3
1988	458.4	528.8	15.4	380.8	−16.9	71.5	73.8	3.2	59.0	−17.4
1989	414.2	418.0	0.9	301.0	−27.3	54.0	79.7	47.7	63.8	18.1
1990	379.8	516.2	35.9	371.6	−2.2	71.9	79.0	9.9	63.2	−12.1
1991	399.0	492.9	23.5	354.9	−11.1	81.3	93.8	15.4	75.0	−7.7
1992	367.8	382.0	3.9	275.1	−25.2	75.3	78.0	3.6	62.4	−17.1
1993	461.0	704.5	52.8	507.2	10.0	82.9	97.5	17.6	78.0	−5.9
1994	429.1	638.6	48.8	459.8	7.2	81.5	96.1	17.9	76.9	−5.6
1995	398.3	569.9	43.1	410.3	3.0	84.2	102.56	21.8	82.1	−2.6
1996	424.7	639.3	50.5	460.3	8.4	82.4	83.4	1.3	66.7	−19.0
1997	460.9	695.1	50.8	500.4	8.6	88.5	111.25	25.7	89.0	0.6
1998	454.5	655.9	44.3	472.3	3.9	82.0	102.68	25.3	82.2	0.2
1999	432.6	632.0	46.1	455.0	5.2	93.5	102.42	9.5	81.9	−12.4
2000	496.3	582.3	17.3	419.3	−15.5	87.7	98.5	12.2	84.8	−3.4
2001	489.7	619.3	26.5	445.9	−8.9	95.4	107.67	12.8	91.8	−3.8
2002	470.7	650.1	38.1	468.1	−0.6	76.4	117.61	54.0	99.3	30.0
2003	375.2	441.1	17.6	317.6	−15.3	78.8	88.2	11.9	73,44	−6.8
2004	480.6	722.4	50.3	520.2	8.2	89.9	125.68	39.7	103.8	15.4
2005	482.2	711.5	47.6	512.3	6.2	84.8	121.74	43.6	99.5	17.3
2006	421.5	581.0	37.9	418.3	−0.7	84.3	102.66	21.8	93.0	−1.6
2007	537.7	771.4	43.5	555.4	3.3	77.5	107.48	38.7	86.0	10.9
	Mean	Mean	Mean	Mean	Mean	Mean	Mean	Mean	Mean	Mean
	445.1	624.3	39.8	449.5	0.63	77.2	95.4	24.8	77.3	1.0
	Factor adjust		0.72			Factor adjust		0.80		
	RMSE (dt/ha)		42.9			RMSE (dt/ha)		9.6		
	% error		9.6			% error		11.5		
	WIA		0.89			WIA		0.74		

Data sources: BioSTAR data and LSN. obs., observed yield; pred., predicted yield; e%, percent error of the predicted; adjust, adjusted value of the predicted; ae%, percent error of the adjusted value of prediction. Values are in deci-tons per hectare (dt/ha) fresh mass for maize and dt/ha grain weight for winter wheat. Mean, mean value of column; RMSE, root-mean-square error; WIA, Willmott index of agreement.

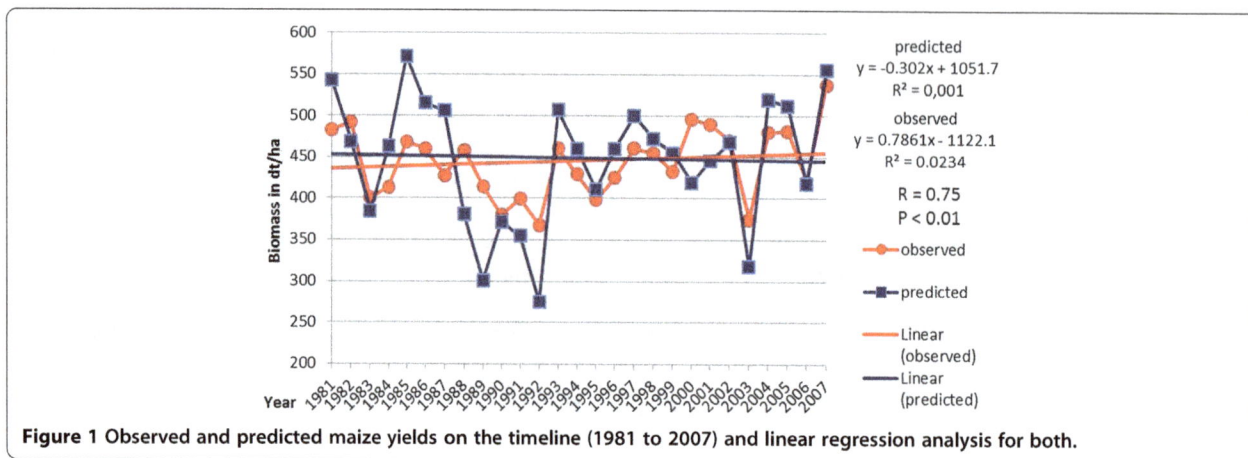

Figure 1 Observed and predicted maize yields on the timeline (1981 to 2007) and linear regression analysis for both.

Below the individual year listings, mean values for each column, the adjustment factor, the root-mean-square error (RMSE), the overall percent error (calculated from the RMSE) and the *Willmott index of agreement* are given for both crops.

The adjusted yield value is a product of the predicted yield value and the adjustment factor (factor adjust).

The adjustment factor is needed here to adjust the output of the model (which has been calibrated with field trial data) to realistically achievable yields of actual agriculture. The yields of actual agriculture can be up to 30% lower than those achieved in controlled field trials. Possible explanations for this are more homogenous soils and optimal crop care (pest and weed control and fertilizer application) on small trail parcels of several square metres compared to large agricultural lots of many hectares.

To derive this adjustment factor, the mean of the observed yield has been divided by the mean of the predicted yield. For maize, the resulting adjustment factor is 0.72 (72%), and for winter wheat, it is 0.80 (80%). Four years, 1989 and 1992 for maize and 1981

and 2002 for winter wheat, show higher deviations of the predicted from the observed yield values.

For winter wheat, the yield was overestimated by 31.7% in 1981 (after the adjustment) and 64.7% (before the adjustment). A possible explanation for this is an unusually wet spring/early summer with rainfall amounts of 108 and 146 mm in May and June. Compared to the long-term (1981 to 2010) averages for these months in the region of Hannover (56 and 65 mm), this is a surplus of 93% and 125% of rainfall. The same holds true for the year 2002 (overestimation of 30%) where July was extremely wet with average rainfall amounts of 171 mm. This is a surplus of 160% in comparison to the average (67 mm) from the period 1981 to 2010 for this region and month.

The model uses these additional litres of water to generate an increased crop growth. Possibly adverse effects due to continuous rain, like water-logged soils (not modelled in this case due to limited soil data), or humidity-related fungus problems are not accounted for. Strongest deviations for maize show up for the years 1989 and 1992, where yields have been underestimated by

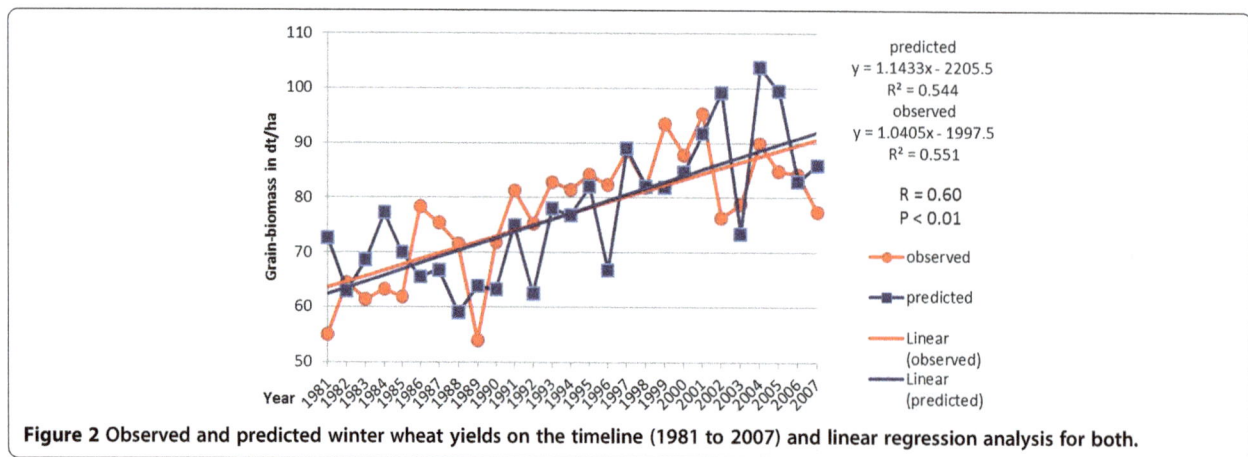

Figure 2 Observed and predicted winter wheat yields on the timeline (1981 to 2007) and linear regression analysis for both.

27.3% and 25.2% after the adjustment factor (0.72) has been applied to the simulation data.

Since winter wheat is not grown on low-quality soils, only model results from soil data with *nFK* values >90 mm have been used for the comparison of the observed and predicted winter wheat yields. For maize, which is usually grown on a big span of soil types, only very poor quality soils with field capacity values below 40 mm have been omitted from the comparison.

In Figures 1 and 2, yield trends (observed and predicted) across the analysed 27 years for maize and winter wheat and their linear regressions are displayed. The regression lines for maize more or less show stagnating yields over the years, and the regressions for winter wheat show a trend of increasing yields in the time slot. For the observed winter wheat data, the explanation lies in advances in breeding. For the simulated yields, the advances in breeding have been accounted for after the simulation run. Deductions for all years before 2000 following the regression line of the observed yields have been made. This is necessary because the model is calibrated to fit with yield data from the period 2005 to 2010 (current state breeding). After 2000, no statistically significant yield gains through breeding can be accounted for. For both maize and winter wheat, the correlation (observed vs. predicted values) is significant with $P < 0.01$ and $R = 0.75$ for maize and $R = 0.60$ for winter wheat. The overall fit of the predicted yield data (after the adjustment) with the observed data for all years combined is at a good level for both cultures with satisfactory statistical error values. The RMSE values for maize and winter wheat are at 42.9 dt/ha (deci-tons per hectare) fresh mass (maize) and 9.6 dt/ha grain weight (winter wheat); the percent errors are at 9.36% (maize) and 11.5% (winter wheat), and the Willmott index of agreement is at 0.89 for maize and 0.74 for winter wheat.

Biomass potentials in the region of Hannover

Biomass potentials (dry matter yields per hectare) for maize, triticale and cup plant have been calculated with the crop model BioSTAR. In a second step, these yields have been joined with a geographical information system (GIS) map of the soil dataset (*Bodenschätzungsdaten*/soil evaluation data) of the region of Hannover [23]. This

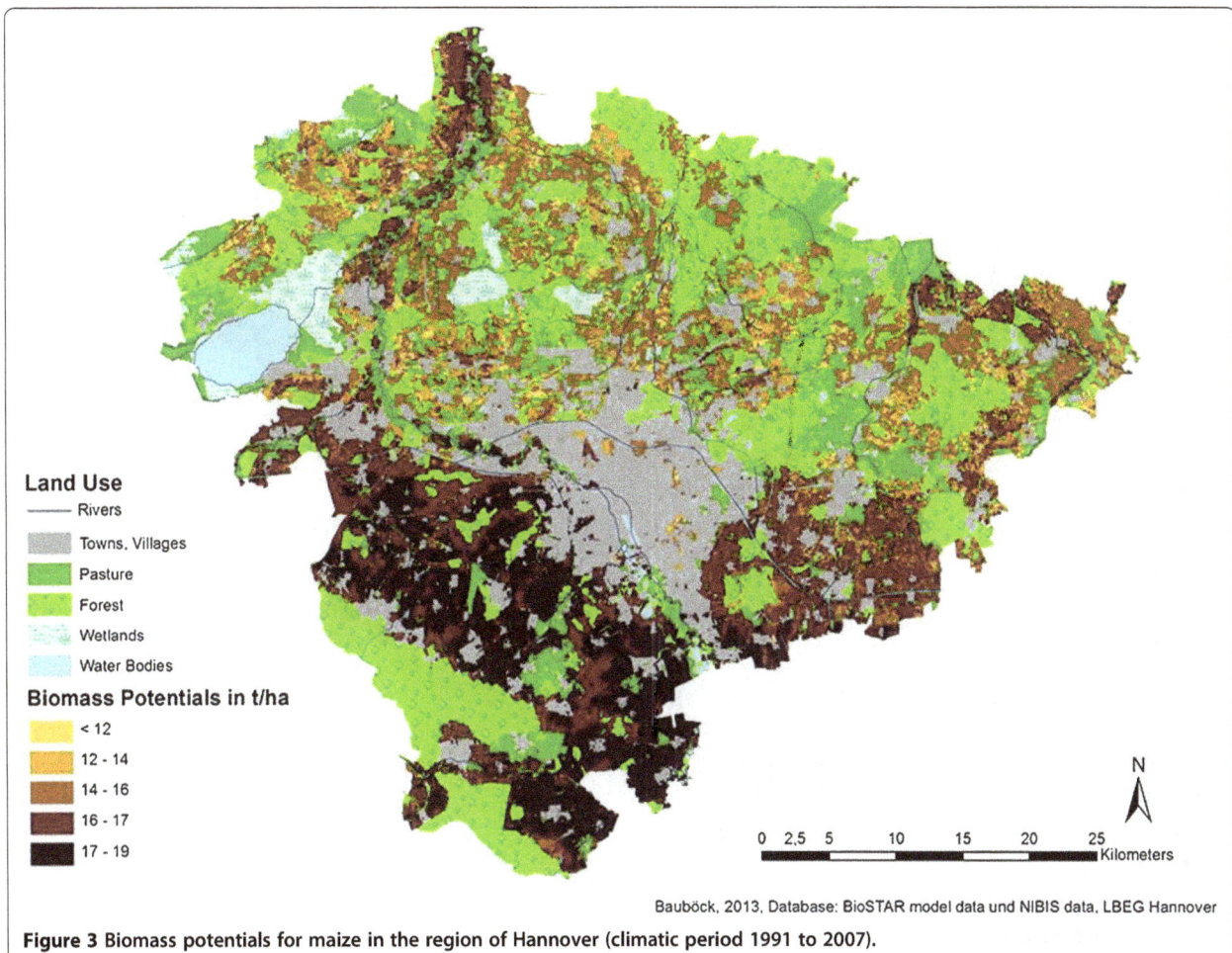

Baubök, 2013, Database: BioSTAR model data und NIBIS data, LBEG Hannover

Figure 3 Biomass potentials for maize in the region of Hannover (climatic period 1991 to 2007).

map has a total of 114,357 individual soil units, divided into eight different soil types and five different climate regions (see 'Methods' section). The yield data has been joined with the procedure described in the 'Methods' section. Because the years 1990 to 2007 are more representative for the current state climate (dryer summers, higher temperatures), only these years have been used for the biomass potential calculation of the three cultures.

In Figures 3, 4, and 5, the results of the joined data are cartographically displayed. Figure 6 shows the spatial distribution of the *nFK* values in the region. For triticale and cup plant, no adjustment factor (yields from field trials vs. actual yields) could be determined due to lack of statistical data. Hence, for triticale and for cup plant, the same factor was used as for winter wheat (0.8). The distribution of the biomass potentials for all three cultures shows a similar pattern roughly following the distribution of the *nFK* values (and thus the soil type distribution) in the region. All three cultures profit from the occurrence of the loamy and silty type soils and the higher *nFK* values (150 to 260 mm) in the southern part of the region and produce the highest yields there. The northern part of the region is mostly dominated by sandy type soils (*nFK* values below 150 mm), with an exception of the 'Leine' river valley in the northwest. Maize, a C_4-culture with a high yield potential, does show up with the highest biomass potentials of the three crops. In the southern part of the region, maize has average potentials between 16 and 19 t dry mass/ha. In the northern part (except for the Leine valley), the potentials are lower and range between 12 and 16 t/ha. Soils with potentials below 12 t/ha are only few.

For cup plant, the biomass potentials are not as high as for maize. However, yields between 16 and 17 t/ha are frequent in the south, and on some soils, yields of up to 19 t/ha can be reached. The northern part, dominated by the sandy type soils, has average biomass potentials around 12 to 14 t/ha (except for the Leine valley, where it is higher), but on some soils, the potentials can be lower than 12 t/ha.

Triticale shows a different biomass potential distribution pattern than the other two cultures. The division into north and south can still be recognized here, but the differences are less pronounced.

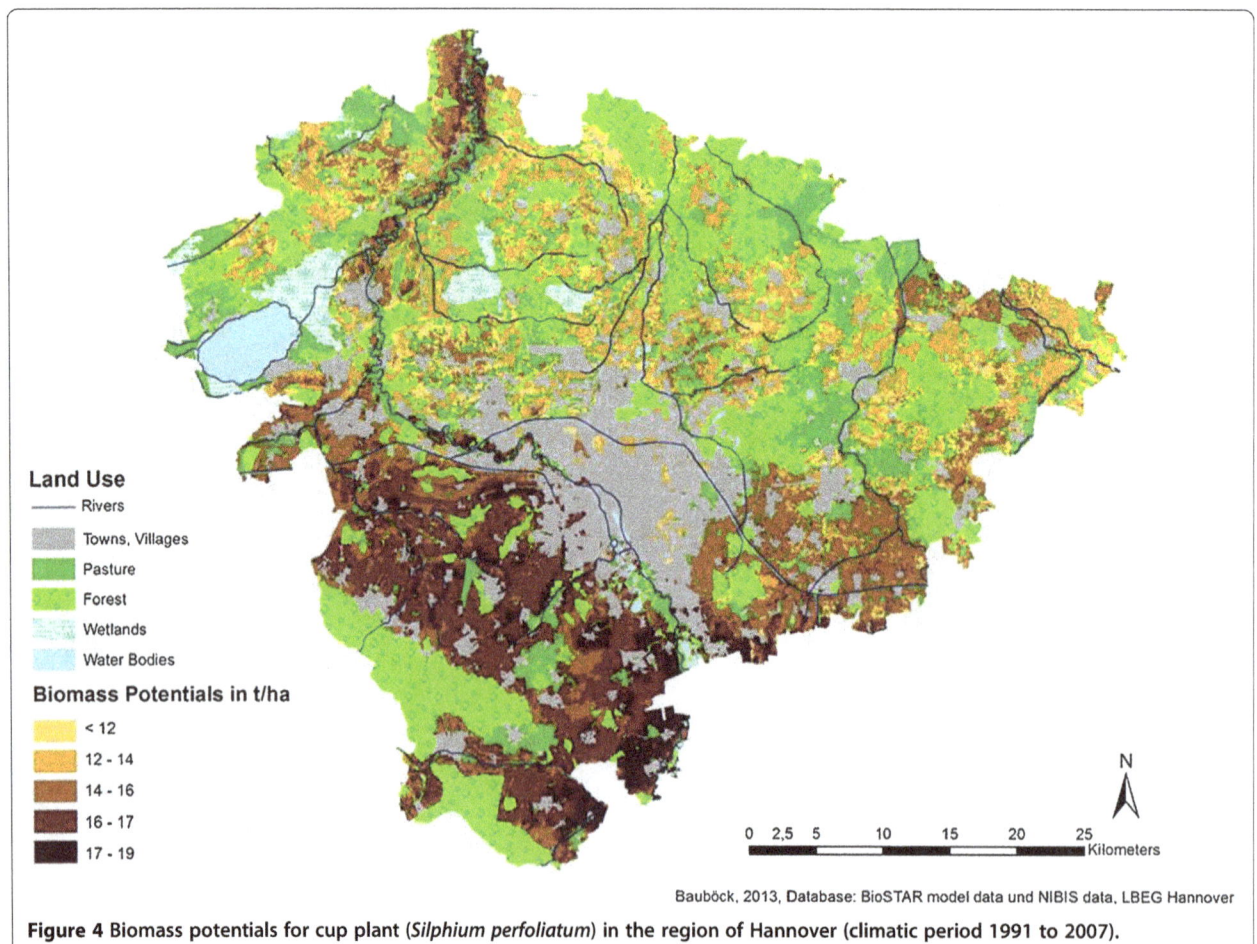

Land Use
— Rivers
Towns, Villages
Pasture
Forest
Wetlands
Water Bodies

Biomass Potentials in t/ha
< 12
12 - 14
14 - 16
16 - 17
17 - 19

Baubӧck, 2013, Database: BioSTAR model data und NIBIS data, LBEG Hannover

Figure 4 Biomass potentials for cup plant (*Silphium perfoliatum*) in the region of Hannover (climatic period 1991 to 2007).

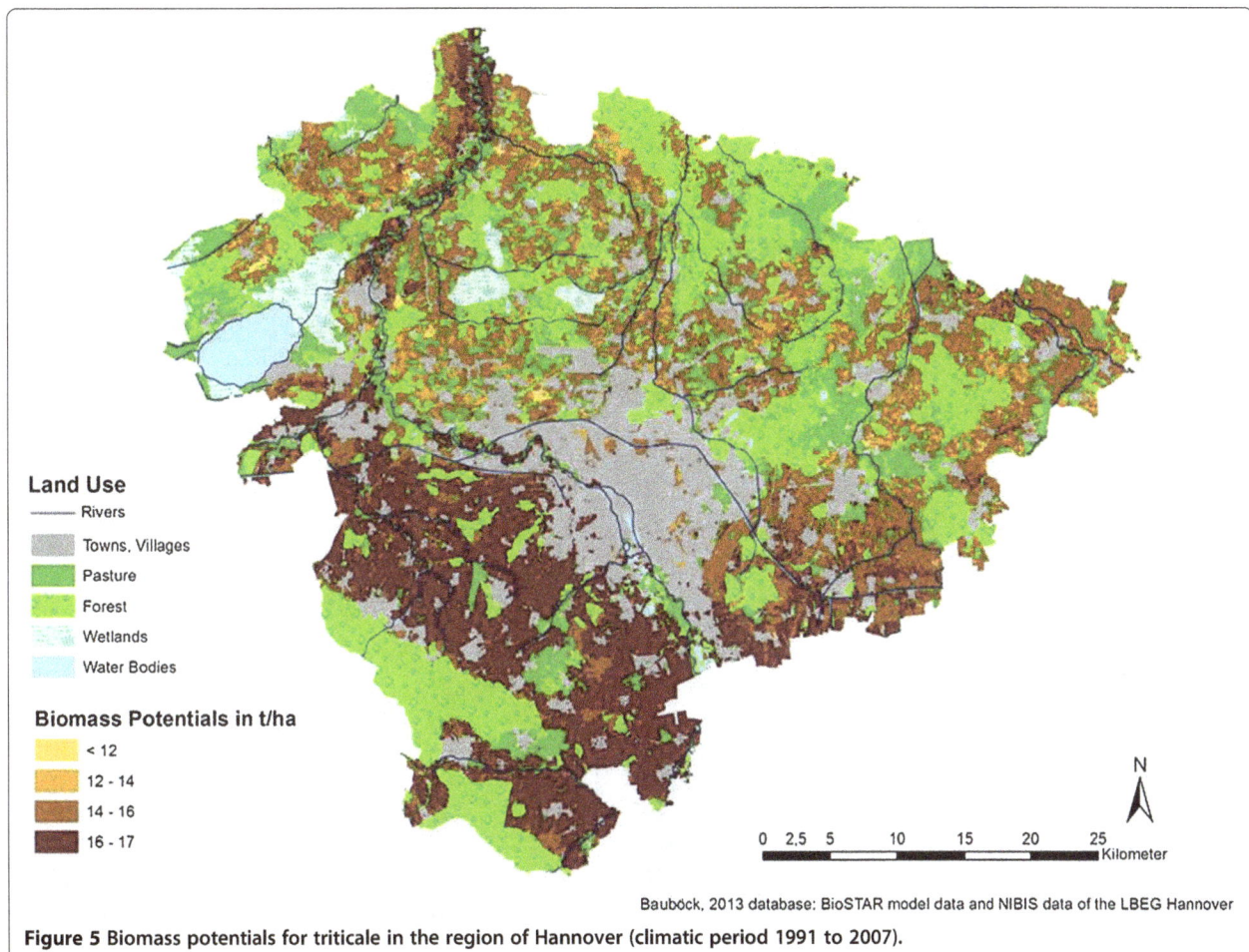

Figure 5 Biomass potentials for triticale in the region of Hannover (climatic period 1991 to 2007).

The high yield stability of triticale on a wide span of soil qualities has been found in previous research as well (see [16]). Triticale has average potentials between 16 and 17 t/ha in most parts of the southern region and yield potentials between 14 and 16 t/ha in the north. Soils with yield potentials below 12 t/ha and even below 14 t/ha are few.

As an overall conclusion for the three cultures, it can be said that all of them are interesting crops for biomass production in the region due to their overall high biomass potentials. For the high-quality soils in the southern part of the region, all three cultures have high potentials of 16 t/ha or more. In the north, cup plant appears to have the lowest potentials and maize and triticale are very similar.

Conclusions

The calculation of the spatial distribution of the biomass potentials for two alternative energy crops (triticale and cup plant) and maize in the region of Hannover has been broken down into two steps. In the first step, the crop model BioSTAR (see [22]) was used to calculate the biomass potentials for maize and winter wheat

with climate data (period 1981 to 2007 and five different climate stations) and soil data (35 soil and *nFK* classes). The generated biomass yields have then been compared with statistical yield data (obtained from the LSN), and a correction factor was deduced from this comparison. In the second step, the biomass yields for triticale and cup plant have been calculated with the model, and the generated yields for maize, triticale and cup plant (after correction) have been connected to a GIS-soil map of the region to visualize the spatial distribution of the biomass potentials for the three cultures.

The first step of the modelling has shown that BioSTAR was able to predict the aggregated yearly yields for winter wheat and maize in the region at a satisfactory level of accuracy, but individual years stand out with stronger than average (9.6% for maize and 11.2% for winter wheat) deviation of up to 31.7% from the observed yield. Possible explanations for these model deviations lie in the resolution of the input data. Both the climate and the soil data used for the simulations are strongly aggregated datasets. The climate data is a monthly aggregation from five representative climate stations in the region, and the

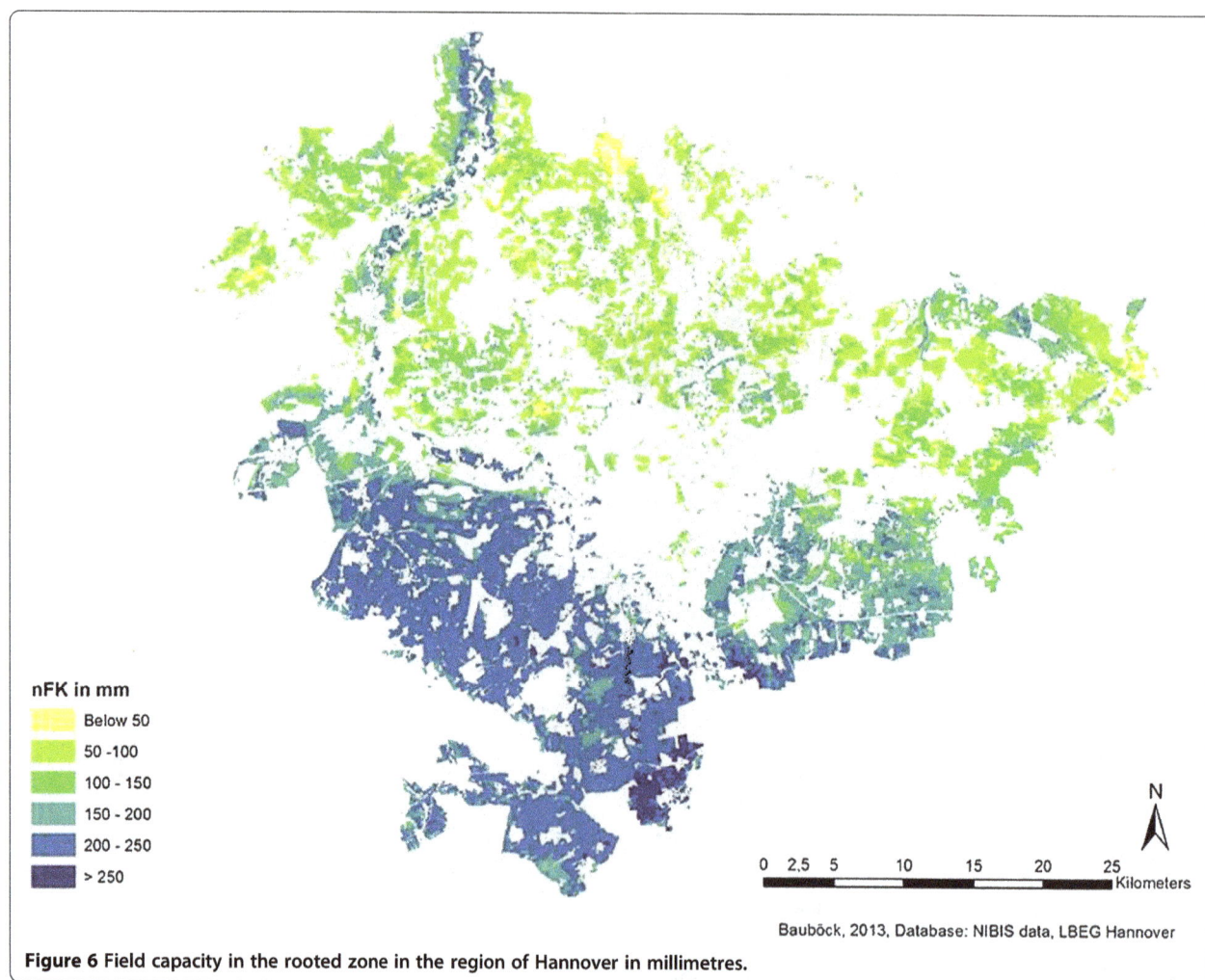

Figure 6 Field capacity in the rooted zone in the region of Hannover in millimetres.

soil data has been aggregated into 35 soil types and *nFK* classes. Additionally, only homogenous soil profiles have been assumed as inputs for the model. All these aggregations make it possible to predict biomass potentials and their spatial distribution for a large region, like the region of Hannover, with a reasonable amount of time and work input. Previous BioSTAR validation procedures with non-aggregated climate and soil data have produced better fits of observed and predicted yield data [24]. The biomass potential calculation presented in this article serves as a demonstration of the method itself and to give an overview and comparison of the relative biomass potentials of maize and the two modelled alternative energy crops, triticale and cup plant. Depending on the exactness and resolution of the input data, a higher degree of model prediction accuracy can be achieved.

The method (biomass calculation with the crop model BioSTAR) offers itself as a tool for a wide span of applications concerning a sustainable biomass production and production allocation on agricultural sites.

Based upon existing bioenergy facilities in a landscape and their respective claim of acreage, expansion scenarios can be defined and potential areas for biomass usage can be located.

This has been done in a series of workshops in the region of Hannover. In the first step, bioenergy expansion scenarios were defined in cooperation with the local government and representatives from agriculture institutions and farmers. Based upon sustainability criteria and taking mixed crop rotations with food crops, feed stuff and alternative energy crops into account, existing bioenergy potentials have been identified.

Protected (nature conservation) areas were omitted in this analysis. Through GIS visualization, suitable areas for bioenergy production could be found. Additional criteria like the prevalence of heat sinks, farms with animal manure production or other residual materials could be accounted for in this step.

Additionally, the method offers applications which go beyond the use of biomass in a region. Other potential applications are for instance to show how different crop

Table 2 Reference table for approximation of *nFK* values from soil type and soil number

| | nFK classes [mm] and the corresponding class boundaries of the soil numbers | | | | | | |
	50	90	140	200	250	270	300
S	≤21	>21 ≤ 37	>37				
Sl	≤17	>17 ≤ 31	>31 ≤ 47	>47			
lS	≤18	>18 ≤ 32	>32 ≤ 49	>49 ≤ 71	>71		
SL	≤17	>17 ≤ 31	>31 ≤ 48	>48 ≤ 69	>69 ≤ 86	>86	
sL	≤18	>18 ≤ 33	>33 ≤ 51	>51 ≤ 73	>73 ≤ 92	<92	
L	≤17	>17 ≤ 31	>31 ≤ 48	>48 ≤ 69	>69 ≤ 86	>86 ≤ 95	>95
LC	≤21	>21 ≤ 38	>38 ≤ 59	>59 ≤ 84	>84		
C	≤21	>21 ≤ 38	>38 ≤ 60	>60			

Source: Changed, after Müller et al. [27]. S, sand; Sl, slightly loamy sand; lS, loamy sand; SL, very loamy sand; sL, sandy loam; L, loam; LC, heavy loam; C, clay.

rotations and possible changes in climate will affect the agricultural potential in a region or on a farm. Therefore, the method appears to be a useful tool for the prediction of agricultural potentials under varying environmental and crop management frame conditions.

Methods
Model and model verification
Biomass calculations for maize, winter wheat, winter triticale and cup plant (*S. perfoliatum*) have been performed with the crop model BioSTAR (see [22]). BioSTAR is a new

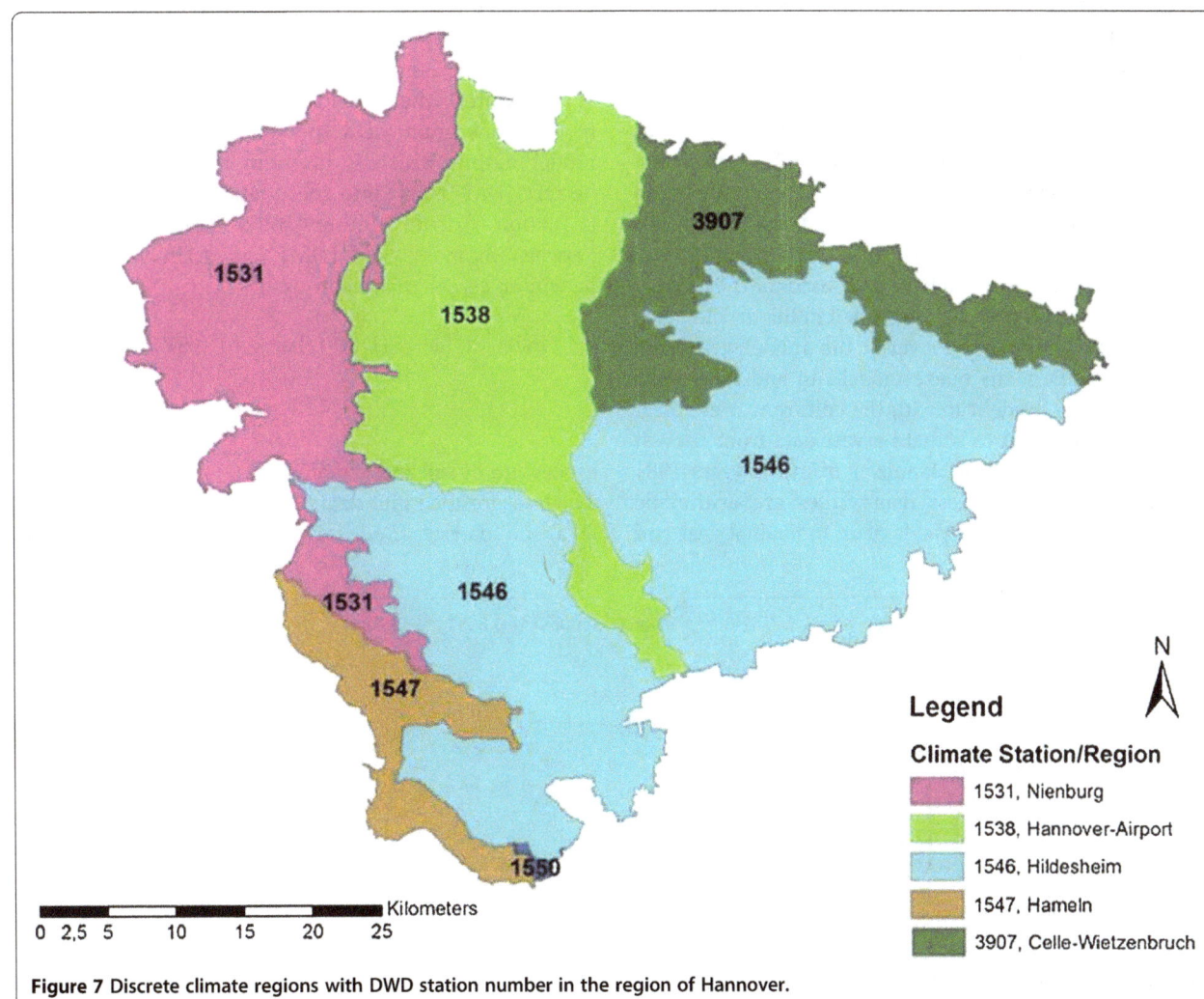

Figure 7 Discrete climate regions with DWD station number in the region of Hannover.

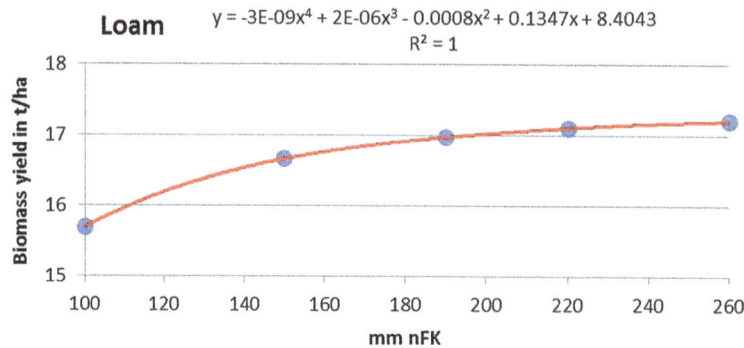

Figure 8 Polynomial fit for *nFK*-dependent yield (triticale on loam).

crop model which was originally developed for the prediction of site-specific biomass potentials for bioenergy crops, mainly maize and cereals for silage. The model is capable of modelling the biomass and yield potentials for a number of agricultural crops, including perennial crops like cup plant (*S. perfoliatum* and short rotation coppices). Due to the software architecture of the model (MS Access® database interface), large numbers of individual sites can be modelled in one single procedure and data editing for large data files, containing soil and climate datasets of whole regions, is easily manageable.

BioSTAR offers four different growth calculation methods (growth engines) and four different ET_0 calculation methods. Because most of the model testing and calibration performed up to now were done with the CO_2-based growth engine and the photosynthesis rate-dependent transpiration method, these two have been used for the calculation of the biomass potentials described in this article.

As a cross-check and to verify the model generated data, the results of the maize model run and the results of an additional model run with the culture winter wheat have been compared to actual harvest data from harvest statistics of the region of Hannover from the years 1981 to 2007 [25]. The harvest data values are arithmetic means of all the reported yields of an individual year and

can be interpreted as representative of a whole region, in this case the region of Hannover.

For this comparison, the biomass yields of maize (model output is in tons dry mass per hectare) had to be converted to deci-tons (dt) fresh mass (unit of the harvest statistic). Because the dry matter content of the yields is probably rarely recorded and, if so, it definitely does not find its way into the highly aggregated statistical data, a value of 35% dry matter content for maize harvested for silage has been assumed.

The winter wheat yields in the statistics are given in deci-tons grain yield (per hectare). To convert the model output which is given in tons dry matter per hectare (DM [t/ha]) into these units, a linear regression (Equation 1) from Moeser [26] describing the relation between biomass yield (DM) and grain yield (Grain) in winter cereals has been used.

$$\text{Grain [dt/ha]} = (\text{DM [t/ha]} \times 10 + 51.377)/2.5188$$

(1)

Processing of soil and climate data
The basis for the input data used with the BioSTAR model is a soil dataset (*Bodenschätzungsdaten*/soil evaluation

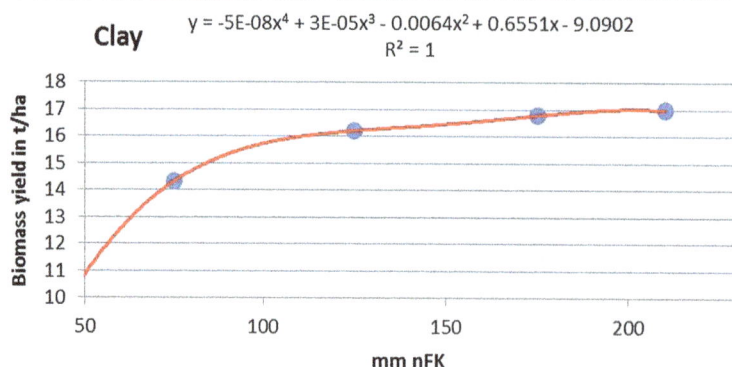

Figure 9 Polynomial fit for *nFK*-dependent yield (triticale on clay).

data) of the region of Hannover with a very high spatial resolution (1:5,000). The relevant information contained in the soil dataset is the dominant soil texture class (soil type) at the individual sites and the German soil classification index number, the *Bodenzahl* (in the following referred to as the *soil number*). The soil number is a quality measure for agricultural soils ranging from below 20 (poor-quality soil) to 100 (highest quality soil).

Following Müller et al. [27], a calculation procedure has been applied to calculate the plant available field capacity (in the following referred to as *nFK*) for each individual soil evaluation site in the region of Hannover. With the knowledge of the soil number and the soil type of a site, the field capacity value can be taken from a reference table. The tabulated *nFK* values (Table 2) were then used to generate linear and polynomial fit equations to approximate *nFK* values for any given soil number and soil type. In total, there are eight soil types and seven *nFK* classes.

The soil evaluation map of the region of Hannover contains more than 114,000 sites. To shorten the calculation procedure of the BioSTAR model and to avoid redundancy, only the biomass yields of three to five *nFK* classes (depending on the *nFK* span of the soil type) for each soil type have been calculated with the model. Additionally, the calculation was differentiated by using climate data from five different DWD climate stations in the region of Hannover (Figure 7), each representing a homogenous climate region (on the basis of long-range climate measurements) (data taken from NIBIS®). From each climate station, the daily climate values from 1981 through 2007 have been converted into monthly mean values. On the basis of these monthly mean values of the five stations, the BioSTAR calculations have been performed. As a result, a very large number of sites (>114,000) with 8 dominant soil types and a climatic variability expressed in the data of 5 climate stations (Figure 7) was converted into 175 different data units (35 soil and *nFK* classes × 5 climate zones) to be processed for each individual year (1981 to 2007). Using this method allows the aggregation of the >114,000 soil sites into 175 data units and thus eliminates redundancies and shortens the calculation procedure.

The resulting biomass yields (175 data units × 27 years) for the three modelled crops (maize, triticale, cup plant) have then been averaged over the whole time period (1981 to 2007).

For the transfer of the aggregated yield data (35 *nFK* classes, 5 climate regions) to the actual GIS (geographical information system) soil type data table, polynomial equations describing the relation between the *nFK* class of each soil type and the corresponding yield for each crop have been generated. These polynomial fit curves thus serve as data interpolation curves and are not meant to describe the accuracy of the fit, and only *nFK* values within an upper and a lower threshold have

been used (thus no *x* values below zero can occur). As an example, the polynomial fit curves for triticale on loam and clay are displayed (Figures 8 and 9).

Loamy soils (these include silts in this classification system), which are generally of a high quality, start at much higher *nFK* values than clays (100 mm vs. <50 mm). Yield gains on loam are strong between 100 and 150 mm of available soil water and then decline. The curve for clay type soils shows a strong increase in productivity for *nFK* values between 50 and 100 mm and then levels of an inclination similar to that of the loam curve.

Abbreviations
BioSTAR: Biomass Simulation Tool for Agricultural Resources; C_4 culture: crop with a C_4 carbon pathway; dt: deci-tons; DWD: Deutscher Wetterdienst; ET_0: potential (reference) evapotranspiration rate; GDR: German Democratic Republic; GIS: geographical information system; LWK: Landwirtschaftskammer (Niedersachsen); LSN: Lower Saxony Department of Statistics; *nFK*: available field capacity; NIBIS®: Lower Saxony soil information system; RMSE: root-mean-square error; WIA: Willmott index of agreement.

Competing interests
The authors declare that they have no competing interests.

Authors' contributions
RB carried out the biomass calculations, the statistical analysis and the implementation of the data into the GIS and participated in the structuring of the article. MKM contributed the scientific and farming practice background for the investigated cultures, helped with the statistical analysis and participated in the structuring of the article. MK conceived the study and helped with the implementation of the data into the GIS. All authors read and approved the final manuscript.

Acknowledgements
The research of this study was made possible by the funding from the Lower Saxony Ministry of Science and Culture (MWK), interdisciplinary research project: 'Sustainable use of bioenergy: bridging climate protection, nature conservation and society' sub-projects 2.1 and 2.2.

Author details
[1]Department of Cartography, GIS and Remote Sensing, Research Project 'BIS', University of Göttingen, Goldschmidtstraße 5, Göttingen 37077, Germany.
[2]Interdisciplinary Centre of Sustainable Development, Research Project 'BIS', University of Göttingen, Goldschmidtstraße 1, Göttingen 37077, Germany.

References
1. FNR: *Basisdaten Bioenergie. Fachagentur für Nachwachsende Rohstoffe*; 2013. http://mediathek.fnr.de/broschuren/basisdaten-bioenergie.html.
2. ML: *Biogas in Niedersachsen. Entwicklung, Stand und Perspektiven.* Ministerium für Ernährung, Landwirtschaft und Verbraucherschutz; 2012. www.ml.niedersachsen.de/download/72747.
3. LWK: *Niedersachsen ist Agrarland Nr. 1.* Landwirtschaftskammer Niedersachsen; 2013. http://www.lwk-niedersachsen.de/index.cfm/portal/6/nav/355/article/16175.html.
4. FNR: *Energiepflanzen für Biogasanlagen – Regionalbroschüre Niedersachsen.* Fachagentur für Nachwachsende Rohstoffe; 2013. http://mediathek.fnr.de/energiepflanzen-fur-biogasanlagen-regionalbroschure-niedersachsen.html.
5. FNR: *Energiepflanzen.* Fachagentur Nachwachsende Rohstoffe; 2012. http://energiepflanzen.fnr.de/pflanzen/mehrjaehrige/durchwachsene-silphie/.
6. Maiskomitee: *Bedeutung des Maisanbaus in Deutschland.* Deutsches Maiskomitee; 2013. http://www.maiskomitee.de/web/intranetHomepages.aspx?hp=30a01c5a-cb8c-9314-9398-742c9d12a03e.
7. MU: *Biogas in Niedersachsen. Stand der Entwicklung und Perspektiven.* Niedersächsisches Ministerium für Umwelt, Energie und Klimaschutz; 2012. http://www.umwelt.niedersachsen.de/download/73203/.

8.   Rode M, Kanning H: *Natur- und raumverträglicher Ausbau energetischer Biomassepfade.* Stuttgart: Ibidem; 2010.

9.   Knickel K, Janssen B, Schramek J, Käppel K: **Naturschutz und Landwirtschaft: Kriterienkatalog zur "Guten fachlichen Praxis".** *Schriftenreihe für Angewandte Landschaftsökologie* 2001, **41**:152.

10. German Environmental Protection Law: **Gesetz über Naturschutz und Landschaftspflege (Bundesnaturschutzgesetz - BNatSchG).** http://www.gesetze-im-internet.de/bundesrecht/bnatschg_2009/gesamt.pdf.

11. Karpenstein-Machan M: **Implementation of integrative energy crop cultivation concepts on biogas farms.** In *International Nordic Bioenergy. Book of Proceedings*, Volume 52. Edited by Mia S. Finbio publication; 2011:127–133. ISBN 978-952-5135-51-0.

12. Karpenstein-Machan M: **Integrativer Energiepflanzenbau als Baustein der regionalen Energiewende.** *Ländlicher Raum* 2013, **3**:26–28.

13. Karpenstein-Machan M: *Konzepte für den Energiepflanzenbau.* DLG -Verlag: Frankfurt; 1997.

14. Karpenstein-Machan M: *Energiepflanzenbau für Biogasanlagenbetreiber.* DLG -Verlag: Frankfurt; 2005.

15. Karpenstein-Machan M: **Sustainable cultivation concepts for domestic energy production from biomass.** *Critical Reviews of Plant Science, Special Issue on bio energy* 2001, **20**:1–14.

16. Karpenstein-Machan M: **Neue Perspektiven für den Naturschutz durch einen ökologisch ausgerichteten Energiepflanzenbau.** *Naturschutz und Landschaftsplanung* 2004, **2/36**:58–64.

17. Karpenstein-Machan M, Honermeier B, Hartmann F: *Produktion aktuell, Triticale.* DLG -Verlag: Frankfurt; 1994.

18. Conrad M, Biertümpfel A: **Optimierung des Anbauverfahrens für Durchwachsende Silphie (Silphium perfoliatum L.) als Kofermentpflanze in Biogasanlagen sowie Überführung in die landwirtschafliche Praxis.** *Project Report* 2010. http://www.tll.de/ainfo/pdf/silp0111.pdf.

19. Biertümpfel A, Götz R, Reinhold G, Zorn W: **Leitlinie zur effizienten und umweltverträglichen Erzeugung von Durchwachsender Silphie.** In *Thüringer Landesanstalt für landwirtschaft [Ed]*; 2013. http://www.tll.de/ainfo/pdf/ll_silphie.pdf.

20. Karpenstein-Machan M: **Integrative energy crop cultivation as a way to a more nature-orientated agriculture.** In *Sustainable Bioenergy Production - An Integrated Approach*. Edited by Ruppert H, Kappas M, Ibendorf J. Heidelberg: Springer; 2013:143–180.

21. Karpenstein-Machan M: **Pflanzenbauliche Optimierung und Umsetzung eines integrativen Energiepflanzenbaus.** In *Project Report "Nachhaltige Nutzung von Energie aus Biomasse im Spannungsfeld von Klimaschutz, Landschaft und Gesellschaft".* Edited by Ruppert; 2014. In Press.

22. Bauböck R: **GIS-gestützte Modellierung und Analyse von Agrar- Biomassepotenzialen in Niedersachsen – Einführung in das Pflanzenmodell BioSTAR.** In University of Göttingen: PhD thesis; 2013. http://hdl.handle.net/11858/00-1735-0000-000E-0ABB-9.

23. Engel N, Mithöfer K: **Bearbeitung, Übersetzung und Auswertung digitaler Bodenschätzungsdaten.** *Arbeitshefte Boden* 2003, **1**:5–43.

24. Bauböck R: **Simulating the yields of bioenergy and food crops with the crop modeling software BioSTAR: the carbon-based growth engine and the BioSTAR ET0 method.** *Environ Sci Eur* 2014, **26**:1.

25. Niedersächsisches Landesamt für Statistik: *Yield statistics for Lower Saxony, received from the statistical department (Niedersächsisches Landesamt für Statistik) by personal request in 2009.*

26. Moeser J: **3-n.Info: Alternativkulturen zur Produktion von Biomasse.** http://www.3-n.info/pdf_files/Vortraege/100914_06_moeser_alternativkulturen.pdf.

27. Müller U, Engel N, Heidt L: **Ableiten der potenziellen Beregnungswassermenge aus verfügbaren Boden und Klimadaten.** *Geoberichte* 2012, **52**:32–48.

8

Republished study: long-term toxicity of a Roundup herbicide and a Roundup-tolerant genetically modified maize

Gilles-Eric Séralini[1*], Emilie Clair[1], Robin Mesnage[1], Steeve Gress[1], Nicolas Defarge[1], Manuela Malatesta[2], Didier Hennequin[3] and Joël Spiroux de Vendômois[1]

Abstract

Background: The health effects of a Roundup-tolerant NK603 genetically modified (GM) maize (from 11% in the diet), cultivated with or without Roundup application and Roundup alone (from 0.1 ppb of the full pesticide containing glyphosate and adjuvants) in drinking water, were evaluated for 2 years in rats. This study constitutes a follow-up investigation of a 90-day feeding study conducted by Monsanto in order to obtain commercial release of this GMO, employing the same rat strain and analyzing biochemical parameters on the same number of animals per group as our investigation. Our research represents the first chronic study on these substances, in which all observations including tumors are reported chronologically. Thus, it was not designed as a carcinogenicity study. We report the major findings with 34 organs observed and 56 parameters analyzed at 11 time points for most organs.

Results: Biochemical analyses confirmed very significant chronic kidney deficiencies, for all treatments and both sexes; 76% of the altered parameters were kidney-related. In treated males, liver congestions and necrosis were 2.5 to 5.5 times higher. Marked and severe nephropathies were also generally 1.3 to 2.3 times greater. In females, all treatment groups showed a two- to threefold increase in mortality, and deaths were earlier. This difference was also evident in three male groups fed with GM maize. All results were hormone- and sex-dependent, and the pathological profiles were comparable. Females developed large mammary tumors more frequently and before controls; the pituitary was the second most disabled organ; the sex hormonal balance was modified by consumption of GM maize and Roundup treatments. Males presented up to four times more large palpable tumors starting 600 days earlier than in the control group, in which only one tumor was noted. These results may be explained by not only the non-linear endocrine-disrupting effects of Roundup but also by the overexpression of the EPSPS transgene or other mutational effects in the GM maize and their metabolic consequences.

Conclusion: Our findings imply that long-term (2 year) feeding trials need to be conducted to thoroughly evaluate the safety of GM foods and pesticides in their full commercial formulations.

Keywords: Genetically modified; GMO; Roundup; NK603; Rat; Glyphosate-based herbicides; Endocrine disruption

* Correspondence: criigen@criigen.info
[1]Institute of Biology, EA 2608 and CRIIGEN and Risk Pole, MRSH-CNRS, Esplanade de la Paix, University of Caen, Caen, Cedex 14032, France
Full list of author information is available at the end of the article

Empirical natural and social sciences produce knowledge (in German: Wissenschaften schaffen Wissen) which should describe and explain past and present phenomena and estimate their future development. To this end quantitative methods are used. Progress in science needs controversial debates aiming at the best methods as basis for objective, reliable and valid results approximating what could be the truth. Such methodological competition is the energy needed for scientific progress. In this sense, ESEU aims to enable rational discussions dealing with the article from G.-E. Séralini et al. (Food Chem. Toxicol. 2012, 50:4221–4231) by re-publishing it. By doing so, any kind of appraisal of the paper's content should not be connoted. The only aim is to enable scientific transparency and, based on this, a discussion which does not hide but aims to focus methodological controversies. -Winfried Schröder, Editor of the Thematic Series "Implications for GMO-cultivation and monitoring" in Environmental Sciences Europe.

Background

There is an ongoing international debate as to the necessary length of mammalian toxicity studies, including metabolic analyses, in relation to the consumption of genetically modified (GM) plants [1]. Currently, no regulatory authority requires mandatory chronic animal feeding studies to be performed for edible genetically modified organisms (GMOs), or even short-term studies with blood analyses for the full commercial formulations of pesticides as sold and used, but only for the declared active principle alone. However, several 90-day rat feeding trials have been conducted by the agricultural biotechnology industry. These investigations mostly concern GM soy and maize that are engineered either to be herbicide-tolerant (to Roundup (R) in 80% of cases), or to produce a modified Bt toxin insecticide, or both. As a result, these GM crops contain new pesticide residues for which new maximum residue levels (MRL) have been established in some countries.

Though the petitioners conclude in general that no major physiological changes is attributable to the consumption of the GMO in subchronic toxicity studies [2-5], significant disturbances have been found and may be interpreted differently [6,7]. A detailed analysis of the data in the subchronic toxicity studies [2-5] has revealed statistically significant alterations in kidney and liver function that may constitute signs of the early onset of chronic toxicity. This may be explained at least in part by pesticide residues in the GM feed [6,7]. Indeed, it has been demonstrated that R concentrations in the range of 10^3 times below the MRL can induce endocrine disturbances in human cells [8] and toxic effects thereafter [9]. This may explain toxic effects seen in experiments in rats *in vivo* [10] as well as in farm animals [11]. After

several months of consumption of an R-tolerant soy, the liver and pancreas of mice were affected, as highlighted by disturbances in sub-nuclear structure [12-14]. Furthermore, this toxic effect was reproduced by the application of R herbicide directly to hepatocytes in culture [15].

More recently, long-term and multi-generational animal feeding trials have been performed, with some possibly providing evidence of safety, while others conclude on the necessity of further investigation because of metabolic modifications [16]. However, in contrast with the study we report here, none of these previous investigations have included a detailed follow-up of the animals, including multiple (up to 11) blood and urine sampling over 2 years, and none has investigated either the GM NK603 R-tolerant maize or Roundup.

Furthermore, evaluation of long-term toxicity of herbicides is generally performed on mammalian physiology employing only their active principle, rather than the complete formulations as used in agriculture. This was the case for glyphosate (G) [17], the declared active chemical constituent of R. It is important to note that G is only able to efficiently penetrate target plant organisms with the help of adjuvants present in the various commercially used R formulations [18]. Even if G has shown to interact directly with the active site of aromatase at high levels [19], at low contaminating levels, adjuvants may be better candidates than G to explain the toxicity or endocrine disruptive side effects of R on human cells [8,20] and also *in vivo* for acute toxicity [21]. In this regard, it is noteworthy that the far greater toxicity of full agricultural formulations compared to declared supposed active principles alone has recently been demonstrated also for six other major pesticides tested *in vitro* [22]. When G residues are found in tap water, food, or feed, they arise from the total herbicide formulation although little data is available as to the levels of the R adjuvants in either the environment or food chain. Indeed, adjuvants are rarely monitored in the environment, but some widely used adjuvants (surfactants) such as nonylphenol ethoxylates, another ethoxylated surfactant like POEA present in R, are widely found in rivers in England and are linked with disruption of wildlife sexual reproduction [23]. Adjuvants are found in groundwater [24]. The half-life of POEA (21 to 42 days) is even longer than for G (7 to 14 days) in aquatic environments [25]. As a result, the necessity of studying the potential toxic effects of total chemical mixtures rather than single components has been strongly emphasized [26-28]. On this basis, the regular measurement of only G or other supposed active ingredients of pesticides in the environment constitute at best markers of full formulation residues. Thus, in the study of health effects, exposure to the diluted whole formulation may be more representative of environmental pollution than exposure to G alone.

With a view to address this lack of information, we performed a 2-year detailed rat feeding study. Our study was designed as a chronic toxicity study and as a direct follow-up to a previous investigation on the same NK603 GM maize conducted by the developer company, Monsanto [3]. A detailed critical analysis of the raw data of this sub-chronic 90-day rat feeding study revealed statistically significant differences in multiple organ function parameters, especially pertaining to the liver and kidneys, between the GM and non-GM maize-fed group [3,7]. However, Monsanto's authors dismissed the findings as not 'biologically meaningful' [3], as was also the case with another GM corn [29]. The European Food Safety Authority (EFSA) accepted Monsanto's interpretation on NK603 maize [30], like in all other cases.

Our study is the first and to date the only attempt to follow up Monsanto's investigation and to determine whether the differences found in the NK603 GM maize-fed rats, especially with respect to liver and kidney function, were not biologically meaningful, as claimed, or whether they developed into serious diseases over an extended period of time.

The Monsanto authors adapted Guideline 408 of the Organization for Economic Co-operation and Development (OECD) for their experimental design [3]. Our study design was based on that of the Monsanto investigation in order to make the two experiments comparable, but we extended the period of observation from Monsanto's 90 days to 2 years. We also used three doses of GMOs (instead of Monsanto's two) and Roundup to determine treatment dose response, including any possible non-linear as well as linear effects. This allowed us to follow in detail the potential health effects and their possible origins due to the direct or indirect consequences of the genetic modification itself in the NK603 GM maize, or due to the R herbicide formulation used on the GM maize (and not G alone), or both. Because of recent reviews on GM foods indicating no specific risk of cancer [2,16], but indicating signs of hepatorenal dysfunction within 3 months [1,7], we had no reason to adopt a carcinogenesis protocol using 50 rats per group. However, we prolonged to 2 years the biochemical and hematological measurements and measurements of disease status, as allowed, for example, in OECD protocols 453 (combined chronic toxicity and carcinogenicity) and 452 (chronic toxicity). Both OECD 452 and 453 specify 20 rats per sex per group but require only 50% (ten per sex per group, the same number that we used in total) to be analyzed for biochemical and hematological parameters. Thus, these protocols yield data from the same number of rats as our experiment. This remains the highest number of rats regularly measured in a standard GM diet study, as well as for a full formulated pesticide at very low environmentally relevant levels.

We used the Sprague-Dawley strain of rat, as recommended for chronic toxicology tests by the National Toxicology Program in the USA [31], and as used by Monsanto in its 90-day study [3]. This choice is also consistent with the recommendation of the OECD that for a chronic toxicity test, rats of the same strain should be used as in studies on the same substance but of shorter duration [32]. We then also tested for the first time three doses (rather than the two usually employed in 90-day protocols) of the R-tolerant NK603 GM maize alone, the GM maize treated with R, and R alone at very low environmentally relevant doses, starting below the range of levels permitted by regulatory authorities in drinking water and in GM feed.

Overall, our study is the first in-depth life-long toxicology study on the full commercial Roundup formulation and NK603 GM maize, with observations on 34 organs and measurement of 56 parameters analyzed at 11 time points for most organs, and utilizing 3 doses. We report here the major toxicological findings on multiple organ systems. As there was no evidence in the literature on GM food safety evaluation to indicate anything to the contrary, this initial investigation was designed as a full chronic toxicity and not a carcinogenicity study. Thus, we monitored in details chronologically all behavioral and anatomical abnormalities including tumors. A full carcinogenicity study, which usually focuses only on observing incidence and type of cancers (not always all tumors), would be a rational follow-up investigation to a chronic toxicity study in which there is a serious suspicion of carcinogenicity. Such indications had not been previously reported for GM foods.

Our findings show that the differences in multiple organ functional parameters seen from the consumption of NK603 GM maize for 90 days [3,7] escalated over 2 years into severe organ damage in all types of test diets. This included the lowest dose of R administered (0.1 ppb, 50 ng/L G equivalent) of R formulation administered, which is well below permitted MRLs in both the USA (0.7 mg/L) [33] and European Union (100 ng/L) [34]. Surprisingly, there was also a clear trend in increased tumor incidence, especially mammary tumors in female animals, in a number of the treatment groups. Our data highlight the inadequacy of 90-day feeding studies and the need to conduct long-term (2 years) investigations to evaluate the life-long impact of GM food consumption and exposure to complete pesticide formulations.

Results
Biochemical analyses of the maize feed
Standard biochemical compositional analysis revealed no particular differences between the different maize types and diets, the GM and non-GM maize being classified as substantially equivalent, except for transgene DNA

quantification. For example, there was no difference in total isoflavones. In addition, we also assayed for other specific compounds, which are not always requested for establishing substantial equivalence. This analysis revealed a consistent and statistically significant ($p < 0.01$) decrease in certain phenolic acids in treatment diets, namely ferulic and caffeic acids. Ferulic acid was decreased in both GM maize and GM maize + R diets by 16% to 30% in comparison to the control diet (889 ± 107, 735 ± 89, respectively, vs. control 1,057 ± 127 mg/kg) and caffeic acid in the same groups by 21% to 53% (17.5 ± 2.1, 10.3 ± 1.3 vs. control 22.1 ± 2.6 mg/kg).

Anatomopathological observations and liver parameters

All rats were carefully monitored during the experiment for behavior, appearance, palpable tumors, and infections. At least ten organs per animal were weighed and up to 34 analyzed postmortem, at the macroscopic and/ or microscopic levels (Table 1). Due to the large quantity of data collected, it cannot all be shown in one report, but we present here the most important findings. There was no rejection by the animals of the diet with or without GM maize, nor any major difference in body weight (data not shown).

The most affected organs in males were the liver, hepatodigestive tract, and kidneys (Table 2; Figure 1A,B,C,D,E, F,G,H,I). Liver abnormalities such as hepatic congestions and macroscopic and microscopic necrotic foci were 2.5 to 5.5 times more frequent in all treatments than in control groups, where only two rats out of ten were affected with one abnormality each. For instance, there were 5 abnormalities in total in the GMO 11% group (2.5 times higher than controls) and 11 in the GMO 22% group (5.5 times greater). In addition, by the end of the experiment, Gamma GT hepatic activity was increased, particularly in the GMO + R groups (up to 5.4 times higher), this probably being reflective of liver dysfunction. Furthermore, cytochrome P450 activity generally increased in the presence of R (either in drinking water or in the GM maize-containing diet) according to the dose and up to 5.7 times greater at the highest dose.

Transmission electron microscopic observations of liver samples confirmed changes for all treated groups in relation to glycogen dispersion or appearance in lakes, increase of residual bodies and enlargement of cristae in mitochondria (Figure 2, panels 2 to 4). The GM maize-fed groups either with or without R application showed a higher heterochromatin content and decreased nucleolar dense fibrillar components, implying a reduced level of mRNA and rRNA transcription. In the GMO + R group (at the highest dose), the smooth endoplasmic reticulum was drastically increased and nucleoli decreased in size, becoming more compact. In the R alone treatment groups, similar trends were

observed, with a partial resumption of nucleolar activity at the highest dose.

Degenerating kidneys with turgid inflammatory areas demonstrated the increased incidence of marked and severe chronic progressive nephropathies, which were up to two fold higher in the 33% GM maize or lowest dose R treatment groups (Table 2; Figure 1, first line).

Biochemical analyses of blood and urine samples

Biochemical measurements of blood and urine were focused on samples taken at the 15th month time point, as this was the last sampling time when most animals were still alive (in treated groups 90% males, 94% females, and 100% controls). Statistical analysis of results employed OPLS-DA 2-class models built between each treated group per sex and controls. Only models with an explained variance $R^2(Y) \geq 80\%$, and a cross-validated predictive ability $Q^2(Y) \geq 60\%$, were used for selection of the discriminant variables (Figure 3), when their regression coefficients were significant at a 99% confidence level. Thus, in treated females, kidney failures appeared at the biochemical level (82% of the total disrupted parameters). Levels of Na and Cl or urea increased in urine with a concomitant decrease of the same ions in serum, as did the levels of P, K, and Ca. Creatinine and creatinine clearance decreased in urine for all treatment groups in comparison to female controls (Table 3). In GM maize-treated males (with or without R), 87% of discriminant variables were kidney-related, but the disrupted profiles were less obvious because of advanced chronic nephropathies and deaths. In summary, for all treatments and both sexes, 76% of the discriminant variables versus controls were kidney-related.

Furthermore, in females (Table 3), the androgen/estrogen balance in serum was modified by GM maize and R treatments (at least 95% confidence level, Figure 3). For male animals at the highest R treatment dose, levels of estrogens were more than doubled.

Tumor incidence

Tumors are reported in line with the requirements of OECD chronic toxicity protocols 452 and 453, which require all 'lesions' (which by definition include tumors) to be reported. These findings are summarized in Figure 4. The results are presented in the form of real-time cumulative curves (each step corresponds to an additional tumor in the group). Only the growing largest palpable growths (above a diameter of 17.5 mm in females and 20 mm in males) are presented (for example, see Figure 5A,B,C). These were found to be in 95% of cases non-regressive tumors (Figure 5D,E,F,G,H,I,J) and were not infectious nodules. These arose from time to time; then, most often disappeared and were not different from controls after bacterial analyses. The real tumors were recorded

Table 1 Protocol used and comparison to existing assessment and to non-mandatory regulatory tests

Treatments and analyses	In this work	Hammond et al. 2004	Regulatory tests
Animals measured/group/sex	10/10 SD rats (200 rats measured)	10/20 SD rats (200 rats measured/total 400)	At least 10 rodents
Duration in months	24 (chronic)	3 (subchronic, 13 weeks)	3
Doses by treatment	3	2	At least 3
Treatments + controls	GMO NK603, GMO NK603 + Roundup, Roundup, and closest isogenic maize	GMO NK603 + Roundup, closest isogenic maize, and 6 other maize lines non substantially equivalent	GMOs or Chemicals (in standard diet or water)
Animals by cage (same sex)	1 to 2	1	1 or more
Monitoring/week	2	1	1 or more
Organs and tissues studied			For high dose and controls
Organs weighted	10	7	At least 8
Histology/animal	34	17/36	At least 30
Electronic microscopy	Yes	No	No
Feed and water consumptions	Measured	For feed only	At least feed
Behavioral studies (times)	2	1 (no protocol given)	1
Ophthalmology (times)	2	0	2
Blood parameters	31 (11 times for most)	31 (2 times)	At least 25 (at least 2 times)
Plasma sex steroids	Testosterone, estradiol	No	No, except if endocrine effects suspected
Number of blood samples/animal	11, each month (0 to 3) then every 3 months	2, weeks 4 and 13	1, at the end
Urine parameters studied	16	18	7 if performed
Number of urine samples	11	2	Optional, last week
Liver tissue parameters	6	0	0
Roundup residues in tissues	Studied	Not studied	Not mandatory
Microbiology in feces or urine	Yes	Yes	No
Transgene in tissues	Studied	Not studied	Not studied

The protocol used in this work was compared to the regulatory assessment of NK603 maize by the company (Hammond et al. 2004), and to non-mandatory regulatory *in vivo* tests for GMOs, or mandatory for chemicals (OECD 408). Most relevant results are shown in this paper.

independently of their grade, but dependent on their morbidity, since non-cancerous tumors can be more lethal than those of cancerous nature, due to internal hemorrhaging or compression and obstruction of function of vital organs, or toxins or hormone secretions. These tumors progressively increased in size and number, but not proportionally to the treatment dose, over the course of the experiment (Figure 4). As in the case of rates of mortality (Figure 6), this suggests that a threshold in effect was reached at the lower doses. Tumor numbers were rarely equal but almost always more than in controls for all treated groups, often with a two- to threefold increase for both sexes. Tumors began to reach a large size on average 94 days before controls in treated females and up to 600 days earlier in two male groups fed with GM maize (11 and 22% with or without R).

Table 2 Summary of the most frequent anatomical pathologies observed

Organs and associated pathologies	Controls	GMO 11%	GMO 22%	GMO 33%	R (A)	R (B)	R (C)	GMO 11% + R	GMO 22% + R	GMO 33% + R
Males										
Kidneys, CPN	3 (3)	4 (4)	5 (5)	7 (7)	6 (6)	5 (5)	3 (3)	5 (5)	4 (4)	4 (4)
Liver	2 (2)	5 (4)	11 (7)	8 (6)	11 (5)	9 (7)	6 (5)	5 (4)	7 (4)	6 (5)
Hepatodigestive tract	6 (5)	10 (6)	13 (7)	9 (6)	23 (9)	16 (8)	9 (5)	9 (6)	13 (6)	11 (7)
Females										
Pituitary	9 (6)	23 (9)	20 (8)	8 (5)	22 (8)	16 (7)	13 (7)	19 (9)	9 (4)	19 (7)
Mammary glands	10 (5)	22 (8)	10 (7)	16 (8)	26(10)	20(10)	18 (9)	17 (8)	16 (8)	15 (9)
Mammary tumors	8 (5)	15 (7)	10 (7)	15 (8)	20 (9)	16(10)	12 (9)	10 (6)	11 (7)	13 (9)

After the number of pathological abnormalities, the number of rats affected out of the initial ten is indicated in parentheses. Only marked or severe chronic progressive nephropathies (CPN) are listed in male animals, excluding two nephroblastomas in groups consuming GMO 11% and GMO 22% + Roundup. Hepatodigestive pathological signs in males concern the liver, stomach, and small intestine (duodenum, ileum, or jejunum). Pathological signs in liver are mostly congestions, macroscopic spots, and microscopic necrotic foci. In females, pituitary dysfunctions include adenomas, hyperplasias, and hypertrophies. Mammary fibroadenomas and adenocarcinomas are the major tumors detected; galactoceles and hyperplasias with atypia were also found and added to the pathological signs in mammary glands.

In female animals, the largest tumors were in total five times more frequent than in males after 2 years, with 93% of these being mammary tumors. Adenomas, fibroadenomas, and carcinomas were deleterious to health due to their very large size (Figure 5A,B,C) rather than the grade of the tumor itself. Large tumor size caused impediments to either breathing or digestion and nutrition because of their thoracic or abdominal location and also resulted in hemorrhaging (Figure 5A,B,C). In addition, one metastatic ovarian cystadenocarcinoma and two skin tumors were identified. Metastases were observed in only two cases; one in a group fed with 11% GM maize and another in the highest dose of R treatment group.

Up to 14 months, no animals in the control groups showed any signs of palpable tumors, whilst 10% to 30% of treated females per group developed tumors, with the exception of one group (33% GMO + R). By the beginning of the 24th month, 50% to 80% of female animals had developed tumors in all treatment groups, with up to three tumors per animal, whereas only 30% of controls were affected. A summary of all mammary tumors at the end of the experiment, independent of size, is presented in Table 2. The same trend was observed in the groups receiving R in their drinking water (Figure 4, R treatment panels). The R treatment groups showed the greatest rates of tumor incidence, with 80% of animals

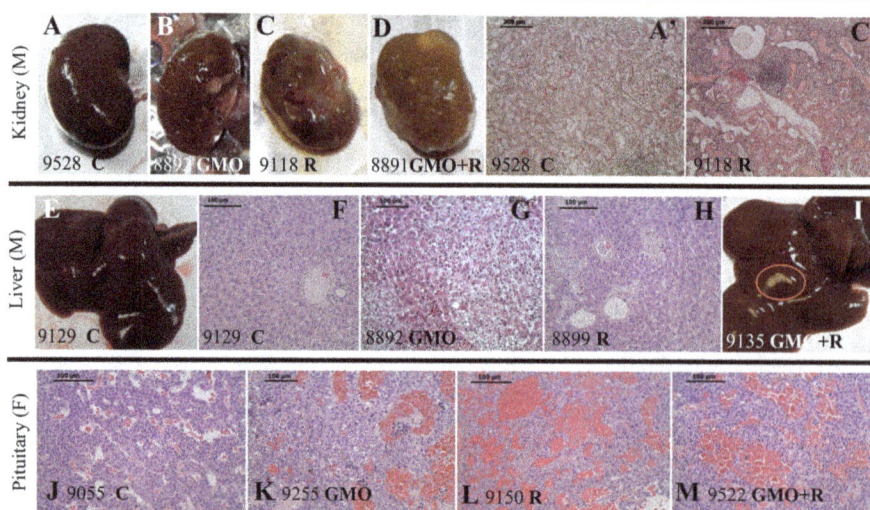

Figure 1 Anatomopathological observations in rats fed GMO treated or not by Roundup and effects of Roundup alone. Macroscopic (**A** to **D**) and microscopic (**A'** and **C'**) photographs show male left kidneys and livers (**E** to **I**) and female pituitaries (**J** to **M**), in accordance to Table 2. The number of each animal and its treatment is specified. Macroscopic pale spots (**I**) and microscopic necrotic foci in liver (**G** clear-cell focus, **H** basophilic focus with atypia), and marked or severe chronic progressive nephropathies, are illustrated. In females, pituitary adenomas (**K** to **M**) are shown and compared to control (**J**, rat number and **C** for control). Apostrophes after letters indicate organs from the same rat.

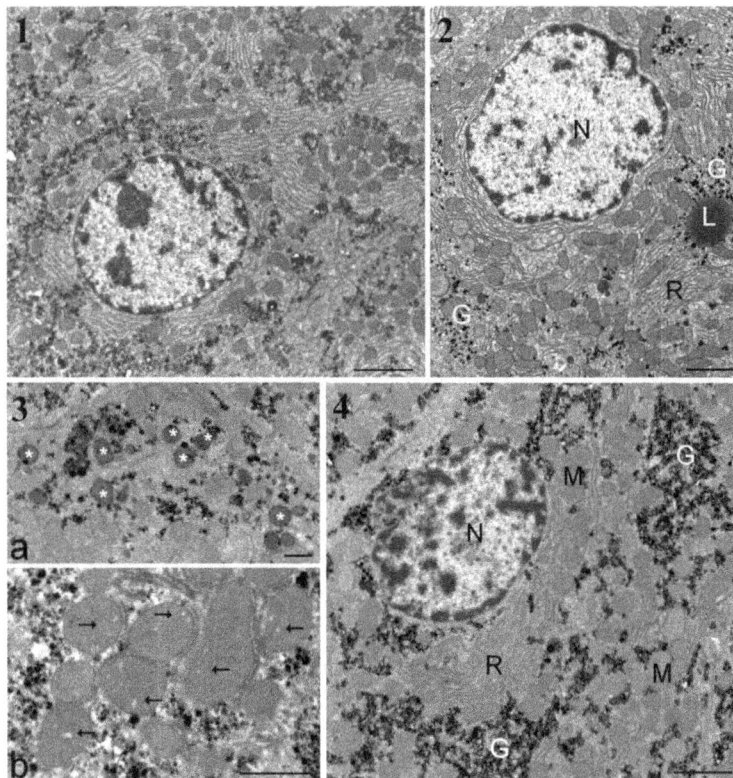

Figure 2 Ultrastructure of hepatocytes in male rats from groups presenting the greatest degree of liver pathology. (1) Typical control rat hepatocyte (bar 2 μm except in 4). **(2)** Effects with Roundup at the lowest dose. Glycogen (G) is dispersed in the cytoplasm. L, lipid droplet; N, nucleus; R, rough endoplasmic reticulum. **(3)** Details of treatment effects with 22% dietary GMO (bar 1 μm). a, cluster of residual bodies (asterisks); b, mitochondria show many enlarged cristae (arrows). **(4)** Hepatocytes of animal fed GM maize (GMO) at 22% of total diet. Large lakes of glycogen occur in the cytoplasm. M, mitochondria.

affected (with up to three tumors for one female), in each group. Using a non-parametric multiple comparison analysis, mammary tumor incidence was significantly increased at the lowest dose of R compared to controls ($p < 0.05$, Kruskal-Wallis test with *post hoc* Dunn's test). All females except one (with metastatic ovarian carcinoma) presented in addition mammary hypertrophies and in some cases hyperplasia with atypia (Table 2).

The second most affected organ in females was the pituitary gland, in general around two times more than in controls for most treatments (Table 2; Figure 1J,K,L,M). Again, at this level of examination, adenomas and/or hyperplasias and hypertrophies were noticed. For all R treatment groups, 70% to 80% of animals presented 1.4 to 2.4 times more abnormalities in this organ than controls.

The large palpable tumors in males (in kidney and mostly skin) were by the end of the experimental period on average twice as frequent as in controls, in which only one skin fibroma appeared during the 23rd month. At the end of the experiment, internal non-palpable tumors were added, and their sums were lower in males than in females. They were not significantly different

from controls, although slightly increased in females (Figure 4, histogram insets).

Mortality

The rates of mortality in the various control and treatment groups are shown as raw data in Figure 6. Control male animals survived on average 624 ± 21 days, whilst females lived for 701 ± 20 days during the experiment, plus in each case, a 5-week starting age at reception of animals and a 3-week housing stabilization period. After mean survival time had elapsed, any deaths that occurred were considered to be largely due to aging. Before this period, 30% control males (three in total) and 20% females (only two) died spontaneously, while up to 50% males and 70% females died in some groups on diets containing the GM maize (Figure 6, panels GMO, GMO + R). However, the rate of mortality was not proportional to the treatment dose, reaching a threshold at the lowest (11%) or intermediate (22%) amounts of GM maize in the equilibrated diet, with or without the R application on the crop. It is noteworthy that the first two male rats that died in both GM maize-treated groups had to be euthanized due to Wilms' kidney tumors that had grown by this time to over

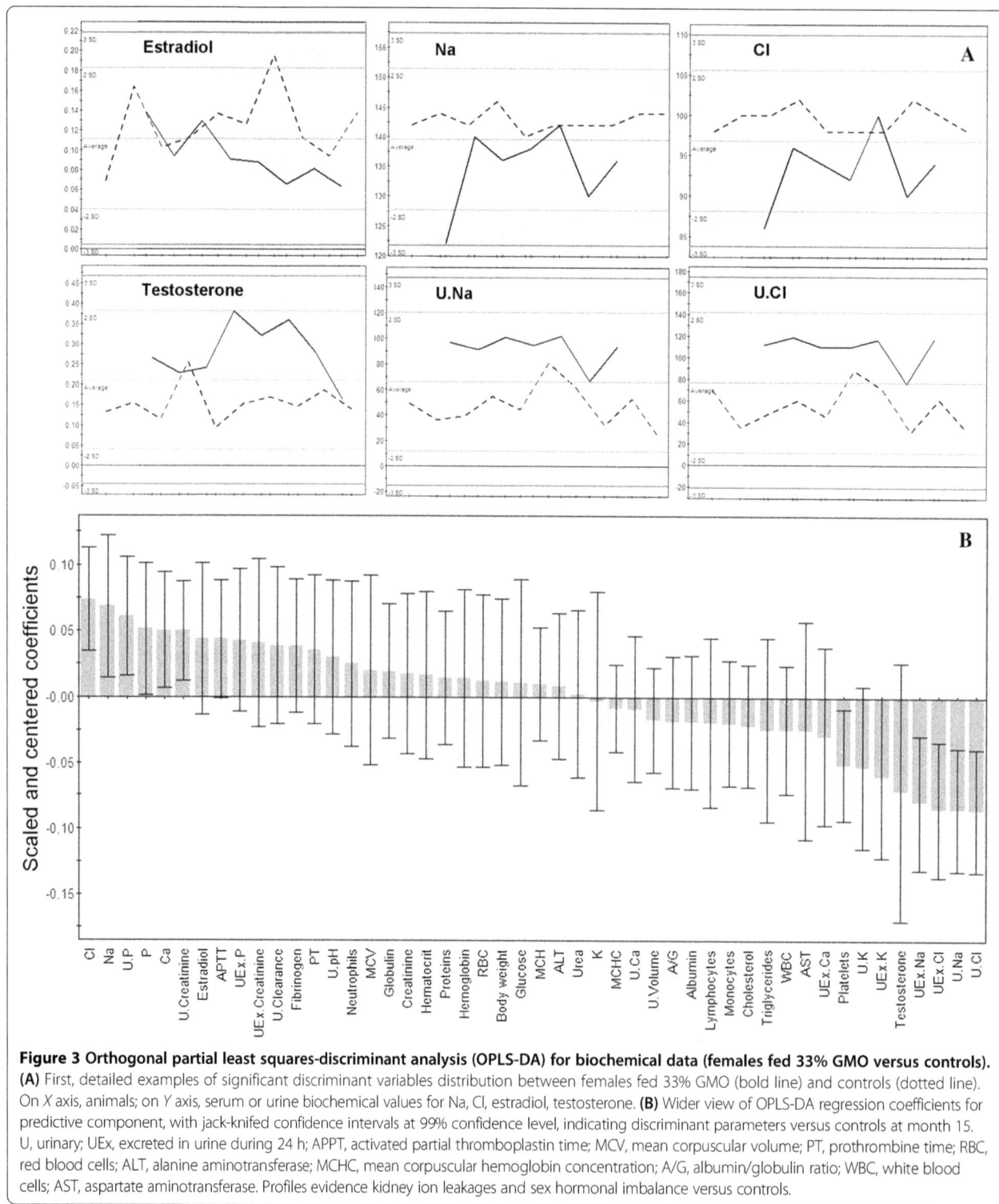

Figure 3 Orthogonal partial least squares-discriminant analysis (OPLS-DA) for biochemical data (females fed 33% GMO versus controls).
(A) First, detailed examples of significant discriminant variables distribution between females fed 33% GMO (bold line) and controls (dotted line). On X axis, animals; on Y axis, serum or urine biochemical values for Na, Cl, estradiol, testosterone. (B) Wider view of OPLS-DA regression coefficients for predictive component, with jack-knifed confidence intervals at 99% confidence level, indicating discriminant parameters versus controls at month 15. U, urinary; UEx, excreted in urine during 24 h; APPT, activated partial thromboplastin time; MCV, mean corpuscular volume; PT, prothrombine time; RBC, red blood cells; ALT, alanine aminotransferase; MCHC, mean corpuscular hemoglobin concentration; A/G, albumin/globulin ratio; WBC, white blood cells; AST, aspartate aminotransferase. Profiles evidence kidney ion leakages and sex hormonal imbalance versus controls.

25% of body weight. This was approximately a year before the first control animal died. The first female death occurred in the 22% GM maize feeding group and resulted from a mammary fibroadenoma 246 days before the first control female death. The maximum difference in males was five times more deaths occurring by the 17th month in the group consuming 11% GM maize and in females six times greater mortality by the 21st month on the 22% GM maize diet with and without R. In the female cohorts, there were two to three times more deaths in all treated groups compared with controls by the end of the experiment and deaths occurred earlier in general. Females were

Table 3 Percentage variation of parameters indicating kidney failures of female animals

Discriminant variables		GMO 11%	GMO 22%	GMO 33%	GMO 11% + R	GMO 22% + R	GMO 33% + R	R (A)	R (B)	R (C)
Gonadal hormones	Estradiol	5	−2	−25	8	−1	2	−26	−73[a]	39
	Testosterone	56[a]	17	81	5	−9	27	97[a]	−72[a]	10
Serum decrease or increase	Na	−1	−4[a]	−6[a]	2	1	1	−7	0	−3
	Cl	−5	−7	−6[a]	−1	−2	−2	−8[a]	−1	−4
	P	−17	−18[a]	−20[a]	−6	−11	−13	−32[a]	−9	−13
	K	2	−4	0	4	5	10	−4	8	−5[a]
	Ca	2[a]	−2	−5[a]	4	3	3	−6	3	−6[a]
Urinary increase	Urea	15	12	−1	12	18[a]	15	0	13	32[a]
	Na	52	−2	95[a]	25	33	30	62[a]	65	91[a]
	Na ex	50	24	125[a]	24	50	68	108[a]	51	7
	Cl	46	5	101[a]	14	35	28	67[a]	56	94[a]
	Cl ex	51	31	138[a]	20	63	70	121[a]	48	13
Urinary decrease	Clearance	−20[a]	−20[a]	−19	−4	−11	−20	−20[a]	−24[a]	−40[a]
	Creatinine	−19	−37	−36[a]	−5	−32[a]	−37[a]	−43	−23	−1
	Creatinine ex	−18	−17[a]	−21	−5	−11	−19[a]	−21[a]	−22[a]	−39[a]

OPLS-DA was performed on 48 variables at month 15. Here, we show mean differences (%) of variables ([a]discriminant at 99% confidence level) indicating kidney parameters of female animals, together with sex hormones. Male kidney pathologies are already illustrated in Figure 1.

more sensitive to the presence of R in drinking water than males, as evidenced by a shorter lifespan (Figure 6, panels R). The general causes of death represented in histogram format within each of the panels in Figure 6, are linked mostly to mammary tumors in females and to problems in other organ systems in males.

Discussion

This report describes the first long-term (2-year) rodent (rat) feeding study investigating possible toxic effects arising from consumption of an R-tolerant GM maize (NK603) and a complete commercial formulation of R herbicide. The aims of this investigation were essentially twofold. First, to evaluate whether the signs of toxicity, especially with respect to liver and kidney functions, seen after 90 days' consumption of a diet containing NK603 R-tolerant GM maize [3,7] escalated into serious ill health or dissipated over an extended period of time. Second, to determine if low doses of full commercial R formulation at permitted levels were still toxic, as indicated by our previous *in vitro* studies [8,9]. The previous toxicity study with NK603 maize employed only this GM crop that had been sprayed with R during cultivation [3]. However, in our study presented here, in addition to extending the treatment period from 90 days to 2 years and in order to better ascertain the source of any ill health observed, we included additional test feeding groups. These consisted of NK603 maize grown without as well as with R application and R alone administered via drinking water. Furthermore, we used three levels of dosing in all cases rather than the two

previously used [3], in order to highlight any dose response effects of a given treatment. It is also important to note that our study is the first to conduct blood, urine, and organ analyses from animals treated with the complete agricultural formulation of R and not just G alone, as measured by the manufacturer [35].

Our data show that the signs of liver and kidney toxicity seen at 90 days from the consumption of NK603 GM maize [3,7] do indeed escalate into severe disease over an extended period. Furthermore, similar negative health effects were observed in all treatment groups (NK603 GM maize with or without R application and R alone).

What is also evident from our data is that ill effects were not proportional to the dose of either the NK603 GM maize ± R or R alone. This suggests that the observed disease may result from endocrine disruptive effects, which are known to be non-monotonic. Similar degrees of pathological symptoms occurred from the lowest to the highest doses, suggesting a threshold effect [36]. This corresponds to levels likely to arise from consumption or environmental exposure, such as either 11% GM maize in food, or 50 ng/L G equivalent of R-formulation, a level which can be found in some contaminated drinking tap waters and which falls within authorized limits.

Death in male rats was mostly due to the development of severe hepatorenal insufficiencies, confirming the first signs of toxicity observed in 90-day feeding trials with NK603 GM maize [7]. In females, kidney ion leakage was evident at a biochemical level at month 15, when severe nephropathies were observed in dead male animals at postmortem, at the anatomopathological level. Early

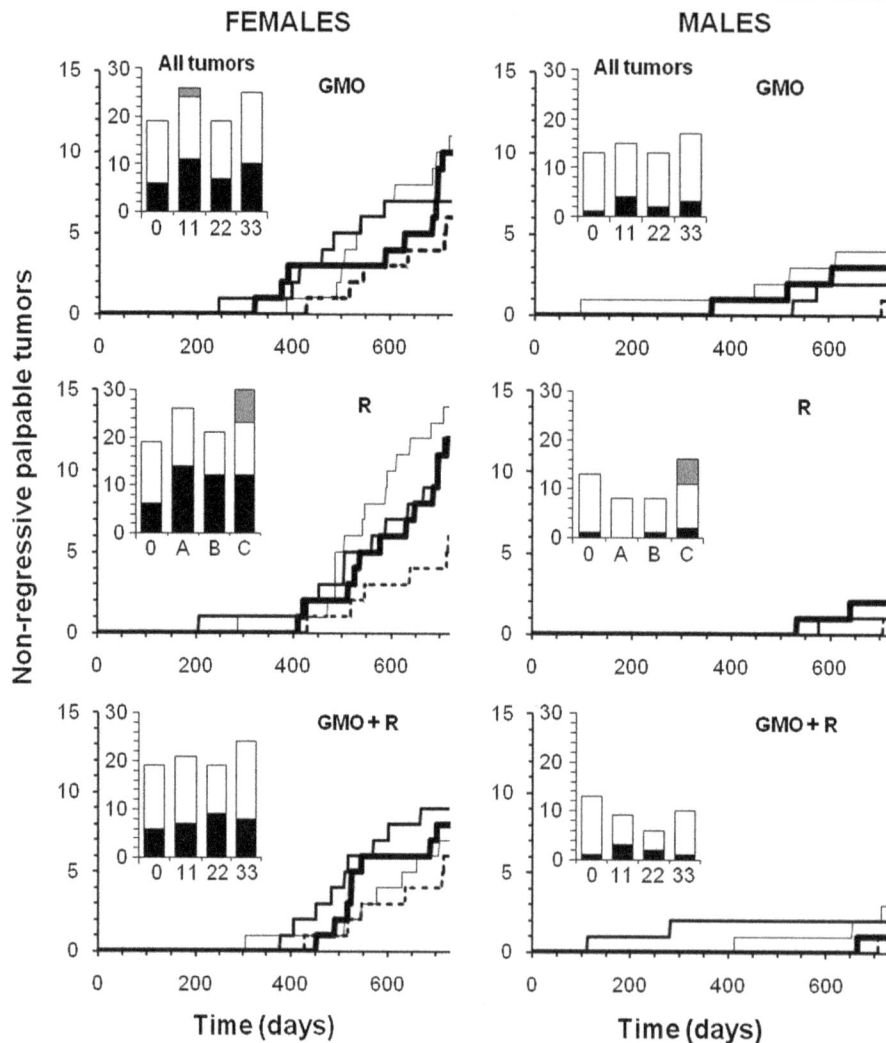

Figure 4 Largest non-regressive tumors in rats fed GMO treated or not by Roundup and effects of Roundup alone. Rats were fed with NK603 GM maize (with or without application of Roundup) at three different doses (11%, 22%, and 33% in their diet; thin, medium, and bold lines, respectively) compared to the substantially equivalent closest isogenic non-GM maize (control, dotted line). Roundup was administered in drinking water at three increasing doses, same symbols, environmental **(A)**, MRL in some agricultural GMOs **(B)**, and half of minimal agricultural levels **(C)**, see 'Methods'). The largest tumors were palpable during the experiment and numbered from 20 mm in diameter for males and 17.5 mm for females. Above this size, 95% of growths were non-regressive tumors. Summary of all tumors are shown in the bar histograms: black, non-regressive large tumors; white, small internal tumors; grey, metastases.

signs of toxicity at month 3 in kidney and liver were also observed for 19 edible GM crops containing pesticide residues [1]. It is known that only elderly male rats are sensitive to chronic progressive nephropathies [37]. Therefore, the disturbed kidney functional parameters may have been induced by the reduced levels of phenolic acids in the GM maize feed used in our study, since caffeic and ferulic acids are beneficial to the kidney as they prevent oxidative stress [38,39]. This possibility is consistent with our previous observation that plant extracts containing ferulic and caffeic acids were able to promote detoxification of human embryonic kidney cells after culture in the presence of R [40]. It is thus possible that NK603 GM maize consumption,

with its reduced levels of these compounds, may have provoked the early aging of the kidney physiology, similarly to R exposure causing oxidative stress [41]. Disturbances in global patterns of gene expression leading to disease via epigenetic effects cannot be excluded, since it has been demonstrated that numerous pesticides can cause changes in DNA methylation and histone modification, thereby altering chromatin compaction and thus gene expression profiles [42].

Disturbances that we found to occur in the male liver are characteristic of chronic toxicity, confirmed by alterations in biochemical liver and kidney function parameters. The observation that liver function in female animals

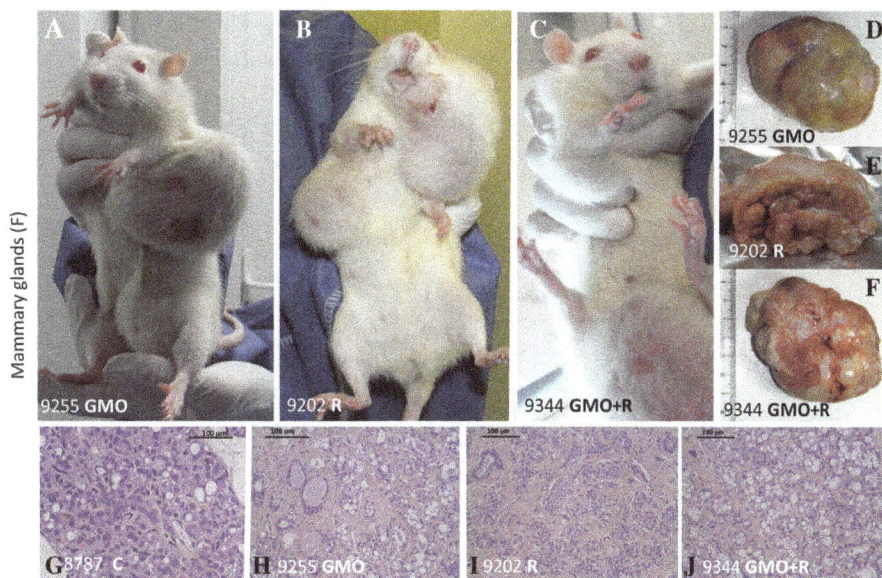

Figure 5 Examples of female mammary tumors observed. Mammary tumors are evidenced (**A**, **D**, **H**, representative adenocarcinoma, from the same rat in a GMO group) and in Roundup and GMO + Roundup groups, two representative rats (**B**, **C**, **E**, **F**, **I**, **J** fibroadenomas) are compared to controls. A normal representative rat in controls is not shown, only a minority of them having tumors up to 700 days, in contrast with the majority affected in all treated groups. (**G**) The histological control.

was less negatively affected may be due to the known protection from oxidative stress conferred by estrogen [43]. Estrogen can induce expression of genes such as superoxide dismutase and glutathione peroxidase via the MAP kinase-NF-kB signaling pathway, thus providing an antioxidant effect [43]. Furthermore, liver enzymes have been clearly demonstrated as sex-specific in their expression patterns, including in a 90-day rat feeding trial of NK603 GM maize [7]. However, in a long-term study, evidence of early liver aging was observed in female mice fed with R-tolerant GM soy [12]. In the present investigation, deeper analysis at an ultrastructural level revealed evidence of impediments in transcription and other defects in cell nuclear structure that were comparable in both sexes and dose-dependent in hepatocytes in all treatments. This is consistent with the well-documented toxic effect of very low dilutions of R on apoptosis, mitochondrial function, and cell membrane degradation, inducing necrosis of hepatocytes, and in other cell lines [8,9,44,45].

The disruptions of at least the estrogen-related pathways and/or enhancement of oxidative stress by all treatments need further confirmation. This can be addressed through the application of transcriptomic, proteomic, and metabolomic methods to analyze the molecular profile of kidneys and livers, as well as the GM NK603 maize [46-48]. Other possible causes of observed pathogenic effects may be due to disturbed gene expression resulting from the transgene insertional, general mutagenic, or metabolic effects [49,50] as has been shown for MON810 GM maize [51,52]. A consequent disruption of general metabolism in the GMO cannot be excluded, which could lead, for example, to the production of other potentially active compounds such as miRNAs [53] or leukotoxin diols [54].

The lifespan of the control group of animals corresponded to the mean for the strain of rat used (Harlan Sprague-Dawley), but as is frequently the case with most mammals, including humans [55], males on average died before females, except for some female treatment groups. All treatments in both sexes enhanced large tumor incidence by two- to threefold in comparison to our controls and also the number of mammary tumors in comparison to the Harlan Sprague-Dawley strain [56] and overall around threefold in comparison to the largest study with 1,329 Sprague-Dawley female rats [57]. This indicates that the use of historical data to compare our tumor numbers is not relevant, first, since we studied the difference with concurrent controls chronologically (and not only at the end of the experiment, as is the case in historical data), and second, since the diets of historical reference animals may have been contaminated with several non-monitored compounds including GMOs and pesticides at levels used in our treatments. In our study, the tumors also developed considerably faster than in controls, even though the majority of tumors were observed after 18 months. The first large detectable tumors occurred at 4 and 7 months into the study in males and females, respectively, further underlining the inadequacy of the standard 90-day feeding trials for evaluating GM crop and food toxicity [1]. Future studies employing larger cohorts of animals providing

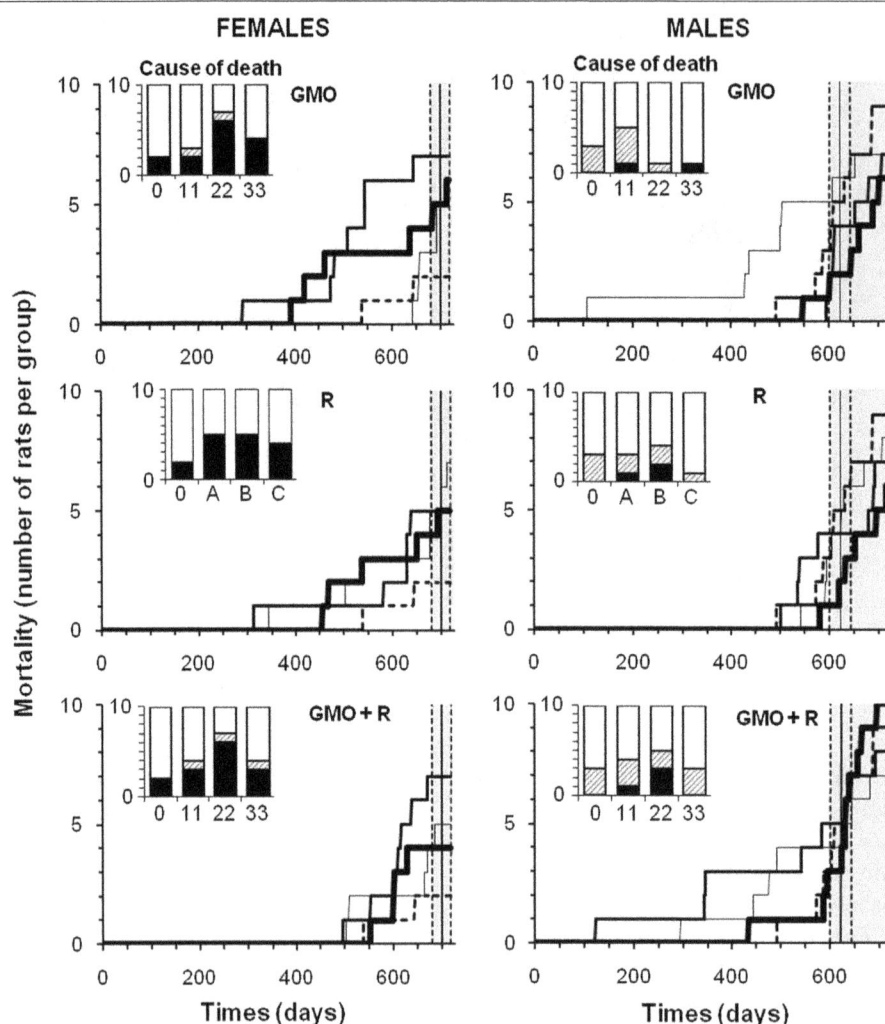

Figure 6 Mortality of rats fed GMO treated or not with Roundup and effects of Roundup alone. The symbols of curves and treatments are explained in the caption of Figure 4. Lifespan during the experiment for the control group is represented by the vertical bar ± SEM (grey area). In bar histograms, the causes of mortality before the grey area are detailed in comparison to the controls (0). In black are the necessary euthanasia because of suffering in accordance with ethical rules (tumors over 25% body weight, more than 25% weight loss, hemorrhagic bleeding, etc.); and in hatched areas, spontaneous mortality.

appropriate statistical power are required to confirm or refute the clear trend in increased tumor incidence and mortality rates seen with some of the treatments tested in this study. As already stated, our study was not designed as a carcinogenicity study that would have required according to OECD the use of 50 rats per sex per group. However, we wish to emphasize that the need for more rats to provide sufficient statistical power may be biased by the presence of contaminants in the diets used in gathering historical control data, increasing artificially the background of tumors, which would inappropriately be called in this case 'spontaneous' or due to the genetic strain. For instance, toxic, hormonal disrupting or carcinogenic levels of pesticides, PCBs, plasticizers, dioxins, or heavy metals may contaminate the diets or drinking water

used for the establishment of 'spontaneous' tumors in historical data [58-62].

In females, induced euthanasia due to suffering and deaths corresponded mostly to the development of large mammary tumors. This was observed independently of the cancer grade but according to impact on morbidity. These appeared to be related to the various treatments when compared to the control groups. These tumors are generally known to be mostly estrogen-dependent [63]. We observed a strikingly marked induction of mammary tumors in groups administered R alone, even at the very lowest dose (50 ng/L G equivalent dilution in adjuvants). At this concentration *in vitro*, G alone is known to induce human breast cancer cell growth via estrogen receptors [64]. In addition, R with adjuvants has been shown to disrupt aromatase, which synthesizes estrogen [19], and to interfere with estrogen and androgen receptors in cells [8]. Furthermore, R appears to be a sex endocrine disruptor *in vivo* in males [10]. Sex steroid levels were also modified in treated rats in our study. These hormone-dependent phenomena are confirmed by enhanced pituitary dysfunction in treated females. An estrogen-modified feedback mechanism may act at this level [65,66]. The similar pathological profiles provoked by the GM maize + R diet may thus be explained at least in part by R residues present in this feed. In this regard, it is noteworthy that the medium dose of the R treatment tested (400 mg/Kg G equivalent) corresponds to acceptable residue levels of this pesticide in some edible GMOs.

Interestingly and perhaps surprisingly, in the groups of animals fed with the NK603 GM maize without R application, similar effects with respect to enhanced tumor incidence and mortality rates were observed. For instance, comparing the 11% GMO-treated female group to the controls, the assumption that the tumors are equally distributed is rejected with a level of significance of 0.54% with the Westlake exceedance test [67]. The classical tests of Kolmogorov-Smirnov (one-sided) and Wilcoxon-Mann-Whitney reach α values of significance, which are respectively of 1.40% and 2.62%.

A possible explanation for this finding is the production of specific compound(s) in the GM feed that are either directly toxic and/or cause the inhibition of pathways, which in turn generates toxic effects. This is despite the fact that the variety of GM maize used in this study was judged by industry and regulators as being substantially equivalent to the corresponding non-GM closest isogenic line [3,30]. As the total chemical composition of the GM maize has not been measured in detail, the use of substantial equivalence as a concept in risk assessment is insufficient to highlight potential unknown toxins and therefore cannot replace long-term animal feeding trials for GMOs.

A cause of the ill effects resulting from NK603 GM maize alone observed in this study could be the fact that it is engineered to overexpress a modified version of the *Agrobacterium tumefaciens* 5-enolpyruvylshikimate-3-phosphate synthase (EPSPS-CP4) [3], which confers R tolerance. The modified EPSPS is not inhibited by G, in contrast to the wild-type enzyme in the crop. This enzyme is known to drive the first step of aromatic amino acid biosynthesis in the plant shikimate pathway. In addition, estrogenic isoflavones and their glycosides are also products of this pathway [68]. A limited compositional analysis showed that these biochemical pathways were not disturbed in the GM maize used in our study. However, our analysis did reveal that the levels of caffeic and ferulic acids in the GM diet, which are also secondary metabolites of the plant shikimate pathway, but not

always measured in regulatory tests, were significantly reduced. This may lower their protective effects against carcinogenesis and mammalian tumor formation [69,70]. Moreover, these phenolic acids, and in particular ferulic acid, may modulate estrogen receptors or the estrogenic pathway in mammalian cells [71]. This does not exclude the possibility of the action of other unknown metabolites. This explanation also corresponds to the fact that the observed effects of NK603 GM maize and R were not additive but reached a threshold. This implies that both the NK603 maize and R may cause hormonal disturbances in the same biochemical and physiological pathways.

Conclusions

In conclusion, the consumption of NK603 GM maize with or without R application or R alone gave similar pathologies in male and female rats fed over a 2-year period. It was previously known that G consumption in water above authorized limits may provoke hepatic and kidney failure [33]. The results of the study presented here clearly indicate that lower levels of complete agricultural G herbicide formulations, at concentrations well below officially set safety limits, can induce severe hormone-dependent mammary, hepatic, and kidney disturbances. Similarly, disruption of biosynthetic pathways that may result from overexpression of the EPSPS transgene in the GM NK603 maize can give rise to comparable pathologies that may be linked to abnormal or unbalanced phenolic acid metabolites or related compounds. Other mutagenic and metabolic effects of the edible GMO cannot be excluded. This will be the subject of future studies, including analyses of transgene, G and other R residue presence in rat tissues. Reproductive and multigenerational studies will also provide novel insight into these problems. This study represents the first detailed documentation of long-term deleterious effects arising from consumption of a GMO, specifically a R-tolerant maize, and of R, the most widely used herbicide worldwide.

Taken together, the significant biochemical disturbances and physiological failures documented in this work reveal the pathological effects of these GMO and R treatments in both sexes, with different amplitudes. They also show that the conclusion of the Monsanto authors [3] that the initial indications of organ toxicity found in their 90-day experiment were not 'biologically meaningful' is not justifiable.

We propose that agricultural edible GMOs and complete pesticide formulations must be evaluated thoroughly in long-term studies to measure their potential toxic effects.

Methods
Ethics

The experimental protocol was conducted in an animal care unit authorized by the French Ministries of Agriculture and Research (Agreement Number A35-288-1). Animal experiments were performed according to ethical guidelines of animal experimentations (CEE 86/609 regulation), including the necessary observations of all tumors, in line with the requirements for a long-term toxicological study [32], up to a size where euthanasia on ethical grounds was necessary.

Concerning the cultivation of the maize used in this study, no specific permits were required. This is because the maize was grown (MON-00603-6 commonly named NK603) in Canada, where it is authorized for unconfined release into the environment and for use as a livestock feed by the Canadian Food Inspection Agency (Decision Document 2002-35). We confirm that the cultivation did not involve endangered or protected species. The GM maize was authorized for import and consumption into the European Union (CE 258/97 regulation).

Plants, diets, and chemicals

The varieties of maize used in this study were the DKC 2678 R-tolerant NK603 (Monsanto Corp., USA), and its nearest isogenic non-transgenic control DKC 2675. These two types of maize were grown under similar normal conditions, in the same location, spaced at sufficient distance to avoid cross-contamination. The genetic nature, as well as the purity of the GM seeds and harvested material, was confirmed by qPCR analysis of DNA samples. One field of NK603 was treated with R at 3 L ha^{-1} (WeatherMAX, 540 g/L of G, EPA Reg. 524-537), and another field of NK603 was not treated with R. Corn cobs were harvested when the moisture content was less than 30% and were dried at a temperature below 30°C. From these three cultivations of maize, laboratory rat chow was made based on the standard diet A04 (Safe, France). The dry rat feed was made to contain 11%, 22%, or 33% of GM maize, cultivated either with or without R, or 33% of the non-transgenic control line. The concentrations of the transgene were confirmed in the three doses of each diet by qPCR. All feed formulations consisted of balanced diets, chemically measured as substantially equivalent except for the transgene, with no contaminating pesticides over standard limits. All secondary metabolites cannot be known and measured in the composition. However, we measured isoflavones and phenolic acids including ferulic acid by standard HPLC-UV. All reagents used were of analytical grade. The herbicide diluted in the drinking water was the commercial formulation of R (GT Plus, 450 g/L of G, approval 2020448, Monsanto, Belgium). Herbicide levels were assessed by G measurements in the different dilutions by mass spectrometry.

Animals and treatments

Virgin albino Sprague-Dawley rats at 5 weeks of age were obtained from Harlan (Gannat, France). All animals were kept in polycarbonate cages (820 cm^2, Genestil, France) with two animals of the same sex per cage. The litter (Toplit classic, Safe, France) was replaced twice weekly. The animals were maintained at 22 ± 3°C under controlled humidity (45% to 65%) and air purity with a 12 h-light/dark cycle, with free access to food and water. The location of each cage within the experimental room was regularly changed. This 2-year life-long experiment was conducted in a Good Laboratory Practice (GLP) accredited laboratory according to OECD guidelines. After 20 days of acclimatization, 100 male and 100 female animals were randomly assigned on a weight basis into ten equivalent groups. For each sex, one control group had access to plain water and standard diet from the closest isogenic non-transgenic maize control; six groups were fed with 11%, 22%, and 33% of GM NK603 maize either treated or not treated with R. The final three groups were fed with the control diet and had access to water supplemented with respectively 1.1 × 10^{-8}% of R (0.1 ppb or 50 ng/L of G, the contaminating level of some regular tap waters), 0.09% of R (400 mg/kg G, US MRL of 400 ppm G in some GM feed), and 0.5% of R (2.25 g/L G, half of the minimal agricultural working dilution). This was changed weekly. Twice-weekly monitoring allowed careful observation and palpation of animals, recording of clinical signs, measurement of any tumors, food and water consumption, and individual body weights.

Anatomopathology

Animals were sacrificed during the course of the study only if necessary because of suffering according to ethical rules (such as 25% body weight loss, tumors over 25% body weight, hemorrhagic bleeding, or prostration) and at the end of the study by exsanguination under isoflurane anesthesia. In each case, detailed observations and anatomopathology was performed and the following organs were collected: brain, colon, heart, kidneys, liver, lungs, ovaries, spleen, testes, adrenals, epididymis, prostate, thymus, uterus, aorta, bladder, bone, duodenum, esophagus, eyes, ileum, jejunum, lymph nodes, lymphoreticular system, mammary glands, pancreas, parathyroid glands, Peyer's patches, pituitary, salivary glands, sciatic nerve, skin, spinal cord, stomach, thyroid, and trachea. The first 14 organs (at least ten per animal depending on the sex, Table 1) were weighted, plus any tumors that arose. The first nine were divided into two parts and one half was immediately frozen in liquid nitrogen/carbonic ice. The remaining parts including other organs were rinsed in PBS and stored in 4% formalin before anatomopathological study. These samples were used for further paraffin-embedding, slides, and HES histological staining. For transmission electron microscopy, the kidneys, livers, and tumors were cut into 1 mm^3 fragments. Samples were fixed in pre-chilled 2% paraformaldehyde/2.5% glutaraldehyde in 0.1 M PBS pH 7.4 at 4°C for 3 h and processed as previously described [13].

Biochemical analyses

Blood samples were collected from the tail vein of each rat under short isoflurane anesthesia before treatment and after 1, 2, 3, 6, 9, 12, 15, 18, 21, and 24 months: 11 measurements were obtained for each animal alive at 2 years. It was first demonstrated that anesthesia did not impact animal health. Two aliquots of plasma and serum were prepared and stored at $-80°C$. Then, 31 parameters were assessed (Table 1) according to standard methods including hematology and coagulation parameters, albumin, globulin, total protein concentration, creatinine, urea, calcium, sodium, potassium, chloride, inorganic phosphorus, triglycerides, glucose, total cholesterol, alanine aminotransferase, aspartate aminotransferase, gamma glutamyltransferase (GT), estradiol, and testosterone. In addition, at months 12 and 24, the C-reactive protein was assayed. Urine samples were collected similarly 11 times, over 24 h in individual metabolic cages, and 16 parameters were quantified including creatinine, phosphorus, potassium, chloride, sodium, calcium, pH, and clearance. Liver samples taken at the end made it possible to perform assays of CYP1A1, 1A2, 3A4, 2C9 activities in S9 fractions, with glutathione S-transferase and gamma-GT.

Statistical analysis

In this study, multivariate analyses were more appropriate than pairwise comparisons between groups because the parameters were very numerous, with samples of ten individuals. Kaplan-Meyer comparisons, for instance, were not used because these are better adapted to epidemiological studies. Differences in the numbers of mammary tumors were studied by a non-parametric multiple comparisons Kruskal-Wallis test, followed by a *post hoc* Dunn's test with the GraphPad Prism 5 software.

Biochemical data were treated by multivariate analysis with the SIMCA-P (V12) software (UMETRICS AB Umea, Sweden). The use of chemometrics tools, for example, principal component analysis (PCA), partial least squares to latent structures (PLS), and orthogonal PLS (OPLS), are robust methods for modeling, analyzing, and interpreting complex chemical and biological data. OPLS is a recent modification of the PLS method. PLS is a regression method used in order to find the relationship between two data tables referred to as X and Y. PLS regression [72] analysis consists in calculating by means of successive iterations, linear combinations of the measured X-variables (predictor variables). These linear combinations of X-variables give PLS components (score vectors t). A PLS component can be thought of as a new variable - a latent variable - reflecting the information in the original X-variables that is of relevance for modeling and predicting the response Y-variable by means of the maximization of the square of covariance (Max $cov^2(X,Y)$). The number of components is determined by cross validation. SIMCA software uses the nonlinear iterative partial least squares algorithm (NIPALS) for the PLS regression. Orthogonal partial least squares discriminant analysis (OPLS-DA) was used in this study [73,74].

The purpose of discriminant analysis is to find a model that separates groups of observations on the basis of their X variables. The X matrix consists of the biochemical data. The Y matrix contains dummy variables which describe the group membership of each observation. Binary variables are used in order to encode a group identity. Discriminant analysis finds a discriminant plan in which the projected observations are well separated according to each group. The objective of OPLS is to divide the systematic variation in the X-block into two model parts, one linearly related to Y (in the case of a discriminant analysis, the group membership), and the other one unrelated (orthogonal) to Y. Components related to Y are called predictive, and those unrelated to Y are called orthogonal. This partitioning of the X data results in improved model transparency and interpretability [75]. Prior to analysis, variables were mean-centered and unit variance scaled.

Competing interests

The author(s) declare that they have no competing interests, and that, in contrast with regulatory assessments for GMOs and pesticides, they are independent from companies developing these products.

Authors' contributions

GES directed and with JSV designed and coordinated the study. EC, RM, SG, and ND analyzed the data, compiled the literature, and participated in the drafting of the manuscript and final version. MM performed transmission electron microscopy. DH performed OPLS-DA statistical analysis. All authors read and approved the final manuscript.

Acknowledgements

We thank Michael Antoniou for English assistance, editing, and constructive comments on the manuscript. We gratefully acknowledge the Association CERES, for research on food quality, representing more than 50 companies and private donations, the Foundation 'Charles Leopold Mayer pour le Progrès de l'Homme', the French Ministry of Research, and CRIIGEN for their major support.

Author details

[1]Institute of Biology, EA 2608 and CRIIGEN and Risk Pole, MRSH-CNRS, Esplanade de la Paix, University of Caen, Caen, Cedex 14032, France. [2]Department of Neurological, Neuropsychological, Morphological and Motor Sciences, University of Verona, Verona 37134, Italy. [3]Risk Pole, MRSH-CNRS, Esplanade de la Paix, University of Caen, Caen, Cedex 14032, France.

References

1. Seralini G-E, Mesnage R, Clair E, Gress S, de Vendomois J, Cellier D: Genetically modified crops safety assessments: present limits and possible improvements. *Environ Sci Eur* 2011, **23**:10.
2. Domingo JL, Gine Bordonaba J: A literature review on the safety assessment of genetically modified plants. *Environ Int* 2011, **37**:734–742.
3. Hammond B, Dudek R, Lemen J, Nemeth M: Results of a 13 week safety assurance study with rats fed grain from glyphosate tolerant corn. *Food Chem Toxicol* 2004, **42**:1003–1014.

4. Hammond B, Lemen J, Dudek R, Ward D, Jiang C, Nemeth M, Burns J: Results of a 90-day safety assurance study with rats fed grain from corn rootworm-protected corn. *Food Chem Toxicol* 2006, 44:147–160.

5. Hammond BG, Dudek R, Lemen JK, Nemeth MA: Results of a 90-day safety assurance study with rats fed grain from corn borer-protected corn. *Food Chem Toxicol* 2006, 44:1092–1099.

6. Seralini GE, Cellier D, de Vendomois JS: New analysis of a rat feeding study with a genetically modified maize reveals signs of hepatorenal toxicity. *Arch Environ Contam Toxicol* 2007, 52:596–602.

7. Spiroux de Vendômois J, Roullier F, Cellier D, Seralini GE: A comparison of the effects of three GM corn varieties on mammalian health. *Int J Biol Sci* 2009, 5:706–726.

8. Gasnier C, Dumont C, Benachour N, Clair E, Chagnon MC, Seralini GE: Glyphosate-based herbicides are toxic and endocrine disruptors in human cell lines. *Toxicology* 2009, 262:184–191.

9. Benachour N, Seralini GE: Glyphosate formulations induce apoptosis and necrosis in human umbilical, embryonic, and placental cells. *Chem Res Toxicol* 2009, 22:97–105.

10. Romano MA, Romano RM, Santos LD, Wisniewski P, Campos DA, de Souza PB, Viau P, Bernardi MM, Nunes MT, de Oliveira CA: Glyphosate impairs male offspring reproductive development by disrupting gonadotropin expression. *Arch Toxicol* 2012, 86:663–673.

11. Krüger M, Schrödl W, Neuhaus J, Shehata A: Field investigations of glyphosate in urine of Danish dairy cows. *J Environ Anal Toxicol* 2013, 3:5.

12. Malatesta M, Boraldi F, Annovi G, Baldelli B, Battistelli S, Biggiogera M, Quaglino D: A long-term study on female mice fed on a genetically modified soybean: effects on liver ageing. *Histochem Cell Biol* 2008, 130:967–977.

13. Malatesta M, Caporaloni C, Gavaudan S, Rocchi MB, Serafini S, Tiberi C, Gazzanelli G: Ultrastructural morphometrical and immunocytochemical analyses of hepatocyte nuclei from mice fed on genetically modified soybean. *Cell Struct Funct* 2002, 27:173–180.

14. Malatesta M, Caporaloni C, Rossi L, Battistelli S, Rocchi MB, Tonucci F, Gazzanelli G: Ultrastructural analysis of pancreatic acinar cells from mice fed on genetically modified soybean. *J Anat* 2002, 201:409–415.

15. Malatesta M, Perdoni F, Santin G, Battistelli S, Muller S, Biggiogera M: Hepatoma tissue culture (HTC) cells as a model for investigating the effects of low concentrations of herbicide on cell structure and function. *Toxicol In Vitro* 2008, 22:1853–1860.

16. Snell C, Bernheim A, Berge JB, Kuntz M, Pascal G, Paris A, Ricroch AE: Assessment of the health impact of GM plant diets in long-term and multigenerational animal feeding trials: a literature review. *Food Chem Toxicol* 2012, 50:1134–1148.

17. Williams GM, Kroes R, Munro IC: Safety evaluation and risk assessment of the herbicide Roundup and its active ingredient, glyphosate, for humans. *Regul Toxicol Pharmacol* 2000, 31:117–165.

18. Cox C: Herbicide factsheet - glyphosate. *J Pesticide Reform* 2004, 24:10–15.

19. Richard S, Moslemi S, Sipahutar H, Benachour N, Seralini GE: Differential effects of glyphosate and roundup on human placental cells and aromatase. *Environ Health Perspect* 2005, 113:716–720.

20. Mesnage R, Bernay B, Seralini GE: Ethoxylated adjuvants of glyphosate-based herbicides are active principles of human cell toxicity. *Toxicology* 2013, 313:122–128.

21. Adam A, Marzuki A, Abdul Rahman H, Abdul Aziz M: The oral and intratracheal toxicities of ROUNDUP and its components to rats. *Vet Hum Toxicol* 1997, 39:147–151.

22. Mesnage R, Defarge N, Spiroux De Vendômois J, Séralini GE: Major pesticides are more toxic to human cells than their declared active principles. *Biomed Res Int* 2014, Vol 2014:Article ID 179691.

23. Jobling S, Burn RW, Thorpe K, Williams R, Tyler C: Statistical modeling suggests that antiandrogens in effluents from wastewater treatment works contribute to widespread sexual disruption in fish living in English rivers. *Environ Health Perspect* 2009, 117:797–802.

24. Krogh KA, Vejrup KV, Mogensen BB, Halling-Sørensen B: Liquid chromatography-mass spectrometry method to determine alcohol ethoxylates and alkylamine ethoxylates in soil interstitial water, ground water and surface water samples. *J Chromatogr A* 2002, 957:45–57.

25. Giesy J, Dobson S, Solomon K: Ecotoxicological risk assessment for Roundup® herbicide. In *Reviews of Environmental Contamination and Toxicology*, Volume Volume 167. Edited by Ware G. New York: Springer; 2000:35–120. Reviews of Environmental Contamination and Toxicology.

26. Cox C, Surgan M: Unidentified inert ingredients in pesticides: implications for human and environmental health. *Environ Health Perspect* 2006, 114:1803–1806.

27. Mesnage R, Clair E, Séralini GE: Roundup in genetically modified plants: regulation and toxicity in mammals. *Theorie in der Ökologie* 2010, 16:31–33.

28. Monosson E: Chemical mixtures: considering the evolution of toxicology and chemical assessment. *Environ Health Perspect* 2005, 113:383–390.

29. Doull J, Gaylor D, Greim HA, Lovell DP, Lynch B, Munro IC: Report of an Expert Panel on the reanalysis by of a 90-day study conducted by Monsanto in support of the safety of a genetically modified corn variety (MON 863). *Food Chem Toxicol* 2007, 45:2073–2085.

30. EFSA: Opinion of the scientific panel on genetically modified organisms on a request from the commission related to the safety of foods and food ingredients derived from herbicide-tolerant genetically modified maize NK603 for which a request for placing on the market was submitted under Article 4 of the Novel Food Regulation (EC) No 258/97 by Monsanto (QUESTION NO EFSA-Q-2003-002). *EFSA J* 2003, 9:1–14.

31. King-Herbert A, Sills R, Bucher J: Commentary: update on animal models for NTP studies. *Toxicol Pathol* 2010, 38:180–181.

32. OECD: *OECD guideline no. 452 for the testing of chemicals: Chronic toxicity studies: Adopted 7 September 2009.* OECD Publishing. Paris, France: 2009.

33. EPA: Basic information about glyphosate in drinking water. 2014, http://waterepagov/drink/contaminants/basicinformation/glyphosatecfm (last access March).

34. Union E: COUNCIL DIRECTIVE 98/83/EC of 3 November 1998 on the quality of water intended for human consumption. *Off J Eur Commun L* 1998, 330(32):51298.

35. German Federal Agency CPFS: Monograph on glyphosate by the German federal agency for consumer protection and food safety. *Annex B-5: Toxicol Metabol* 1998.

36. Vandenberg LN, Colborn T, Hayes TB, Heindel JJ, Jacobs DR Jr, Lee DH, Shioda T, Soto AM, Vom Saal FS, Welshons WV, Zoeller RT, Myers JP: Hormones and endocrine-disrupting chemicals: low-dose effects and nonmonotonic dose responses. *Endocr Rev* 2012, 33:378–455.

37. Hard GC, Khan KN: A contemporary overview of chronic progressive nephropathy in the laboratory rat, and its significance for human risk assessment. *Toxicol Pathol* 2004, 32:171–180.

38. Srinivasan M, Rukkumani R, Ram Sudheer A, Menon VP: Ferulic acid, a natural protector against carbon tetrachloride-induced toxicity. *Fundam Clin Pharmacol* 2005, 19:491–496.

39. M UR, Sultana S: Attenuation of oxidative stress, inflammation and early markers of tumor promotion by caffeic acid in Fe-NTA exposed kidneys of Wistar rats. *Mol Cell Biochem* 2011, 357:115–124.

40. Gasnier C, Laurant C, Decroix-Laporte C, Mesnage R, Clair E, Travert C, Seralini GE: Defined plant extracts can protect human cells against combined xenobiotic effects. *J Occup Med Toxicol* 2011, 6:3.

41. El-Shenawy NS: Oxidative stress responses of rats exposed to Roundup and its active ingredient glyphosate. *Environ Toxicol Pharmacol* 2009, 28:379–385.

42. Collotta M, Bertazzi PA, Bollati V: Epigenetics and pesticides. *Toxicology* 2013, 307:35–41.

43. Vina J, Borras C, Gambini J, Sastre J, Pallardo FV: Why females live longer than males? Importance of the upregulation of longevity-associated genes by oestrogenic compounds. *FEBS Lett* 2005, 579:2541–2545.

44. Benachour N, Sipahutar H, Moslemi S, Gasnier C, Travert C, Seralini GE: Time- and dose-dependent effects of roundup on human embryonic and placental cells. *Arch Environ Contam Toxicol* 2007, 53:126–133.

45. Peixoto F: Comparative effects of the Roundup and glyphosate on mitochondrial oxidative phosphorylation. *Chemosphere* 2005, 61:1115–1122.

46. Jiao Z, Si XX, Li GK, Zhang ZM, Xu XP: Unintended compositional changes in transgenic rice seeds (Oryza sativa L.) studied by spectral and chromatographic analysis coupled with chemometrics methods. *J Agric Food Chem* 2010, 58:1746–1754.

47. Zhou J, Ma C, Xu H, Yuan K, Lu X, Zhu Z, Wu Y, Xu G: Metabolic profiling of transgenic rice with crylAc and sck genes: an evaluation of unintended effects at metabolic level by using GC-FID and GC-MS. *J Chromatogr B Analyt Technol Biomed Life Sci* 2009, 877:725–732.

48. Zolla L, Rinalducci S, Antonioli P, Righetti PG: Proteomics as a complementary tool for identifying unintended side effects occurring in transgenic maize seeds as a result of genetic modifications. *J Proteome Res* 2008, 7:1850–1861.

Republished study: long-term toxicity of a Roundup herbicide and a Roundup-tolerant...

99

49. Latham JR, Wilson AK, Steinbrecher RA: The mutational consequences of plant transformation. *J Biomed Biotechnol* 2006, 2006:25376.

50. Wilson AK, Latham JR, Steinbrecher RA: Transformation-induced mutations in transgenic plants: analysis and biosafety implications. *Biotechnol Genet Eng Rev* 2006, 23:209–237.

51. Rosati A, Bogani P, Santarlasci A, Buiatti M: Characterisation of 3' transgene insertion site and derived mRNAs in MON810 YieldGard maize. *Plant Mol Biol* 2008, 67:271–281.

52. Abdo E, Barbary O, Shaltout O: Feeding study with Bt corn (MON810: ajeeb YG) on rats: biochemical analysis and liver histopathology. *Food Nutri Sci* 2014, 5:185–195.

53. Zhang L, Hou D, Chen X, Li D, Zhu L, Zhang Y, Li J, Bian Z, Liang X, Cai X, Yin Y, Wang C, Zhang T, Zhu D, Zhang D, Xu J, Chen Q, Ba Y, Liu J, Wang Q, Chen J, Wang J, Wang M, Zhang Q, Zhang J, Zen K, Zhang CY: Exogenous plant MIR168a specifically targets mammalian LDLRAP1: evidence of cross-kingdom regulation by microRNA. *Cell Res* 2012, 22:107–126.

54. Markaverich BM, Crowley JR, Alejandro MA, Shoulars K, Casajuna N, Mani S, Reyna A, Sharp J: Leukotoxin diols from ground corncob bedding disrupt estrous cyclicity in rats and stimulate MCF-7 breast cancer cell proliferation. *Environ Health Perspect* 2005, 113:1698–1704.

55. WHO: World health statistics. 2012, Geneva, Switzerland: WHO press <http://whoint> (Last access August).

56. Brix AE, Nyska A, Haseman JK, Sells DM, Jokinen MP, Walker NJ: Incidences of selected lesions in control female Harlan Sprague–Dawley rats from two-year studies performed by the National Toxicology Program. *Toxicol Pathol* 2005, 33:477–483.

57. Chandra M, Riley MG, Johnson DE: Spontaneous neoplasms in aged Sprague–Dawley rats. *Arch Toxicol* 1992, 66:496–502.

58. Hayes TB: There is no denying this: defusing the confusion about atrazine. *Biosciences* 2004, 54:1139–1149.

59. Desaulniers D, Leingartner K, Russo J, Perkins G, Chittim BG, Archer MC, Wade M, Yang J: Modulatory effects of neonatal exposure to TCDD, or a mixture of PCBs, p, p'-DDT, and p-p'-DDE, on methylnitrosourea-induced mammary tumor development in the rat. *Environ Health Perspect* 2001, 109:739–747.

60. Schecter AJ, Olson J, Papke O: Exposure of laboratory animals to polychlorinated dibenzodioxins and polychlorinated dibenzofurans from commercial rodent chow. *Chemosphere* 1996, 32:501–508.

61. Kozul CD, Nomikos AP, Hampton TH, Warnke LA, Gosse JA, Davey JC, Thorpe JE, Jackson BP, Ihnat MA, Hamilton JW: Laboratory diet profoundly alters gene expression and confounds genomic analysis in mouse liver and lung. *Chem Biol Interact* 2008, 173:129–140.

62. Howdeshell KL, Peterman PH, Judy BM, Taylor JA, Orazio CE, Ruhlen RL, Vom Saal FS, Welshons WV: Bisphenol A is released from used polycarbonate animal cages into water at room temperature. *Environ Health Perspect* 2003, 111:1180–1187.

63. Harvell DM, Strecker TE, Tochacek M, Xie B, Pennington KL, McComb RD, Roy SK, Shull JD: Rat strain-specific actions of 17beta-estradiol in the mammary gland: correlation between estrogen-induced lobuloalveolar hyperplasia and susceptibility to estrogen-induced mammary cancers. *Proc Natl Acad Sci USA* 2000, 97:2779–2784.

64. Thongprakaisang S, Thiantanawat A, Rangkadilok N, Suriyo T, Satayavivad J: Glyphosate induces human breast cancer cells growth via estrogen receptors. *Food Chem Toxicol* 2013, 59C:129–136.

65. Popovics P, Rekasi Z, Stewart AJ, Kovacs M: Regulation of pituitary inhibin/activin subunits and follistatin gene expression by GnRH in female rats. *J Endocrinol* 2011, 210:71–79.

66. Walf AA, Frye CA: Raloxifene and/or estradiol decrease anxiety-like and depressive-like behavior, whereas only estradiol increases carcinogen-induced tumorigenesis and uterine proliferation among ovariectomized rats. *Behav Pharmacol* 2010, 21:231–240.

67. Deheuvels P: On testing stochastic dominance by exceedance, precedence and other distribution-free tests, with applications. In *Chapter 10 in Statistical Models and Methods for Reliability and Survival Analysis John Wiley & Sons*; 2013.

68. Duke SO, Rimando AM, Pace PF, Reddy KN, Smeda RJ: Isoflavone, glyphosate, and aminomethylphosphonic acid levels in seeds of glyphosate-treated, glyphosate-resistant soybean. *J Agric Food Chem* 2003, 51:340–344.

69. Kuenzig W, Chau J, Norkus E, Holowaschenko H, Newmark H, Mergens W, Conney AH: Caffeic and ferulic acid as blockers of nitrosamine formation. *Carcinogenesis* 1984, 5:309–313.

70. Baskaran N, Manoharan S, Balakrishnan S, Pugalendhi P: Chemopreventive potential of ferulic acid in 7,12-dimethylbenz[a]anthracene-induced mammary carcinogenesis in Sprague–Dawley rats. *Eur J Pharmacol* 2010, 637:22–29.

71. Chang CJ, Chiu JH, Tseng LM, Chang CH, Chien TM, Wu CW, Lui WY: Modulation of HER2 expression by ferulic acid on human breast cancer MCF7 cells. *Eur J Clin Invest* 2006, 36:588–596.

72. Eriksson L, Johansson E, Kettaneh-Wold N, Wold S: *Multi and Megavariate Data Analysis Part I - Principles and Applications.* Umea, Sweden: Umetrics AB; 2006.

73. Weljie AM, Bondareva A, Zang P, Jirik FR: (1)H NMR metabolomics identification of markers of hypoxia-induced metabolic shifts in a breast cancer model system. *J Biomol NMR* 2011, 49:185–193.

74. Wiklund S, Johansson E, Sjostrom L, Mellerowicz EJ, Edlund U, Shockcor JP, Gottfries J, Moritz T, Trygg J: Visualization of GC/TOF-MS-based metabolomics data for identification of biochemically interesting compounds using OPLS class models. *Anal Chem* 2008, 80:115–122.

75. Eriksson L, Johansson E, Kettaneh-Wold N, Trygg J, Wikström C, Wold S: *Multi- and Megavariate Data Analysis Part II. Advanced Applications and Method Extensions.* Umea, Sweden: Umetrics; 2006.

9

Modelling and mapping of plant phenological stages as bio-meteorological indicators for climate change

Winfried Schröder[*], Gunther Schmidt and Simon Schönrock

Abstract

Background: In Hesse, a federal state in central Germany, the average air temperature of the period 1991 to 2009 was by 0.9°C higher compared to that of 1961 to 1990. A further rise in air temperature of up to 3.7°C compared to the reference period 1971 to 2000 is expected until the end of the twenty-first century. This may affect the beginning and length of phenological stages of plants. Hence, this project should analyse and model spatiotemporal trends of plant phenology as being an indicator for climate change-related biological effects. Meteorological data together with data on 35 phenological phases of plants indicating different phenological seasons and observed at 6,500 sites in Germany (553 in Hesse) between 1961 and 2009 were analysed in a GIS. Estimations on phenological developments in the future periods 2031 to 2060 and 2071 to 2100 were based on data from four regional climate models.

Results: Thirty-one out of 35 phases started earlier in the years 1991 to 2009 compared with 1961 to 1990. These shifts were stronger in Hesse (8 days) than in Germany (6 days). As winter phases tend to shift towards the end of the year, a prolongation of the vegetation period of up to 3 weeks was observed. More than 70% of the phases were correlated with air temperature by $r \geq 0.5$, more than 50% even by $r \geq 0.7$. Since the 1990s, phenological shifts and regional differences in phase onsets amplified. In many cases, the shifts between 2071 to 2100 and 1961 to 1990 are expected to be at least twice as high as those between 1991 to 2009 and 1961 to 1990.

Conclusions: The presented approach allows revealing statistical relationships between air temperatures and phenological onsets. Thus, shifts in plant phenology are an appropriate bioindicator to map early signs of ecosystem transitions under climate change. The phenological records allow estimating future trends of plant phenological development. Using phenological maps as presented in this article, efficient adaption strategies may be planned and implemented in terms of, e.g. adjusting delineation, shape and allocation of protected areas.

Keywords: Bioindicator; Climate projections; Geostatistics; Phenological shift

Background

Climate change impacts on individual species comprise shifts in phenology, productivity, distribution and, thus, biodiversity. Therefore, phenological phases are used as indicators for detecting ecological impacts of climate change on flora and fauna such as plants, migratory birds or fishes and, consequently, on ecosystems reflecting the results of manifold combinations of environmental interactions [1]. Changes in the timing of phenological stages of plants such as foliation, flowering, fruit ripening, colour changing and leaf fall are recognized as globally coherent ecological fingerprints of climate change. Plant development in terms of phenological stages as exemplified is a rather sensitive bio-meteorological response to environmental variation [2], which is ecologically meaningful since changes in phenophases serve as both forcing and inhibiting ecological processes [3] across spatial scales from individuals to landscapes [4]. Hence, shifts in phenology might be of importance for the implementation of mitigation and adaptation measures in compliance with regulations such as the European Habitat

* Correspondence: wschroeder@iuw.uni-vechta.de
Landscape Ecology, University of Vechta, P.O. Box 1553, Vechta 49364, Germany

Directive on the one hand [5] and for crop yields and food security issues on the other hand [6,7].

According to the etymologic origin of the term phenology, i.e. the ancient Greek word 'phainestai' meaning 'to appear', which was introduced by [8], plant phenology examines annually and periodically reappearing events in growth and development of plants [9-11]. The study of the timing of recurring biological events encompasses the causes of their timing with regard to biotic and abiotic drivers. The interrelation among the phases of the same or different species is called phenology [12]. Phenology is, thus, an integrative environmental science [13]. Phenological observations corroborated on the one hand that phenological phases can exhibit remarkable interannual variability and large spatial differences due to individual characteristics such as genes and age and environmental factors such as meteorological conditions at the micro- and macro-scale, soil conditions, water supply, diseases, and competition. On the other hand, the seasonal development of plants is, however, mainly influenced by air temperature, photoperiod and precipitation. In particular, spring development in the Northern Hemisphere mid-latitudes mainly depends on the temperatures in winter and spring [14-17].

Recent studies corroborated that the beginning of phenological phases, such as blooming or foliation, is closely related to air temperature [1,2,18-25]. As higher temperatures advance the course of phenological events [26], phenological data reflect biological response to this feature of climate change [27] and, therefore, can be used for climate bio-monitoring [28]. Regarding the sensitivity of spring phenology of plants to warming across temporal and spatial climate gradients in independent databases, [19] found good congruence, despite significant differences in species richness and geographic distribution and concluded that this should encourage 'to move beyond basic statistical diagnosis of trends towards explicit predictions into the future'.

Between 1906 and 2005, the global mean air temperature increased by 0.74°C [29]. In Germany, the long-term mean air temperature of the period 1991 to 2009 was 0.9°C higher compared to that of the climate reference period of 1961 to 1990. The same holds true for the German federal state Hesse. According to different climate projections and emission scenarios, a further rise of the long-term mean air temperature in Hesse from 9.1°C in the period of 1991 to 2009 to 12°C at the end of the twenty-first century is expected. Given that background, within a GIS environment, this investigation was focused on(1) the development of measured (1961 to 2009) and projected (2031 to 2060 and 2071 to 2100) air temperatures and phenological observations (1961 to 2009), (2) the spatial variation of the beginning of phenological phases in different natural landscapes, (3) the correlation between air temperatures and the onset

of phenological phases and the (4) calculation of maps depicting the spatial patterns of measured (1961 to 2009) and projected (2031 to 2060 and 2071 to 2100) phenological phases.

Results and discussion
Development of measured (1961 to 2009) and projected (2031 to 2060 and 2071 to 2100) air temperatures and phenological observations (1961 to 2009)

The annual mean air temperatures increased from 8.2°C (1961 to 1990) to 9.1°C (1991 to 2009) in both Germany and Hesse. Regarding the natural landscapes of Hesse, the regional yearly means range between 7.7°C and 10.1°C (1961 to 1990) and 8.6°C and 10.9°C (1991 to 2009). Air temperatures were strongly associated with elevation patterns (Figure 1). Depending on the model used for temperature projection, the long-term annual mean air temperature rise, comparing the reference periods 1971 to 2000 (8.5°C according to measurements and 8.6°C according to projections) and 2071 to 2100, varies between 3.2°C (ECHAM5/COSMO-CLM) and 3.7°C (WETTREG2010).

Reflecting the measured temperature increase between the periods 1961to 1990 and 1991 to 2009 (Figure 1), 31 of the examined 35 phenological phases advanced towards the beginning of the year as illustrated in Figure 2 for the ten indicator phases. On average, of all the 35 phases, the shifts in Hesse are even stronger (about 8 days) than those in Germany (about 6 days). A lot of phases, especially in Hesse, showed even shifts of more than 10 days. The strongest shifts were detected for phases in spring and early summer. In the further course of the year, some phases, especially in late summer and autumn, show weaker shifts. At the end of the phenological year in late autumn and winter, respectively, some phases even showed a reverse shift towards the end of the year. Along with these phenological shifts, the vegetation period extended. In some Hessian natural land units (Lahn Valley, Westerwald and Odenwald, and Spessart and South Rhoen), the prolongation lasted up to 3 weeks (Figure 3).

Correlation between air temperature and plant phenology

The development of the measured (1961 to 2009) air temperatures and phenological observations as exemplarily depicted in Figure 4 for the beginning of flowering of *Malus domestica* (phase 62) was quantified by the use of Pearson's correlation coefficients (Table 1) quantifying the strength of the statistical association between the onset of the phenophases and air temperatures during those months with highest correlations. For phase 62 (apple bloom), the maximum correlation coefficient was calculated for the months March to June in the years 1971 to 2000, amounting to −0.87.

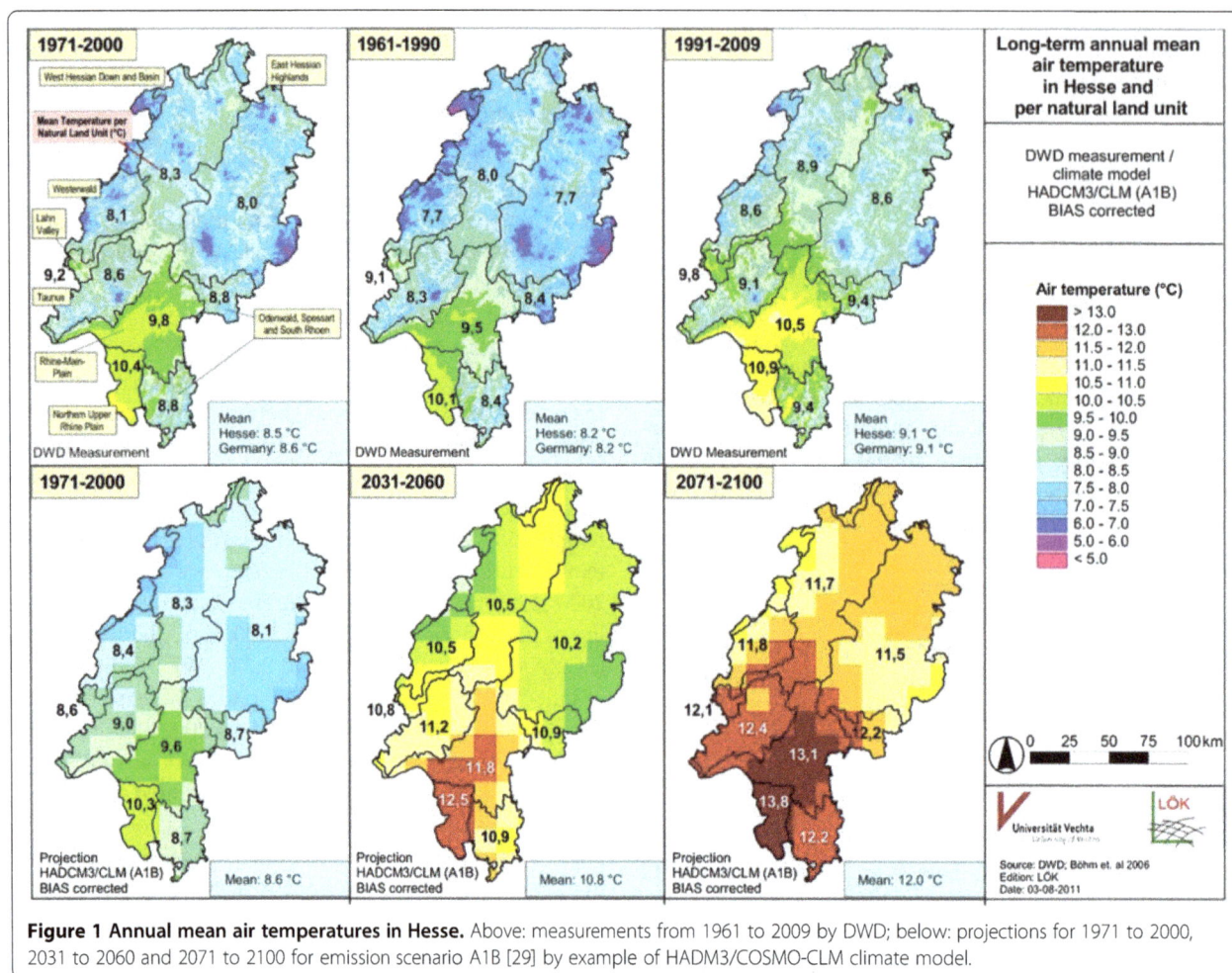

Figure 1 Annual mean air temperatures in Hesse. Above: measurements from 1961 to 2009 by DWD; below: projections for 1971 to 2000, 2031 to 2060 and 2071 to 2100 for emission scenario A1B [29] by example of HADM3/COSMO-CLM climate model.

The respective regression model (Figure 5) explains 76% of the variance.

The bivariate statistical analysis revealed the statistical associations of at least medium strength ($r \geq 0.5$) for more than 70% of the analysed phases in each of the three considered periods. More than 50% even showed a high correlation ($r \geq 0.7$) between air temperatures and phenological onset. The result of the analysis corresponds with the findings for the past phenological development described in the 'Development of measured (1961 to 2009) and projected (2031 to 2060 and 2071 to 2100) air temperatures and phenological observations (1961 to 2009)' section: Almost all phases indicating an earlier beginning showed negative correlation coefficients. These findings corroborate, spatially differentiated, the hypothesis that air temperature is a significant driver for phenological development. However, phases with less intense shifts in the further course of the year ('Development of measured (1961 to 2009) and projected (2031 to 2060 and 2071 to 2100) air temperatures and phenological observations (1961 to 2009)' section) showed only weak correlation coefficients, especially in autumn. Eventually, two of those

phases with shifts towards the end of the year (phases 73 and 226) revealed positive coefficients. This implies that high temperatures in autumn have reverse effects on these phases. Whereas high temperatures stimulate the beginning of spring and summer phases, they retard the onset of late autumn and winter phases [11].

The correlation analyses in this study considered spatial auto-correlation. The results of the respective computations according to [30] corroborated that auto-correlation considerably reduced the degrees of freedom. However, the correlations remained statistically significant ($p < 0.01$). For instance, no relevant differences were found for the significance of Spearman's ($r_S = -0.858$) and Person's ($r_P = -0.873$) correlation coefficients regarding the statistical association of apple bloom and air temperature from 1971 to 2000.

Spatial patterns of measured (1961 to 2009) and projected (2031 to 2060 and 2071 to 2100) phenological phases

Based on the results of the regression analysis ('Correlation between air temperature and plant phenology'

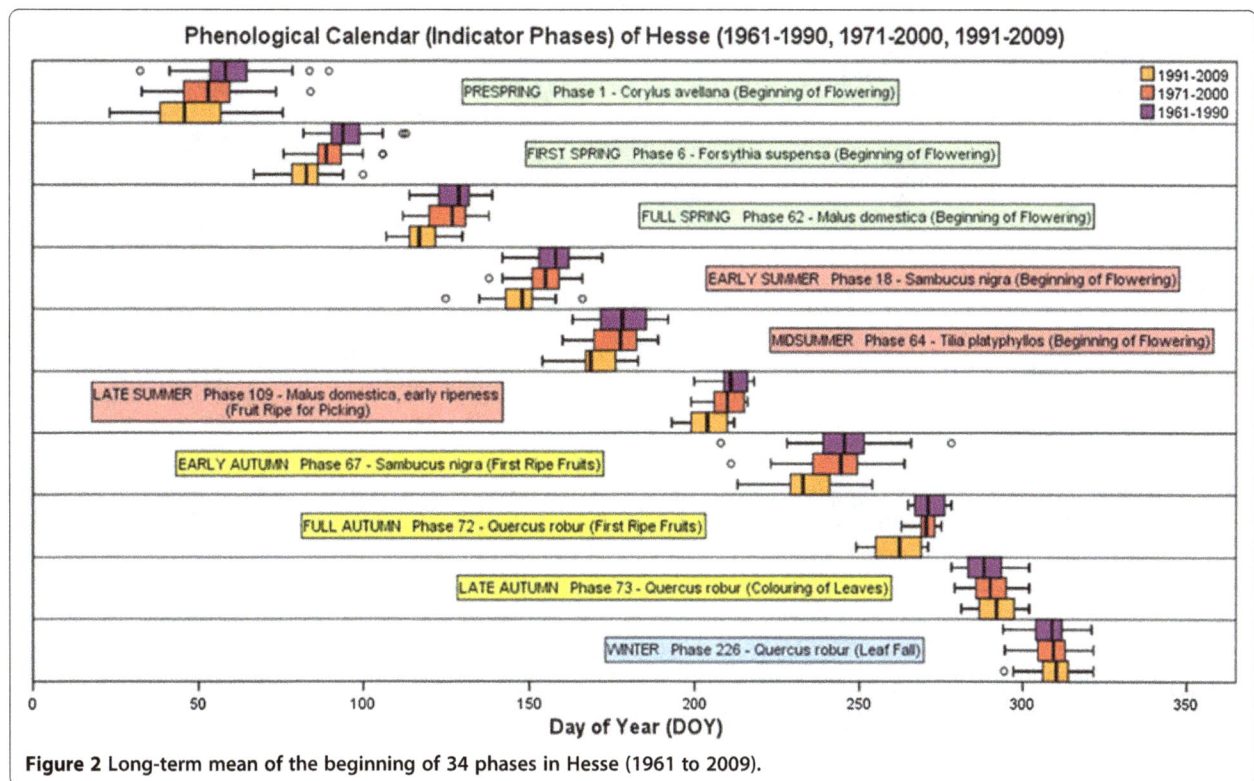

Figure 2 Long-term mean of the beginning of 34 phases in Hesse (1961 to 2009).

section), regression kriging was applied to 23 of the 35 investigated phases, indicating a correlation coefficient of at least 0.5 between air temperatures and phenological onset. The results of the phenological mapping are illustrated by the example of apple *(Malus domestica)* flowering. The upper row of Figure 6 depicts long-term phenological maps of Hesse for three past periods (1961 to 1990, 1971 to 2000 and 1991 to 2009). Regarding these three maps, in Hesse, apple flowering began 8 days earlier in the period 1991 to 2009 compared to the period 1961 to 1990 (even 10 days when comparing only the averaged observations in these periods). As could be expected, topographical patterns were reflected in the phenological maps: Lower regions indicating rather warm temperatures (e.g. *Rhine-Main river valleys*) were characterized by early apple bloom. In comparison, mountainous regions (e.g. the *Westerwald*) showed late-phase beginnings. These observations coincide with the regionally differentiated analysis based on the respective natural land units. On average, of the period 1971 to 2000, for instance, apple flowering occurred nearly 3 weeks earlier (April 21) in the *Northern Upper Rhine Plain* than that in the *Eastern Hessian Highlands* (May 10). Furthermore, the phenological development reflects the rise of air temperatures, as long-term phase shifts between the periods 1971 to 2000 and 1991 to 2009 are more distinct than those between the periods 1961 to

1990 and 1971 to 2000. The maximum shift between 1961 to 1990 and 1971 to 2000 lasted for 4 days (*Taunus*), whereas between the periods 1971 to 2000 and 1991 to 2009, the maximum shift lasted for 9 days (*Westerwald*). In summary, most regions of the Hessian uplands are affected by stronger shifts of the phase onset (e.g. *Westerwald, Western Hessian Down and Basin, East Hessian Highlands*), whereas valleys and lowlands in the south of Hesse are less affected.

The lower three maps of Figure 6 show, by example of one (HADCM3/CLM) of the four climate models considered, the projected future phenological development of apple bloom in Hesse for the future periods 2031 to 2060 and 2071 to 2100 as well as for the reference period 1971 to 2000. According to this climate model, apple flowering in Hesse will shift on average about 11 days towards the beginning of the year (from May 5 to April 24) between the periods 1971 to 2000 and 2031 to 2060, and another 7 days (April 17) compared to the period 2071 to 2100. For some parts of Hesse, especially in the lowlands, apple flowering should begin even before April 10 in the period 2031 to 2060, and an average onset at the end of March and the very beginning of April is projected for the period 2071 to 2100. Depending on the respective climate model used in the examination at hand, the long-term mean shifts of apple bloom in Hesse range between 13 and

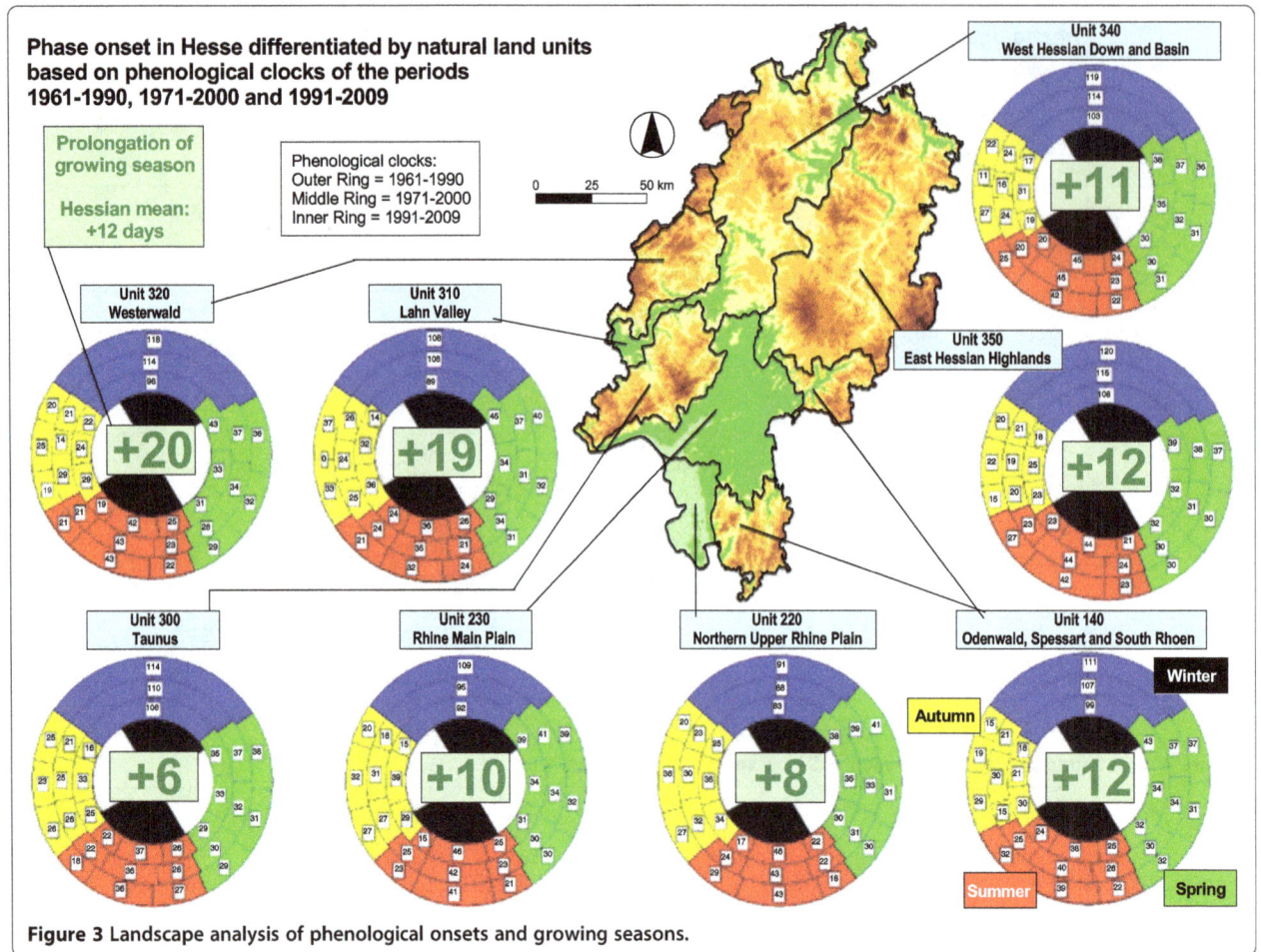

Figure 3 Landscape analysis of phenological onsets and growing seasons.

18 days when comparing the periods 1971 to 2000 and 2071 to 2100.

Table 2 contains the shifts for 22 additional phases indicated as differences between the respective long-term means of the period 1961 to 1990 and the future period 2071 to 2100. The differences are all negative. Consequently, according to the results of the four climate models, the observed tendency of the past phenological development ('Development of measured (1961 to 2009) and projected (2031 to 2060 and 2071 to 2100) air temperatures and phenological observations (1961 to 2009)' section) will obviously continue until the end of the twenty-first century. The ECHAM5/COSMO-CLM-model (ECLM_K) and the REMO/UBA-Model (RUBA_K) are rather conservative, projecting mostly less-intense shifts, whereas shifts projected by HADM3/COSMO-CLM (HCLM_K) and WETTREG2010 (WETTR) are more distinct. Nevertheless, with only few exceptions, the shifts of the assessed phases for all the models between the periods 2071 to 2100 and 1961 to 1990 are at least twice as high as they are between 1991 to 2009 and 1961 to 1990. For many phases, they are even three times higher or more. The

most affected phases are due to occur about 1 month earlier and more (hazel flowering, 40 days earlier).

Conclusions

For analysing spatial patterns of the plant phenological development the geostatistical estimation of phenological surface maps is of remarkable importance, especially for those natural land units with small extents and a low number of observation sites. Several statistical values could prove the quality of the surface maps. Referring to the bivariate statistical analysis, the applied approach of using air temperature data of only those months that showed strong correlation between air temperature and phase onset instead of using annual mean air temperature data enabled powerful regression models.

The four different climate models used for projecting the future phase onsets showed different characteristics. As two of them projected rather moderate shifts until the end of the twenty-first century (ECHAM5/CLM and REMO/UBA), the two others (HADCM3/CLM and WETTREG 2010) projected stronger shifts. The

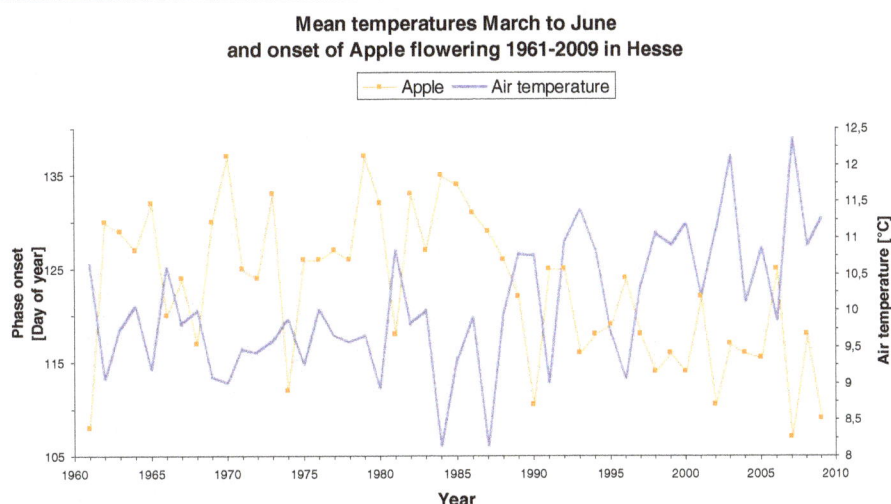

Figure 4 Mean temperatures (March to June) and onset of flowering of *Malus domestica* (1961 to 2009).

phenological monitoring should be complemented by a standardized metadata acquisition as done in the European moss survey [31] to promote the interpretation of phenological data. One further measure to support correctness, objectivity and reliability of plant phenological observations as well as the spatial density of monitoring networks is the use of digital repeat photography [32].

The application of regression models calculated for the years 1961 and 2009 to project changes in 2031 to 2100 may be misleading, since these constitute an extrapolation that may bias future impacts of climate change on the timing of phenological phases, given that the range of temperatures projected for the future would be significantly higher than those observed in the past and, thus, were not covered by the range of observational data used to derive the regression models. However, it should be recognized that, without any exception, according to basic and widely accepted knowledge of the philosophy of science, spatial and temporal extrapolations beyond time and space of measurements never can be fully justified in a strict sense of logical, inductive, reasoning *ex ante* [33-37]. Their validation needs empirical investigations ex post which are, in case of temporal extrapolations in terms of forecasts, prognoses and projections, impossible during the respective presence.

The presented approach allows revealing statistical relationships between air temperatures and phenological onsets and, by calculating residuals, indicates the extent of influence of other drivers affecting phenological development. Additionally, even though the residuals could not be explained in terms of identifying latent factors, the spatial structure of the residuals was considered within the regression kriging approach used for mapping

recent and potential future spatial and temporal trends of phenological developments.

In this investigation, only one emission scenario (A1B) was considered. To establish a more comprehensive range of projections of the impacts of climate change, the study could have included other scenarios, particularly those that are more moderate compared to A1B as for instance B1 and B2. The results presented in this paper rely on a study which was conducted in the framework of the research program INKLIM-A founded by the federal state Hesse which brought together experts of several environmental sciences, including experts for climate modelling. Based on comprehensive knowledge represented in the research consortium, it was a unanimous decision that it would make no sense to consider the B1 scenario since the assumptions made for this scenario are meanwhile outdated. However, a wide range of temperature variation was considered by applying the four climate models reflecting, both, more conservative and more extreme temperature developments.

As documented in this investigation and in many studies across the world (e.g. America, Asia, Europe), the phenological records allow to estimate future trends of plant phenological development and related ecological processes for agriculture, forestry, human health and the global economy [22]. Thus, bio-monitoring of climate change using plant phenology as an indicator should further be refined [38]. We did not consider the potential impact of the geographic distribution of the phenological observations across Hesse, assuming that the observations were randomly distributed [39]. This assumption could be proved in a succeeding study. A refinement of the regression models presented in this study might be achieved by additional calculations for spatially and/or

Table 1 Correlation coefficients (Pearson) between air temperature and phenological onset (1961 to 2009)

Phase	r value		
	1961 to 1990	1971 to 2000	1991 to 2005
1	−0.79	−0.79	−0.71
2	−0.75	−0.73	−0.64
6	−0.85	−0.88	−0.82
52	−0.70	−0.69	−0.62
7	−0.73	−0.77	−0.72
115	−0.52	0.61	−0.51
62	−0.82	−0.87	−0.83
13	−0.73	−0.77	−0.70
15	−0.83	−0.86	−0.83
19	−0.46	−0.53	−0.45
18	−0.75	−0.77	−0.74
123	−0.60	−0.73	−0.70
20	−0.56	−0.65	−0.63
64	−0.71	−0.76	−0.63
100	−0.73	−0.73	−0.61
109	−0.76	−0.74	−0.62
65	−0.11	−0.07[a]	−0.06[a]
67	−0.55	0.59	−0.57
177	−0.39	−0.38	−0.33
72	−0.30	−0.34	−0.34
68	−0.38	−0.40	−0.43
73	0.12	0.14	0.13
94	0.12	0.16	0.23
226	-	-	0.15
54	−0.82	−0.86	−0.84
56	−0.83	−0.85	−0.81
60	−0.84	−0.88	−0.84
102	−0.69	−0.76	−0.68
103	−0.65	−0.74	−0.60
104	−0.63	−0.63	−0.58
107	−0.51	−0.63	−0.51
108	−0.30	−0.39	−0.28
171	−0.44	−0.51	−0.62
172	−0.62	−0.78	−0.85
205	-	−0.78	−0.36

[a]Correlation is not significant.

timely defined clusters in Hesse/Germany [40] could prove that such specified models could be more effective than the general models and, thus, might be the basis for better plant phenology projections.

Climate change impacts on ecosystems are widely recorded in terms of phenological shifts of organisms worldwide. As shown in this investigation, changes in plant phenology evidently reflect a warming trend. Thus, phenology is useful as a primary tool for mapping early signs of ecosystem transitions under climate change across areas of large spatial extend. However, besides shifts in timing of phenological events, plants may also shift their geographical distribution towards more favourable climates or adapt to the altered local conditions. Shifts through phenotypic plasticity occur prior to and more rapidly than the more profound changes in species distribution and genetics [41] corroborated that different temperate plant genotypes require varying amount of heat energy for starting annual growth and reproduction due to adaptation and other ecological and evolutionary processes along climatic gradients, which is quantitatively reflected in the timing of phenophases. Accordingly, earlier timing indicates higher efficiency, i.e. less heat energy needed to trigger phenophase transitions.

Since temperature increase is expected to continue until the end of the twenty-first century, distinct effects on flora distribution are, thus, very likely and will result in changes in ecological processes, such as species migration or extinction, and in agricultural management. Cryophilic species are expected to move into northern latitudes and high altitudes, whereas thermophilic species are due to emigrate from southern regions into the north, repressing probably domestic species [42].

The phenological development of plants influences the mass and energy cycle of the biosphere [43]: Plant vegetative cycles determine the flows of matter, e.g. carbon dioxide and water, and energy between land surface and atmosphere. Additionally, canopy development and senescence are linked to seasonal changes in surface resistance and roughness, as well as the turbulent exchange of water and energy. Ma et al. [43] calculated that the leaf onset days for wheat, barley and rapeseed in Germany advanced by 1.6, 3.4 and 3.4 days per decade, respectively, during 1961 to 2000. This modelled trend of advanced onset days could be corroborated by observations from the International Phenology Gardens in Europe [44] reported that the temperature increase since 1980 has already lowered the worldwide wheat yields by 5.5%, without considering the effect of increasing CO_2 levels, and by 2.5% when considering C-fertilization. Between 1981 and 2002, the average global wheat yield decreased by 88 kg/ha. Wheat in the developing countries is expected to suffer most among major crops from rising temperatures in low-latitude countries. Up to 2050, the wheat yield levels are expected to decrease by 5% to 9% for rain-fed systems. Thus, plant phenological observations are crucial not only for nature protection but also for agricultural management.

Using phenological maps as presented in this article, appropriate adaption strategies may be planned and implemented [45] in terms of, e.g. adjusting delineation,

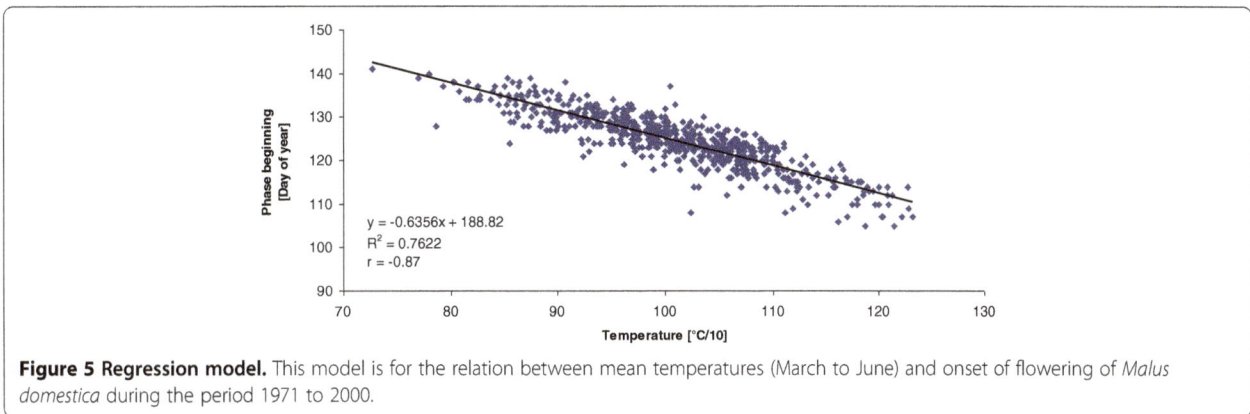

Figure 5 Regression model. This model is for the relation between mean temperatures (March to June) and onset of flowering of *Malus domestica* during the period 1971 to 2000.

shape and geographical position of protected areas [46,47]. In terms of agricultural management, the selection of crops and cultivars must be adjusted to the changed climatic conditions. Furthermore, farmers should cope with increased problems with insect pests potentially occurring due to increased air temperatures [39,48]. Another threat is the increasing risk of frost damage due to the earlier occurrence of phenological events [49].

Referring to resources management, irrigation in the summer will be necessary for larger areas and longer periods as precipitation is expected to decrease in Hesse during the vegetation period [50,51]. On the other hand, there are also positive effects of climate change: prolongation of the growing season [52] might lead to increased yields in some regions, and cultivation of new fruit varieties might be possible [48,53,54].

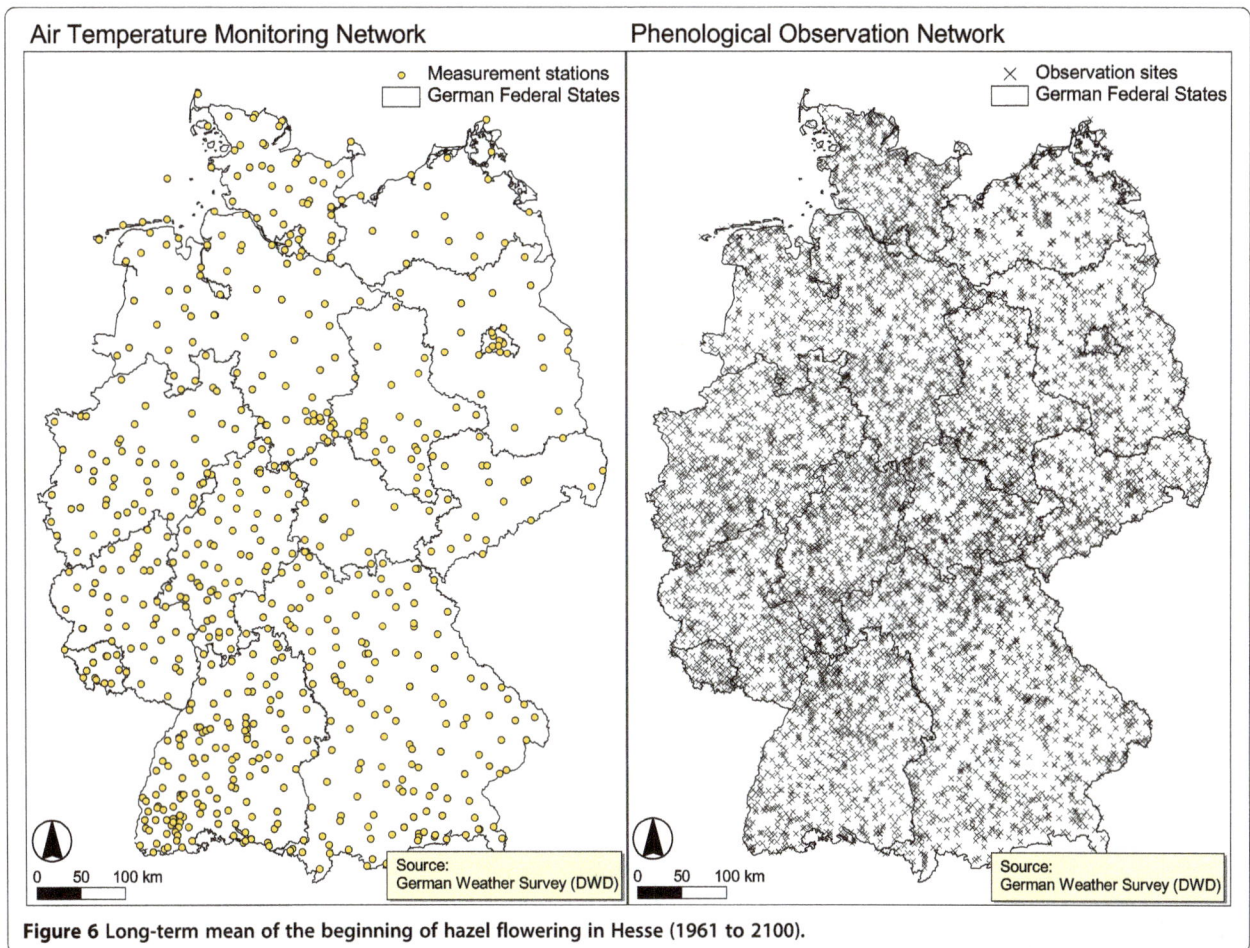

Figure 6 Long-term mean of the beginning of hazel flowering in Hesse (1961 to 2100).

Table 2 Projected long-term mean shifts of 23 phases between 2071 to 2100 and 1961 to 1990

Phase	1	2	6	52	7	115	62	13	15	19	18	123	20	64	100	109	67	54	56	60	102	103	104
Differences (days) between 1991 to 2009 and 1961 to 1990 (DWD-observation)																							
Observed	−14	−9	−11	−10	−8	−6	−10	−8	−9	−7	−10	−4	−7	−9	−8	−5	−11	−6	−7	−7	−12	−9	−10
Differences (days) between 2071 to 2100 (projection) and 1961 to 1990 (DWD-observation)																							
ECLM_K	−30	−22	−23	−13	−16	−7	−16	−14	−15	−12	−17	−12	−15	−19	−25	−26	27	−15	−15	−19	−26	−21	−24
HCLM_K	−32	−23	−26	−18	−19	−11	−20	−17	−18	−14	−19	−12	−15	−20	−25	−27	29	−17	−18	−21	−25	−21	−24
RUBA_K	−33	−24	−25	−13	−17	−7	−15	−14	−15	−12	−16	−11	−14	−17	−22	−23	25	−15	−14	−19	−23	−18	−20
WETTR_00	−40	−28	−30	−18	−21	−11	−21	−18	−19	−15	−21	−13	−16	−22	−25	−27	27	−20	−19	−25	−27	−21	−25
WETTR_05	−40	−28	−30	−16	−20	−10	−19	−18	−17	−13	−19	−11	−15	−21	−26	−26	27	−19	−17	−23	−26	−21	−25

ECLM_K, HCLM_K, and RUBA_K WETTR are the climate models applied for the SRES scenario A1B [29]. ECHAM5/COSMO-CLM (bias corrected), HADM3/COSMO-CLM (bias corrected), REMO/UBA (bias corrected) and WETTREG2010 run_00 and run_55 [56-59]. DWD, Deutscher Wetterdienst (National Meteorological Service of the Federal Republic of Germany).

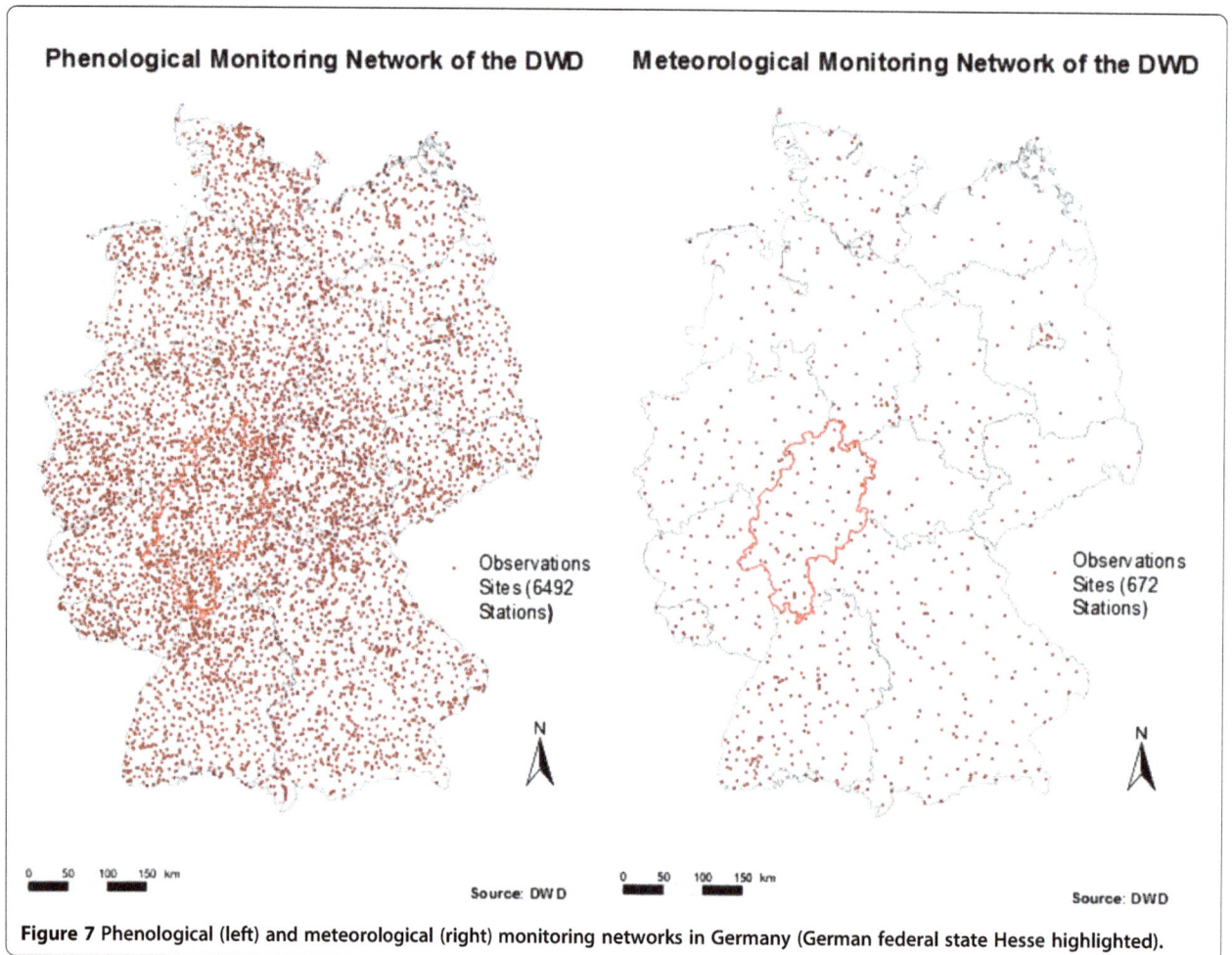

Figure 7 Phenological (left) and meteorological (right) monitoring networks in Germany (German federal state Hesse highlighted).

Methods

Data

For analysing and mapping the past phenological development, observations on plant phenology as well as temperature measurements collected by the German Weather Survey (Deutscher Wetterdienst (DWD)) from 1961 to 2009 (phenological data up to 2005) were used (Figure 7). The German phenological monitoring network comprises almost 6,500 sites (553 in Hesse), observing more than 270 phenophases of wild-growing plants and crop plants (agricultural plants, fruits and vine).

Table 3 Phenological phases investigated (IDs, plant, phase and phenological season)

Index	Plant	Phase	ID	Vegetation layers	Phase rank	Phenological season
	Wild growing plants					
1	*Corylus avellana*	Beginning of flowering	B	Shrub/tree layer	Indicator phase	Prespring
2	*Galanthus nivalis*	Beginning of flowering	B	Herb layer	Alternative Phase	Prespring
6	*Forsythia suspensa*	Beginning of flowering	B	Shrub layer	Indicator phase	First spring
52	*Ribes uva-crispa*	Beginning of unfolding of leaves	BO	Shrub layer	Alternative phase	First spring
7	*Aesculus hippocastanum*	Beginning of unfolding of leaves	BO	Tree layer		First spring
115	*Anemone nemorosa*	Beginning of flowering	B	Herb layer		First spring
62	*Malus domestica*	Beginning of flowering	B	Tree layer	Indicator phase	Full spring
13	*Quercus robur*	Beginning of unfolding of leaves	BO	Tree layer	Alternative phase	Full spring
15	*Syringa vulgaris*	Beginning of flowering	B	Shrub layer		Full spring
19	*Alopecurus pratensis*	Flowering general blossom	AB	Herb layer		Full spring
18	*Sambucus nigra*	Beginning of flowering	B	Shrub layer	Indicator phase	Early summer
123	*Robinia pseudoacacia*	Beginning of flowering	B	Tree layer	Alternative phase	Early summer
20	*Dactylis glomerata*	Flowering general blossom	AB	Herb layer		Early summer
64	*Tilia platyphyllos*	Beginning of flowering	B	Tree layer	Indicator phase	Midsummer
100	*Ribes rubrum*	Fruit ripe for picking	F	Shrub layer	Alternative phase	Midsummer
109	*Malus domestica (early ripeness)*	Fruit ripe for picking	F	Tree layer	Indicator phase	Late summer
65	*Calluna vulgaris*	Beginning of flowering	B	Herb layer		Late summer
67	*Sambucus nigra*	First ripe fruits	F	Shrub layer	Indicator phase	Early autumn
177	*Rosa canina*	First ripe fruits	F	Shrub layer		Early autumn
72	*Quercus robur*	First ripe fruits	F	Tree layer	Indicator phase	Full autumn
68	*Aesculus hippocastanum*	First ripe fruits	F	Tree layer	Alternative phase	Full autumn
73	*Quercus robur*	Colouring of leaves	BV	Tree layer	Indicator phase	Late autumn
226	*Quercus robur*	Leaf fall	BF	Tree layer	Indicator phase	Winter
94	*Triticum aestivum*	Emergence	AU	Herb layer	Indicator phase	Winter
	Fruits					
54	*Prunus avium*	Beginning of flowering	B	Tree layer		First spring
56	*Prunus cerasus*	Beginning of flowering	B	Tree layer		First spring
60	*Pyrus communis*	Beginning of flowering	B	Tree layer		First spring
102	*Prunus avium (early ripening)*	Fruit ripe for picking	F	Tree layer		Midsummer
103	*Prunus avium (late ripening)*	Fruit ripe for picking	F	Tree layer		Midsummer
104	*Prunus cerasus*	Fruit ripe for picking	F	Tree layer		Midsummer
107	*Pyrus communis (early ripening)*	Fruit ripe for picking	F	Tree layer		Early autumn
108	*Pyrus communis (late ripening)*	Fruit ripe for picking	F	Tree layer		Full autumn
	Vine					
171	vine (Muller-Thurgau)	Beginning of sprouting	A	Shrub layer		First spring
172	vine (Muller-Thurgau)	Beginning of flowering	B	Shrub layer		Early summer
205	vine (Muller-Thurgau)	Grape gathering	L	Shrub Layer		Full autumn

The phenological observations were conducted two or three times a week within a defined area by volunteers according to a guideline [55]. For the analysis at hand, 35 phases were selected. They comprise so-called indicator phases for different phenological seasons and respective alternative phases [55]. Additional phases were selected to describe the phenology of fruits and vine plants representing high economic importance for fruit growers (Table 3).

The phenological data sets were provided as vector data (point layer); climate data were provided as grid data sets with a spatial resolution of 1×1 km^2. For mapping the potential future phenological development during the years 2031 to 2060 and 2071 to 2100, the results for the SRES A1B emission scenario [29] processed by four models (REMO/UBA, ECHAM5/COSMO-CLM, HADM3/COSMO-CLM and WET-TREG2010 (two runs)) were used. The modelled temperature maps were provided as grid data with a spatial resolution of 20×20 km [56-59]. Before statistical analysis ('Statistical analyses' section), the phenology data were checked for quality by use of three approaches, testing whether (1) each observation site covers at least 90% of the respective long-term period (1961 to 1990, 1971 to 2000 and 1991 to 2005) (temporal representativeness), (2) the onset at one site differed conspicuously from the average beginning at the surrounding the observation sites (neighbourhood analysis) and (3) the onset was remarkably early or late compared to the average (outlier analysis). This quality control complemented the quality check conducted routinely by the DWD before offering the data. The plausibility analysis finally resulted in the exclusion of 100 long-term datasets [60].

Statistical analyses

After quality control as described in the 'Data' section, the meteorological and phenological data were analysed by *descriptive statistics* to detect the trends in the past phenological development. After having proved the normal distribution of the data, the strength of the statistical association between air temperature and each of the respective phenophase for the periods 1961 to 1990, 1971 to 2000 and 1991 to 2009 was computed by the use of *bivariate correlation analysis* (Pearson's product-moment correlation coefficients) and modelled by the use of *linear regression analysis* [61]. To enhance the reliability of the results, the computations comprised temperature and phenological data covering the whole territory of Germany instead of only those of Hesse. The phenological point data and temperature grids were intersected within a GIS to estimate the air temperature values for each phenological observation site.

In environmental systems, *auto-correlation* is a widespread phenomenon [62,63] which, in statistics, is defined as the similarity of, or correlation between, the values of a process at neighbouring points in time or space. Positive auto-correlation means that the individual observations contain information which is part of the other, timely or spatial neighbouring, observations. Subsequently, the effective sample size is less than the number of the realized observations. Negative auto-correlation can have the opposite effect, thus making the effective sample larger than the realized sample. Therefore, auto-correlation can have several implications for, e.g. statistical inference testing and regression analysis [64]. It could be shown that positive spatial auto-correlation enhances type-I errors, so parametric statistics such as Pearson correlation coefficients are declared significant when they should not be [65]. Thus, in this investigation, spatial auto-correlation was considered in the correlation analyses according to [30].

The regression analysis was based only on those months showing the strongest correlation between air temperatures and the respective phenological onset instead of only using annual or monthly mean air temperatures. Accordingly, long-term mean temperatures were calculated for each sequence of these related months (e.g. averaged temperatures from March to June for the period 1971 to 2000) and then related to the respective phenological observations for each of the three periods. The resulting r values were classified as follows: $r < 0.20$, very low correlation; $0.20 \geq r \geq -0.49$, low; $0.50 \geq r \geq 0.69$, medium; $0.70 \geq r \geq 0.89$, high; and $r \geq 0.9$, very high correlation [66].

For those phases showing a significant and at least medium correlation ($r \geq 0.5$), phenological maps for each period (1961 to 1990, 1971 to 2000, 1991 to 2009, 2031 to 2060 and 2071 to 2100) were calculated in a GIS by *regression kriging* [67,68]. The regression equation derived for each phase and period was thereby applied to the long-term mean temperature grids to calculate a surface map for the respective phenophase. Then, these regression maps were added by kriging maps depicting the spatial structure of the residuals of the regression models based on auto-correlation functions determined by use of variogram analysis. The potential future phenological development was estimated by applying the regression equations of the reference period 1971 to 2000 to the air temperature grids of the periods 2031 to 2060 and 2071 to 2100 for each of the four climate projections as calculated by the use of REMO/UBA, ECHAM5/COSMO-CLM, HADM3/COSMO-CLM and WET-TREG2010 [56-59]. Additionally, the past and potential future phenological developments were spatially differentiated by intersection of the respective phenological maps with a map on the natural land units [69] of Hesse.

Competing interests

The authors declare that they have no competing interests.

Authors' contributions

WS drafted the manuscript and led the project. SS carried out the statistical analyses, and GS participated in its design and coordinated the investigations. All authors read and approved the final manuscript.

References

1. Ahas R, Aasa A: The effects of climate change on the phenology of selected Estonian plant, bird and fish populations. *Int J Biometeorol* 2006, 51:17–26.
2. Holopainen J, Helama S, Lappalainen H, Gregow H: Plant phenological records in northern Finland since the 18th century as retrieved from databases, archives and diaries for biometeorological research. *Int J Biometeorol* 2013, 57:423–435.
3. Parmesan C: Influences of species, latitudes and methodologies on estimates of phenological response to global warming. *Global Change Biol* 1860–1872, 2007:13.
4. Rosenzweig C, Casassa G, Karoly DJ, Imeson A, Liu C, Menzel A, Rawlins S, Root TL, Seguin B, Tryjanowski P: Assessment of observed changes and responses in natural and managed systems. In *Climate Change 2007: Impacts, Adaptation and Vulnerability. Contribution of Working Group II to the Fourth Assessment Report of the Intergovernmental Panel on Climate Change.* Edited by Parry ML, Canziani OF, Palutikof JP, van der Linden PJ, Hanson CE. Cambridge: Cambridge University Press; 2007:79–131.
5. van Bodegom PM, Verboom J, Witte JPM, Vos CC, Bartholomeus RP, Geertsema W, Cormont A, van der Veen M, Aerts R: Synthesis of ecosystem vulnerability to climate change in the Netherlands shows the need to consider environmental fluctuations in adaptation measures. *Reg Environ Change* 2013. doi:10.1007/s10113-013-0511-x.
6. Brown I: Influence of seasonal weather and climate variability on crop yields in Scotland. *Int J Biometeorol* 2012, 57:605–614.
7. Li Z, Yang P, Tang H, Wu W, Yin H, Liu Z, Zhang L: Response of maize phenology to climate warming in Northeast China between 1990 and 2012. *Reg Environ Change* 2014, 14(1):39–48. doi:10.1007/s10113-013-0503-x.
8. Morren C: Le globe, le temps et la vie. *Bulletins de l'Académieroyale des Sciences, des Lettres et des Beaux-Arts de Belgique* 1849, 2:660–684.
9. Demarée GR, Rutishauser T: From "periodical observations" to "anthochronology" and "phenology"—the scientific debate between AdolpheQuetelet and Charles Morren on the origin of the word "phenology". *Int J Biometeorol* 2012, 55:753–761.
10. Schnelle F: *Pflanzen-Phänologie.* Leipzig: Geest & Portig; 1955.
11. Seyfert F: *Phänologie.* Wittenberg: Ziemsen; 1960.
12. Lieth H: (Ed): *Phenology and Seasonality Modelling.* New York: Springer; 1974.
13. Schwartz MD: *Phenology. An Integrative Environmental Science.* Dordrecht: Kluwer; 2003.
14. Clealand EE, Allen JM, Crimmins TM, Dunne JA, Pau S, Travers SE, Zavaleta ES, Wolkovich EM: Phenological tracking enables positive species responses to climate change. *Ecology* 2012, 93(8):1765–1771.
15. Cook BI, Wolkovichc EM, Parmesan C: Divergent responses to spring and winter warming drive community level flowering trends. *Proc Natl Acad Sci USA* 2012, 109(23):9000–9005.
16. Ellwood ER, Temple SA, Primack RB, Bradley NL, Davis CC: Record-breaking early flowering in the Eastern United States. *PLoS ONE* 2013, 8(1):e53788. doi:10.1371/journal.pone.0053788.
17. Pau S, Wolkovich EM, Cook BI, Davies JT, Kraft NJB, Blomgren K, Betancourt JL, Clealand EE: Predicting phenology by integrating ecology, evolution and climate science. *Glob Chang Biol* 2011, 17:3633–3643.
18. Chmielewski FM, Müller A, Küchler W: Possible impacts of climate change on natural vegetation in Saxony (Germany). *Int J Biometeorol* 2005, 50:96–104.
19. Cook BI, Wolkovich EB, Davies TJ, Ault TR, Betancourt JL, Allen JM, Bolmgren K, Cleland EE, Crimmins TM, Kraft NJB, Lancaster LT, Mazer SJ, McCabe GJ, McGill BJ, Parmesan C, Pau S, Regetz J, Salamin N, Schwartz MD, Travers SE: Sensitivity of spring phenology to warming across temporal and spatial climate gradients in two independent databases. *Ecosystems* 2012, 15:1283–1294.
20. Črepinšek Z, Štampar F, Kajfež-Bogataj L, Solar A: The response of *Corylus avellana* L. phenology to rising temperature in north-eastern Slovenia. *Int J Biometeorol* 2012, 56:681–694.
21. Chuine I, Cambon G, Comtois P: Scaling phenology from local to the regional level: advances from species-specific phenological models. *Global Change Biol* 2000, 6(8):943–952.
22. Khanduri VP, Sharma CM, Singh SP: The effects of climate change on plant phenology. *Environmentalist* 2008, 28:143–147.
23. Schmidt G, Holy M, Pesch R, Schröder W: Changing plant phenology in Germany due to the effects of global warming. *Int J Climate Change* 2010, 2(2):73–84.
24. Schröder W, Pesch R, Schmidt G, Englert C: Analysis of climate change affecting German forests by combination of meteorological and phenological data within a GIS environment. *Sci World J* 2007, 7(S1):84–89.
25. Schröder W, Schmidt G, Hasenclever J: Geostatistical analysis of data on air temperature and plant phenology from Baden-Württemberg (Germany) as a basis for regional scaled models of climate change. *Environ Monit Assess* 2006, 130(1–3):27–43.
26. Kreeb KH: *Methoden zur Pflanzenökologie und Bioindikation.* Stuttgart: Fischer; 1990.
27. Braun P, Brügger R, Bruns E, Clever J, Estreguil C, Flechsig M, De Groot RS, Grutters M, Harrewijn J, Jeanneret F, Martens P, Menne B, Menzel A, Sparks T: *European Phenology Network. Nature's Calendar on the Move.* Wageningen University: Wageningen; 2003.
28. Gebhardt H, Rammert U, Schröder W, Wolf H: Klima-Biomonitoring: Nachweis des Klimawandels und dessen Folgen für die belebte Umwelt. *Umweltwiss Schadst Forsch* 2010, 22:7–19.
29. Intergovernmental Panel on Climate Change (IPCC): *Climate Change 2007: Synthesis Report.* Geneva: IPCC; 2007.
30. Dutilleul P: Modifying the t-test for assessing the correlation between two spatial processes. *Biometrics* 1993, 49:305–314.
31. Pesch R, Schröder W: Assessment of metal accumulation in mosses by combining metadata, statistics and GIS. *Nova Hedwigia* 2006, 82(3–4):447–466.
32. Crimmins MA, Crimmins TM: Monitoring plant phenology using digital repeat photography. *Environ Manage* 2008, 41:949–958.
33. Gardner RH, Kemp WM, Kennedy VS, Petersen JE: (Eds): *Scaling Relations in Experimental Ecology.* New York: Columbia University Press; 2001.
34. Hume D: *An Enquiry Concerning Human Understanding.* London; 1748.
35. *Ecological Scale: Theory and Applications.* Edited by Peterson DL, Parker VT. New York: Columbia University Press; 1998.
36. Schneider DC: *Quantitative Ecology: Spatial and Temporal Scaling.* San Diego: Academic; 1994.
37. Wiens JA: Spatial scaling in ecology. *Funct Ecol* 1989, 3:385–397.
38. Englert C: Plant phenology in Germany between 1951 and 2005: data quality assurance and bioindication of climate change. In *PhD thesis.* University of Vechta; 2010.
39. van Vliet AJH, Bron WA, Mulder S, van der Slikke W, Ode B: Observed climate-induced changes in plant phenology in the Netherlands. *Reg Environ Change* 2013. doi:10.1007/s10113-013-0493-8.
40. Oteros J, García-Mozo H, Hervás-Martínez C, Galán C: Year clustering analysis for modelling olive flowering phenology. *Int J Biometeorol* 2013, 57:545–555.
41. Liang L, Schwartz MD: Testing a growth efficiency hypothesis with continental-scale phenological variations of common and cloned plants. *Int J Biometeorol* 2014. doi:10.1007/s00484-013-0691-6.
42. Theurillat JP, Guisan A: Potential impact of climate change on vegetation in the European Alps: a review. *Clim Change* 2001, 50:77–109.
43. Ma S, Churkina G, Trusilova K: Investigating the impact of climate change on crop phenological events in Europe with a phenology model. *Int J Biometeorol* 2012, 56:749–763.
44. Shiferaw B, Smale M, Braun H-J, Duveiller E, Reynolds M, Muricho G: Crops that feed the world 10. Past successes and future challenges to the role played by wheat in global food security. *Food Sec* 2013, 5:291–317.
45. Elith J, Leathwick J: Species distribution models: ecological explanation and prediction across space and time. *Annu Rev Ecol Evol* 2009, 40:677–697.
46. Mawdsley JR, O'Malley R, Ojima DS: A review of climate-change adaptation strategies for wildlife management and biodiversity conservation. *Conserv Biol* 2009, 23(5):1080–1089.

47. Milad M, Schaich H, Bürgi M, Konold W: **Climate change and nature conservation in Central European forests: a review of consequences, concepts and challenges.** *Forest Ecol Manage* 2011, 261:829–843.

48. Bindi M, Olesen JE: **The responses of agriculture in Europe to climate change.** *Reg Environ Change* 2011, 11:151–158.

49. Inouye DW: **The ecological and evolutionary significance of frost in the context of climate change.** *Ecol Lett* 2000, 3:457–463.

50. Enke W: *Anwendung eines Statistischen Regionalisierungsmodells auf das Szenario B2 des ECHAM4 OPYC3 Klima-Simulationslaufes bis 2050 zur Abschätzung Regionaler Klimaänderungen für das Bundesland Hessen.* Hessisches Landesamt für Umwelt und Geologie: Wiesbaden; 2003.

51. Enke W: *Erweiterung des Simulationszeitraumes der Wetterlagenbasierten Regionalisierungsmethode auf der Basis des ECHAM4-OPYC3 Laufes für die Dekaden 2011/2020 und 2051/2100, Szenario B2.* Hessisches Landesamt für Umwelt und Geologie: Wiesbaden; 2004.

52. Henniges Y, Danzeisen H, Zimmermann R-D: **Regionale Klimatrends mit Hilfe der Phänologischen Uhr, Dargestellt am Beispiel Rheinland-Pfalz.** *Umweltwiss Schadst Forsch* 2005, 17(1):28–34.

53. Priess JA, Heistermann M, Schaldach R, Onigkeit J, Mimler M, Trinks D, Alcamo J: **Klimawandel und Landwirtschaft in Hessen: Mögliche Auswirkungen des Klimawandels auf Landwirtschaftliche Erträge.** In *Abschlussbericht für den Bereich Landwirtschaft, InKlim 2012—Integriertes Klimaschutzprogramm Baustein II: Klimawandel und Klimafolgen in Hessen.* Kassel: Universität Kassel; 2005. http://klimawandel.hlug.de/fileadmin/dokumente/klima/inklim/endberichte/landwirtschaft.pdf.

54. Streitfert A, Grünhage L, Jäger H-J: *Klimawandel und Pflanzenphänologie in Hessen.* Giessen: InstitutfürPflanzenökologie, Justus-Liebig-Universität Giessen; 2005.

55. Deutscher Wetterdienst (DWD): *Anleitung für die Phänologischen Beobachter des Deutschen Wetterdienstes (BAPH).* Offenbach am Main: DWD; 1991.

56. Böhm U, Kücken M, Ahrens W, Block A, Hauffe D, Keuler K, Rockel B, Will A: **CLM—the climate version of LM: brief description and long-term applications.** *COSMO Newsletter* 2006, 6:225–235.

57. Jacob D, Göttel H, Kotlarski S, Lorenz P, Sieck K: *Klimaauswirkungen und Anpassung in Deutschland—Phase 1: Erstellung regionaler Klimaszenarien für Deutschland. Forschungsbericht 204 41 138, UBA-FB 000969.* Regionaler Klimaatlas Deutschland: Dessau; 2008.

58. Keuler K, Lautenschlager M: *ClimateSimulationswith CLM. Climate of the 20th Century run No. 1, 1960-2000, Data Stream 2 und Scenario A1B run No. 1, 2001-2100, European Region, MPI-M/MaD.* Hamburg: Max-Planck Institut für Meteorologie; 2006. http://cera-www.dkrz.de/WDCC/ui/BrowseExperiments.jsp?proj=CLM_regional_climate_model_runs].

59. Kreienkamp F, Spekat A, Enke W: *Ergebnisse eines Regionalen Szenarienlaufs für Deutschland mit dem Statistischen Modell WETTREG2010. Report.* CEC Potsdam: Potsdam; 2010.

60. Schröder W, Schmidt G, Schönrock S: **Landesweite untersuchungen zu beobachteten und zukünftig zu erwartenden änderungen der phänologie von wild- und kulturpflanzen in Hessen und deren implikationen für die forst- und landwirtschaft - klimawandel und pflanzenphänologie in Hessen.** In *Abschlussbericht für das Fachzentrum Klimawandel Hessen, Hessisches Landesamt für Umwelt und Geologie.* Vechta, Wiesbaden; 2012.

61. Bahrenberg G, Giese E: *Statistische Methoden und ihre Anwendung in der Geographie.* Stuttgart: Teubner; 1975.

62. Brown DG, Aspinall T, Bennett DA: **Landscape models and explanation in landscape ecology—a space for generative landscape science?** *Prof Geograph* 2006, 58:369–382.

63. Legendre P: **Spatial autocorrelation: trouble or new paradigm?** *Ecology* 1993, 74:1659–1673.

64. Dale MRT, Fortin M-J: **Spatial autocorrelation and statistical tests: some solutions.** *J Agr Biol Environ Stat* 2009, 14:188–206.

65. Fortin J-M, Payette S: **How to test the significance of the relation between spatially autocorrelated data at the landscape scale: a case study using fire and forest maps.** *Ecosci* 2001, 9:213–218.

66. Hagl S: *Schnelleinstieg Statistik: Daten Erheben, Analysieren, Präsentieren.* Haufe-Lexware: München; 2008.

67. Odeh IOA, McBratney AB, Chittleborough DJ: **Further results on prediction of soil properties from terrain attributes: heterotopic cokriging and regression-kriging.** *Geoderma* 1995, 67:215–226.

68. Zirlewagen D, Raben G, Weise M: **Zoning of forest health conditions based on a set of soil, topographic and vegetation parameters.** *Forest Ecol Manage* 2007, 248:43–55.

69. Meynen E, Schmithüsen J: *Handbuch der naturräumlichen Gliederung Deutschlands.* Selbstverlag der Bundesanstaltfür Landeskunde: Remagen; 1953–1962.

10

Why are nanomaterials different and how can they be appropriately regulated under REACH?

Kathrin Schwirn*, Lars Tietjen and Inga Beer

Abstract

Background: For nanomaterials, not only their chemical composition but also their morphological properties and surface properties determine their characteristics. These properties do not only differ in comparison to the corresponding bulk material but also between different nanoforms of the same substance. Changes in these physico-chemical characteristics can cause changes in chemical properties, reactivity, (photo-) catalytic activities and energetic properties and in turn alter their (eco-) toxicity, fate and behaviour in environmental media and toxico-kinetics. Registration, Evaluation, Authorisation and Restriction of Chemicals (REACH) deals with chemical substances in general and although there are no special provisions that explicitly refer to nanomaterials, they are principally covered by REACH. In October 2012, the European Commission published the Second Regulatory Review on Nanomaterials. In February 2013, the REACH Review from the European Commission was published. Both papers address questions about the regulation of nanomaterials in REACH. The Commission proposes to improve the future situation by adaptation of the REACH Regulation. However, the European Commission plans to revise the annexes only and not the main text of the regulation.

Results and conclusions: In this publication, the authors present their considerations and recommendations on how REACH can adequately be adapted to nanomaterials. In the author's view, the bulk form and nanoforms of the same chemical composition should be treated as the same substance in the context of REACH. However, the regulation of nanomaterials under REACH has to meet specific requirements. Taking into account the plurality of physico-chemical characteristics and resulting changes in the hazard profile, an approach must be found to adequately cover nanomaterials under REACH. Accordingly, the REACH information requirements have to be adapted. This includes lower tonnage thresholds for different REACH obligations (e.g. registration, chemical safety report) which are justified by highly dispersed use together with low mass application, linked with the uncertainties regarding (eco-) toxicity, environmental fate and exposure. If the physico-chemical characteristics of different nanoforms of the same substance differ in a relevant manner they have to be considered separately for further test performance and REACH requirements.

Keywords: REACH; Chemicals regulation; Nanomaterials

Background

Speciality of nanomaterials and the challenges regarding their assessment

According to the European Commission (EU COM), the global quantity of nanomaterials is around 11.5 million tonnes with a market value of roughly 20 bn € per year [1]. Nanomaterials cover a heterogeneous range of materials including *inter alia* inorganic metal and metal oxide nanomaterials, carbon-based nanomaterials and polymeric particulate materials in a variety of forms. A wide range of nanomaterials is already available on the market, and nanomaterials for future applications like e.g. targeted drug delivery systems, novel robotic devices, molecule-by-molecule design and self-assembly structures are in development. Now then, what makes nanomaterials, in particular manufactured nanomaterials, so special that they should be explicitly addressed under the European Chemicals Regulation REACH [2].

For nanomaterials, not only their chemical composition but also their morphological properties like size, shape and surface properties determine their characteristics.

* Correspondence: Kathrin.Schwirn@uba.de
Federal Environment Agency, WoerlitzerPlatz 1, Dessau-Rosslau 06844, Germany

These parameters can affect the chemical properties, reactivity, (photo-) catalytic activities of a substance as well as energetic properties and their confinement. The properties of nanomaterials do not only differ from their bulk counterpart but also between different nanoforms of the same chemical substance and hence do their effects and behaviour. The property changes can lead to e.g. differences in cell penetration, in the mode of action and in the toxicity level, and may also vary in dose-response relationships describing toxicity. Subsequently, the question arises how these specific properties influence behaviour and effects in the environment and whether existing risk assessment and mitigation methods can be applied to nanomaterials without further a do.

Studies show that often nanomaterials can penetrate biological systems, like e.g. cell walls and membranes, and remain within the cell [3]. It is assumed that in this way a higher amount of a substance in comparison to the corresponding bulk form or substances that normally would not pass through barriers can enter the cell and may lead to a variety of toxic impacts. The following processes mediating cell toxicity can be possible: nanomaterials can excite coupled effects by causing ion toxicity together with particle toxicity; nanomaterials can delay and prolong the occurrence of toxic effects resulting from depot properties, or enable the uptake and impact of other toxic substances within the cell by a carrier effect [4-8]. Thus, the toxic consequence of cellular uptake of nanomaterials is hard to predict.

Due to the size dependence of the energy band gaps of materials, the energy band gaps of various nanomaterials correspond with the energy levels of important biological reactions, resulting *inter alia* in an increased formation of reactive oxygen species and subsequently oxidative stress in organisms [9-11]. Additionally, histological analysis upon nanomaterial exposure showed abnormalities in tissues of different organs of fish [12-14]. Furthermore, scientists observed mechanical stress onto aquatic organisms as well as influences on the reproduction of invertebrates like earth worms [15,16]. Photocatalytic active nanomaterials show a clearly higher toxicity under UV irradiation to aquatic organisms than observed for same test conditions without UV irradiation [17,18]. In our view, observations like these cannot be neglected for environmental risk assessment.

Previous studies on the effects of nanomaterials on the environment mainly focused on toxicity for aquatic organisms and the behaviour in the aquatic environment. However, nanomaterials tend to exhibit agglomeration and sedimentation. Therefore, an increased exposure of sediment and soil organisms is expected, but less information is available on the effects on soil and sediment organisms. This is due to the methodical challenges, like the absence of the possibility to detect and observe the nanomaterial and its behaviour in a complex system. Moreover, data on long-term effects of nanomaterials to organisms in the environment, effects on populations and communities (e.g. as determined in mesocosm studies), environmental monitoring as well as data regarding bioaccumulation and biomagnification are rare. Data on effect monitoring is not available so far.

A further obstacle in the assessment of nanomaterials regarding environmental risks is the low comparability and reproducibility of available study results. This is caused by the high variety of existing and in particular investigated nanomaterials, which differ in composition and their production process, variations in sample preparation and characterisation, which can influence the results significantly. Overall, in many cases, the investigated nanomaterials as produced as well as in media are neither sufficiently described nor characterised. This is of importance for instance for nanosilver where the release of silver ions depends on the intrinsic stability against dissolution or other alteration and the composition of the surrounding media [19-21].

The environmental fate of nanomaterials is generally very complex, but this complexity cannot be sufficiently investigated with the currently available standard tests. For instance, test guidelines on dissolution and determination of the partitioning coefficient log K_{OW} as well as adsorption/desorption in soil compartment are not applicable. There also is a lack of guidelines for the determination of dispersion behaviour and transformation in environmental media, respectively. These shortcomings were identified amongst others by the Organisation for Economic Co-operation and Development (OECD) Expert Meeting in the framework of the OECD Working Party of Manufactured Nanomaterials, held in January 2013 in Berlin [22]. Moreover, tools for exposure estimation cannot be sufficiently applied, as existing input parameters are not suitable for the calculation of the environmental distribution [23,24]. Furthermore, a validation of the results of the predicted environmental concentration (PEC) calculation is currently not possible due to the lack of appropriate techniques and devices in order to determine environmental concentrations.

Due to the particular uncertainties concerning evaluation of the possible risks of nanomaterials for human health and the environment, as well as on the grounds of the precautionary principle, the chemicals legislation needs to be revised. The most important aspect right now is the generation of information. This can be achieved by relatively mild measures. The adaptation of existing test methods and, in particular, the creation of nanospecific information requirements for the REACH registration are two possible options. In the following chapters, it is described how nanomaterials are currently regulated under REACH. Subsequently, the authors present their considerations and recommendations on how REACH can adequately be adapted to nanomaterials.

Current regulation of nanomaterials under REACH

Although there are no provisions in REACH that explicitly refer to nanomaterials, it covers nanomaterials since it deals with substances in general. Therefore, also the basic principle that is laid down in Article 1 (3) 'This Regulation is based on the principle that it is for manufacturers, importers and downstream users to ensure that they manufacture, place on the market or use such substances that do not adversely affect human health or the environment' applies to nanomaterials.

Since nanomaterials are substances within the meaning of REACH, they also have to be registered if the yearly manufactured or imported quantity reaches 1 tonne. As it is not regulated otherwise, a manufacturer/importer who manufactures/imports not only a nanomaterial but also the corresponding bulk material with the same chemical identity registers the nanomaterial and the bulk material together in one registration. Nanomaterials are not regarded as separate substances within the meaning of the current REACH rules, but as substances in a certain form. The sum of the quantities (nanomaterial(s) + bulk) is decisive for the calculation of the aforementioned tonnage threshold. If no corresponding bulk material is manufactured or imported by the same registrant, the nanomaterial has to be registered on its own accordingly. However, the registration dossier of the nanomaterial can be a part of the joint submission for the chemically identical substance. For yearly quantities of 10 tonnes or more, a chemical safety report has to be submitted.

Generally, each registrant has the obligation to accurately describe the material that he manufactures/imports or uses and to ensure the safe use of this material. The manufacturer/importer/downstream user that places the bulk material and the nanomaterial(s) on the market is, e. g., obliged to classify and label them differently in case the available data give reason to do so. The questions arise in the details: How does one know if the material is a nanomaterial at all? How does one know if the nanoform on hand has to be treated separately from or jointly with another nanoform?

Many experts are of the opinion that the testing requirements, test strategies and test methods under REACH are in principle applicable to nanoscale substances. However, some adaptations to the specificities of nanomaterials are needed [25,26].

The explanations above show that obligations regarding nanomaterials exist. Nevertheless, e.g. the registration dossiers that ECHA has received so far reveal that there is an obvious deficit. Hardly any information about nanomaterials was contained or retrievable in the dossiers. In addition for the few information on nanomaterials provided, the responsible registrants did mostly not indicate which nanoform was tested or which nanoform the information referred to, respectively [27]. This shows that the obligations regarding nanomaterials are either not clear enough or the current regulation offers too many loopholes, or both. Also, for the second registration period which ended 31 May 2013, only four substances were registered as nanomaterials [28].

ECHA is working on clarifications by adapting guidance documents, forming advisory groups, conducting projects, organising trainings and webinars and much more. Nevertheless, the discussion is still ongoing whether this is enough or whether the regulation itself has to be adapted.

Activities in the EU

On the European level, the discussion about nanomaterials has been ongoing for several years. In May 2004, the European Commission adopted the Communication 'Towards a European Strategy for Nanotechnology' [29]. The Communication proposed actions to promote a strong role of Europe in nanoscience und nanotechnology. The communication also covered the need to address potential risks for health and environment. In June 2005, the European Commission presented an action plan 'Nanosciences and nanotechnologies' for 2005 to 2009.

In 2008 the European Commission published a (first) regulatory review of EU legislation with respect to nanomaterials. The main conclusion was 'Current legislation covers in principle the potential health, safety and environmental risks in relation to nanomaterials. The protection of health, safety and the environment needs mostly to be enhanced by improving implementation of current legislation'.

In 2009 the European Parliament responded to the European Commission Communication in a resolution [30] and did the following:

1. Called for a regulatory and policy framework that explicitly addresses nanomaterials
2. Called on the Commission to review all relevant legislation
3. Called for an inventory and product labelling
4. Called specifically on the Commission to evaluate the need to review REACH concerning *inter alia*:

(a) Simplified registration for nanomaterials manufactured or imported below 1 tonne
(b) Consideration of all nanomaterials as new substances

(c) A chemical safety report with exposure assessment for all registered nanomaterials

(d) Notification requirements for all nanomaterials placed on the market on their own, in preparations or in articles

While first nanospecific provisions were integrated in the Regulation (EC) no. 1333/2008 on food additives, specific provisions for nanomaterials were also created in some other legislation, which was approved in the last years such as biocidal products (Regulation (EU) no. 528/2012) and cosmetics regulation (Regulation (EC) no. 1223/2009). The central chemical regulation REACH, however, was not adapted to better address nanomaterial.

Indeed, the European Commission started projects to investigate the needs for additional provisions. Based on the outcomes of the REACH Implementation Projects [Information Requirements (RIP-oN 2); Chemical Safety Assessment (RIP-oN 3)], ECHA prepared appendices to the Guidance on Information Requirements and Chemical Safety Assessment (IR & CSA). In the project on Substance Identification of Nanomaterials (RIP-oN 1), no consensus between the stakeholders was reached on whether parameters like size, shape and surface treatment of a nanomaterial can be regarded as a so-called identifier (changes in parameters trigger a new substance) or 'characterizer' (changes in parameters do not trigger a new substance, but another form of a substance).

The European Commission published a Second Regulatory Review on Nanomaterials in October 2012 and the REACH Review in February 2013 [1,31]. These reports depict the Commission's main conclusions regarding nanomaterials: According to the European Commission, the REACH registration and proof of safe use for nanomaterials should be based on a case by case approach, and each type of nanomaterial should be clearly described. Since only very limited information about nanomaterials was provided in the first registration period by December 2010, the Commission proposes to improve the situation in future by adaption of the REACH Regulation. The Commission initiated a public consultation on how the annexes of REACH could be amended to ensure that nanomaterials are registered more clearly under REACH and that the safe use of nanomaterials is adequately demonstrated within the registration dossiers, as part of an impact assessment in May 2013 [32]. The consultation comprised five potential policy options which were measured against a baseline that assumes no new policy actions. The consultation asked how respondents consider the potential impact of the options on cost, safety and overall efficiency of the regulatory process based on 37 main questions and 183 subquestions. The consultation closed mid of September 2013. The initial results of the consultation were presented at the Member State experts'

meeting in October 2013. In total, 142 responses to the questionnaire were submitted. Forty-nine percent of the replies were made by industry, while 12% were from national authorities, and 7% from consumer associations and non-governmental associations. The results were split into two opposites. While 38% prefer a loosening of REACH for nanomaterials, 31% called for tougher information requirements. The final results of the impact assessment are now expected for early 2014 [33].

However, the Commission plans to revise the annexes only and not the main text of the regulation. The Commission justifies this approach by being able to use the faster and lighter comitology procedure and by avoiding to re-open the general REACH discussion. This approach was supported by a majority of Member States, at least as a first step. In our view one reason for the Commission to choose the comitology procedure is that the Commission has a stronger position than during the ordinary legislative procedure.

Similar discussions on regulatory frameworks take place in other countries and in academia [34-36]. However, this publication focuses on the ongoing discussion in Europe.

In our opinion, the planned amendments and, in particular, the refusal regarding the adaptation of the main text of REACH are not sufficient in order to receive adequate, meaningful and relatable information on nanomaterials. In the interest of legal clarity and certainty, we propose that the definition for nanomaterials should be integrated in Article 3. Furthermore, additional amendments of the REACH regulation should be conducted which will be presented in the following chapters.

Results and discussion

There is a high variety of existing nanomaterials which differ in chemical composition, size, shape, crystallinity and surface modification. Figure 1 shows exemplarily a small excerpt of the variety of possible nanomaterials. Taking into account this plurality of physico-chemical characteristics and resulting changes in the hazard profile, an approach must be found to adequately cover nanomaterials under REACH. There are two approaches conceivable to cover nanomaterials under REACH- treating them as substances on its own or as specific forms of a substance.

Nanomaterial as a substance on its own

The bulk form and nanoforms of the same chemical composition could be treated as different substances within the meaning of REACH. In the context of RiP-oN 1, this approach was called 'size as an identifier'. To be more precise, such an approach could use the change of different parameters (e.g. differences in size, shape, surface) to decide on the regulatory substance identity.

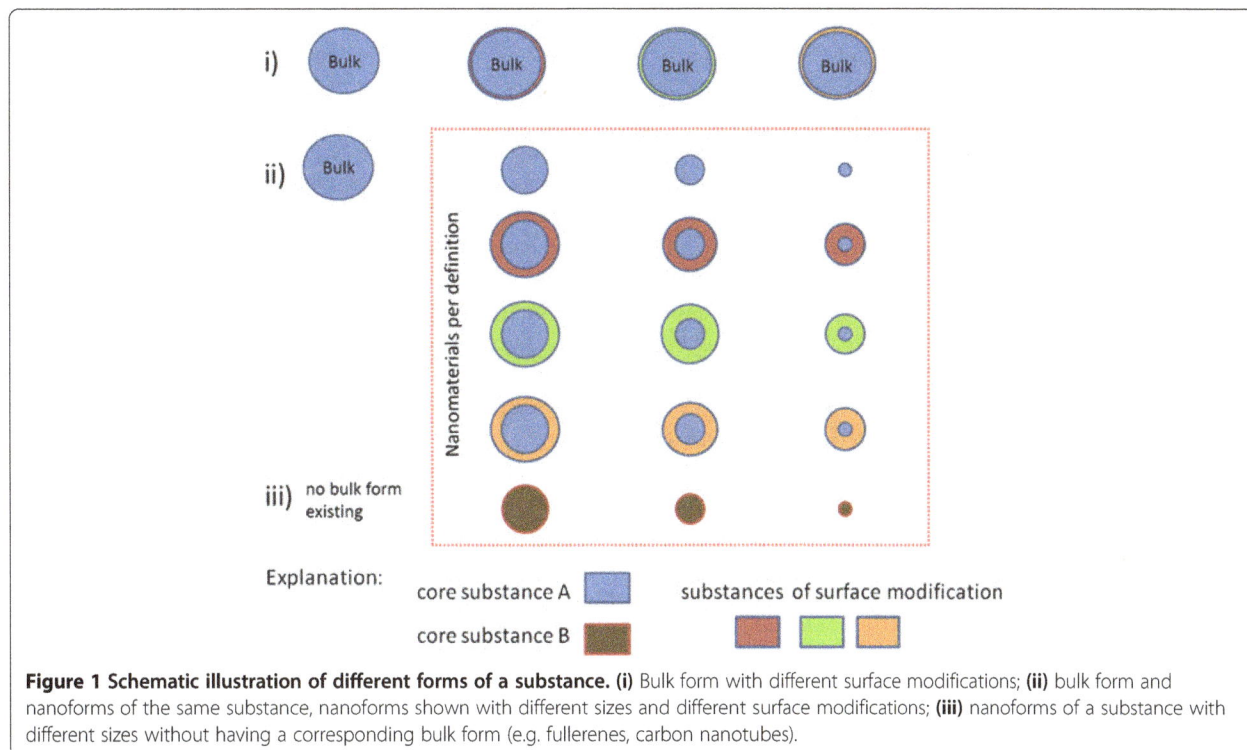

Figure 1 **Schematic illustration of different forms of a substance. (i)** Bulk form with different surface modifications; **(ii)** bulk form and nanoforms of the same substance, nanoforms shown with different sizes and different surface modifications; **(iii)** nanoforms of a substance with different sizes without having a corresponding bulk form (e.g. fullerenes, carbon nanotubes).

REACH and in particular the classification, labelling, packaging (CLP) Regulation [37] base in general on the assumption that a substance has an intrinsic hazard profile independent from the manufacturing process. Thus, REACH information can be shared between the registrants of the substance, and the hazard classification based on the CLP Regulation should correspond (e.g. agreed entry, harmonised classification). However, this approach has limitations already for bulk materials. For example, bulk materials of the same substance can differ in their hazard profiles based on impurities or macroscopic particle size, leading to different classifications under the CLP Regulation. As already mentioned above, different nanomaterials of the same chemical composition often have very different properties and, subsequently, can differ in their hazard profile. This condition for instance can be utilised to define nanomaterials as substances on their own. In a paper from Competent Authorities for REACH and CLP (CARACAL), it is described that indeed in some cases the nanoform could be treated as a substance on its own: 'In the case of substances at nanoscale, it is possible that some substances which in the past have been identified by the same EINECS number may have to be considered as different substances for the purpose of REACH' [38]. However, the paper does not provide criteria for this decision.

In the context of this approach, there is a need for clear criteria to avoid a split-up in many substances with small tonnages below the REACH triggers for registration requirements due to the variability of nanomaterials. The argument of the CARACAL paper based on the guidance for data sharing is 'whether or not data sharing would give a meaningful result' [39]. Some stakeholders are in favour of considering nanomaterials as substances on their own. If even the different nanoforms of a substance were treated as different substances, such an approach would need very low tonnage triggers for the registration and data obligations in REACH. A proposal from KemI gives an impression of some of the necessary changes [40]. KemI proposes to regulate nanomaterials by an individual regulation, which supports the existing REACH regulation. Nanomaterials shall be considered as substances of its own with a registration starting from 10 kg/a. However, the KemI proposal does not include an approach how to differentiate between different nanomaterials of the same chemical composition.

In summary, for the legal implementation of such an approach, clear criteria to decide if two nanomaterials of the same chemical composition are different substances in the meaning of REACH are necessary. It must be carefully considered what the consequences of the aforementioned change of the substance definition are for other pieces of legislation that address substances as such. Furthermore, such an approach would need a comprehensive review of the diverse instruments of the REACH regulation in order to ensure the workability of the instruments like data sharing, substance evaluation, information requirements and chemical safety assessment.

Nanomaterials as specific forms of a substance

Basically the same requirements for nanomaterials can be implemented with a different approach which is more in line with the substance definition and structure of the REACH regulation. The bulk form and nanoforms of the same chemical composition could be treated as the same substance in the context of REACH. This approach was called 'size as a characterizer' in the discussion of RIP-oN1.

Also in this case, the regulation of nanomaterials under REACH has to meet specific requirements. This includes a differentiated consideration of the bulk form and nanoform and the different nanoforms of the same substance respectively. Separate risk assessments shall be performed for the different nanoforms. An adequate handling of surface-treated nanomaterials has to be defined. Tonnage bands and information requirements need to be adjusted, and even the role of the downstream user has to be reconsidered. For all these requirements, the burden of proof has to be on the side of the registrant. In the following sections, we will present some corner stones of the proposal published by the German federal authorities responsible for REACH [41]. The concept presents considerations how REACH could be amended to adequately cover nanomaterials.

One important aspect is the substance identity. Generally, for a well-defined substance under REACH, the substance identity is defined solely by the molecular structure and chemical composition. Bulk and nanomaterial with the same molecular structure are chemically identical. This means that the bulk form and nanoform of a substance generally have to be registered in the same dossier. Therefore, the concept follows the characteriser approach.

Nevertheless, special characteristics concerning (eco-) toxicology, toxico-kinetics and environmental fate, together with the existing uncertainties and special features with regard to mode of action, necessitate requirements which go beyond those laid down in REACH to date. The information requirements under REACH therefore have to be adapted.

What should be the requirements for nanomaterials under REACH?

Nanomaterials have a low bulk density. This comes along with a typically high technical effectiveness caused by a high specific surface area and changes of reactivity, respectively. These characteristics together allow a wide dispersive use by a low mass application of the substance. Therefore and because of the uncertainties regarding (eco-) toxicology, environmental fate and exposure information requirements should already apply at lower tonnage bands. Following the structure of the REACH regulation for tonnage-based triggers, it would be reasonable to stipulate a simplified registration for nanomaterials starting from 100 kg/a, which requires information about

substance identity, characterisation and use. Starting at a tonnage band of 1 t/a, requirements regarding (eco-) toxicology listed in a new nanospecific annex have to be fulfilled. This nanospecific annex covers information requirements for the different tonnage levels. With respect to environment, these nanospecific information requirements subject chronic tests instead of acute tests at lower tonnage levels. Regarding the presumable partitioning of nanomaterials within the environment, appropriate target organisms have to be taken into account. That means information requirements must cover toxicity to sediment and soil organisms at lower tonnages. Furthermore, low water solubility as the exclusive waiving criterion for aquatic testing is not appropriate for nanomaterials, since also insoluble nanomaterials can show effects in the environment. Moreover, a chemical safety report, which considers every nanoform separately, has to be submitted if the sum of all nanoforms of a substance reaches the quantity of 1 t/a. Future adjustments regarding assessment concepts and test guidelines have to be taken into account.

Information requirements should first of all include a comprehensive characterisation of the nanoforms. This means that for each nanoform within a substance registration in addition to the identification of the chemical composition for substance identity, a characterisation of morphological parameters (e.g. size, shape, and crystal structure), surface properties (e.g. charge, surface reactivity, functional group, dispersability) and its solubility in different media would be necessary. However, it has to be noted that the further development and standardisation of reference methods for characterisation are still ongoing.

The registrant can use the information on the nanoforms' characteristics to ascertain if different nanoforms of a substance can be considered jointly or separately for the fulfilment of information requirements. This decision should be based on the aforementioned physical and chemical parameters and whether these differ or equal in a relevant way. A difference should be considered as relevant if it is likely that it leads to a change of the hazard profile. If nanoforms of a substance differ in a relevant way, information requirements have to be fulfilled separately for the individual forms. In a further step, an endpoint-specific waiving and read across between different nanoforms of the same substance should be possible on a scientific basis. The process of examination whether different nanoforms can be considered jointly or separately for information requirements is illustrated in Figure 2.

How can surface-treated nanomaterials be handled under REACH?

For nanomaterials, the surface to volume ratio dramatically increases with decreasing size and thus the surface plays a major role in the interaction with its surroundings.

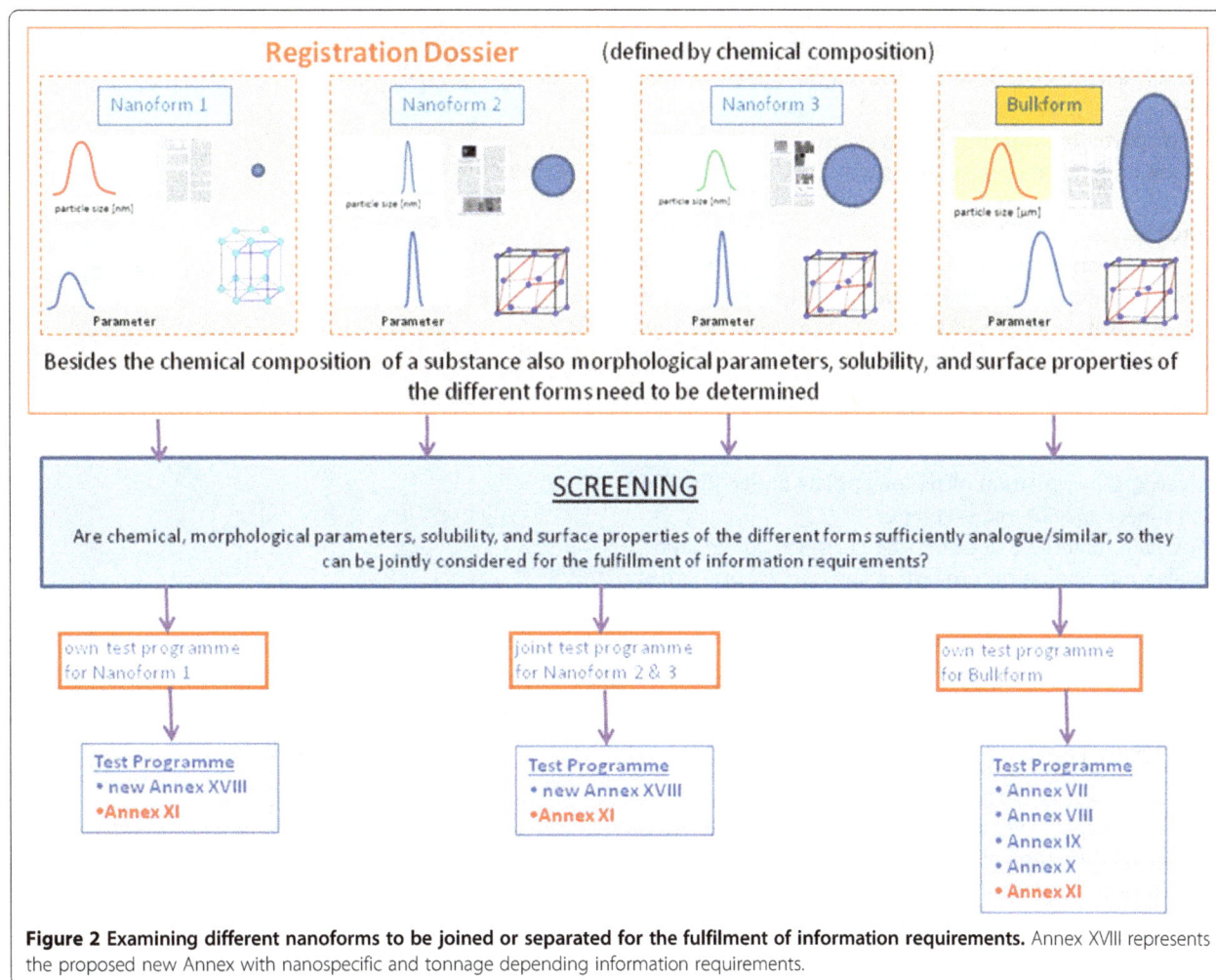

Figure 2 Examining different nanoforms to be joined or separated for the fulfilment of information requirements. Annex XVIII represents the proposed new Annex with nanospecific and tonnage depending information requirements.

Changes of surface properties (e.g. by modification with other substances) would result in the change of the nanomaterial's behaviour and resulting effects. Therefore, surface properties and their changes must not be neglected in the consideration of different forms of nanoscaled substances.

On the one hand, the conventional approach is to register surface-treated and surface-treating substances separately. This approach can be used for substances for which surface is of minor importance, but it is not appropriate for nanomaterials since nanomaterials have a very high surface to value ratio which increases with decreasing particle size. On the other hand, it is discussed that each surface-treated substance is regarded as a substance on its own. However, there are diverse substances existing as nanomaterials. Additionally, these can be surface treated with a variety of different substances leading to a countless number of new substances. Thus, the question appears how surface-treated nanomaterials can be considered under REACH while avoiding the aforementioned splitting into numerous new substances?

A possibility could be to define a surface-treated nanomaterial as a separate nanoform of the untreated substance. We propose to apply the substance identity approach with the 80 wt.% criterion. If the surface-treated nanomaterial consists of at least 80 wt.% of the core material, it fulfils this criterion and is regarded as a separate nanoform of the core material. If the surface-treated nanomaterial consists of less than 80 wt.% of the core material, it does not fulfil the criterion and has to be defined as a new substance. Following this approach the registrant has to demonstrate that the different nanoforms can be jointly considered or have to be separately considered for further test performances and REACH requirements.

In general, the implementation of this approach requires the adaptation of the articles and the annexes of REACH. Especially, demands like a simplified registration for nanomaterials starting from 100 kg/a and a chemical safety report for quantities of 1 t/a or more can only be realised by amending the main text of REACH. Other demands like adapted information requirements, the separate consideration of different nanoforms for testing obligations and for the chemical safety report, the possibility of scientifically justified waiving and read across, and the handling of surface-treated nanomaterials could at least partly be implemented by amending the REACH annexes and guidance documents only.

Conclusions

On the European level, there are ongoing discussions whether nanomaterials should be treated as substances on their own or as a specific form of a substance and what requirements for nanomaterials under REACH should look like. In our view, the bulk form and nanoforms of the same chemical composition should be treated as the same substance in the context of REACH. However, the regulation of nanomaterials under REACH has to meet specific requirements.

For nanomaterials, not only the chemical composition but also morphological properties and surface properties determine the special characteristics. These properties do not only differ in comparison to the corresponding bulk material but also between different nanoforms of the same substance. Changes in the physico-chemical characteristic can cause changes in chemical properties, reactivity, (photo-) catalytic activities and energetic properties and in turn alter their (eco-) toxicity, fate, behaviour in environment media and toxico-kinetics. Taking into account this plurality of physico-chemical characteristics and resulting changes in the hazard profile, an approach must be found to adequately cover nanomaterials under REACH. This would mean that the REACH information requirements have to be adapted.

Regarding environmental information, chronic tests instead of acute tests and toxicity to sediment and soil organisms are necessary at lower tonnage thresholds. This is justified by potentially wide dispersive use even by a low mass application linked with the uncertainties regarding (eco-) toxicity, environmental fate and exposure. If the physico-chemical characteristics of different nanoforms of the same substance differ in a relevant manner, they have to be considered separately for the further test performance and REACH requirements. A possibility to handle surface-treated nanomaterials would be to consider them as a separate nanoform of the untreated substance.

Methods

This study utilized description of the results of the regulatory work regarding the proposal for the adaptation of the REACH regulation to nanomaterials including literature analysis and taking into account the experiences from regulatory activities.

Abbreviations

CARACAL: Competent Authorities for REACH and CLP; CLP: Classification, Labelling, Packaging; COM: European Commission; ECHA: European

Chemicals Agency; EINECS: European Inventory of Existing Commercial Chemical Substances; EU: European Union; OECD: Organisation for Economic Co-operation and Development; PEC: predicted environmental concentration; REACH: European Chemicals Regulation EC no. 1907/2006 on Registration, Evaluation, Authorization and Restrictions of Chemicals.

Competing interests
The authors declare that they have no competing interests.

Authors' contributions
KS, LT and IB contributed in equal parts to this publication. All authors read and approved the final manuscript.

Acknowledgements
We would like to thank our colleagues from the German Federal Institute for Risk Assessment and the German Federal Institute for Occupational Safety and Health for the productive discussions and joint development of the proposal published by the federal authorities responsible for REACH.

References
1. European Commission: *Communication from the Commission to the European Parliament, the Council and the European Economic and Social Committee - second regulatory review on nanomaterials.* http://eur-lex.europa.eu/LexUriServ/LexUriServ.do?uri=COM:2012:0572:FIN:en:PDF.
2. European Parliament Council: **Regulation (EC) no 1907/2006 of the European Parliament and of the Council of 18 December 2006 concerning the Registration, Evaluation, Authorisation and Restriction of Chemicals (REACH), establishing a European Chemicals Agency, amending Directive 1999/45/EC and repealing Council Regulation (EEC) no 793/93 and Commission Regulation (EC) no 1488/94 as well as Council Directive 76/769/EEC and Commission Directives 91/155/EEC, 93/67/EEC, 93/105/EC and 2000/21/EC.** *Official J EU* 2006, **L353**:1–849.
3. Navarro E, Baun A, Behra R, Hartmann NB, Filser J, Miao AJ, Quigg A, Santschi PH, Sigg L: **Environmental behavior and ecotoxicity of engineered nanoparticles to algae, plants, and fungi.** *Ecotoxicology* 2008, **17**:372–386.
4. Poynton HC, Lazorchak JM, Impellitteri CA, Smith ME, Rogers K, Patra M, Hammer K, Allen JH, Vulpe CD: **Differential gene expression in *Daphnia magna* suggests distinct modes of action and bioavailability for ZnO nanoparticles and Zn ions.** *Environ Sci Technol* 2011, **45**:762–768.
5. Neal AL, Kabengi N, Grider A, Bretsch PM: **Can the soil bacterium *Cupriavidus necator* sense ZnO nanomaterials and aqueous Zn^{2+} differentially?** *Nanotoxicology* 2012, **6**(Suppl 4):371–380.
6. Bilberg K, Hovgaard MB, Besenbacher F, Baatrup E: **In vivo toxicity of silver nanoparticles and silver ions in zebrafish (*Danio rerio*).** *J Toxicol* 2012, **2012**:784:9.
7. Sun H, Zhang X, Niu Q, Chen Y, Crittenden JC: **Enhanced accumulation of arsenate in carp in the presence of titanium dioxide nanoparticles.** *Water Air Soil Pollut* 2007, **178**:245–254.
8. Misra SK, Dybowska A, Berhanu D, Luoma SN, Valsami-Jones E: **The complexity of nanoparticle dissolution and its importance in nanotoxicological studies.** *Sci Total Environ* 2012, **438**:225–232.
9. Burello E, Worth AP: **A theoretical framework for predicting the oxidative stress potential of oxide nanoparticles.** *Nanotoxicology* 2010, **5**:228–235.
10. George S, Xia T, Rallo R, Zhao Y, Ji Z, Lin S, Wang X, Zhang H, France B, Schoenfeld D, Damoiseaux R, Liu R, Lin S, Bradley KA, Cohen Y, Nel AE: **Use of a high-throughput screening approach coupled with in vivo zebrafish embryo screening to develop hazard ranking for engineered nanomaterials.** *ACS Nano* 2011, **5**:1805–1817.
11. Zhang H, Ji Z, Xia T, Meng H, Low-Kam C, Liu R, Pokhrel S, Lin S, Wang X, Liao YP, Wang M, Li L, Rallo R, Damoiseaux R, Telesca D, Mädler L, Cohen Y, Zink JI, Nel A: **Use of metal oxide nanoparticle band cap to develop a predictive paradigm for oxidative stress and acute pulmonary inflammation.** *ACS Nano* 2012, **6**:4349–4368.
12. Frederici G, Shaw BJ, Handy RD: **Toxicity of titanium dioxide nanoparticles to rainbow trout (*Oncorhynchus mykiss*): gill injury, oxidative stress, and other physiological effects.** *Aquat Toxicol* 2007, **84**:415–430.

13. Smith CJ, Shaw BJ, Handy RD: Toxicity of SWCNT to rainbow trout: respiratory toxicity, organ pathologies, and other physiological effects. *Aquat Toxicol* 2007, **82**:94–109.

14. Ramsden CS, Smith TJ, Shaw BJ, Handy RD: Dietary exposure to titanium dioxide nanoparticles in rainbow trout, (*Oncorhynchus mykiss*): no effect on growth, but subtle biochemical disturbances in the brain. *Ecotoxicology* 2009, **18**:939–951.

15. Dabrunz A, Duester L, Prasse C, Seitz F, Rosenfeldt R, Schilde C, Schaumann GE, Schulz R: Biological surface coating and molting inhibition as mechanisms of TiO2 nanoparticle toxicity in *Daphnia magna*. *PloS One* 2011, **6**(Suppl 5):e20112.

16. Schlich K, Terytze K, Hund-Rinke K: Effects of TiO2 nanoparticles in the earthworm reproduction test. *Environ Sci Eur* 2012, **24**:5.

17. Ma H, Bennan A, Diamond S: Phototoxicity of TiO2 nanoparticles under solar radiation to two aquatic species: *Daphnia magna* and Japanese medaka. *Environ Toxicol Chem* 2012, **31**(7):1621–1629.

18. Miller RJ, Bennett S, Keller AA, Pease S, Lenihan HS: TiO2 nanoparticles are phototoxic to marine phytoplankton. *PLoS One* 2012, **7**(Suppl 1):e30321.

19. Gondikas AP, Morris A, Reinsch BC, Marinakos SM, Lowry GV, Hsu-Kim H: Cysteine-induced modifications of zero-valent silver nanomaterials: implications for particle surface chemistry, aggregation, dissolution, and silver speciation. *Environ Sci Technol* 2012, **46**:7037–7704.

20. Kennedy AJ, Chappell MA, Bednar AJ, Ryan AC, Laird JG, Stanley JK, Steevens JA: Impact of organic carbon on the stability and toxicity of fresh and stored silver nanoparticles. *Environ Sci Technol* 2012, **46**:10772–10780.

21. Tejamaya M, Römer I, Merrifield RC, Lead JR: Stability of citrate, PVP, and PEG coated silver nanoparticles in ecotoxicology media. *Environ Sci Technol* 2012, **46**:7011–7017.

22. Kuehnel D, Nickel C: The OECD expert meeting on ecotoxicology and environmental fate - towards the development of improved OECD guidelines for testing nanomaterials. *Sci Total Environ* 2014, **472**:347–353.

23. Arvidsson R, Molander S, Sandén BA, Hasselöv M: Challenges in exposure modelling of nanoparticles in aquatic environments. *Hum Ecol Risk Assess* 2011, **17**:245–262.

24. Hansen SF, Baun A, Tiede K, Gottschalk F, Van der Meent D, Peijnenburg W, Fernandes T, Riediker M: *Consensus report based on the Nano Impact Net workshop: environmental fate and behaviour of nanoparticles - beyond listing of limitation*. http://www.nanoimpactnet.eu/uploads/Deliverables/D2.4.pdf.

25. Hankin SM, Peters SAK, Poland CA, Hansen SF, Holmqvist J, Ross BL, Varet J, Aitken RJ: *Specific advice on fulfilling information requirements for nanomaterials under REACH (RIP-oN 2) - final project report*. http://ec.europa.eu/environment/chemicals/nanotech/pdf/report_ripon2.pdf.

26. Organisation for Economic Co-operation and Development: *Environment Directorate: Preliminary review of OECD test guidelines for their applicability to manufactured nanomaterials*. http://search.oecd.org/officialdocuments/displaydocumentpdf/?doclanguage=en&cote=env/jm/mono%282009%2921.

27. European Commission: *NANO SUPPORT Project: scientific technical support on assessment of nanomaterials in REACH registration dossiers and adequacy of available information*. http://ec.europa.eu/environment/chemicals/nanotech/pdf/jrc_report.pdf.

28. European Chemical Agency: *The outcome of the second REACH registration deadline*. http://echa.europa.eu/documents/10162/13126357/press_memo_reach_2013_en.pdf.

29. European Commission: *Towards a European strategy for nanotechnology*. http://ec.europa.eu/nanotechnology/pdf/nano_com_en_new.pdf.

30. European Parliament: *European Parliament resolution of 24 April 2009 on regulatory aspects of nanomaterials*. http://www.europarl.europa.eu/sides/getDoc.do?type=TA&reference=P6-TA-2009-0328&language=EN.

31. European Commission: *General report on REACH in accordance with Article 117(4) of REACH and Article 46(2) of CLP, and a review of certain elements of REACH in line with Articles 75(2), 138(2), 138(3) and 138(6) of REACH*. http://eur-lex.europa.eu/LexUriServ/LexUriServ.do?uri=COM:2013:0049:FIN:EN:PDF.

32. European Commission: *Public consultation on the modification of the REACH annexes on nanomaterials*. http://ec.europa.eu/enterprise/sectors/chemicals/reach/nanomaterials/index_en.htm.

33. European Commission: *13th Meeting of Competent Authorities for REACH and CLP (CARACAL): document CA/36/2013*. https://circabc.europa.eu/sd/d/1e829ba0-9450-4a08-8b8a-64695a2f6e70/04%20-%20CA_36_2013_Nano%20IA_v2.doc.

34. Organisation for Economic Co-operation and Development, Environment Directorate: *Publications in the series on the safety of manufactured nanomaterials*. http://www.oecd.org/science/nanosafety/publicationsintheseriesonthesafetyofmanufacturednanomaterials.htm.

35. Linkov I, Satterstorm FK, Monica JC Jr, Hansen F, Davis TA: Nano risk governance: current developments and future perspectives. *Nanotechnology* 2009, **6**:203–220.

36. Grieger KD, Linkov I, Hansen F, Baun A: Environmental risk analysis for nanomaterials: review and evaluation framework. *Nanotoxicology* 2012, **6**:196–212.

37. European Parliament, Council: Regulation (EC) no 1272/2008 of the European Parliament and of the Council of 16 December 2008 on classification, labelling and packaging of substances and mixtures, amending and repealing Directives 67/548/EEC and 1999/45/EC, and amending Regulation (EC) no 1907/2006. *Official J EU* 2008, **L353**:1–1355.

38. European Commission: *Nanomaterials in REACH; paper from the meeting of the REACH competent authorities for the implementation of regulation (EC) 1907/2006 (REACH)*. http://ec.europa.eu/enterprise/sectors/chemicals/files/reach/nanomaterials_en.pdf.

39. European Chemical Agency: *Guidance on data sharing*. http://echa.europa.eu/documents/10162/13631/guidance_on_data_sharing_en.pdf.

40. Swedish Chemical Agency: *Draft proposal for a regulation on nanomaterials*. http://www.kemi.se/Documents/Forfattningar/Reach/Draft-proposal-regulation-nanomaterials.pdf.

41. Umweltbundesamt, Bundesinstitut für Risikobewertung, Bundesamt für Arbeitsschutz und Arbeitsmedizin: *Nanomaterials and REACH - background paper on the position of German Competent Authorities*. http://www.umweltbundesamt.de/publikationen/nanomaterials-reach.

11

Laboratory leaching tests on treated wood according to different harmonised test procedures

Ute Schoknecht[1*], Ute Kalbe[2], André van Zomeren[3] and Ole Hjelmar[4]

Abstract

Background: Laboratory leaching tests on treated wood were performed during a European robustness study in the framework of the validation of a tank leaching test procedure that has been proposed for construction products in order to determine the potential release of dangerous substances which can be transferred to soil and groundwater. The release of substances has to be determined also for materials treated with biocidal products according to the requirements of the European Biocidal Products Regulation. A similar leaching test procedure was already harmonised for treated wood for this purpose. Both test procedures were applied in parallel to wood treated with the same preservative to investigate whether the results of these tests can replace each other. Additional experiments were performed to further investigate unexpected effects of L/A ratio on leaching of copper and duration of storage of treated test specimens.

Results: Both procedures generate similar results concerning cumulative emissions of tebuconazole, copper, dissolved organic carbon and total nitrogen. The emission rates with time are in comparable ranges for both leaching protocols. Emissions of copper increased with decreasing L/A ratios. Strong correlation of copper concentrations and dissolved organic carbon as well as total nitrogen concentrations in eluates indicates that this observation is caused by co-elution of copper with organic substances. Duration of storage of treated test specimens affected emissions for the investigated wood preservative.

Conclusions: Based on these findings, results from both test procedures can be used to describe leaching characteristics and avoid double testing of treated wood to fulfil the requirements of the European regulations for either biocides or construction products. Leaching of substances from treated wood is a complex process that depends on its chemical composition and ageing processes.

Keywords: Construction products; Biocidal product; Emission; Leaching; Tank leaching test; Treated wood; European regulation

Background

The EU member states require that construction works are designed and executed so as not to endanger the safety of persons, domestic animals or property nor damage the environment within the Construction Products Regulation [1]. This includes concern about possible emissions of harmful substances due to contact of construction products with water.

CEN/TC 351/WG 1 'Release from construction products into soil, ground water and surface water' was mandated to develop harmonised laboratory leaching tests and to test the applicability of these methods for a series of different construction products within a robustness study as a part of the validation process [2]. The results of this study were reported by Hjelmar et al. [3].

A dynamic surface leaching test (DSLT) has been proposed for testing of monolithic materials and described in a Draft Technical Specification (FprCEN/TS 16637-2:2013 [4], temporarily named TS-2). This procedure is intended to also apply for structural timber since it is considered a 'monolithic product'. Therefore, timber was

* Correspondence: ute.schoknecht@bam.de
[1]Division 4.1 Biodeterioration and Reference Organisms, BAM Federal Institute for Materials Research and Testing, Unter den Eichen 87, D-12205 Berlin, Germany
Full list of author information is available at the end of the article

included in the robustness study. It was decided to use wood specimens treated with a wood preservative to assure that methods for the quantification of inorganic as well as organic substances in the eluates were available and that the experiments provide data for the evaluation of the test procedure.

Leaching of inorganic and to some extent also organic components of wood preservatives from treated wood has been investigated for many years with respect to the type of preservative, leaching mechanisms, environmental impact and test conditions [5-13]. Reduction of leaching has been a driving force for the development of new types of wood preservatives. However, leaching of active substances from treated wood can be intended to some extent to avoid growth of microorganisms in a water film on wet surfaces or soil in the surrounding area of wood. Therefore, a precautionary environmental risk assessment remains relevant for treated wood.

During the European robustness study, tebuconazole and copper from the applied wood preservative were analysed in the eluates. The effects of temperature as well as different ratios between water volume and exposed surface area (L/A ratio) on the test results were investigated.

The robustness study revealed that tebuconazole emissions decreased slightly with decreasing L/A ratios, whereas the emissions of copper from treated wood - surprisingly - increased with decreased amounts of water. The experiments did not indicate dependency on the temperature between 10°C and 27°C since the differences in the results were in the range of the relative standard deviations between parallel tests. There were hints that the results were also affected by the duration of storage of the treated test specimens.

Treated and untreated timber was investigated according to a water renewal scheme lasting for 36 days. However, based on experiments on other construction products, a 64-day scheme was selected for the final TS-2 draft [3].

In additional experiments, the results of the TS-2 test for construction products were compared to results from tests that were developed for treated wood by CEN/TC 38 'Durability of wood and wood-based products' and by the OECD Task Force on Biocides in order to fulfil the requirements of the regulations on biocidal products [14]. These tests were designed for different use conditions, i.e. permanent and occasional contact with water [15-18]. The test procedures for wood in permanent contact with water - OECD 313 [17] and CEN/TS 15119-2 [16] - correspond to a DSLT. The test parameters are similar for all three procedures with only a few deviations, e.g. different L/A ratios (see Table 1). The required L/A ratio and sampling scheme are the same in OECD 313 and CEN/TS 15119-2:2012. Therefore, in this paper, the procedures for treated wood will be referred to as 'OECD 313'.

Parallel experiments according to TS-2 (preliminary name, intended final name is CEN/TS 16637-2) and OECD 313 were performed to investigate whether the results of both procedures can replace each other to fulfil the requirements of both the European regulations in order to avoid double testing. Additional experiments at different L/A ratios and after a long period of storage of the treated test specimens were performed to further investigate unexpected effects that were observed during the robustness study.

This report is directed at the comparison of TS-2 and OECD 313 tests, variability of test results from leaching experiments on treated wood and confirmation of effects of the variation of L/A ratios and ageing of the treated test specimens on emissions of the selected substances.

Results and discussion
Comparison of TS-2 and OECD 313 experiments
Emission curves for tebuconazole, copper, dissolved organic carbon (DOC) and total nitrogen (TN) from parallel TS-2 and OECD 313 experiments are presented in Figure 1. The replicates investigated at conditions stipulated in the current documents for both leaching tests were selected for this comparison. Both TS-2 and OECD 313 experiments yield similar leaching curves for each of the considered substances despite of slight deviations of parameters.

It has to be considered that similar emission rates in TS-2 and OECD 313 experiments mean higher concentrations of these components in the eluates from the OECD 313 experiments, since the standardised L/A ratios applied for the comparison were 8 and 2.5 cm^3/cm^2, respectively (see Tables 1 and 2).

Standard deviations between parallel test assemblies are usually lower for test assemblies using sets of five test specimens as required in OECD 313 than for tests using a single test specimen as allowed for TS-2 experiments since differences in surfaces of test specimens are compensated to a certain extent if wood samples are combined. Relative standard deviations of tebuconazole concentrations in the eight eluate fractions (i.e. samples obtained after the defined immersion periods) from parallel experiments were 11% (3% to 24%) in TS-2 experiments using two test specimens per test assembly and 6% (3% to 10%) in OECD 313 experiments. Relative standard deviations of copper concentrations in the corresponding eluate fractions of parallel experiments were 14% (6% to 42%) in TS-2 experiments and 2% (1% to 5%) in OECD 313 experiments. The maximum standard deviations for the TS-2 experiments were observed for the first eluate fractions that were obtained after 2 h of immersion. This observation probably reflects concentration differences on the surface of the test specimens at the beginning of the test.

Table 1 Test setup for the applied version of TS-2 and CEN/TS 15119-2 as well as OECD 313 experiments using treated wood

Parameter	TS-2			CEN/TS 15119-2 and OECD 313		
Test specimens and treatment with wood preservatives	Representative of the product to be tested			Quality parameters for wood are defined (e.g. 100% sapwood), reference to EN 113		
	Preparation procedure to be developed by the respective product Technical Committees			Treatment and conditioning of test specimens as in accordance with either the recommendations made by the supplier, commercial treatment practices or EN 252		
				Test specimens with a retention within 5% of the mean are selected and end-sealed to prevent leaching via the end grain		
Test assemblies	At least one test piece per test assembly			TS 15119-2: exposed surface area: ≥ 200 cm^2, number of test specimens per test assembly: ≥ 3		
	Monolithic products: all dimensions >40 mm					
	Flat products (sheet-like and plate-like): minimum of exposed surface area: 100 cm^2, one dimension <40 mm			OECD 313: sets of five test specimens according to EN 113 size blocks are recommended (yield 200-cm^2 exposed surface area)		
Water	Demineralised water or deionised water or water of equivalent purity with a conductivity <0.5 mS/m according to grade 3 specified in ISO 3696			CEN/TS 15119-2: water complying with grade 3 of EN ISO 3696, water especially designed for environmental investigations or deionised water, pH 5 to 7 (not adjusted)		
				OECD 313: deionised water is recommended, synthetic seawater has to be used when wood exposed to seawater is to be evaluated		
L/A ratio	8 ml/cm^2			2.5 ml/cm^2		
Temperature	20°C[a]			20°C ±2°C		
Sampling schemes[a]	Day	Sample number (eluate fraction)	Total duration of the test [days]	Day	Sample number (eluate fraction)	Total duration of the test [days]
	1	1	0.08	1	1	0.25
	2	2[b]	1	2	2	1
	3	3	2.25	3	3	2
	9	4[b]	8	5	4	4
	15	5	14	9	5	8
	16	6	15	16	6	15
	29	7[b]	28	23	7	22
	37	8	36	30	8	29

[a]FprCEN/TS 16637-2 defines a sampling scheme including eight immersion periods (steps) within 64 days and temperature of 22°C ±3°C as a result of the robustness study [3,4]; [b]eluate fractions selected for chemical analysis from experiments at ECN.

The log-log graphs in Figure 1 include lines with a slope of −0.5 that indicate periods during the experiments when leaching can be assumed to be mainly controlled by diffusion. Diffusion-controlled processes are expected to be proportional to the square root of time (i.e. exponent 0.5). This exponent becomes factor 0.5 in logarithmic functions of cumulative emissions and −0.5 in logarithmic functions of emission rates with time. During the early stage of the experiments, the emission curves of all components follow the lines with a slope of −0.5 almost parallel, i.e. leaching is mainly controlled by diffusion. Steeper graphs indicate depletion after longer periods of permanent immersion. The applied TS-2 sampling scheme includes a short immersion step from the 14th to the 15th day. This caused relatively high emission rates for tebuconazole during this immersion period (see Figure 1a), but not for Cu, DOC and TN. For tebuconazole and copper, these observations are in agreement with former experiments [9,11]. It is assumed that leaching of tebuconazole

is restricted due to its low water solubility (29 mg/l at 20°C, independent of pH [19]). Providing fresh water strongly supports leaching of this substance.

The cumulative emissions of three parallel experiments at standardised conditions during the first series of experiments exhibited a very good repeatability (see Table 2 and Figure 2). The cumulative emissions of tebuconazole, copper and DOC were in similar ranges for TS-2 and OECD 313 experiments (see results for the first test series in Table 2). The relative standard deviations of the cumulative emissions during the experiments were similar for TS-2 and OECD 313 experiments.

Higher emissions of tebuconazole in the TS-2 experiment were probably caused by the higher amount of water per unit surface area, whereas slightly higher emissions of copper in OECD 313 experiments were probably caused by higher DOC concentrations in the eluates from this test.

The total emissions during the tests according to OECD 313 differed slightly between experiments using

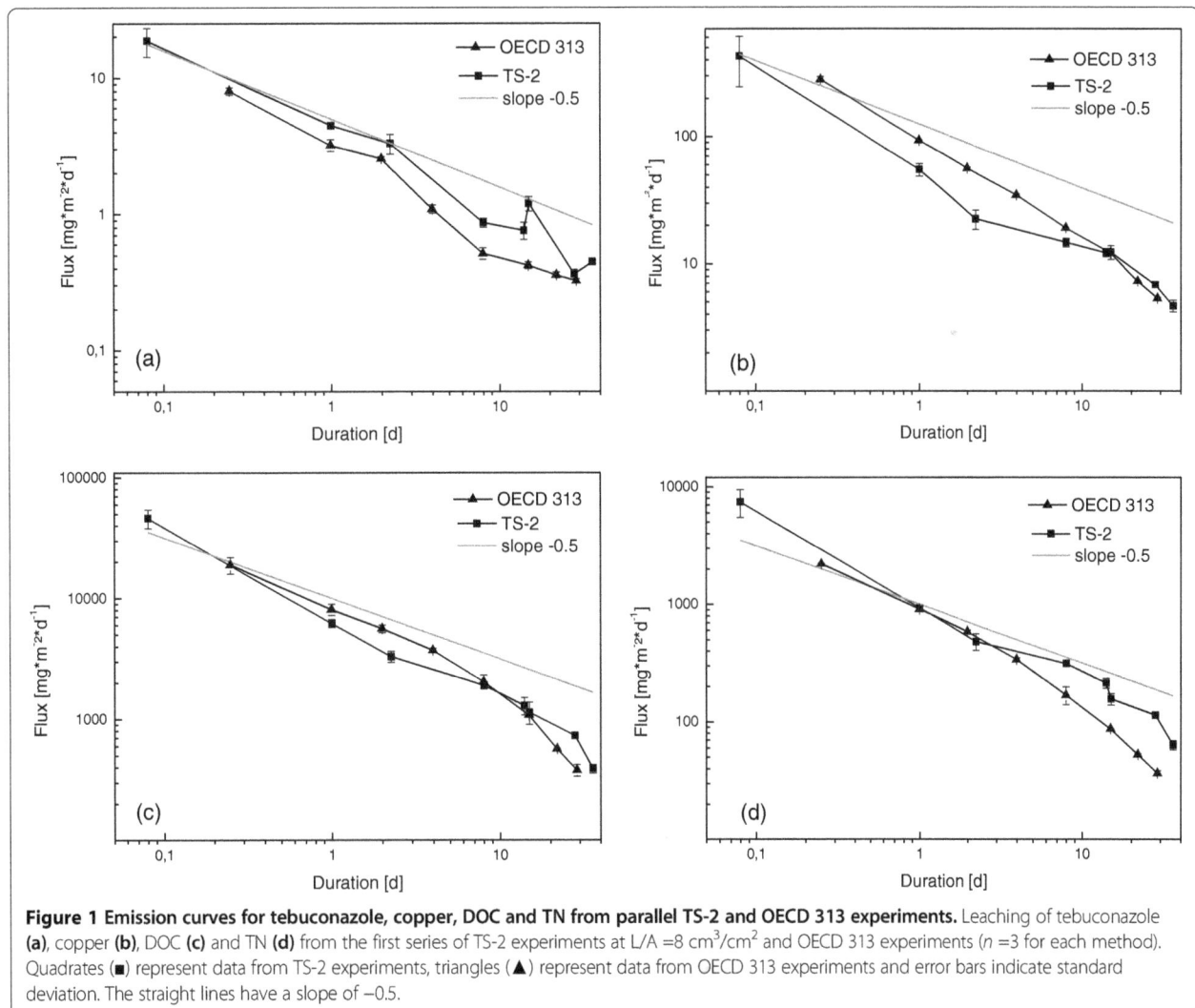

Figure 1 Emission curves for tebuconazole, copper, DOC and TN from parallel TS-2 and OECD 313 experiments. Leaching of tebuconazole **(a)**, copper **(b)**, DOC **(c)** and TN **(d)** from the first series of TS-2 experiments at L/A =8 cm³/cm² and OECD 313 experiments (n =3 for each method). Quadrates (■) represent data from TS-2 experiments, triangles (▲) represent data from OECD 313 experiments and error bars indicate standard deviation. The straight lines have a slope of −0.5.

sets of five test specimens and the test using a single test specimen from the same origin as the specimens for the TS-2 test (see data for experiments of series 1 in Table 2 and Figure 2). This is probably caused by different origins and ages of the wood.

Cumulative DOC emissions from experiments on treated wood are about 20% higher than DOC emissions from the control experiments using untreated wood specimens (Table 2). This difference is obviously caused by components of the wood preservative.

Most of the TN in the eluates is related to treatment with the wood preservative that includes amines (see Table 2). It was shown by Lupsea et al. [20] that nitrogen is not only leached as the added amine but also as a series of substances that originate from reactions of nitrogen (amino group) contained in the wood preservative with wood components. Higher TN emissions in TS-2 experiments can be caused by different evaporations of ethanolamine during the 39-day period of conditioning of

the treated test specimens prior to the leaching experiment. The amount of evaporated ethanolamine was probably higher from the small test specimens used for OECD 313 experiments having a ratio between surface area and volume of 2.50 cm²/cm³ compared to a ratio of 1.24 cm²/cm³ for the 'stakes' used for TS-2 experiments.

The mass of the wood specimens increased until 29 days of immersion in the OECD 313 procedure. The wood moisture content, as calculated from the mass data, increased from about 12% to 50%. This is close to the upper values that have been reported to range between 30% and 55% for wood exposed to precipitation [13].

The pH of the eluates from TS-2 and OECD 313 experiments ranged between pH 4.7 and 5.8 for untreated wood and usually between pH 6.5 and 7.5 (up to pH 8.6 in one experiment) for treated wood (see Table 2). An increase of the pH in the eluates from the treated wood specimens has to be expected. It is caused by leaching of alkaline ethanolamine that is part of the wood preservative.

Table 2 Test data from TS-2 (mg/m^2 in 36 days) and OECD 313 (mg/m^2 in 29 days) experiments

Duration of storage[a] [weeks]	Name of the experiment	Repetition	Cumulative emission								pH range	
			Cu [mg/m^2]	SD	Tebuconazole [mg/m^2]	SD	DOC [mg/m^2]	SD	TN [mg/m^2]	SD	Max	Min
7	L/A 8 control		<LOD		<LOD		37,445		126		5.83	4.76
	L/A 2.2	1	591		19.5		44,678		7,955		7.21	6.46
	L/A 5	1	510		28.5		47,214		7,603		7.37	6.34
	L/A 8	*1*	*426*		*29.3*		*46,852*		*7,497*		*7.40*	*6.37*
		2	*399*		*27.9*		*44,744*		*7,160*		*7.96*	*6.24*
		3	*401*		*29.4*		*46,797*		*7,218*		*7.27*	*6.45*
	L/A 9	1	450		31.2		45,464		7,572		7.44	6.55
	L/A 8	*Mean*	*409*	*15*	*28.9*	*0.8*	*46,131*	*1,201*	*7,292*	*180*		
	All experiments	Mean	463	75	27.6	4.1	45,958	1,135	7,501	289		
39	L/A 8 rep control		<LOD		<LOD		13,550		352		5.86	4.75
	L/A 2.2 rep	2	1,036		20.0		34,654		6,139		6.96	6.24
		3	1,000		19.9		34,024		5,706		7.29	6.10
	L/A 5 rep	2	867		28.7		34,746		5,968		7.52	6.28
	L/A 8 rep	*4*	*764*		*39.2*		*36,443*		*6,047*		*7.81*	*6.34*
		5	*890*		*39.0*		*36,470*		*6,126*		*8.61*	*6.35*
	L/A 8	*Mean*	*827*		*39.1*		*36,457*		*6,087*			
	All experiments	Mean	911	109	29.4	9.6	35,267	1,121	5,997	177		
8	OECD 313 control		<LOD		<LOD		32,405		139		5.03	4.72
	OECD 313	*1*	*499*		*18.8*		*44,670*		*4,337*		*6.80*	*6.40*
		2	*520*		*19.6*		*45,876*		*4,320*		*6.78*	*6.50*
		3	*503*		*17.6*		*46,134*		*4,569*		*6.75*	*6.45*
	OECD 313	*Mean*	*507*	*11*	*18.7*	*1.0*	*45,273*	*782*	*4,409*	*139*		
	OECD 313[a] control						22,326		76		4.60	4.22
	OECD 313[a]	1	425		22.5		37,550		6,308		7.12	6.60
68	OECD 313[a] rep	2	834		16,9		[b]		[b]		6.04	5.59

Experiments at conditions stipulated in TS-2 and OECD 313 are indicated by italics. Mean data of parallel experiments according to these conditions are indicated by italic entries.
[a]'Duration of storage' refers to treated test specimens. SD, standard deviation (mg/m^2); 'control', experiment with untreated wood; OECD 313, test specimens $(1.5 \times 2.5 \times 5 \text{ cm}^3)$; OECD 313[a], test specimen as in TS-2 $(2.5 \times 5 \times 15 \text{ cm}^3)$.
[b]During the first four immersion periods (4 days), DOC emissions were about 80%, and TN emissions were about 50% of the amounts detected during the same period of time in the related experiment with test specimens after 8 weeks of storage of the treated test specimens.

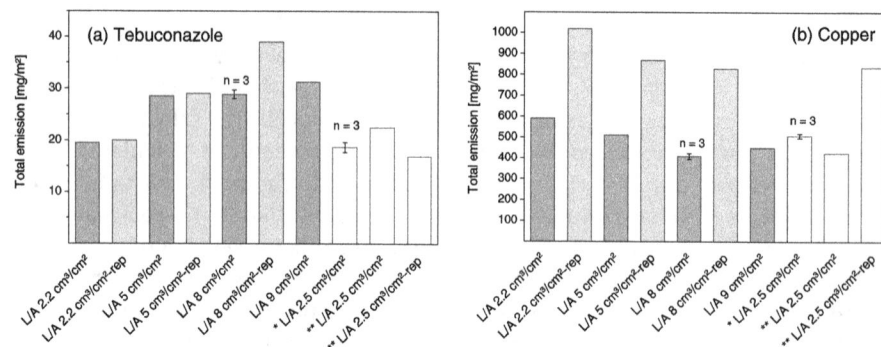

Figure 2 Cumulative emission of tebuconazole and copper in both series of TS-2 experiments and OECD 313 experiments. Cumulative emission of tebuconazole **(a)** and copper **(b)** in both series of TS-2 experiments (first series: dark grey columns, repeated experiments: grey columns) and OECD 313 experiments using small test specimens (*) and test specimens as in TS-2 tests (**) (white columns). Error bars indicate standard deviation. The numbers of replicates are given in Table 2.

Variability of test results from leaching experiments on treated wood

TS-2 experiments at standardised conditions (L/A =8 cm^3/cm^2, 20°C ±2°C; later defined to be 22°C ±3°C in FprCEN/TS 16637-2:2013) were performed at BAM and ECN with three parallel test assemblies in both laboratories (see Table 3). The test samples originate from the same batch but had to be shipped to ECN Petten. The experiments at BAM were started 7 and 39 weeks after treatment, and the experiments at ECN were started 17 weeks after treatment due to the schedule of all experiments within the robustness study at ECN. The eluate fractions 2, 4 and 7 from the ECN experiments were shipped to BAM immediately after sampling. All samples were analysed at BAM for tebuconazole and copper. The measured concentrations are summarised in Figure 3.

Results that fall within the average ±2.8 SD range for the standardised conditions were regarded to be in the same range in the robustness study [3]. If this criterion is applied, and the results of the first experiments performed at BAM (without shipment of test specimens and eluate samples, shortest period of storage after treatment) were regarded as reference values, some of the tebuconazole and copper concentrations of the eluate fractions from the ECN experiment just fit. The tebuconazole concentrations tend to be lower, whereas the copper concentrations tend to be higher for the ECN experiments compared to the results from the BAM experiment. Relatively high deviation of tebuconazole concentrations in the second fractions and copper concentrations in the fourth fractions were observed for the three parallel experiments at ECN Petten. This is probably not caused by uncertainty of the analytical methods or analytical outliers. Relative standard deviations

of tebuconazole concentrations from three parallel injections were 1% to 2% for all samples. Tebuconazole concentrations in two samples of eluate fraction 2 were considerably lower than the data from the experiment at BAM, whereas the concentration of the third sample was similar to the data from the BAM experiment. Variation of copper concentrations was generally higher in the experiment at ECN than at BAM - perhaps due to the fact that a single test specimen per test assembly was used in the experiments at ECN whereas two test specimens were combined in the test assemblies at BAM (see Table 3).

Several parameters cause differences of analytical results within and between the tests at the participating laboratories. The subsamples were taken from the same parts of the treated stakes and randomly distributed to ECN and BAM. However, samples of treated wood cannot be homogenous. In addition, the wood samples as well as the eluates were shipped to and from ECN Petten, whereas the wood samples were stored under controlled conditions and the eluates were analysed directly after sampling at BAM. This should not affect copper concentrations, but shipment might potentially reduce the concentration of tebuconazole. Possibly, this is an explanation for lower tebuconazole concentrations in the samples from ECN.

Variation of L/A ratios

The concentrations of tebuconazole, copper, DOC and TN in the eluates from TS-2 experiments at different L/A ratios decrease in the following order: L/A 2.2 cm^3/cm^2 > L/A 5 cm^3/cm^2 > L/A 8 cm^3/cm^2 ~ L/A 9 cm^3/cm^2. A repeated series of experiments using different L/A ratios confirmed that tebuconazole emissions increase with L/A ratios (see Table 2 and Figure 2a). This is in agreement with

Table 3 Overview on the parameters of the leaching tests

Test procedure	Name of the experiment	Number	Test specimens (dimensions) [cm^3]	Test specimens per test assembly [number]	Duration of storage after treatment [weeks]	L/A [cm^3/cm^2]
TS-2	L/A 8 (ECN)	3	2.5 × 5 × 15	1	17	8
	L/A 8	3		2	7	8
	L/A 2.2	1		3	7	2.2
	L/A 5	1		3	7	5
	L/A 9	1		2	7	9
	L/A 8 rep	2		2	39	8
	L/A 2.2 rep	2		3	39	2.2
	L/A 5 rep	1		3	39	5
OECD 313	OECD 313	3	1.5 × 2.5 × 5	5	8	2.5
	OECD 313[a]	1	2.5 × 5 × 15	1	8	2.5
	OECD 313[a]	1		1	68	2.5

Tests were performed at BAM unless otherwise stated in the name of the experiment; 'rep' stands for repeated experiments. The temperature was 20°C ±2°C in all experiments.
[a]'Duration of storage' refers to treated test specimens. SD, standard deviation (mg/m^2); 'control', experiment with untreated wood; OECD 313, test specimens (1.5 × 2.5 × 5 cm^3); OECD 313[a], test specimen as in TS-2 (2.5 × 5 × 15 cm^3).

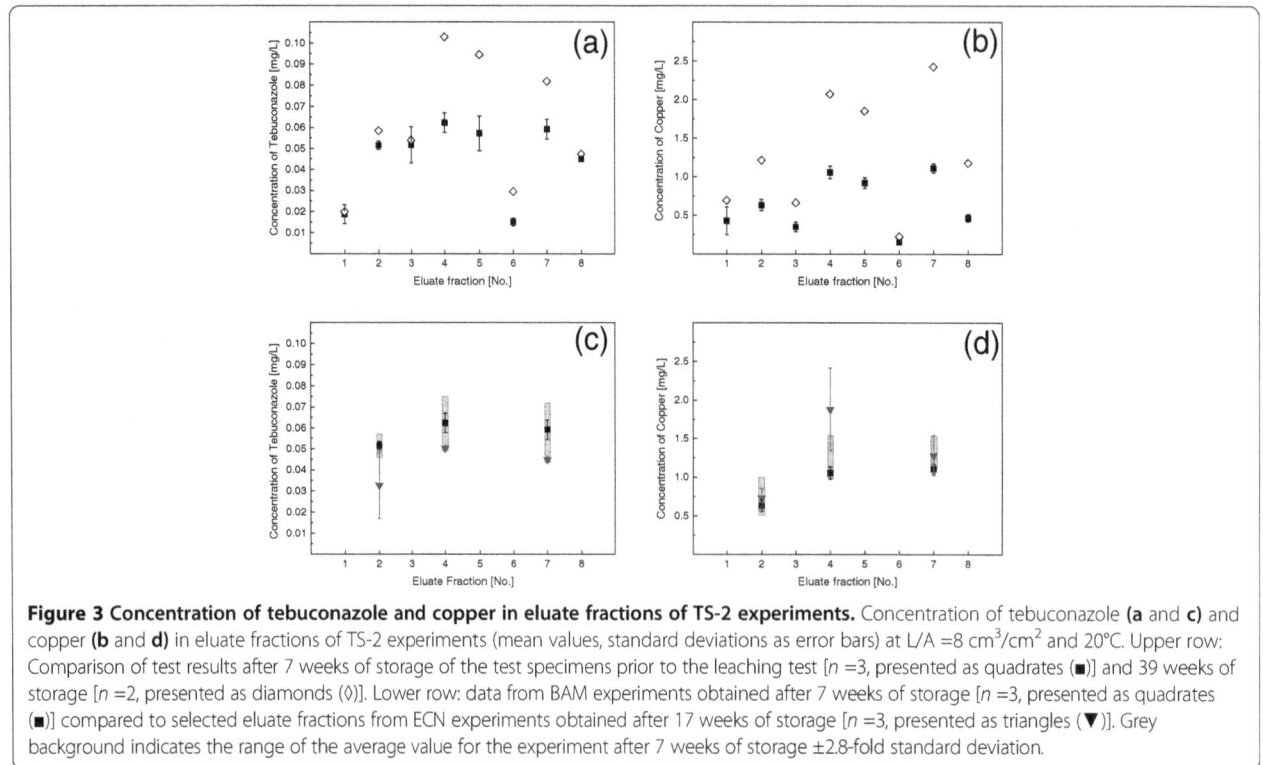

Figure 3 Concentration of tebuconazole and copper in eluate fractions of TS-2 experiments. Concentration of tebuconazole (**a** and **c**) and copper (**b** and **d**) in eluate fractions of TS-2 experiments (mean values, standard deviations as error bars) at L/A =8 cm^3/cm^2 and 20°C. Upper row: Comparison of test results after 7 weeks of storage of the test specimens prior to the leaching test [n =3, presented as quadrates (■)] and 39 weeks of storage [n =2, presented as diamonds (◊)]. Lower row: data from BAM experiments obtained after 7 weeks of storage [n =3, presented as quadrates (■)] compared to selected eluate fractions from ECN experiments obtained after 17 weeks of storage [n =3, presented as triangles (▼)]. Grey background indicates the range of the average value for the experiment after 7 weeks of storage ±2.8-fold standard deviation.

previous experiments [11]. Cumulative emissions of DOC and TN were in similar ranges for all test assemblies of treated wood of each test series rather independent from the L/A ratio. Although leaching of copper differs considerably between the series of experiments, the emissions of copper decreased with increasing L/A ratios in both series of experiments (see Figure 2b). This has not been described so far, and it is assumed to be related to higher concentrations of organic substances as indicated by DOC and TN concentrations of the eluates at low L/A ratios. Leaching of copper is strongly related to the concentrations of DOC and TN in the eluates, whereas co-elution with organic substances is less important for tebuconazole (see Table 4). The normalised emission curves, i.e. emission rates given as percentage of the concentration in the first eluate versus duration of the test, are rapidly decreasing and very similar for copper, DOC and TN, whereas the progression of the curves for tebuconazole is more gently dipping (not shown). Similar curve progression also supports the assumption that copper and organic substances are co-eluted. This is in agreement with the

results reported by Lupsea et al. [20,21]. Increased emission of copper due to co-elution with DOC has been observed by van Zomeren et al. [22], Tiruta-Barna and Schiopu [23] and Lupsea et al. [20]. Co-elution of copper and organic substances, including nitrogenous substances, is supported by binding reactions as described by Lupsea et al. [20,21]. Therefore, the observation on copper emissions is related to the actual composition of the wood preservative that was used for the experiments.

Ageing of treated test specimens
The first series of TS-2 experiments was performed after 7 and 17 weeks of storage of the treated test specimens at BAM and ECN, respectively. A second series of experiments was performed at BAM using test specimens after 39 weeks of storage (see Table 3). The emissions of tebuconazole were in similar ranges in both series of TS-2 experiments with the exception of the data from the repeated experiments at L/A =8 cm^3/cm^2 (see Figure 2a). It was noticed that the copper emissions were higher in the repeated experiments at all L/A ratios tested. This

Table 4 Correlation of concentrations of copper and tebuconazole with DOC and TN in eluates from TS-2 experiments

Storage of treated test specimens	Number of eluates	Correlation coefficients for concentrations in eluates			
		Cu/DOC	Cu/TN	Tebuconazole/DOC	Tebuconazole/TN
7 weeks	48	0.973	0.971	0.561	0.553
39 weeks	40	0.982	0.911	0.639	0.673

observation was confirmed by OECD 313 experiments that were performed after 8 and 68 weeks of storage (see Figure 2b). Emissions of DOC and TN were lower in the second series of both types of experiments (see Table 2).

Possibly, the observed differences in the leachability of compounds were caused by ageing of the treated test specimens. However, higher emissions of copper in the second series of experiments were unexpected, since binding of copper in wood, so-called 'fixation', is expected to proceed with time and is known to reduce its leachability [5]. This result also contradicts the assumption that copper emissions are related to the DOC concentrations that were lower in the second series of experiments. The results cannot be explained by any deviation during the experiments or analysis of the eluates. There are two options to explain this result: the constituents of the wood preservative were not equally distributed in the stakes, e.g. copper fixation was less effective if ethanolamine did not penetrate into the middle sections of the stakes that were investigated in the second series of the experiments. Another option is that the composition of DOC originating from the wood and the wood preservative, i.e. the availability of functional groups for binding of copper, is changing with time. This is a question for further research.

The observed effect of ageing of test specimens is probably related to the applied wood preservative and its chemical reactions with wood components.

Chemical changes like binding reactions of the components of the wood preservative and evaporation of organic substances are supposed to take place during storage. Experiments of Lupsea et al. [20] revealed different leachability of substances possessing phenolic and carboxylic groups for untreated wood specimens that had been stored for different periods of time. Changes in the leachability of copper and tebuconazole were also related to these differences.

Leaching of components from treated wood is a complex process that is influenced by the chemical composition of the wood [20,21]. This composition does not only depend on the origin of the wood (e.g. species and growth conditions) and the components of wood preservatives but also on ageing of the wood.

Conclusions

Results from leaching experiments with treated wood are likely to vary considerably due to the inhomogeneity of wood that has to be used as test specimens. This can only slightly be compensated by using combinations of several test specimens in a test assembly, but cannot be circumvented. In addition, differences in the test results are acceptable to a certain degree if the aim of the test is a general description of leaching processes. Test specimens of the same origin and storage time have to

be used if minor differences between wood preservatives have to be detected.

Despite the slight differences in the test conditions and test duration, the results of TS-2 and OECD 313 experiments are similar with regard to the total cumulative emission after 36 and 29 days, respectively. Both procedures indicate diffusion-controlled leaching of the target substances.

The robustness validation within CEN has led to a total test duration of 64 days in the final version of the TS-2 method while the TS-2 experiments during this study lasted 36 days. That is supposed to cause slightly higher total emissions due to the longer test duration, but the cumulative emission and the flux will remain comparable at comparable test duration.

Both the TS-2 and the OECD 313 test procedures are suitable to describe leaching characteristics of substances emitted from treated wood. The results can be used to fulfil the requirements of the Construction Products Regulation as well as the Biocidal Products Regulation for estimations for which laboratory data are considered appropriate, i.e. data from TS-2 experiments can replace data from OECD 313 experiments and vice versa. It is expected that this applies also for other types of wood preservatives than the one included in this study.

Certain results are probably related to the applied wood preservative, i.e. additional components in the preservative. The effect of changes of L/A ratios on copper emission can be related to the amount and composition of co-eluting DOC. Changes of the chemical composition of the test specimens during ageing of the treated test specimens are assumed to cause increased leachability of copper. Therefore, a detailed description on preparation and conditioning of test specimens is required for leaching tests with treated wood. It has to be considered that leaching processes also depend on the actual composition of the test specimens when experimental leaching data are used to derive leaching characteristics.

Methods

Wood preservative

A water-based wood preservative of the Cu/triazole type containing 10% copper, 0.39% tebuconazole and ethanolamine was provided by a company for this study. The content of the active substances was confirmed by chemical analysis of the product. Tebuconazole (CAS 107534-96-3) is approved as an active substance for wood preservatives according to the European regulations on biocidal products. It acts as a fungicide. The molecular mass amounts to 307.8 g/mol, the water solubility of tebuconazole (technical) is 0.029 g/l (pH 7), the log K_{OW} is 3.5 (20°C) (K_{OW}: octanol/water partition coefficient) and the vapour pressure is 1.7×10^{-6} Pa (20°C). The substance is stable towards photolysis in aqueous solution and stable towards hydrolysis between pH 5 and 9 at 25°C [19].

Preparation of test specimens

Test specimens of *Pinus sylvestris* (sapwood) (50 × 5 × 2.5 cm^3, referred to as 'stakes') were treated by vacuum pressure impregnation with a 2.12% (w/w) solution of the wood preservative in water in a pilot plant. The retention of the preservative was 15.1 ± 0.2 kg/m^3 which is the amount required for use class 4 according to EN 335 [24] (i.e. wood intended to be in permanent contact with the ground or water). The test specimens were conditioned according to prEN 252 [25] at 20°C ±2°C, 65% ±5% relative humidity for a minimum of 7 weeks. Thirty-nine days after treatment, the specimens were cut to pieces of 15 × 5 × 2.5 cm^3 and sealed at the cut faces using a mixture of Sigillon I and II (polyurethane-based varnish) since these areas are actually not exposed to water in a construction work. The OECD 313 experiments were performed using one test specimen from this batch, but mainly using test specimens of 5 × 2.5 × 1.5 cm^3 as defined in EN 113 [26] that were vacuum-impregnated with the same preservative. The stakes and the test specimens according to EN 113 originate from different trees.

Leaching experiments

After conditioning, the treated test specimens were used to test the influence of the ambient temperature at ECN Petten (NL), whereas the tests under variation of L/A were performed at BAM. For the first series of experiments, the two 'outer' subsamples from the ends of identical stakes were allocated to the laboratories. The subsamples from the middle sections were kept as a reserve and were used for repeated experiments at BAM. The reported results originate from experiments at BAM if not specified otherwise.

The test specimens were placed in either glass or stainless steel containers and covered with the water volume required to achieve the specified L/A ratios. It was necessary to use different numbers, i.e. one to three test specimens per test assembly to achieve the required L/A ratios for the TS-2 tests in the test vessels. Certain test parameters are illustrated in Figure 4. The OECD 313 tests were performed in parallel to the first series of TS-2 tests. Water renewal schemes for TS-2 and OECD 313 experiments are given in Table 1. Mass data were recorded for the test specimens in OECD 313 experiments. Further details on experimental setup parameters are given in Table 3.

Analysis of eluates

The eluates from the experiments at BAM were analysed for DOC and TN by catalytic combustion and non-dispersive infrared (NDIR) detection according to EN 1484:1997 [27] using a Shimadzu TOC-VCPH analyzer (Shimadzu Corporation, Nakagyo-ku, Kyoto, Japan). Eluate samples from all laboratories were analysed at BAM for tebuconazole by liquid chromatography coupled with MS detection (Agilent 1100 Series with a 6130 Quadrupole LC/MS, Agilent Technologies, Santa Clara, CA, USA) without further sample preparation. The samples were separated on a reversed phase column (Phenomenex Luna C18(2) 3 µm, 50 × 2 mm, Phenomenex, Aschaffenburg, Germany) and eluted by a gradient of 0.2% acetic acid in water and acetonitrile. Quantification of tebuconazole was based on the signal at m/z 308. Each sample was analysed twice at minimum. The LOD was 0.0035 mg/l and relative standard deviation of parallel injections of samples was ≤2%. The samples for the analysis of copper were stabilised with a few drops of nitric acid and analysed at BAM by atomic absorption spectrometry using a UNICAM 696 AA spectrometer (Thermo Scientific, Waltham, MA, USA) (LOD 0.17 mg/l, relative standard deviation 2%).

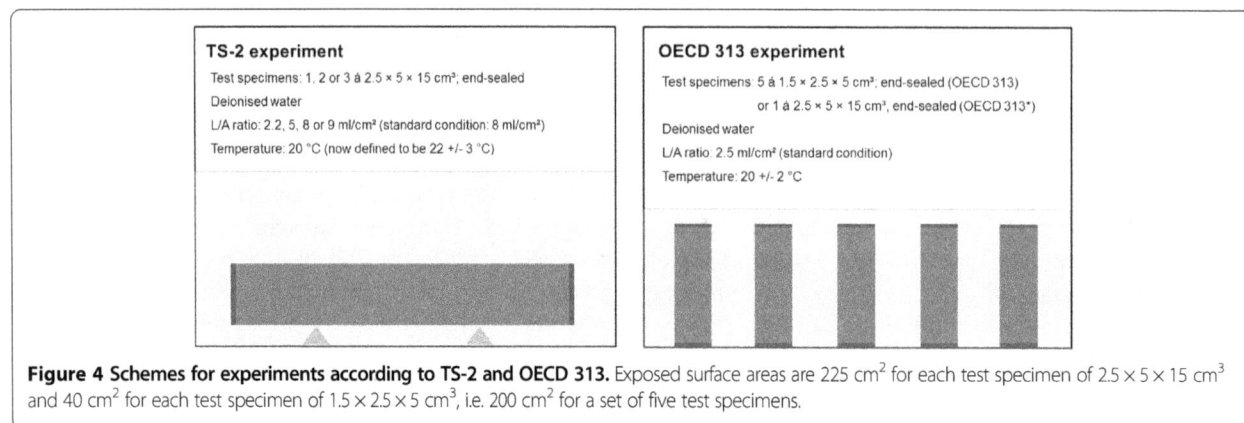

TS-2 experiment

Test specimens: 1, 2 or 3 á 2.5 × 5 × 15 cm³; end-sealed

Deionised water

L/A ratio: 2.2, 5, 8 or 9 ml/cm³ (standard condition: 8 ml/cm²)

Temperature: 20 °C (now defined to be 22 +/- 3 °C)

OECD 313 experiment

Test specimens: 5 á 1.5 × 2.5 × 5 cm³; end-sealed (OECD 313)
or 1 á 2.5 × 5 × 15 cm³, end-sealed (OECD 313*)

Deionised water

L/A ratio: 2.5 ml/cm² (standard condition)

Temperature: 20 +/- 2 °C

Figure 4 Schemes for experiments according to TS-2 and OECD 313. Exposed surface areas are 225 cm^2 for each test specimen of 2.5 × 5 × 15 cm^3 and 40 cm^2 for each test specimen of 1.5 × 2.5 × 5 cm^3, i.e. 200 cm^2 for a set of five test specimens.

Abbreviations
BAM: Bundesanstalt für Materialforschung und -prüfung; DOC: dissolved organic carbon; ECN: Energy Research Centre of the Netherlands); L/A: liquid (water) volume/exposed surface area; LOD: limit of detection; SD: standard deviation; TN: total nitrogen.

Competing interests
The authors declare that they have no competing interests.

Authors' contributions
All authors were substantially involved in the conception and design of the study and interpretation of the data. UK, AZ and US were responsible for the acquisition of the data. US performed the analysis of the data and drafted the manuscript. UK, AZ and OH critically revised the manuscript. All authors read and approved the final manuscript.

Acknowledgements
This work was partly funded by the EU Commission (Project No. 11810594). The authors thank the producer for the support with the wood preservative and T. Sommerfeld, Y. de Laval, K. Nordhauß and P.A. Bonouvrie who carefully performed the leaching tests and analysis of the eluates.

Author details
[1]Division 4.1 Biodeterioration and Reference Organisms, BAM Federal Institute for Materials Research and Testing, Unter den Eichen 87, D-12205 Berlin, Germany. [2]Division 4.3, Contaminant Transfer and Environmental Technologies, BAM Federal Institute for Materials Research and Testing, Unter den Eichen 87, D-12205 Berlin, Germany. [3]Department of Environmental Assessment, ECN, Westerduinweg 3, 1755 LE Petten, the Netherlands. [4]Waste and Soil, DHI, Agern Allé 5, DK-2970 Hørsholm, Denmark.

References
1. Regulation (EU) No. 305/2011 of the European Parliament and of the Council of 9 March 2011 laying down harmonised conditions for the marketing of construction products and repealing Council Directive 89/106/EEC. http://eur- lex.europa.eu/LexUriServ/LexUriServ.do?uri= OJ: L:2011:088:0005:0043:EN:PDF.
2. European Commission (EC): Mandate M/366. Horizontal complement to the mandates to CEN/CENELEC, concerning the execution of standardization work for the development of horizontal standardized assessment methods for harmonized approaches relating to dangerous substances under the Construction Products Directive (CPD). Emissions to indoor air, soil, surface water and ground water. Brussels: European Commission; 2005
3. Hjelmar O, Hyks J, Wahlström M, Laine-Ylijoki J, van Zomeren A, Comans R, Kalbe U, Schoknecht U, Krüger O, Grathwohl P, Wendel T, Abdelghafour M, Méhu J, Schiopu N, Lupsea M: Robustness validation of TS-2 and TS-3 developed by CEN/TC351/WG1 to assess release from products to soil, surface water and groundwater. Report 2013. https://www.nen.nl/web/file?uuid=9355006a-a1bc-416e-bd8c.
4. FprCEN/TS 16637-2:2013: Construction Products - Assessment of Release of Dangerous Substances – Part 2: Horizontal Dynamic Surface Leaching Test (Temporarily Named TS-2). CEN/TC 351: 2013. CEN-CENELEC Management Centre Brussels.
5. Lebow S: Leaching of Wood Preservative Components and Their Mobility in the Environment - Summary of Pertinent Literature. Madison: Gen. Tech. Rep. FPL-GTR-93. Forest Service, Forest Products Laboratory, U.S. Department of Agriculture; 1996.
6. Hingston JA, Moore J, Bacon A, Lester JN, Murphy RJ, Collins CD: The importance of the short-term leaching dynamics of wood preservatives. Chemosphere 2002, 47(5):517–523.
7. Hingston JA, Bacon A, Moore J, Collins CD, Murphy RJ, Lester JN: Influence of leaching protocol regimes on losses of wood preservative biocides. Bull Environ Contam Toxicol 2002, 68(1):118–125.
8. Lebow ST, Cooper P, Lebow PK: Variability in Evaluating Environmental Impacts of Treated Wood. Madison: Res. Pap. FPL-RP-620. Forest Service, Forest Products Laboratory, U.S. Department of Agriculture; 2004.
9. Schoknecht U, Mathies H, Wegner R, Melcher E, Seidel B, Kussatz C, Maletzki D: The Influence of Test Parameters on the Emission of Biocides from Preservative-Treated Wood in Leaching Tests. UFOPLAN 203 67 441. BAM Berlin: Research report; 2004.
10. Waldron L, Cooper PA, Ung TY: Prediction of long-term leaching potential of preservative-treated wood by diffusion modelling. Holzforschung 2005, 50(5):81–588.
11. Schoknecht U, Mathies H, Morsing N, Lindegaard B, van der Sloot HA, van Zomeren A, Deroubaix G, Legay S, Tadeo JL, Garcia-Valcárcel AI, Gigliottti G, Zadra C, Hajšlová J, Tomaniová M, Wegner R, Bornkessel C, Fürhapper C: Inter-laboratory Evaluation of Laboratory Test Methods to Estimate the Leaching from Treated Wood. BAM Berlin: European Grant Agreement no. 04/375757/C4, Report; 2005.
12. Townsend TG, Solo-Gabriele H: Environmental Impacts of Treated Wood. Boca Raton: CRC Press; 2006.
13. Lebow S, Lebow P, Foster D: Estimating preservative release from treated wood exposed to precipitation. Wood Fiber Sci 2008, 40(4):562–571.
14. Regulation (EU) No 528/2012 of the European Parliament and of the Council of 22 May 2012 concerning the making available on the market and use of biocidal products. http://eur-lex.europa.eu/legal-content/EN/TXT/PDF/?uri=CELEX:32012R0528&qid=1413179620144&from=EN.
15. Wood Held in the Storage Yard After Treatment and Wooden Commodities Exposed in Use Class 3 (Not Covered, Not In Contact with the Ground) - Laboratory Method. CEN/TC 38: 2008. CEN-CENELEC Management Centre Brussels.
16. CEN TS 15119-2:2012: Durability of Wood and Wood-Based Products, Determination of Emissions from Preservative Treated Wood to the Environment - Part 2: Wooden Commodities Exposed in Use Class 4 or 5 (In Contact with the Ground, Fresh Water or Sea Water) - Laboratory Method. CEN/TC 38: 2012. CEN-CENELEC Management Centre, Brussels; 2007.
17. OECD 313:2007: Estimation of Emissions from Preservative-Treated Wood to the Environment: Laboratory Method for Wooden Commodities That Are Not Covered And Are In Contact with Fresh Water or Seawater. OECD; 2007.
18. OECD Guidance for industry data submissions on the estimation of emissions from wood preservative-treated wood to the environment: for wood held in storage after treatment and for wooden commodities that are not covered and are not in contact with ground (July 2009), Series on Testing and Assessment No. 107. http://www.oecd.org/dataoecd/42/31/43411595.pdf.
19. Assessment report: tebuconazole (PT 8). http://ec.europa.eu/environment/biocides/.
20. Lupsea M, Mathies H, Schoknecht U, Tiruta-Barna L, Schiopu N: Biocide leaching from CBA treated wood - a mechanistic interpretation. Sci Total Environ 2013, 444:522–530.
21. Lupsea M, Tiruta-Barna L, Schiopu N, Schoknecht U: Modelling inorganic and organic biocide leaching from CBA-amine (copper-boron-azole) treated wood based on characterisation leaching tests. Sci Total Environ 2013, 461–462:645–654.
22. van Zomeren A, Meeussen HCL, van der Sloot HA: Characterisation of the leaching behaviour of preserved wood and results of modelling release. In Schoknecht et al.: Interlaboratory Evaluation of Laboratory Test Methods to Estimate the Leaching from Treated Wood. European Grant Agreement no. 04/375757/C4. 2005.
23. Tiruta-Barna L, Schiopu N: Modelling inorganic biocide emission from treated wood. J Hazard Mater 2011, 192:1476–1483.
24. EN 335:2013: Durability of Wood and Wood-Based Products - Use Classes: Definitions, Application to Solid Wood and Wood-Based Products. CEN/TC 38: 2013. CEN-CENELEC Management Centre Brussels.
25. prEN 252:2012: Field Test Method for Determining the Relative Protective Effectiveness of a Wood Preservative in Ground Contact. CEN/TC 38: 2012. CEN-CENELEC Management Centre Brussels.
26. EN 113:1996: Wood Preservatives - Method of Test for Determining the Protective Effectiveness Against Wood Destroying Basidiomycetes - Determination of the Toxic Values. CEN/TC 38: 1996. CEN-CENELEC Management Centre Brussels.
27. EN 1484:1997: Water Analysis - Guidelines for the Determination of Total Organic Carbon (DOC) and Dissolved Organic Carbon (DOC). CEN/TC 230: 1997. CEN-CENELEC Management Centre Brussels.

12

Microinjection into zebrafish embryos (*Danio rerio*) - a useful tool in aquatic toxicity testing?

Sophia Schubert*, Nadia Keddig, Reinhold Hanel and Ulrike Kammann

Abstract

Background: Microinjection was tested as a potentially powerful tool to introduce natural and anthropogenic pollutants directly into fish eggs to determine their toxicological impact on fish. With this technique, parental transfer of lipophilic contaminants may be mimicked. Here, we investigated the applicability of pollutant injection into the yolk of early zebrafish (*Danio rerio*) eggs with special regard on survival after vehicle injection. Tested vehicles were autoclaved tap water, dimethyl sulfoxide (DMSO), methanol, and triolein.

Results: Highest mortality occurred after the injection of DMSO and methanol. The lethality rates were up to 40% higher than under control conditions. Best survival rates were obtained after triolein and water injections. However, the triolein droplet was not assimilated by the embryo within 96 h post fertilization suggesting an incomplete uptake of triolein-solubilized chemicals. Technical aspects concerning microinjection in zebrafish eggs are discussed with special emphasis on quantitative injection.

Conclusions: Microinjection into the yolk cell of zebrafish eggs is feasible, but the application of exact volumes appears problematic. However, microinjection is a powerful tool for studies without the demand for high volume accuracy. Adopting microinjection for pollutant research requires further investigation.

Keywords: Toxicity testing; Microinjection; Zebrafish embryo; Vehicle injection; Triolein; Dimethyl sulfoxide; Methanol

Background

Early life stages often show a greater sensitivity towards contaminants than adults [1,2]. Hence, for environmental risk assessment, it is specifically important to determine the influence of contaminants during embryonic development. Early life stages of aquatic species including fish face different pollutant exposure routes. Besides maternal transfer, in which contaminants are mobilized during game to genesis together with parental fat reserves to build up ovaries [3-5], they experience waterborne exposure or get in direct contact to the sediment, immediately after the embryo is released into the environment.

In wild fish, significant pollutant concentrations have been found in both, oocytes [3,6] and spawned eggs [1,7] giving evidence that maternal pollutant transfer in fish cannot be neglected. The pollutant transfer from adults to offspring was also investigated in laboratory studies with medaka [5] and zebrafish [8,9]. In these studies, fish were exposed to pollutants and observed for several weeks. A potent way to mimic maternal pollutant transfer with the advantage to shorten the experimental duration offers the direct substance administration into the early fish egg via microinjection.

During the last decades, microinjection has been widely used in experimental biology. Microinjection allows, e.g., the production of transgenic cell lines or animals. It also offers the direct administration of supporting or harmful substances into cells to investigate their mode of action or toxic potential [10,11]. Unlike the classical Fish Embryo Toxicity Test, both polar and nonpolar substances can be administered and natural barriers, i.e., the chorion [12] and embryonic envelope, can be overcome. However, microinjection as a tool for the administration of contaminants has only rarely been tested. In the early

* Correspondence: sos-publication@gmx.net
Institute of Fisheries Ecology, Thünen Institute (TI), Palmaille 9, 22767 Hamburg, Germany

1990s, xenobiotics [13] and different organochlorines [14-16] were injected into rainbow trout eggs. This studies resulted in relatively high mortalities even for the control groups (≥30%) questioning the reliability of the method [17-19]. Since the year 2000, Japanese medaka became the most favorable species for toxicant application via microinjection [20-28].

Medaka is a common model organism for many laboratory purposes. It is easy to maintain, comprise a fast development, and its embryos are moderately transparent and therefore an adequate choice for developmental studies [17]. The yolk of medaka embryos has additional oil droplets comprising lipid reserves necessary as nutritional resources during development [10,11,29,30]. Prior to injection, pollutants were diluted into triolein to mimic one of the natural oil droplets. Other fish species were rarely tested in microinjection studies linked to toxicity evaluation so far. However, investigating the effects of injected toxicants to species other then medaka is important to develop more generalized toxicity levels.

Even though D. rerio is a common vertebrate model, it was so far not used for direct pollutant administration directly into the egg yolk to mimic maternal transfer. Due to the phylogenetic divergence of medaka and zebrafish, a comparative approach seems appropriate to test, whether the injection procedure itself is equally harmless to the embryos of zebrafish as compared to medaka. Thereby, we aim for basic parameters that are essential for a reliable microinjection procedure into the yolk cell of one-cell staged zebrafish embryos.

In a second step, we tested the effect of four different vehicle substances, i.e., water, dimethyl sulfoxide (DMSO), methanol, and triolein on zebrafish embryos. Water, DMSO, and methanol are regularly used as solvents in bioassays like the Fish Embryo Toxicity (FET) Test. Salts, alcohols, and acids are easily solved in water. Methanol is often used to extract other polar substances as secondary metabolites of plants and other biomaterials [31]. DMSO is known to improve the solubility of less polar contaminants

and is accepted not to be harmful to the developing individuals as long as their concentrations were kept below 2.5% v/v in the FET [31]. In contrast, triolein is presumably an excellent carrier for lipophilic substances. It was found to be the most promising vehicle administered in medaka embryos causing a very low mortality among them. Identifying a variety of vehicle substances offering the possibility to administer a wide range of polar and nonpolar substances directly into the yolk of eggs with one-cell staged zebrafish embryo is one major aim of our study. This will add basic knowledge to the field of environmental toxicity testing.

Results

The survival after vehicle injection ranged from 100% to 60% depending on the injected substance as well as the injection volume within 24 h post fertilization (hpf). Between 24 and 96 hpf, survival of injected embryos decreased not significantly (Table 1). Individuals injected with 4.2 nL DMSO or methanol showed significantly lower survival rates than zebrafish eggs after injection with similar volumes of autoclaved tap water or triolein (Figure 1A, B, Table 2). In general, the strength of an effect depends on injection volumes. Smaller injection volumes caused comparably less, but not significantly less mortality.

A DMSO injection of 0.5 nL resulted in 12% mortality 24 hpf. All other vehicle substances caused less mortality. However, high visible volume alterations occurred especially during DMSO and methanol injections. No secure statement of the real volume in place of the nominal volume can be made especially for methanol. Highest volume constancy was obtained after triolein injection. A distinct oil droplet appeared in the yolk sac, which was still visible 96 hpf (Figure 2).

Discussion

The experiments show that the microinjection of substances into the yolk of early egg stages of the zebrafish

Table 1 Survival rates (SR) of zebrafish (*Danio rerio*) eggs after vehicle injection

Substance	Eggs	Replicates	Volume [nL]	24 h SR [%]	SD	96 h SR [%]	SD
Untreated	718	6	-	100	0	98	3.6
Autoclaved water	121	3	0.5	100	0	95	0
	155	3	4.2	84	2.9	84	2.9
DMSO	158	3	0.5	88	5.9	88	8.7
	90	3	4.2	60	5.0	55	6.2
Methanol	125	3	0.5	97	0.9	96	1.6
	92	4	4.2	63	10.8	59	9.9
Triolein	94	3	0.5	97	2.4	96	0.4
	110	3	4.2	94	3.5	94	3.5

Untreated (pooled data) and after the pure injection of autoclaved tap water, DMSO, methanol, and triolein 24 and 96 h post fertilization (hpf), given with standard deviation (SD), respectively, total number of injected eggs, number of replicates, and injection volumes.

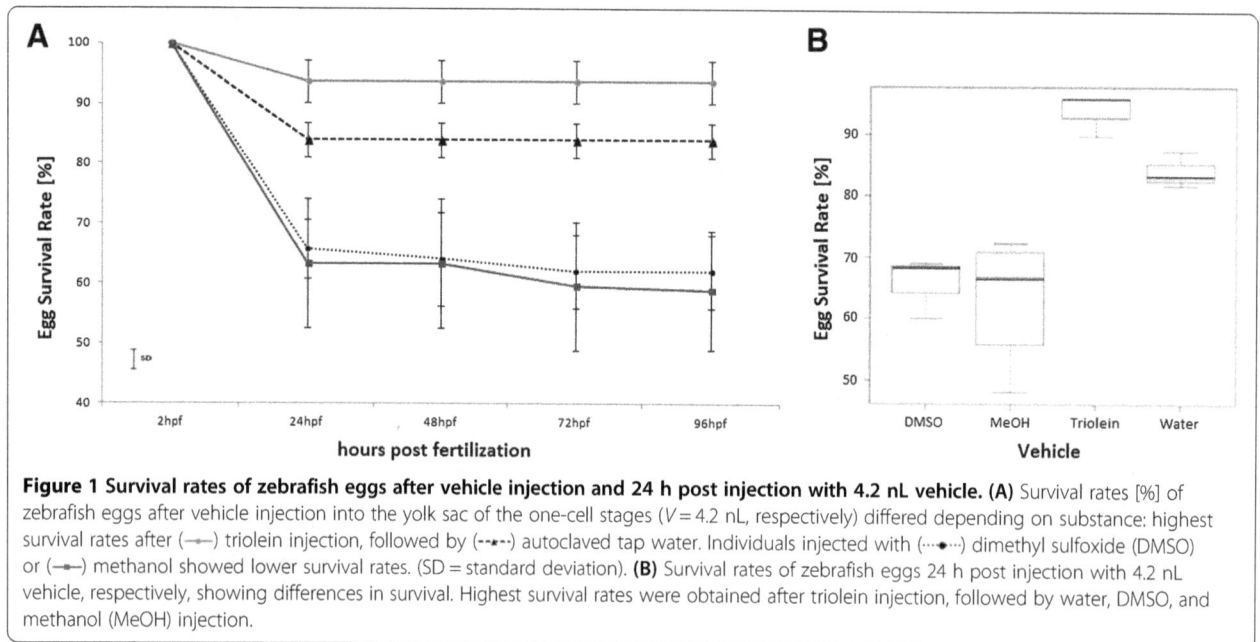

Figure 1 Survival rates of zebrafish eggs after vehicle injection and 24 h post injection with 4.2 nL vehicle. **(A)** Survival rates [%] of zebrafish eggs after vehicle injection into the yolk sac of the one-cell stages ($V = 4.2$ nL, respectively) differed depending on substance: highest survival rates after (—•—) triolein injection, followed by (--•--) autoclaved tap water. Individuals injected with (···•···) dimethyl sulfoxide (DMSO) or (—•—) methanol showed lower survival rates. (SD = standard deviation). **(B)** Survival rates of zebrafish eggs 24 h post injection with 4.2 nL vehicle, respectively, showing differences in survival. Highest survival rates were obtained after triolein injection, followed by water, DMSO, and methanol (MeOH) injection.

embryo is a feasible tool even though survival rates differ depending on the injected vehicle. Methanol and DMSO caused higher mortalities after injection into egg yolk than triolein or water. From this result, DMSO and methanol cannot be recommended as putative vehicles for microinjection into the yolk of one- to two-cell staged zebrafish eggs. The high mortality rate after DMSO injection matches previous results for rainbow trout (*Oncorhynchus mykiss*) eggs [15,16]. For *O. mykiss* mortality rates after the injection of DMSO, acetone, and dioxane were found to be higher than 60%. Even though dose-dependent survival rates could be shown, the high mortalities in the control groups challenged the reliability of the method [17-19].

Triolein was chosen for the application of lipophilic substances to mimic a maternal pollutant transfer. It has already been successfully used for microinjection into early egg stages of medaka (*Oryzias latipes*) carrying ciguatoxin [21,25], type B brevetoxin [24], and azaspiracid-1 [19] as well as anthropogenic substances as tributyltin [26], a DDT metabolite [23], pharmaceuticals [28], crude oil extracts [26], and polychlorinated naphthalenes [22]. In

contrast to zebrafish embryos, eggs of medaka naturally contain oil droplets which are involved in embryo developmental processes [10,11,29,30]. Mortality rates in vehicle controls tend to zero after the pure injection of triolein [20-28]. Our results concerning low mortalities post oil injection into the embryos of *D. rerio* are in accordance with the results for medaka. However, triolein was not assimilated by the zebrafish embryo until the termination of the experiment 96 hpf. We assume that a lipophilic pollutant dissolved in such a triolein droplet will not substantially affect the development of a zebrafish embryo during this time. It may be possible that an injected triolein droplet is assimilated later than 96 hpf, as zebrafish embryos completely consume their yolk sac during 165 ± 12 hpf [32]. To clarify whether the injected oil droplet is generally assimilated at a later point in development of zebrafish embryos, a prolongation of the experiment may be considered. However, in this case, the whole experiment needs permission by the Animal Welfare Act which would be contradictory to the general idea of the micro-injection to be an alternative to experiments with adult animals concerning maternal transfer.

Water could be shown to be an adequate vehicle causing mortalities below 20% in zebrafish embryos when low volumes were injected (≤ 4.2 nL). The injection volume seems to be an important factor for egg survival post injection. High injection volumes (≥ 4.2 nL) cause yolk sac swelling. Consequently, yolk sac content leaks through the injection piercing if it is not sealed. Survival rates can be enhanced by 14% by sealing immediately after vehicle injection as has been shown for salmonid eggs [15]. However, sealing of any injection hole is a time consuming procedure. Instead of sealing, we decided to

Table 2 Pairwise comparison of survival rates of zebrafish eggs 24 h after vehicle injection

	DMSO	MeOH	Triolein
MeOH	0.65795	-	-
Triolein	0.00084*	0.00029*	-
Water	0.01097**	0.00379*	0.12088

One-cell staged eggs were injected with 4.2 nL of either dimethyl sulfoxide (DMSO), methanol (MeOH), autoclaved tap water, or triolein. Differences between survival rates were *significant after Bonferroni correction ($p \leq 0.008$) and **significant without Bonferroni correction ($p \leq 0.05$).

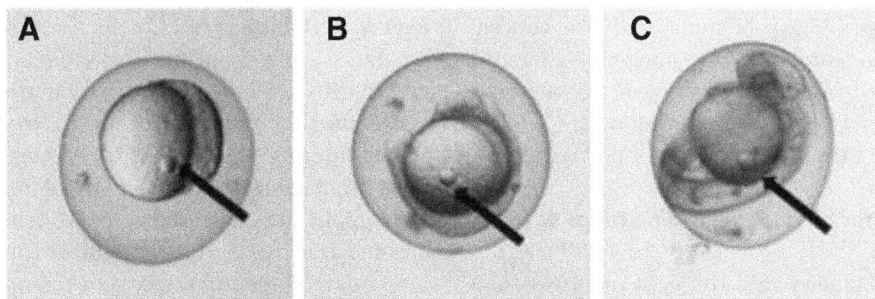

Figure 2 Developmental stages of a zebrafish embryo after injection of 4.2 nL triolein into the one-cell stage. Injected triolein droplet remains visible inside the yolk sac at **(A)** 2 hpf, **(B)** 24 hpf, and **(C)** 48 hpf.

use small injection volumes. The chosen volume of 4.2 nL does not exceed 10% of zebrafish egg volume, which was recommended to avoid egg trauma by Walker et al. [15]. In our experiments, no yolk sac leaking post egg injection was observed after the injection of 4.2 nL.

Survival of triolein-injected individuals showed no remarkable difference between the tested injection volumes. For water, methanol, and DMSO, we found higher survival rates for smaller injection volumes. In general, the injection volume is directly linked to the applied concentration. An increase in volume resulted in a simultaneous increase in dose. However, this is only true for methanol and DMSO, which are known to be toxic to the fish embryo in higher concentrations [31], but cannot be assumed for water. Water is generally nontoxic to the fish embryo. Hence, smaller injection volumes seem to cause less mortality [14]. However, a smaller injection volume needs less injection pressure and/or a shorter injection time, which appears advantageous at the first sight. But the risk of needle clogging is enhanced at the same time. Volume constancy over a series of injections may therefore not be warranted when low volumes are injected. We observed this effect especially during the injection of low methanol volumes. The injection needle was successively clogging, and an injection droplet was not always visible. Hence, the high survival rate after the injection of 0.5 nL methanol is possibly overestimated. Methanol as a putative vehicle was not tested before in any of the prior studies concerning microinjection as tool for pollutant administration. However, it is used in a few cases as substance carrier for, e.g., perfluorooctane sulfonate (PFOS) [33] and was shown to be tolerated up to a concentration of 2% by early zebrafish embryos in the Fish Embryo Toxicity Test [31].

Generally, volume fluctuations may be due to changes on needle tip diameters or depend on the viscosity of the cytoplasm of the injected cell [34]. During the experiments, volume variations appeared on a regular basis. As these volume differences result in concentration shifts, the nominal and the real concentrations differ

from each other by an unknown dimension. In ecotoxicology, high-quality data are essential and concentration variation as a consequence of volume oscillations affect data reliability. According to the OECD guideline for the Fish Embryo Toxicity Test, nominal and real concentrations should not differ more than ±20% from each other to assure data reliability [18]. As volume variations occur regularly and partly unnoticed throughout a series of injections, it is not warranted that the real concentration deviates maximal ±20% from the nominal concentration. Hence, the use of data derived from microinjection is presumably not feasible for the risk assessment of pollutants.

Furthermore, volume determination in general is difficult as injection volumes are in pico- to nanoliter range. Here, an object micrometer was used for the volume identification. Depending on the micrometer scale, volume determination is more or less accurate. It comprises a high-error risk. Differences in droplet diameter of 10 μm lead to approximately 10% differences in calculated injection volumes. To further reduce the high-error risk, it may be reasonable to use a micrometer with an even smaller scale. Due to the rapid development of a zebrafish egg [35] and therefore for time saving as well as practical reasons, we determined the injection volume prior and past a series of 40 to 60 injections of one vehicle in a droplet of mineral oil spotted on the micrometer scale [36]. Hence, volume oscillations between single injections were not observed. Probably the best way to circumvent this effect is to measure the diameter of the injected droplet inside the egg. But this is not feasible for any vehicle. A distinct injection droplet in the shape of a sphere was only visible after triolein administration. The droplets appearing after DMSO, methanol, or water injection were rather diffuse. They had either no distinct shape or appeared as several small injection droplets within the yolk. Volume determination after each injection was not feasible in any of the described scenarios.

Diameters of injected triolein have previously been determined for medaka. Due to the volume oscillations between single injections, the authors reported a volume range instead of a distinct injection volume for dose–response relationships [25,37]. Nominal and real concentrations are not identical for any single injection. As mentioned above, this approach may lead to uncertain data not feasible for any effect concentration (EC) calculation or to estimate the risk potency of a pollutant.

Recommendations for the use of microinjections as tool in toxicity testing

In general, toxicity studies with fish eggs require a sufficient number of healthy and freshly fertilized eggs for each concentration and control treatment to assure the gain of reliable dose–response relationships of the tested contaminant. Therefore, the use of laboratory fish as, e.g., zebrafish or medaka has the major advantage of constant egg supply throughout the year independent from seasons and environmental disturbances. In general, for zebrafish two spawning routes are established within different laboratories. These are mass and group spawning [17,18,38]. During mass spawning, relatively high numbers of eggs are produced once a day. Due to the rapid development of the zebrafish embryos [38,39] gaining a high number of eggs within a single egg stage is rather difficult and, therefore, laborious pre-selecting of egg stages may be necessary. Contrary, group spawning offers a smaller but constant egg supply throughout the day, when groups are assembled consecutively and small fish groups can spawn one after another [17,18]. Due to the rapid development of the zebrafish eggs and the demand to inject in at least similar egg stages to assure similar distribution patterns of the contaminant within the embryo, we advise to favor group spawning over mass spawning for further applications.

The use of laboratory fish assures a high value of reproducibility, even if a constant egg supply may experience variations in egg quality and quantity due to variations in the breed. To identify variations in egg quality, it is recommended to determine the spontaneous lethality rate of the fish breed. This rate is a measure for all mortality events putatively occurring during embryonic development without any influence of a harmful substance or microinjection. The spontaneous lethality rate is needed to keep tests reproducible and reliable throughout parallel and repeated tests, (KNK, SS, Wosniok W, submitted). It can be determined in short examinations prior to experiments. To distinguish already low contaminant effects from background mortality, we propose a correspondingly low spontaneous lethality rate (≤2.5%), which is in line with the results from Kammann et al. [40,41]. In comparison, the OECD recommends to discard datasets with a lethality equal or above 10% under control conditions [18,19]. Ali et al. [42] found a spontaneous lethality for zebrafish ELS test to be 9% in control treatments. Due to the purpose of a study, a small spontaneous lethality

rate may be essential especially when the authors are aiming for EC10 instead of EC50. In this case, distinguishing a substance effect from the background mortality is more crucial, (KNK, SS, Wosniok W, submitted), [43].

We further recommend for *D. rerio* injecting substances during the earliest developmental stages. The most homogeneous distribution of injected substances throughout the addressed compartment was achieved by keeping track of a GFP protein distribution during the initial developmental stages, i.e., the one-cell stage (data not shown). This strategy offers another advantage. Piercing the chorion by an injection needle may become more difficult with ongoing development of fish eggs due to the chorion hardening [44,45]. It is recommended that injection needles should be mostly inflexible and sharp to pierce a chorion even though it is already hardening and to avoid unwanted egg or needle damage [46].

Besides, the usage of appropriate needles, a sufficient supply of healthy and fertilized eggs, as well as the determination of the spontaneous lethality rate, we recommend examining the success rate. It describes the number of times when the contaminant is successfully injected into the addressed compartment. However, we propose that the success rate should be above 90%. Accepting a rate below this recommendation leads to the detection of either false-positive or even no results.

Conclusions

Microinjection is an easy way to administer substances into newly fertilized fish eggs comprising many advantages. Once established in the lab, it can be applied for many fish species with only minor modifications. Effects on embryonic development become visible almost immediately after injection. Even small contaminant effects can be distinguished from background mortality. In contrast to the classical Fish Embryo Toxicity Test, both polar and nonpolar substances can be administered and natural barriers, i.e., the chorion [12] and embryonic envelope, can be overcome. However, preventing volume oscillations across consecutive injections remain difficult. Therefore, the usage of microinjection as a tool for the calculation of dose–response relationships in terms of environmental risk assessment may be problematic.

Methods
Maintenance and egg production of zebrafish

Wild-type zebrafish brood stock was held in breeding groups of about 20 females and 30 males in the facilities of the Thünen Institute of Fisheries Ecology in Hamburg, Germany. Fish were kept in three glass aquaria (160 L) at 26°C ± 2°C and a light/dark period of 14 h/10 h in tap water. Water quality was maintained by external bioactive filter devices. Filter material and aquarium water were changed twice a week. Fish were fed *ad libitum* twice a day with dry flake food (Tetramin, Tetra Werke, Melle, Germany).

Embryos were obtained from mass spawning. Eggs were collected 30 min after the light was switched on and rinsed in aquarium water. Embryos were inspected under an inverted microscope and staged according to Kimmel et al. [47].

Assurance of egg quality/validation criteria

For valid experiments, eggs were obtained only from spawns with a fertilization rate higher than 70% according to the OECD guideline for fish-egg assays with zebrafish embryos [18]. Spontaneous lethality (SL) of the fish breed was determined on a regular basis and used as a measure of egg quality. In sterile 24-well plates, embryos were kept in groups of five eggs per well under standard test conditions without the influence of any toxicant for 96 h. Each well contained 1 mL of autoclaved tap water. The plates were kept at 26°C ± 2°C and a light/dark period of 14 h/10 h. Dissolved oxygen was not measured as no severe oxygen stress for the embryo was expected during the test procedure. Braunbeck et al. [17] reported that zebrafish eggs are capable to tolerate oxygen concentrations of 2 mg/L without the development of malformations. Furthermore, even 100 µL water per egg was tested indicating no appearing oxygen stress. Here, the used water volume per egg was at least 200 µL.

In three independent tests, each with 120 zebrafish eggs, mean SL was determined. Additionally, each microinjection test comprised a negative control containing 60 eggs, which were neither treated with any pollutant nor subjected to the microinjection procedure itself. In general, for valid test procedures the control group needed to have a survival rate higher than 90% [18,19].

Chemicals and substances

Injection vehicles were purchased and prepared as follows: methanol (99,9%) was obtained from Merck KGaA, Darmstadt, Federal State of Hesse, Germany. Triolein (≥99%) and DMSO (CHROMASOLV, 99.7%) were purchased from Sigma-Aldrich, Seelze, Federal State of Lower Saxony, Germany. All substances were used undiluted. High-quality tap water (≤8°dH) was autoclaved prior to experiments. Aliquots of vehicles were kept in the fridge at 4°C prior to experiments.

Microinjection

Preparation

Injection needles (needle types: O.D. = 20 µm; BM100T-10, ends fire polished, beveled, Biomedical Instruments, Germany or Femtotip II, Eppendorf, Germany) were backfilled with 10 µL substance by a Microloader (Eppendorf, Germany). The needle was placed in the microinjection manipulator (Narishige MN-151, Narishige International

Limited, London, United Kingdom) connected to a pneumatic microinjection pump (FemtoJet from Eppendorf, Germany). Injections were made with × 20 magnification under an inverted microscope (Nikon MSZ800, Nikon GmbH, Düsseldorf, Federal State of North Rhine-Westphalia, Germany). Injection volume was determined according to Sive et al. [37] by an object micrometer (Bresser 1/10 mm, Bresser GmbH, Rhede, Federal State of North Rhine-Westphalia, Germany). Briefly, for the three vehicles, DMSO, methanol, and water, a mineral droplet was mounted on the scale of the object micrometer. For triolein, a drop of water was used as matrix for measurements. The arising vehicle sphere was measured with the scale of the object micrometer. Each vehicle was administered several times into the oil droplet until the target injection volume was achieved. According to the sphere volume formula ($V = 1/6\pi d^3$), a sphere diameter of 1 bar corresponded to an injection volume of 0.5 nL. Two bars corresponded to 4.2 nL. Injection volume needed to be measured and adjusted for every solution, concentration as well as for every control prior and past injections due to the putative variations at the needle tip during injection procedure.

Procedure

One-cell stadium zebrafish embryos were stringed at the edge of a microscope slide placed in a petri dish. Surplus water was removed with a paper towel such that the eggs were immobilized during the injection procedure. Per treatment between 40 and 60 eggs was consistently injected with triolein, DMSO, methanol, or autoclaved water, into the yolk. Each treatment was tested in triplicate.

To avoid needle clogging by any of the injection substances, capillaries were cleaned at frequent intervals. Post injection eggs were rinsed with autoclaved tap water (26°C ± 2°C) into a petri dish. After 2 h, viable eggs were separated from coagulated and/or non-fertilized eggs and transferred in groups of two to five individuals into the wells of a 24-well plate. Viable eggs were kept in 1 mL autoclaved tap water at 26°C ± 2°C and a 14 h/10 h-light/dark rhythm. Eggs were checked for coagulation and malformation every 24 h. Experimentation was terminated at latest 96 h post fertilization (hpf).

Statistics

Survival rates of 24 h-old zebrafish embryos injected with 4.2 nL of each vehicle were normally distributed (p value = 0.69). A Fligner-Killeen test of homogeneity of variances yielded not significant (df = 3, p value = 0.50). As a consequence, we performed a one-way analysis of means not assuming equal variances. It ended up in a significant differences between treatment groups (p value = 0.003). To identify the location of these differences, we chose a pairwise t-test which was Bonferroni corrected ($p \leq 0.008$).

The entire statistic was performed with the freeware 'R' [48]. All data are represented as means ± standard deviation (SD).

Abbreviations

ANOVA: analysis of variances; DDT: dichlorodiphenyltrichloroethane; DMSO: dimethyl sulfoxide; EC: effect concentration; hpf: hours post fertilization; MeOH: methanol; SD: standard deviation; SL: spontaneous lethality; SR: survival rate; TCDD: 2,3,7,8-tetrachlorodibenzo-p-dioxin.

Competing interests

The authors declare that they have no competing interests.

Authors' contributions

SS has designed the study, assembled, analyzed, and interpreted the data and wrote the manuscript. NK has made substantial contributions to the conception and design of the study and has been involved in drafting the manuscript and revising it critically for important intellectual content. RH and UK have been involved in drafting the manuscript and revising it critically for important intellectual content. All authors read and approved the final manuscript.

Acknowledgements

The authors thank Manfred Trenk and Marc Willenberg for their skillful technical assistance as well as Malte Damerau, Wolfgang Gerwinski, Michael Haarich, Norbert Theobald, Jochen Trautner, and Werner Wosniok for their thematic support. This study was incorporated into 'MERIT-MSFD: Methods for detection and assessment of risks for the marine ecosystem due to toxic contaminants in relation to implementation of the European Marine Strategy Framework Directive'. It was supported by a grant (grant number 10017012) from the German Federal Ministry of Transport and Digital Infrastructure (BMVI) and the German Maritime and Hydrographic Agency (BSH).

References

1. Russell RW, Gobas FAPC, Haffner GD: **Maternal transfer and in ovo exposure of organochlorines in oviparous organisms: a model and field verification.** *Environ Sci Technol* 1999, 33(Suppl):1–15.
2. Hutchinson TH, Solbe J, Kloepper-Sams PJ: **Analysis of the ECETOC aquatic toxicity (EAT) database III - comparative toxicity of chemical substances to different life stages of aquatic organisms.** *Chemosphere* 1998, 36:129–142.
3. Serrano R, Blanes MA, López FJ: **Maternal transfer of organochlorine compounds to oocytes in wild and farmed gilthead sea bream (*Sparus aurata*).** *Chemosphere* 2008, 70:561–566.
4. Russell RW, Gobas FAPC, Haffner GD: **Maternal transfer and in ovo exposure of organochlorines in oviparous organisms: a model and field verification.** *Environ Sci Technol* 1999, 33:416–420.
5. Van de Merwe JP, Chan AKY, Lei ENY, Yau MS, Lam MHW, Wu RSS: **Bioaccumulation and maternal transfer of PBDE 47 in the marine medaka (*Oryzias melastigma*) following dietary exposure.** *Aquat Toxicol* 2011, 103:199–204.
6. Von Westernhagen H, Rosenthal H: **Bioaccumulating substances and reproductive success in Baltic flounder *Platichthys flesus*.** *Aquat Toxicol* 1981, 1:85–99.
7. Peng H, Wei Q, Wan Y, Giesy JP, Li L, Hu J: **Tissue distribution and maternal transfer of poly- and perfluorinated compounds in Chinese sturgeon (*Acipenser sinensis*): implications for reproductive risk.** *Environ Sci Technol* 2010, 44:1868–1874.
8. Nyholm JR, Norman A, Norrgren L, Haglund P, Andersson PL: **Maternal transfer of brominated flame retardants in zebrafish (*Danio rerio*).** *Chemosphere* 2008, 73:203–208.
9. King Heiden TC, Struble CA, Rise ML, Hessner MJ, Hutz RJ, Carvan MJ III: **Molecular targets of 2,3,7,8-tetrachlorodibenzo-p-dioxin (TCDD) within the zebrafish ovary: insights into TCDD-induced endocrine disruption and reproductive toxicity.** *Reprod Toxicol* 2008, 25:47–57.
10. Dodd A, Curtis PM, Williams LC, Love DR: **Zebrafish: bridging the gap between development and disease.** *Hum Mol Genet* 2000, 9:2443–2449.
11. Spitsbergen J, Kent M: **The state of the art of the zebrafish model for toxicology and toxicologic pathology research - advantages and current limitations.** *Toxicol Pathol* 2003, 31:62–87.
12. Mizell M, Romig ES: **The aquatic vertebrate embryo as a sentinel for toxins: zebrafish embryo dechorionation and perivitelline space microinjection.** *Int J Dev Biol* 1997, 41:411–423.
13. Norrgren L, Andersson T, Björk M: **Liver morphology and cytochrome P450 activity in fry of rainbow trout after microinjection of lipid-soluble xenobiotics in the yolk-sac embryos.** *Aquat Toxicol* 1993, 26:307–316.
14. Zabel E, Cook P, Peterson R: **Toxic equivalency factors of polychlorinated dibenzo-p-dioxin, dibenzofuran and biphenyl congeners based on early life stage mortality in rainbow trout (*Onchorhynchus mykiss*).** *Aquat Toxicol* 1995, 31:315–328.
15. Walker MK, Hufnagle LCJ, Clayton MK, Peterson RE: **An egg injection method for assessing early life stage mortality of polychlorinated dibenzo-p-dioxins, dibenzofurans, and biphenyls in rainbow trout, (*Onchorhynchus mykiss*).** *Aquat Toxicol* 1992, 22:15–38.
16. Walker MK, Spitsbergen JM, Olson JR, Peterson RE: **2, 3, 7, 8-Tetrachlorodibenzo-p-dioxin (TCDD) toxicity during early life stage development of lake trout (*Salvelinus namaycush*).** *Can J Fish Aquat Sci* 1991, 48:875–883.
17. Braunbeck T, Boettcher M, Hollert H, Kosmehl T, Lammer E, Leist E, Rudolf M, Seitz N: **Towards an alternative for the acute fish LC(50) test in chemical assessment: the fish embryo toxicity test goes multi-species - an update.** *ALTEX* 2005, 22:87–102.
18. OECD: *Fish Embryo Acute Toxicity (FET) Test. Test Guideline No. 236. OECD Guidelines for the Testing of Chemicals.* Paris: OECD Publishing; 2013:1–22. [*OECD Guidelines for the Testing of Chemicals, Section 2*]. www.oecd.org.
19. ASTM E729-96: *Standard Guide for Conducting Acute Toxicity Tests on Test Materials with Fishes, Macroinvertebrates, and Amphibians.* West Conshohocken, PA, United States: ASTM International; 2007:1–22. www.astm.org.
20. Colman JR, Twiner MJ, Hess P, McMahon T, Satake M, Yasumoto T, Doucette GJ, Ramsdell JS: **Teratogenic effects of azaspiracid-1 identified by microinjection of Japanese medaka (*Oryzias latipes*) embryos.** *Toxicon* 2005, 45:881–890.
21. Edmunds JS, McCarthy RA, Ramsdell JS: **Ciguatoxin reduces larval survivability in finfish.** *Toxicon* 1999, 37:1827–1832.
22. Villalobos SA, Papoulias DM, Meadows J, Blankenship AL, Pastva SD, Kannan K, Hinton DE, Tillitt DE, Giesy JP: **Toxic responses of medaka, d-rR strain, to polychlorinated naphthalene mixtures after embryonic exposure by in ovo nanoinjection: a partial life-cycle assessment.** *Environ Toxicol Chem* 2000, 19:432–440.
23. Villalobos SA, Papoulias DM, Pastva SD, Blankenship AL, Maedows J, Tillit DE, Giesy JP: **Toxicity of o, p′-DDE to medaka d-rR strain after a one-time embryonic exposure by in ovo nanoinjection: an early through juvenile life cycle assessment.** *Chemosphere* 2003, 53:819–826.
24. Colman JR, Ramsdell JS: **The type B brevetoxin (PbTx-3) adversely affects development, cardiovascular function, and survival in medaka (*Oryzias latipes*) embryos.** *Environ Health Perspect* 2003, 111:1920–1925.
25. Colman JR, Dechraoui M-YB, Dickey RW, Ramsdell JS: **Characterization of the developmental toxicity of Caribbean ciguatoxins in finfish embryos.** *Toxicon* 2004, 44:59–66.
26. Escoffier N, Gaudin J, Mezhoud K, Huet H, Chateau-Joubert S, Turquet J, Crespeau F, Edery M: **Toxicity to medaka fish embryo development of okadaic acid and crude extracts of Prorocentrum dinoflagellates.** *Toxicon* 2007, 49:1182–1192.
27. Hano T, Oshima Y, Kim SG, Satone H, Oba Y, Kitano T, Inoue S, Shimasaki Y, Honjo T: **Tributyltin causes abnormal development in embryos of medaka, *Oryzias latipes*.** *Chemosphere* 2007, 69:927–933.
28. Nassef M, Kim SG, Seki M, Kang IJ, Hano T, Shimasaki Y, Oshima Y: **In ovo nanoinjection of triclosan, diclofenac and carbamazepine affects embryonic development of medaka fish (*Oryzias latipes*).** *Chemosphere* 2010, 79:966–973.
29. Gonzalez-Doncel M, Okihiro MS, Villalobos SA, Hinton DE, Tarazona JV: **A quick reference guide to the normal development of *Oryzias latipes* (Teleostei, Adrianichthyidae).** *J Appl Ichthyol* 2005, 21:39–52.
30. Iwamatsu T: **Stages of normal development in the medaka *Oryzias latipes*.** *Mech Dev* 2004, 121:605–618.
31. Maes J, Verlooy L, Buenafe OE, de Witte PAM, Esguerra CV, Crawford AD: **Evaluation of 14 organic solvents and carriers for screening applications in zebrafish embryos and larvae.** *PLoS One* 2012, 7:e43850.

32. Jardine D, Litvak MK: **Direct yolk sac volume manipulation of zebrafish embryos and the relationship between offspring size and yolk sac volume.** *J Fish Biol* 2003, **63**:388–397.

33. Sharpe RL, Benskin JP, Laarman AH, Macleod SL, Martin JW, Wong CS, Goss GG: **Perfluorooctane sulfonate toxicity, isomer-specific accumulation, and maternal transfer in zebrafish (***Danio rerio***) and rainbow trout (***Oncorhynchus mykiss***).** *Environ Toxicol Chem* 2010, **29**:1957–1966.

34. Minaschek G, Bereiter-Hahn J, Bertholdt G: **Quantitation of the volume of liquid injected into cells by means of pressure.** *Exp Cell Res* 1989, **183**:434–442.

35. Kimmel CB, Law RD: **Cell lineage of zebrafish blastomeres.** *Dev Biol* 1985, **108**:86–93.

36. Sive H, Grainger R, Harland R: **Calibration of the injection volume for microinjection of Xenopus oocytes and embryos.** *Cold Spring Harb Protoc* 2010, **2010**:1382–1383.

37. Edmunds JS, McCarthy RA, Ramsdell JS: **Permanent and functional male-to-female sex reversal in d-rR strain medaka (Oryzias latipes) following egg microinjection of o, p′-DDT.** *Environ Health Perspect* 2000, **108**:219–224.

38. M W: *The Zebrafish Book. A Guide for the Laboratory Use of Zebrafish (Danio Rerio).* 4th edition. Eugene: Univ. of Oregon Press; 2000.

39. Kimmel CB, Spray DC, Bennett MV: **Developmental uncoupling between blastoderm and yolk cell in the embryo of the teleost Fundulus.** *Dev Biol* 1984, **102**:483–487.

40. Kammann U, Vobach M, Wosniok W: **Toxic effects of brominated indoles and phenols on zebrafish embryos.** *Arch Environ Contam Toxicol* 2006, **51**:97–102.

41. Kammann U, Vobach M, Wosniok W, Schäffer A, Telscher A: **Acute toxicity of 353-nonylphenol and its metabolites for zebrafish embryos.** *Environ Sci Pollut Res Int* 2009, **16**:227–231.

42. Ali S, van Mil HGJ, Richardson MK: **Large-scale assessment of the zebrafish embryo as a possible predictive model in toxicity testing.** *PLoS One* 2011, **6**:e21076.

43. Scholz S, Fischer S, Gündel U, Küster E, Luckenbach T, Voelker D: **The zebrafish embryo model in environmental risk assessment - applications beyond acute toxicity testing.** *Environ Sci Pollut Res Int* 2008, **15**:394–404.

44. Henn K, Braunbeck T: **Dechorionation as a tool to improve the fish embryo toxicity test (FET) with the zebrafish (***Danio rerio***).** *Comp Biochem Physiol C Toxicol Pharmacol* 2011, **153**:91–98.

45. Robles V, Cabrita E, de Paz P, Herráez MP: **Studies on chorion hardening inhibition and dechorionization in turbot embryos.** *Aquaculture* 2007, **262**:535–540.

46. Pase L, Lieschke GJ: **Validating microRNA target transcripts using zebrafish Assays.** *Methods Mol Biol* 2009, **546**:227–240.

47. Kimmel CB, Ballard WW, Kimmel SR, Ullmann B, Schilling TF: **Stages of embryonic development of the zebrafish.** *Dev Dyn* 1995, **203**:253–310.

48. R Core Team. **R: a language and environment for statistical computing.** Vienna, Austria: R Foundation for Statistical Computing. 2013; www.R-project.org.

13

The European technical report on aquatic effect-based monitoring tools under the water framework directive

Ann-Sofie Wernersson[1], Mario Carere[2*], Chiara Maggi[3], Petr Tusil[4], Premysl Soldan[4], Alice James[5], Wilfried Sanchez[5], Valeria Dulio[5], Katja Broeg[6], Georg Reifferscheid[7], Sebastian Buchinger[7], Hannie Maas[8], Esther Van Der Grinten[9], Simon O'Toole[10], Antonella Ausili[3], Loredana Manfra[3], Laura Marziali[11], Stefano Polesello[11], Ines Lacchetti[2], Laura Mancini[2], Karl Lilja[12], Maria Linderoth[12], Tove Lundeberg[12], Bengt Fjällborg[1], Tobias Porsbring[1], DG Joakim Larsson[13], Johan Bengtsson-Palme[13], Lars Förlin[13], Cornelia Kienle[14], Petra Kunz[14], Etienne Vermeirssen[14], Inge Werner[14], Craig D Robinson[15], Brett Lyons[16], Ioanna Katsiadaki[16], Caroline Whalley[17], Klaas den Haan[18], Marlies Messiaen[19], Helen Clayton[20], Teresa Lettieri[21], Raquel Negrão Carvalho[21], Bernd Manfred Gawlik[21], Henner Hollert[22], Carolina Di Paolo[22], Werner Brack[23], Ulrike Kammann[24] and Robert Kase[14]

Abstract

The Water Framework Directive (WFD), 2000/60/EC, requires an integrated approach to the monitoring and assessment of the quality of surface water bodies. The chemical status assessment is based on compliance with legally binding Environmental Quality Standards (EQSs) for selected chemical pollutants (priority substances) of EU-wide concern. In the context of the mandate for the period 2010 to 2012 of the subgroup Chemical Monitoring and Emerging Pollutants (CMEP) under the Common Implementation Strategy (CIS) for the WFD, a specific task was established for the elaboration of a technical report on aquatic effect-based monitoring tools. The activity was chaired by Sweden and co-chaired by Italy and progressively involved several Member States and stakeholders in an EU-wide drafting group. The main aim of this technical report was to identify potential effect-based tools (e.g. biomarkers and bioassays) that could be used in the context of the different monitoring programmes (surveillance, operational and investigative) linking chemical and ecological status assessment. The present paper summarizes the major technical contents and findings of the report.

Keywords: Effect-based tools; Priority substances; Aquatic ecosystems; Water Framework Directive; Bioassays; Chemical pollution; Biomarkers

Review

Introduction

The Water Framework Directive (WFD), 2000/60/EC, [1] requires an integrated approach to the monitoring and assessment of the quality of surface water bodies in the European Union. The assessment of ecological status takes into account the effects at the population and community levels, based on the use of specific indices and ecological quality ratios. The chemical status assessment is based on compliance with legally binding Environmental Quality Standards (EQSs) for 53 selected chemical pollutants (priority substances) of EU-wide concern [2].

Chemical analysis generally requires *a priori* knowledge about the type of substances to be monitored as, for technical and economic reasons, it is not possible to analyse, detect and quantify all substances that are present in the aquatic environment. Even for the thousands of unique substances registered under Registration, Evaluation, Authorisation and Restriction of Chemicals (REACH) it would be highly challenging to perform a chemical

* Correspondence: mario.carere@iss.it
[2]Department of Environment and Primary Prevention, ISS-Italian Institute of Health, Viale Regina Elena, 299, 00161 Rome, Italy
Full list of author information is available at the end of the article

monitoring programme on the aquatic environment. Furthermore, to estimate the risk of effects related to the large number of substances that are present and detected in the environment (including pollutants of emerging concern, metabolites and transformation products), it would be necessary to develop a very large number of assessment criteria (EQS). Such assessment criteria for chemicals are generally developed substance by substance, based on laboratory studies, and usually do not consider the consequences of simultaneous exposure to multiple chemicals [3] occurring in the environment, possibly giving rise to cumulative effects [4].

In the mandate for 2010 to 2012 of the European subgroup Chemical Monitoring and Emerging Pollutants (CMEP) under the WG "Chemical Aspects" of the Common Implementation Strategy (CIS) for the WFD, a specific task was foreseen for the elaboration of a technical report on effect-based tools [5]. The activity was chaired by Sweden and co-chaired by Italy and progressively involved several Member States and stakeholders in an EU-wide drafting group (47 experts). According to the mandate from the CMEP, the aim of the report was to identify potential effect-based tools (e.g. bioassays, biomarkers and ecological indicators) that could be used in the context of the different monitoring programmes (surveillance, operational and investigative) linking the chemical and ecological status assessment.

Technical report
The technical report on aquatic effect-based monitoring tools (Additional file 1) [6] aims at presenting the state of the art of aquatic effect-based monitoring tools and describing how these tools can help EU Member States to make monitoring programmes more efficient (including reduction of monitoring costs). The report further contains specific sections on the use of such tools in marine systems such as the Regional Seas Conventions and the Marine Strategy Framework Directive (MSFD) 2008/56/EC. The MSFD has foreseen the use of effect-based tools: in particular, the indicators related to Descriptor 8 (contaminants and pollution effects) of the MSFD should include effects from hazardous substances on ecosystem components.

For reasons of clearness, the tools described in the report are categorised into three main groups, primarily depending on the type of monitoring approach used:

1) Bioassays, both *in vitro* and *in vivo*, which measure the toxicity of environmental samples under defined laboratory conditions, on cellular or individual levels, respectively.
2) Biomarkers, i.e. biological responses at the cellular or individual levels, measured in field-exposed organisms.

3) Ecological methods, measuring changes observed at higher biological organisation levels, i.e. the population and/or community.

In Europe, several of the tools described in the report are already used for both marine and limnic applications [7,8]. Biomarkers are included in the monitoring programmes of Regional Seas Conventions to detect the presence of substances or combinations of substances not previously identified as a concern and to identify regions of decreased environmental quality. Bioassays are used for example to support risk assessment and management of contaminated sediments and provide decision support for reducing the release of toxic substances into the environment (e.g. in the evaluation of dredged sediments that are considered for sea disposal and whole effluent assessments in the permitting process). They are also used in broad screening of different pollutant sources (such as sewage treatment plant effluents). Other applications include, for example, alarm systems directly triggering control measures (e.g. closing drinking water intakes). Effect-based tools support also the ecotoxicological characterisation and classification of hazardous wastes in the context of the Waste Framework Directive (2008/98/EC). Specific sections of the report are dedicated to EDA/TIE approaches and OMICS.

The report was approved by the CMEP subgroup in Gent, Belgium, (October 2012), by the Working Group on Chemical Aspects in Brussel, Belgium (April 2013), by the Strategic Coordination Group (SCG) of the WFD in Brussel (October 2013) and endorsed by the Water Director Meeting in Vilnius, Lithuania (December 2013).

Effect-based tools in the WFD
The WFD mandates three monitoring programmes:

Surveillance monitoring aimed to supplement and validate an impact analysis, support efficient and effective design of future monitoring programmes and assess long-term changes in natural conditions and changes resulting from anthropogenic activity. Monitoring is performed at least once every management cycle (usually every 6 years).

Operational monitoring aimed to establish the status of water bodies identified as being at risk of failing to meet the WFD environmental objectives and assess any changes in the status resulting from the programme of measures.

Investigative monitoring aimed to determine reasons for exceedances or predicted failure to achieve environmental objectives if the reasons are not already known and to determine the magnitude and impacts of accidental pollution.

As with all other components of a monitoring programme, it is important to assess the suitability of different effect-based tools against the specified objectives of the monitoring programmes [9]. The suitability of any particular approach must be evaluated in terms of method, cost, practicality and capability to provide information that can be translated into management practices useful for achieving the monitoring programme objectives.

As it has been already evident from previous CIS guidance documents, it is possible to identify several objectives for the use of effect-based tools in a WFD context, some of them are mentioned below:

- As screening tools, in the framework of the pressures and impacts assessment to aid in the prioritisation of water bodies.
- To establish early warning systems.
- To prioritise further studies in areas that are not identified as being at risk because they are located far from known local sources.
- To take the effects of chemical mixtures or chemicals that are not analysed into account (e.g. to support investigative monitoring where causes of a decline of specific species are unknown).
- To provide additional support in water and sediment quality assessment, though not as a replacement for conventional chemical and ecological monitoring under the WFD.

Effect-based tools are particularly suitable as part of investigative monitoring programmes, for which the regulatory requirements are determined less formally. However, as with any investigative monitoring, the optimum set of tools varies on a case-by-case basis. The optimal approach will frequently involve several effect-based tools as well as chemical analysis, as illustrated by several of the case studies described in the Appendix section of the report. To optimize cost-effectiveness, it is often wise to make use of the same samples for both chemical and effect-based analyses.

Bioassays *in vitro*

The use of *in vitro* assays is increasing for ethical reasons in order to comply with the regulations on animal experimentation. They measure effects at the subcellular level, such as receptor activation and DNA damage, rather than investigate cells or tissues of organisms [10] exposed in the field (as it is the case with biomarkers), the effects are studied in cells after exposure to environmental samples [11]. An advantage of this approach is that *in vitro* bioassays can often be performed on many different matrixes (such as concentrated extracts of surface water, sediment or pore water samples, biological tissues, passive samplers and effluents). Additional advantages are that only small amounts of sample are generally needed and exposure time is short compared to the time needed for an *in vivo* assay to detect a response.

In most cases, *in vitro* assays are considered highly sensitive, because they measure effects at a low organisational level. Many *in vitro* assays are suitable for screening and high-throughput/automated applications and can be added to the analytical tool package at comparatively lower costs (especially if taking into account the number of substances they respond to).

Table 1 presents some *in vitro* assays that were nominated for monitoring purposes in a Swedish workshop [12]. The table also includes information about the mode of action the assay responds to.

The use of *in vitro* assays is also common outside Europe; for example, they are largely used in the context of the U.S. A. Programme US TOX 21. For some *in vitro* bioassays, results are expressed in chemical equivalents, comparing the response induced by the sample to that induced by the reference chemical (positive control). However, before comparing such bioassay results to water quality criteria developed for single chemicals or a number of specific chemicals, it is necessary to consider that the assay can respond to different combined substances that have the same mode of action, e.g. via receptor activation. Generally, for a mixture of agonistic substances with the same mode of action, the biological signal is higher than for a single substance which makes *in vitro* assays highly suitable as screening tools for environmental samples. They integrate

Table 1 *In vitro* assays and their modes of action

Name/s of assay	Mode of action/endpoint
AR CALUX (anti-)	Androgen receptor (activation or blocking)
DR CALUX	AH receptor binding
ER CALUX (anti-)	Alpha and beta/estrogen receptors
GR CALUX (anti-)	Glucocorticoid receptor
PAH CALUX	AH receptor binding
PR CALUX	Progesterone receptor
Acetylcholinesterase inhibition assay	Inhibition of acetylcholinesterase activity
Carboxylesterase inhibition assay	Inhibition of carboxylesterase activity
Ames	Mutagenicity
umuC	Primary DNA damage
TTR-binding	Competition with thyroid hormone for binding to TTR (transport protein)
TRb CALUX	Thyroid receptor beta
EROD	EROD induction
YES	ER receptor
YAS	AR receptor
P-53 accumulation	Genotoxicity
Green screen	Genotoxicity
RYA	ER receptor
ABC assay	Antibiotic activity

effects of all substances with the same mode of action, e.g. estrogen receptor binding. As such integrative detection tools, *in vitro* assays are also able to quantify and distinguish agonistic and antagonistic effects.

Although many *in vitro* bioassays can be used on any matrix/extract, some are more suited for the assessment of certain matrices than others, in part because they have been so far only validated for certain uses but also because relevant substances eliciting certain types of responses are primarily found in certain compartments. When using *in vitro* bioassays, we must be aware that we use highly specific systems to detect chemicals interacting with a specific cellular receptor. These cellular events have been shown to be an important link in the development of specific adverse effects on higher levels of biological organisation. However, it is obvious that simplified *in vitro* systems do not cover the complexity. Some drawbacks with *in vitro* tools are that, as opposed to *in vivo* bioassays and biomarkers, the systems studied are highly simplified when compared to the complexity of whole organisms. Thus, the potential interactions between different receptors, cells and organs are not detected.

In vitro assays are suitable and sometimes necessary when it is needed to conduct follow-up studies using biomarkers. In comparison to most biomarkers, they can easily be used to track local pollution sources by sampling water and sediment in a pollution gradient or effluents from suspected point sources. *In vitro* assays are also valuable in effect-directed analysis (EDA)/toxicity identification evaluation (TIE) approaches to identify toxic fractions and provide guidance for the identification of causative agents.

Bioassays *in vivo*

In vivo bioassays are tests where whole living organisms (including bacteria) are exposed to environmental samples, such as surface water, sediment, wastewater, dredge material or extracts from such samples. Tests are performed in the laboratory or, less frequently, in the field ("*in situ*" bioassays) [13].

The "endpoint" is the type of effect that is measured in a toxicity test, and some examples that are frequently used in this context are mortality, immobilization, fertilisation rate, hatching rate, embryo development, effects on growth of individuals (e.g. weight), effects on growth of populations (e.g. number of individuals), metabolic or physiological changes, reduced swimming activity, bioluminescence and specific molecular/biochemical responses.

Table 2 reports the most common *in vivo* bioassays applied by certain Member States within aquatic monitoring programmes.

In general, *in vivo* bioassays are broad spectrum assays, i.e. an *in vivo* bioassay responds to a variety of substances and different types of toxicity. An example is the "fish embryo acute toxicity (FET) test," recently adopted by the OECD N.236 of 26 July 2013, that is based on individual exposure of eggs to evaluate the embryotoxicity of samples with the aim to detect contaminants (relevant for the WFD) such as industrial chemicals, pesticides, pharmaceuticals and biocides.

Nevertheless, it is important to underline that the evaluation of toxic effects is based on the responses observed in several species, because they can exhibit intrinsic differences in terms of sensitivity to various chemicals; it also depends on the endpoints measured in the test [14]. Both short- and long-term *in vivo* bioassays should preferably be carried out using at least three species from different taxonomic groups and trophic levels (primary producer, decomposer/saprophytic, detritivore/filter feeder and consumer). A battery of ecotoxicological tests should have sufficient sensitivity and discriminatory power and respond to as many contaminant groups as possible.

Biomarkers

Biomarkers are molecular, biochemical, cellular and physiological indicators of contaminant stress measured in organisms resident or exposed *in situ* in a specific location. They are used in the monitoring programmes of Regional Seas Conventions and more recently in the MSFD to identify the impact from substances or combinations of substances not previously identified to be of concern, study trends and identify regions of decreased environmental quality [15]. Contrary to bioassays but similar to the ecological/community-based tools, biomarkers are analysed on field-exposed, usually resident, organisms. The sampling step is therefore primarily focused on the organisms that should to be examined. However, active monitoring based on caged organisms can also be used to measure biomarkers.

It is important to detect deleterious effects due to chemicals before significant effects at the population level occur. Damage at the population and ecosystem level can take a long time to repair. For certain trophic levels, recolonisation may take much longer than the 6-year management cycles considered in the WFD and MSFD. Ecological tools or indices are not predictive of damage, whereas several biomarkers can be used as early warning systems because they can detect effects caused by chemical substances and other environmental stress at an early stage.

Biomarkers are frequently divided into two different categories, depending on the number of substances/ groups of substances they are known to respond to:

- General (integrative) biomarkers that respond to several classes of toxic substances and, frequently, also to other types of stressors.
- Specific biomarkers that respond primarily to only a few/groups of/substances.

Table 2 Examples of *in vivo* bioassays applied by certain member states within aquatic monitoring programmes

Organism	Test item	Endpoint	Species	Exposure
Bacteria	w, ws, e, p	Bioluminescence	*Aliivibrio fischeri* (f/m)	5 to 30 min.
		Enzyme activity	*Arthrobacter globiformis* (f)	2 h
Algae	w, e, p	Growth	*Phaeodactylum tricornutum* (m)	72 h
			Skeletonema costatum (m)	
		Growth	*Desmodesmus subspicatus* (f)	72 h
			Pseudokirchneriella subcapitata (f)	
		Growth	*Ceramium tenuicorne* (m)	7 days
Plants	w, e, p	Growth	*Lemna minor* (f)	7 days
	ws	Growth	*Myriophyllum aquaticum* (f)	10 days
Rotifera	w, e, p	Mortality	*Brachionus plicatilis*	24 to 48 h
Crustacea (amphipods)	ws	Mortality	*Corophium spp.* (and other amphipods) (m)	10 days
	w, e, p	Mortality	*Artemia franciscana* (m)	24 h, 14 days
		Mortality	*Acartia tonsa* (m)	96 h
			Tigriopus fulvus (m)	
		Mobility, mortality, reproduction	*Daphnia magna* (f)	24/48 h, 21 days
			Cerodaphnia dubia (f)	
Nematoda	ws	Mortality, fertility, reproduction	*Caenorhabditis elegans* (f)	96 h
Annelida	ws	Mortality, reproduction	*Lumbriculus variegatus* (f)	28 days
Insecta	w, ws	Mortality, reproduction	*Chironomus riparius* (f)	48 h, 28 days
Bivalvia	w, e, p	Development	*Crassostrea gigas* (m)	24 to 72 h
			Mytilus galloprovincialis (m)	
			Tapes philippinarum (m)	
Echinodermata	w, e, p	Fertilisation	*Paracentrotus lividus* (m)	≤72 h
		Development	*Sphaerechinus granularis* (m)	
Polychaeta	ws	Mortality	*Hediste diversicolor* (m)	10 days
Vertebrata (fishes)	w, e, p	Mortality and genotoxic damage	*Danio rerio* (and embryos of other species) (f)	96 h, 28 days
			Dicentrarchus labrax (m)	

w, water; e, elutriate; p, pore water; ws, whole sediment; f, freshwater; m, brackish and marine.

For example, imposex is considered to be a highly specific effect, responding primarily to organic tin compounds such as tributyltin (TBT), whereas lysosomal stability is a more general biomarker for cellular stress. Both general and specific biomarkers can be useful, depending on the monitoring goals and the level of prior knowledge regarding the type of contaminants present at the given location. Because general biomarkers respond to several classes of compounds, they cover more substances and are therefore valuable in identifying areas of concern in environments exposed to complex mixtures of stressors. Specific biomarkers are, for example, valuable in second tier assessments to detect and identify the effects of specific types of substances in impacted locations.

Table 3 presents biomarkers used in the integrated monitoring approach proposed by International Council for the Exploration of the Sea (ICES) and the Regional Seas Conventions.

Biomarkers of exposure allow statements about the quality and/or quantity of exposure, whereas with biomarkers of effect statements about effects and the health status of exposed organisms can be made.

Exposure biomarkers, such as ethoxyresorufin-O-deethylase (EROD), can provide a sensitive indication of cellular changes at the enzyme level, which often represent the first warning signals of environmental disturbance. EROD can be used to detect exposure to classes of organic pollutants such as co-planar polychlorinated biphenyls (PCB), polycyclic aromatic hydrocarbons (PAH), planar dibenzodioxins (PCDD) and dibenzofurans (PCDF). Metallothioneins, peroxisomal enzymes (e.g. acyl CoA oxidase) and inhibition of acetylcholinesterase activity are other more or less

Table 3 Biomarkers used in the integrated monitoring approach proposed by ICES and the Regional Seas Conventions

Biomarker	Description	Responds to
EROD activity	Biotransformation enzyme induced by planar hydrocarbon	PCBs, PAHs and dioxin-like compounds
Acetylcholinesterase (AChE) activity	Enzyme implicated in nervous transmission	Organophosphates, carbamates and similar molecules
Vitellogenin (VTG) in male fish	A precursor of egg yolk, normally synthesized by female fish	estrogenic endocrine disrupting compounds
Metallothionein (MT)	Metal scavenger implicated in protection against oxidative stress	Heavy metals and inducer of oxidative stress
Amino-levulinic acid deshydratase (ALAD)	Enzyme implicated in amino-acid metabolism	Lead exposure
Lysosomal stability	General health, lysosomes play a key role in liver injury caused by various xenobiotics	Several classes of pollutants, including PAH, inducer of oxidative stress, metals and organochlorines
DNA adducts	Alteration of DNA structure able to disturb DNA function	Genotoxic compounds including PAHs and other synthetic organic compounds
Imposex biomarkers (e.g. VDSI) in molluscs	Imposition of male sex characteristics on female molluscs	TBT
PAH bile metabolites	PAH metabolites in bile/urine represent the final stage of the biotransformation process	Indirect indicator of PAH exposure
Liver histopathology	General indication about liver damage but can be diagnostic depending on the type of lesion	PAHs
Macroscopic liver neoplasms	Visible fish liver tumours	Cancer inducing substances; PAHs
Externally visible fish diseases	Overall organism health External investigations of fish, significant changes indicate chronic stress	Several classes of pollutants and pathogens
Intersex in fish	Presence of ovarian tissue in male fish gonads compromising reproductive capacity	estrogenic endocrine disrupting compounds
Micronucleus	Damage to genetic material of organisms; could affect their health and potentially also their offspring.	Substances causing chromosomal aberrations (clastogens)
Amphipod/fish embryo alterations	Embryo malformations (viviparous organisms)	Overall organism health; strong correlation observed between malformed embryos and concentrations of metals and organic compounds
Stress proteins	Early stage effects, including oxidative stress	Responds to many types of stress factors
Benthic diatom malformations	Malformations; overall organism health	Significant response to metals and several pesticides, but less to other priority substances
Comet assay	Sensitive tool to detect genetic damage	Substances causing DNA strand breaks
Mussel histopathology (gametogenesis)	Histological studies of, e.g. digestive gland and tube	Many groups of substances, including PAHs, PCBs and heavy metals
Stress on stress	Survival in air	Many groups of substances, including crude oil, copper ions and PCBs
Scope for Growth	Measures alterations in the energy available for growth and reproduction.	Many groups of substances, including di(2-ethylhexyl)phthalate (DEHP), aromatics, pentachlorophenol (PCP), copper, TBT and dichlorvos

specific exposure responses towards trace metals, organic chemicals and organophosphate pesticides, respectively.

On the other hand, biomarkers of effect indicate the occurrence of various forms of molecular to cellular/tissue alterations, although the health-related effects may differ in terms of toxicological and ecological relevance. Some effect biomarkers detect effects at early stages (such as genetic changes), whereas others, such as imposex, are related to later stages from a population risk perspective.

Ecological methods

The assessment criteria (primarily based on values of biodiversity indices) selected as the biological quality elements of the WFD do not respond in a specific way to the effects of hazardous substances. Specific tools for the assessment of hazardous substances applied in the context of WFD monitoring are extremely rare. Therefore, ecological status related to hazardous substances is generally based on individual pollutant concentrations, which may or may not be consistent with the ecological quality status assessment.

A new approach within biomonitoring is to consider the ecological role of communities, based on their functional, rather than structural composition, through the identification of species traits [16]. The resistance and resilience characters of individual taxa determine the response of communities to disturbance. Undisturbed communities display a diversity of species traits, whereas the communities downstream of a pollution source consist of those species that have a suite of traits which convey tolerance to the new conditions. Species that do not have these traits cannot survive [17]. The advantage of using functional traits instead of taxonomic composition (an entirely structural approach) of communities is bound to the *a priori* predictable response of traits to individual stressors, because each selection pressure affects different traits.

Spear

The SPEcies At Risk (SPEAR) bioindicator index, based on biological traits, has been shown to be highly sensitive to particular groups of contaminants such as pesticides and relatively independent of confounding factors [18]. The index measures the proportion of sensitive (SPEAR) and less sensitive (SPEnotAR, "SPEcies not At Risk") species and is expressed as a percentage.

The SPEAR concept is applicable for the assessment of the effects of pesticides on invertebrate communities in rivers but not lakes, coastal areas or temporary streams. Sampling must be performed in early summer (around June and July), not too long after peak pesticide application. Sensitivity data and information on other relevant traits for the taxa are included within the database used for the SPEAR online calculator. So far, validation studies have been performed in Finland, Germany, Sweden, France, Spain, Czech Republic and Australia. Nonetheless, there is a need for further validation before the SPEAR index can be used on a routine basis and as part of the WFD classification. In particular, the baseline sensitivity and variability of the method need to be assessed.

Pict

Pollution-induced community tolerance (PICT) has been suggested as a sensitive tool to track changes in community function (and therefore indicative of structural changes) that can be attributed to toxic substances. The PICT approach was developed by Blanck and Wängberg [19]. The approach relies on the assumption that sensitive components of the exposed community (species, genotypes or phenotypes) will be replaced by more tolerant ones during exposure, thus leading to an increase of community tolerance. PICT is measured by a functional test that detects the consequences of selection pressures. Tolerance development, for example,

can be measured as a shift in the effect concentration (usually EC_{50}) that is obtained with a short-term toxicity test based on an ecophysiological endpoint. Such an endpoint is preferably related to community metabolism (photosynthesis, respiration, protein synthesis, nucleic acid synthesis etc.). In recent years, PICT combined with the transplantation of periphyton communities has been suggested as a promising tool to identify impaired sites by detecting an induced tolerance after transplantation. *In situ* PICT assays using transplanted communities have been suggested as a promising tool that can link ecological and chemical status in the WFD context [20]. The PICT approach has the disadvantage that cannot be used to assess the risks for long-lived organisms with complex life cycles (e.g. insects and vertebrates).

EDA and TIE

While effect-based monitoring indicates hazards due to chemical contamination and provides information on toxicological endpoints of concern, tools are required to identify causes and elucidate links between exposure and effects. EDA and TIE are integrated biological and chemical approaches which aim to identify those compounds in an environmental or technical sample (water, soil, sediment, air, food, consumer product and technical mixture) that cause a biological response. Both approaches combine biotesting, physico-chemical fractionation and chemical analysis in a sequential procedure. However, the philosophy behind both approaches is slightly different [21]. The TIE approach has its origin in whole effluent testing, which focuses on the question, whether an effluent will cause adverse effects on aquatic organisms when emitted to the environment. In the case that effects are detected in whole organisms under realistic exposure conditions, TIE should help to characterise and identify the cause of the measured effect. Thus, TIE applies *in vivo* biotesting and avoids extraction and preconcentration steps as far as possible.

EDA is based on the understanding that environmental samples may contain thousands of mostly organic chemicals and that only a fraction of them can be analysed by chemical target analysis. EDA takes a biological effect (typically observed by effect-based monitoring) as the basis to narrow down the huge amount of possible chemical substances and aims to direct chemical analysis to those compounds that contribute significantly to a measurable effect. Thus, in EDA, bioassays are considered as tools to sensitively detect chemicals with similar biological targets or modes of action. The focus of EDA is on unravelling the contamination with organic toxicants representing the most complex group of chemicals. Similar to chemical analysis, there are no restrictions with respect to extraction or pre-concentration. Since the isolation and identification of individual toxicants out of thousands of components in typical environmental mixtures often demand for large numbers of fractions, high-throughput tools are preferred. In addition, the identification of unknown toxicants is very much supported by information on the mode of action. Both criteria are often met best with *in vitro* assays, although small-scale *in vivo* assays may be helpful, too [22]. The sample or an extract thereof is tested with the bioassays of choice depending on the objective of the study. If effects are detectable, the mixture is fractionated according to the physico-chemical properties of the components. The fractions are tested with the same bioassays for prioritisation according to effects. The mixture may undergo several fractionation steps to further reduce complexity. The components of active fractions are identified and quantified by chemical analytical means. Depending on the objective of the study, in a final confirmation step, the contribution of the identified candidate compound to the measured effect should be quantified or estimated in order to exclude that major contributors have been overlooked.

The major components of EDA are (i) separation including extraction, clean up and fractionation, (ii) biotesting, (iii) chemical analysis including computational tools for structure elucidation and (iv) confirmation.

EDA is a tool for investigative monitoring at selected sites of particular interest or with conspicuous effects. EDA aids in linking ecological status to contamination, to establish cause-effect relationships and to target mitigation measures. Although providing enormous progress over present target chemical monitoring, a general limitation of EDA is the requirement to pre-select toxicological endpoints. The combination of integrating whole organism tests with *in vitro* test batteries applying sufficient pre-concentration reduces the risk to overlook important effects and thus toxicants.

OMICS

The recent advances in DNA sequencing and characterisation of genomes have opened up a range of new possibilities. A particular field of molecular studies within biology is called "Omics" and refers to high-throughput molecular profiling technologies, such as genomics, metagenomics, transcriptomics, proteomics, metabolomics and metabonomics. The suffix "-ome" refers to the collection of all genes or gene products such as the genome, proteome or metabolome, respectively. A study of all or a very large number of these genes would fall under the definition of omics. Omics and bioinformatics tools can e.g. be used to:

- Develop molecular biomarkers of exposure as early signals to predict effects (that at a later stage could have an impact on physiological level and further on at population level).

- Provide information about the mode of action (MOA) of chemicals, i.e. the mechanism of toxicity; in turn, reducing the uncertainties involved in chemical risk assessment by providing, for example, a basis for extrapolation of the effects across species.
- Integrate MOA data with a deleterious outcome and in this way, aid towards understanding the impact on the ecosystem instead of just on single organism or species.
- Distinguish the site of origin of organisms, based on the transcriptomics changes in organisms coming from different locations.

Genomics-DNA microarray applications

Genomics can be indicative of the susceptibility of an organism for a certain chemical or group of chemicals [23] and are more frequently also been used for assessment of complex environmental samples, such as sediments [24]. A DNA microarray is a glass slide or a nylon membrane on which part of the organism's gene sequences (probes) is spotted or synthesised. Normally, complementary DNA (cDNA) is made using reverse transcriptase from RNA in the sample to be analysed. Then, the cDNA is thereafter hybridised to the array. After scanning and image analysis, the RNA abundance (amount of RNA molecules bound to the complementary probes on the microarray) is analysed and the relative gene expression of the treated sample can be compared to the untreated control.

Next-generation sequencing

The development of DNA sequencing technology four decades ago was a major scientific hallmark and opened the doors for several breakthrough achievements in all areas of biology. Next-generation sequencing (NGS) is a more recent technology, also named second-generation sequencing (SGS) and has been commercially available since 2004.

Compared to the first-generation capillary electrophoresis (CE)-based Sanger sequencing, NGS has increased sequencing speed, the throughput to millions of sequencing experiments on fragmented DNA run in parallel, has reduced sequencing costs per base pair in some cases more than 10,000 fold. NGS platforms enable a wide variety of applications including the study of the genome or transcriptome of any organism.

RNA-seq

RNA-seq is a recently developed approach, extending the high-throughput sequencing to the profiling of the transcriptome. Instead of capturing transcript molecules by molecular hybridisation, as on microarrays, RNA-seq directly sequences the transcripts present in a sample. Transcript sequences are then mapped back to a reference genome and counted to assess the expression level of that gene or genomic region [25].

Metagenomics

In recent years, a tremendous increase in DNA sequencing capacity, combined with an unprecedented drop in price per obtained nucleotide sequence, has made it possible to study the functional elements of a microbial ecosystem at the levels of the actual genes responsible for these functions. Such sequencing studies of the total DNA content of an environment are generally referred to as metagenomics [26] and could be used to pinpoint the genes or species that cause, for example, a PICT response. A possible relevant use of metagenomics is to monitor the presence and abundance of antibiotic resistance genes in the environment.

Proteomics

Proteomics is the large-scale study of proteins, particularly their structures and functions [27]. After genomics and transcriptomics, high-throughput proteomics is considered the next step in the study of biological systems. It is much more complicated than genomics mostly because while an organism's genome is more or less constant, the proteome differs from cell to cell and from time to time. This is because distinct genes are expressed in distinct cell types. This means that even the basic set of proteins which are produced in a cell needs to be determined.

Metabolomics

Within metabolomics, the endogenic metabolic profile of an organism is studied. The metabolites that are studied can be considered to be the result of the ongoing metabolic activity of the cells. To measure metabolites is considered advantageous since it is well known that metabolites are formed at an early stage of environmental stress [28].

Conclusions

The topic "effect-based tools" highly ranked in the CIS science-policy interface report [29] elaborated on the basis of inputs from the Working Group "Chemicals" of the WFD". The new mandate 2013 to 2015 of the Working Group, approved by the Water Directors, has foreseen the continuation of the activity on effect-based tools, in particular, in relation to the detection and evaluation of effects caused by mixtures of pollutants. This activity will be strongly linked to the work of the WG Ecostat and the implementation needs of the MSFD. Furthermore, a project has been planned with the aim to evaluate the use of bioanalytical methods for the detection of the pharmaceuticals 17α-ethinylestradiol (EE2) and 17βestradiol (E2) which are included in the WFD "Watch List" of emerging pollutants foreseen by the Directive 2013/39/EU. This technical report, elaborated in close collaboration with the scientific community, can already be considered to

provide important support to the managers, assessors and local operators involved in the analysis and monitoring of surface water bodies.

Appendix

The Appendix section of the technical report on aquatic effect-based monitoring tools (Additional file 2) collects 14 case studies (Table 4), which illustrate how these tools can help to achieve the objectives of the WFD and MSFD and a series of fact-sheets (Table 5) that provide technical specifications for selected individual effect-based tools (biomarkers and bioassays) that are either already used on a routine basis or are gaining in popularity.

The Appendix section contains also a list of available standardized effect-based tools (*in vivo* and *in vitro*, bioassays and biomarkers), established assessment criteria for the marine environment and an overview of available DNA microarrays. Other technical issues, such as sampling aspects, standardization and proposed approaches to assess estrogenic effects are also described in more detail in the Appendix section. Finally, a list of definitions, abbreviations and a wide bibliography section on the topic is included.

Table 4 Case studies

	Case studies
1	Laxsjön - investigating sediment contamination using chemical and *in vitro* bioassay approach
2	Deployment of a multi-biomarker approach to identify the origin of wild fish abnormalities reported in a French stream receiving urban and industrial effluents.
3	Endocrine disruptors in the Irish aquatic environment
4	Swedish national monitoring programme of fish health
5	Evaluation of aquatic environmental estrogens with passive sampling - EPSA
6	Contaminated sediments in the River Elbe basin-EDA
7	Monitoring concentrated surface water with in vivo bioassays in the Netherlands
8	Monitoring imposex on water body level
9	Bioassays for monitoring the offshore platform impacts and their main discharges
10	Evaluation of the utility of microarrays as a biomonitoring tool in field study
11	Use of DNA microarray to test the water quality of river East Turkey Creek (bay of watershed of Florida) potentially impacted by treated wastewater from sprayfield area
12	The risk of chronic impact of pollution on the Bílina River
13	Mechanism-specific tools with zebrafish early life stages in EDA of surface waters
14	Multicriteria assessment of human activity effects on water ecosystems: the case study of Tiber River basin

Table 5 Fact-sheets for certain biomarkers and bioassays

	Fact-sheets of biomarkers and bioassays
Biomarker	Metallothionein (MT)
	ALA-D
	Cytochrome P450 1A activity (EROD; CYP 1A activity)
	DNA adducts
	PAH metabolites
	Liver histopathology (LH)
	Macroscopic liver neoplasm (MLN)
	Externally visible fish diseases
	Reproductive success in eelpout
	Vitellogenin
	Intersex (in male fish)
	Lysosomal stability
	Imposex biomarkers
	Micronucleus assay
	Amphipod embryo alterations
	Stress proteins (heat shock protein)
	Acetylcholinesterase (AChE) assay
	Comet Assay
	Mussel histopathology (gametogenesis)
	Stress on stress
	Scope for Growth (SFG)
	Benthic diatom malformation
Bioassay	DR CALUX/DR Luc assay
	PAH CALUX
	ERα CALUX/ER-Luc (agonistic/antagonistic)
	AR CALUX (agonistic/antagonistic)
	YES
	YAS
	Ames fluctuation test
	Micronucleus assay
	Fish embryo acute toxicity (FET) test

Authors' contributions

All authors are responsible for the general design of the Technical Report. ASW and MC wrote the first draft of the manuscript. All authors contributed on specific aspects, read and approved the final manuscript.

Acknowledgements

We acknowledge the Members of WG "Chemicals" (previous "Chemical Aspects") of the WFD for the contribution and the useful comments provided for the revision of the technical report.

Author details

[1]Swedish Agency for Marine and Water Management, Gullbergs Strandgata 15, Göteborg 404 39, Sweden. [2]Department of Environment and Primary Prevention, ISS-Italian Institute of Health, Viale Regina Elena, 299, 00161 Rome, Italy. [3]ISPRA - Institute for Environmental Protection and Research, Via Brancati 48, 00144 Rome, Italy. [4]T.G. Masaryk Water Research Institute, Podbabská 2582/30, 6, 160 00 Praha, Czech Republic. [5]National Institute for Industrial Environment and Risks INERIS, Rue Jacques Taffanel, 60550 Verneuil en Halatte, France. [6]Baltic Eye, Östersjöcentrum, Stockholms Universitet, SE-106 91, Universitetsvägen 10 A, Stockholms, Sweden. [7]Federal Institute of Hydrology, Am Mainzer Tor 1, 56068 Koblenz, Germany. [8]Rijkswaterstaat, Water, Traffic and Environment, Zuiderwagenplein 2, 8224 Lelystad, The Netherlands. [9]National Institute for Public Health and the Environment (RIVM), 3720, Postbus 1BA, Bilthoven, The Netherlands. [10]EPA, Johnstown Castle Estate, PO Box 3000, Wexford, Ireland. [11]IRSA-CNR, Via del Mulino 19, 20047 Brugherio, Italy. [12]Swedish Environmental Protection Agency, Naturvårdsverket, SE-106 48 Stockholm, Sweden. [13]University of Gothenburg, Guldhedsgatan 10, SE-405 30 Gothenburg, Sweden. [14]Swiss Centre for Applied Ecotoxicology Eawag-EPFL, Überlandstrasse 133, CH-8600 Dübendorf, Switzerland. [15]Marine Scotland Science, 375 Victoria Road, AB11 9DB Aberdeen, Scotland. [16]Cefas, Pakefield Road, Lowestoft, Suffolk NR33 0HT, UK. [17]Defra, Area 3D, Nobel House, Smith Square, London SW1P 3JR, UK. [18]CONCAWE, Vorstlaan165, B-1160 Brussels, Belgium. [19]Eurometaux, Avenue de Broqueville 12, B-1150 Brussels, Belgium. [20]DG Environment - European Commission, Avenue de Beaulieu 9, B-1160 Brussels, Belgium. [21]DG Joint Research Centre - European Commission, Via Enrico Fermi 2749, I-21027 Ispra, Italy. [22]Department of Ecosystem Analysis, Institute for Environmental Research, ABBt RWTH Aachen University, Worringerweg 1, 52074 Aachen, Germany. [23]Helmholtz Centre for Environmental Research-UFZ, Permoserstraße 15, 04318 Leipzig, Germany. [24]Thünen-Institut of Fisheries-Ecology, Palmaille 9, 22767 Hamburg, Germany.

References

1. European Union. Directive 2000/60/EC of the European Parliament and of the Council of 23 October 2000 establishing a framework for Community action in the field of water policy. Off J Eur Union. 2000;L327:1–73.
2. European Union. 2013. Off J Eur Union. 2013;L 226:1–17.
3. European Union: communication from the commission to the council. The combination effects of chemicals. COM/2012/0252 final, 31/05/2012, pp. 1–10.
4. Silva E, Rajapakse N, Kortenkamp A. Something from "nothing" - eight weak estrogenic chemicals combined at concentrations below NOECs produce significant mixture effects. Environ Sci Technol. 2002;2002(36):1751–6.
5. Quevauviller P, Carere M, Polesello S. Chemical monitoring activity for the implementation of the Water Framework Directive. Trends Anal Chem. 2012;36:1–184.
6. Wernersson AS, Maggi C, Carere M. Technical report on aquatic effect-based monitoring tools. Technical Report 2014–077. Luxembourg: Office for Official Publications of the European Communities; 2014.
7. Hallare AV, Seiler TB, Hollert H. The versatile, changing, and advancing roles of fish in sediment toxicity assessment-a review. J Soils Sedim. 2011;11:141–73.
8. Green N, Schøyen M, Øxnevad S, Ruus A, Høgåsen T, Beylich B, Håvardstun J, Rogne Å, Tveiten L. Hazardous substances in fjords and coastal waters - 2010. Levels, trends and effects. Long-term monitoring of environmental quality in Norwegian coastal waters. Norwegian State Pollution Monitoring Programme Report no. 1111/2011. TA-no. 2862/2011. 252, 2011.

Competing interests

The authors declare that they have no competing interests.

9.   Chapman D, Jackson J (1996): Biological monitoring. In Water Quality
      Monitoring - A Practical Guide to the Design and Implementation of
      Freshwater Quality Studies and Monitoring Programmes. Edited by Jamie
      Bartram and Richard Balance Published on behalf of United Nations
      Environment Programme and the World Health Organization.
      ISBN 0 419 22320 7. (Hbk) 0419217304 (PBK).

10.   Connon RE, Geist J, Werner I. Effect-based tools for monitoring and predicting
      the ecotoxicological effects of chemicals in the aquatic environments. Sensors.
      2012;12:12741–71.

11.   Leusch FL, De Jager C, Levi Y, Lim R, Puijker L, Sacher F, et al. Comparison of
      five in vitro bioassays to measure estrogenic activity in environmental
      waters. Environ Sci Technol. 2010;44:3853–60.

12.   Wernersson AS: Swedish monitoring of hazardous substances in the aquatic
      environment-current vs required monitoring and potential developments.
      Länsstyrelsen I Västra Götalands län. Rapport 2012:23.

13.   Piva F, Ciaprini F, Onorati F, Benedetti M, Fattorini D, Ausili A, et al.
      Assessing sediment hazard through a weight of evidence approach with
      bioindicator organisms: a practical model to elaborate data from sediment
      chemistry, bioavailability, biomarkers and ecotoxicological bioassays.
      Chemosphere. 2011;2011(83):475–85.

14.   Ahlf W, Heise S. Sediment toxicity assessment: rationale for effect classes. J
      Soils Sedim. 2005;5:16–20.

15.   Sanchez W, Porcher JM. Fish biomarkers for environmental monitoring
      within the water framework directive. Trends Anal Chem. 2009;28:150–8.

16.   Usseglio-Polatera P, Bournaud M, Richoux P, Tachet H. Biological and
      ecological traits of benthic freshwater macroinvertebrates: relationships and
      definition of groups with similar traits. Freshwater Biol. 2000;43:175–205.

17.   Van den Brink PJ, Alexander AC, Desrosiers M, Goedkoop W, Goethals PLM,
      Liess M, et al. Traits-based approaches in bioassessment and ecological risk
      assessment: strengths, weaknesses, opportunities and threats. Integr Environ
      Assess Managem. 2011;7:198–208.

18.   Schäfer R, Vd Ohe P, Rasmussen J, Kefford B, Beketov M, Schulz R, et al.
      Thresholds for the effects of pesticides on invertebrate communities and
      leaf breakdown in stream ecosystems. Environ Sci Technol. 2012;46:5134–42.

19.   Blanck H, Wängberg S-Å. Induced community tolerance in marine periphyton
      established under arsenate stress. Can J Fish Aquat Sci. 1988;45:1816–9.

20.   Pesce S, Lissalde S, Lavieille D, Margoum C, Mazzella N, Roubeix V, et al.
      Evaluation of single and joint toxic effects of diuron and its main
      metabolites on natural phototrophic biofilms using a pollution-induced
      community tolerance (PICT) approach. Aquat Toxicol. 2010;99:492–9.

21.   Burgess RM, Ho KT, Brack W, Lamoree M. Effect-directed analysis (EDA) and
      toxicity identification evaluation (TIE) (2013): complementary but different
      approaches for diagnosing causes of environmental toxicity. Environ Toxicol
      Chem. 2013;32:1935–45.

22.   Brack W. Effect-directed analysis: a promising tool for the identification of
      organic toxicants in complex mixtures (2003). Anal Bioanal Chem.
      2003;377:397–407.

23.   Gunnarsson L, Kristiansson E, Larsson DGJ (2012): Environmental comparative
      pharmacology: theory and application. In B.W. Brooks and D.B. Huggett (eds.),
      Human Pharmaceuticals in the Environment:Current and Future Perspectives,
      Emerging Topics in Ecotoxicology 4, doi:10.1007/978-1-4614-3473-3_5,
      Springer Science + Business Media.

24.   Kosmehl T, Otte JC, Yang L, Legradi J, Bluhm K, Zinsmeister C, et al. A
      combined DNA-microarray and mechanism-specific toxicity approach with
      zebrafish embryos to investigate the pollution of river sediments. Reprod
      Toxicol. 2012;33:245–53.

25.   Wang Z, Gerstein M, Snyder M. RNA-Seq: a revolutionary tool for transcriptomics.
      Nat Rev Genet. 2009;10:57–63.

26.   Riesenfeld CS, Schloss PD, Handelsman J. Metagenomics: genomic analysis
      of microbial communities. Annu Rev Genet. 2004;38:525–52.

27.   Anderson NL, Anderson NG. Proteome and proteomics: new technologies,
      new concepts, and new words. Electrophoresis. 1998;19:1853–61.

28.   Samuelsson LM, Förlin L, Karlsson G, Adolfsson-Erici M, Larsson JDG. Using
      NMR metabolomics to identify responses of an environmental estrogen in
      blood plasma of fish. Aquat Toxicol. 2006;78(4):341–9.

29.   Kase R, Clayton H, Martini F: Science-Policy Interface (SPI) activity on
      prioritisation of research needs, knowledge availability and dissemination
      for the Working Group E (Chemical Aspects) 2010–2012. Open available at
      CIRCABC at: https://circabc.europa.eu/w/browse/5bf63ff3-b24b-4365-8a57-
      38e4d56b941.

14

Differences in biomass yield development of early, medium, and late maize varieties during the 21st century in Northern Germany

Jan F Degener[*] and Martin Kappas

Abstract

Background: Though there exists a general notion on how maize yields might develop throughout Europe during the current century, modeling approaches on a regional level that account for small-scale variations are not yet universally available. Furthermore, many studies only refer to one variety of maize. However, the few studies that include at least two varieties indicate that the respective choice will play a major role in how the yields will develop under a changing climate throughout the 21st century. This study will evaluate how far this choice of variety will affect future yields, identify the main factors to explain potential differences, and determine the magnitude of spatial variability.

Results: The results suggest clearly differentiated development paths of all varieties. All varieties show a significant positive trend until the end of the century, though the medium variety also shows a significant decline of 5% during the first 30 years and only a slight recovery towards +5% around the century's end. The late variety has the clearest and strongest positive trend, with peaks of more than +30% increase of biomass yields and around 25% average increase in the last three decades. The early variety can be seen as in-between, with no negative but also not an as-strong positive development path. All varieties have their strongest increase after the mid of the 21st century. Statistical evaluation of these results suggests that the shift from a summer rain to a winter rain climate in Germany will be the main limiting factor for all varieties. In addition, summer temperatures will become less optimal for all maize crops. As the data suggests, the increasing atmospheric CO_2 concentrations will play a critical role in reducing the crops water uptake, thus enabling yield increases in the first place.

Conclusions: This study clearly shows that maize yields will develop quite differently under the assumed climatic changes of the 21st century when different varieties are regarded. However, the predominant effect is positive for all discussed varieties and expected to be considerably stronger in the second half of the century.

Keywords: Maize varieties; Climate change; Crop yield; Summer drought; Carbon dioxide

Background

With a production of around 875 Mt, maize was the second most grown crop on earth in 2012, only surpassed by sugarcane and surpassing rice (3rd 718 Mt) and wheat (4th 675 Mt). However, in terms of nutrition, rice and wheat provided around 3.8 times more calories to the world's average human [1]. This spread in the data is a clear indicator for the variety of usage that maize allows for, from its first and foremost use as feed for livestock to a raw material for energy purposes.

As of 2012, like most years before, Lower Saxony (LS) constituted Germany's largest maize producer, accounting for more than a quarter of the 94.56 Mt total German production, while extending over merely 13% of Germany's overall territory. This is due to an over-average yield of 50.6 t/ha (avg. Germany 46.4 t/ha, at 35% dry matter) combined with a relatively large cropping area of 27% (avg. Germany 17%) of the total utilized agricultural area [2].

Around the early 1980's the cropping area of LS for silage maize leveled out at around 220,000 ha for several

*Correspondence: jdegene@gwdg.de
Department of Geography, University of Göttingen, Goldschmidtstr.5, 37077 Göttingen, Germany

years. Around 2004, this began to change rapidly. Within 5 years, the area nearly doubled; after less than a decade, the area already amounted to 514,000 ha in 2012 [3]. An early look into the matter [4] did not show any increase in the local livestock nor a dramatic change in livestock diet or related imports or exports. Even more, the maize cropping area for feed receded by 30,000 ha between 2004 and 2007. Energy maize, however, in LS, used predominantly as a regenerative power source, showed an increase in cropping area by 38,000 ha in only 1 year. Therefore, it can be safely assumed that this increase in cropping area was due to reasons other than livestock farming. While there are some propositions for alternatives to this extensive maize cultivation [5,6], its known production strategies and biomass yields will make it hard for any competing crop to replace maize. Thus, it can be assumed that maize will be around for some time, raising the question how changing regional or local conditions will affect its yield potential.

Wolf and vanDiepen [7] did an early estimation of the European grain maize yield potential, basically coming to the conclusion that no large changes are to be expected for the central part of the European Community and thus for LS. This outlook has not changed dramatically in present day studies [8], generally suggesting no trend or seldom a positive trend in rainfed maize yields for most parts of Germany. Spatial surveys covering only Germany in its entirety are relatively rare. However, it is often pointed out that maize already grows near optimal conditions in Germany and is, as of today, already limited through drought stress in its main growing period of July and August [9,10]. The expected further decline in summer precipitation of around 30% for some areas in Germany [11], however uncertain this change might be, would thus strongly limit the growing conditions of maize in these regions.

In part probably owed to the administrative structure in Germany, the assessment of climate change impacts on crops yields was mostly done on a federal state level. A wide variety of approaches (differences in climate model and dataset, crop model, reference period, etc.) make a direct comparison often difficult at least. However, results from regions close to LS are still of special interest for comparison.

The federal state of Hesse, directly to the south of LS, shows a regional differentiated pattern with a positive maize yield trend (up to +15%) in the southern part and a neutral-to-negative (mostly around −10%) northern part in the middle of the 21st century under the SRES B2 scenario [12]. To the south-west of LS lies North Rhine-Westphalia. Fröhlich [13] did show that most of the state will profit from a changing climate from a silage maize yield increase of around 2% to 4% (B1 scenario) or 3% to 7% (A1B scenario) until 2050. Saxony and Thuringia, both to the south-east of LS, have a generally negative

development until 2050 (A1B) with a decline in maize yields of roughly −10% [14,15]. Especially the study for Thuringia does show how wide these results may spread, even if climate model, scenario, and crop model are kept the same. Four alternative approaches, including more or less progress in cultivation and breeding, further differentiated by dry or moist conditions, resulted here in average yield changes from −8.2% to +38.6%.

Furthermore, Buttlar et al. [16] took a closer look at a part of LS, the region connecting the cities of Hanover, Brunswick, Göttingen, and Hildesheim. This study was, however, rather site specific, with biomass yield changes of maize between −3% and +7% (until mid-century) and −4% to +13% (end of century).

While the current study does not expect to diverge largely from these findings, a regional or even local approach was necessary as a probable basis for action of regional decision makers. An important difference to the mentioned studies lies, however, in the selection of different maize varieties. For simplification, many studies omit the use of different varieties that are differentiated only by their required temperature sums to reach their respective development stages. As Southworth et al. [17] could show in a study in the Midwestern United States, this differentiation can indeed make a difference, as heat-resistant late (or long-term) varieties did show a considerably better yield development in a future climate than varieties with less temperature requirements. However rare, if studies do evaluate distinct varieties, the findings are similar as Liu et al. [18] could show for Northeast China. Most studies, however, only hint in a more general way towards the influence of variety choice [19-21].

Results

The results in this study will describe the *change* in biomass yields during the 21st century. Changes are relative to the mean yields of the decade 2001 to 2010 as a representation of the present time.

Mean yield development

The results in this section give the average yield development of all modeled sites. As can be seen in Figure 1, all three varieties visually show a positive yield development throughout the century. This is further underpinned by the actual biomass yields after 2060, where the average yield per decade is always higher than for the reference period. Not as evident is the shared pattern of the decadic coefficient of variability. All three varieties have their lowest value in the present (3% to 4%) with an increase (except for the comparably low variability between 2031 to 2040) towards mid-century (above 8%) and a slight decline towards 6% to 7% at the end of the century. Actual yields will therefore vary more widely around the decadic mean at mid-century.

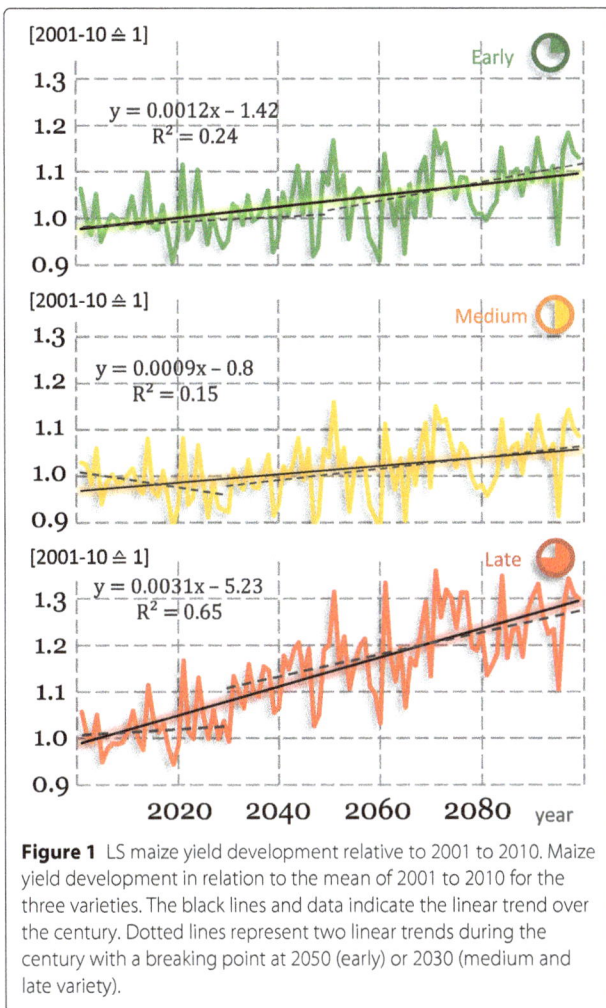

Figure 1 LS maize yield development relative to 2001 to 2010. Maize yield development in relation to the mean of 2001 to 2010 for the three varieties. The black lines and data indicate the linear trend over the century. Dotted lines represent two linear trends during the century with a breaking point at 2050 (early) or 2030 (medium and late variety).

Apart from these shared aspects there are also obvious differences in the overall development. The following description will thus cover each variety on its own. The *early* variety shows an $R^2 = 0.24$ and an average increase in yields of 0.12% p.a. throughout the century. This trend could be slightly better explained through a polynomial model of second or third order with $R^2 = 0.27$; however, no big advantage would be expected from such an approach. The t/n ratio 1.68 shows no significant trend for $\alpha = 0.95$ but would for $\alpha = 0.9$. Mann-Kendall delivers a more unambiguous result with $p < 0.001$ over the century. It is therefore assumed that a significant trend exists throughout the entire time period.

This trend can basically be split into two parts: the period 2001 to 2050 has an $R^2 = 0.003$ in a linear regression model with an average yield development of ±0% p.a. The period 2051 to 2099 has an $R^2 = 0.11$ with an average yield increase of 0.2% p.a. The lack of a trend in the first half of the century is confirmed by its t/n ratio of 0.5 and $p = 0.39$ for MK. For the period 2001 to 2030, there even seems to appear a slight negative development, with a t/n

ratio of -0.63 and a $p = 0.08$ for MK that is, however, not recognized as being significant.

All in all, it seems clear that a change in biomass yields is expected to happen, however, only after the mid of the century and especially after 2070 when there is only 1 year with a critically lower yield than the present average. In these last 30 years, the yields are about 9% higher than in the first decade, with the last decade being the one with the overall highest yields. If total production of early maize would be calculated over the century, 49.1% would be produced during the first half.

The *medium* variety has an $R^2 = 0.15$ and an average increase in yields of 0.09% p.a. over the century. As with the early variety, a slightly better explanation is provided through a polynomial model with $R^2 = 0.19$. A t/n ratio of 1.34 indicates no linear trend while a $p < 0.001$ for MK assumes a significant trend. These numbers represent the lowest indicator for a trend throughout the century of all varieties.

This is due to a different break within the data that occurs around 2030 and is still present when the year is shifted ±10, though weaker. The period 2001 to 2030 shows a linear decline in yields of around -0.2% p.a. and an $R^2 = 0.16$. From 2031 on, this turns towards a positive trend of $+0.13\%$ p.a. and an $R^2 = 0.14$. While the early variety did also show signs for a decline in yields, the data for the medium variety supports it more strongly. While a t/n ratio of -1.3 fails to be significant, the MK with $p = 0.005$ is; therefore, a significant negative trend until 2030 is assumed. On average, this decline will reduce the yields about 5%.

This trend is then reversed towards the end of the century, in such a way that around 2070 the yields are mostly above the present average. However, this happens in a lower magnitude than for the early variety. If only the period 2001 to 2050 is considered, t/n (0.01) and MK ($p = 0.95$) are both highly insignificant, meaning that the average yields are not changing. As this is generally comparable to the early variety, the yield variability is somewhat larger for the medium variety.

The medium variety will thus have the least positive development in the 21st century. Yields after 2070 will on average be 5% above today's. A comparison of both halves of the century has 49.4% of a potential production happening in the first 50 years, again the highest value of all varieties.

The *late* variety is somewhat of an exception. Where early and medium varieties show at least minor comparability, the late variety has a uniquely positive development path. This is evident by just looking at the graph as well as in the numbers of the linear regression model with an $R^2 = 0.65$ and an average yield increase of 0.31% p.a. A change to another regression model does not show any different results. Also the t/n ratio (2.74) and

Mann-Kendall ($p < 0.001$) are more explicit in determining significance than for the other varieties.

However clear, this trend is not entirely constant. While the t/n ratio (2.31) and MK ($p < 0.001$) already show a highly significant trend towards 2050, the data seems to have a break around the year 2030. There appears to be no trend for the period 2001 to 2030 as a t/n ratio of 0.65 and $p = 0.35$ for MK suggest. A linear regression for 2001 to 2030 shows an $R^2 = 0.09$ and a mean yield increase of 0.11% p.a., a non-significant but positive trend, whereas 2031 to 2099 has a $R^2 = 0.37$ and an increase of 0.26% p.a.

The late variety has thus the most positive development throughout the century. The mean yields in the last three decades increase about at least 10% compared to today, with a mean of 25% and a maximum of 36%. The same holds true for a production comparison of both half-centuries, as the first 50 years would only contribute 47.1% to a potential overall production.

Yield development by region
As Figure 2 indicates, there are also certain differences in the regional distribution of potential yield increases or decreases. The late variety does clearly have the most uniform development as it is positive for almost all times and sites. The share of sites with a positive development lies around 53% for the early variety in the period and increases to 87%, 96%, and 95% towards the end of the

century. In a similar fashion, the medium variety starts out at a very low share of 16% positive sites, increasing to 49%, 88%, and 83%.

If the coefficient of variation is calculated regarding all sites and years from the respective periods, all three varieties show an increasing *cov* with progressing time. With 9.3% (2021 to 2040) to 10.7% (2081 to 2099), the late variety does have the least variability. The early (10.4% to 11.5%) and medium variety (10.0% to 11.5%) have a rather similar variation in their yields.

The overall best sites are situated in the west of LS. Two main areas with a below-average yield development can be identified. One is to the north or north-east of LS, the other one to the south. This southern area is, however, not affected uniformly, but rather, quite positive and quite negative sites are alternating. The negative sites are consistently those with an overall shallow profile, situated on the slopes of the hilly landscape. By contrast, the sites with a positive development lie within the fertile river valleys with their good soil quality.

Good soil quality is here defined only through the soil's ability to retain water, basically defined by its field capacity.

Statistical process analysis
To determine the relative influence of certain variables on crop growth, a multivariate linear regression model was

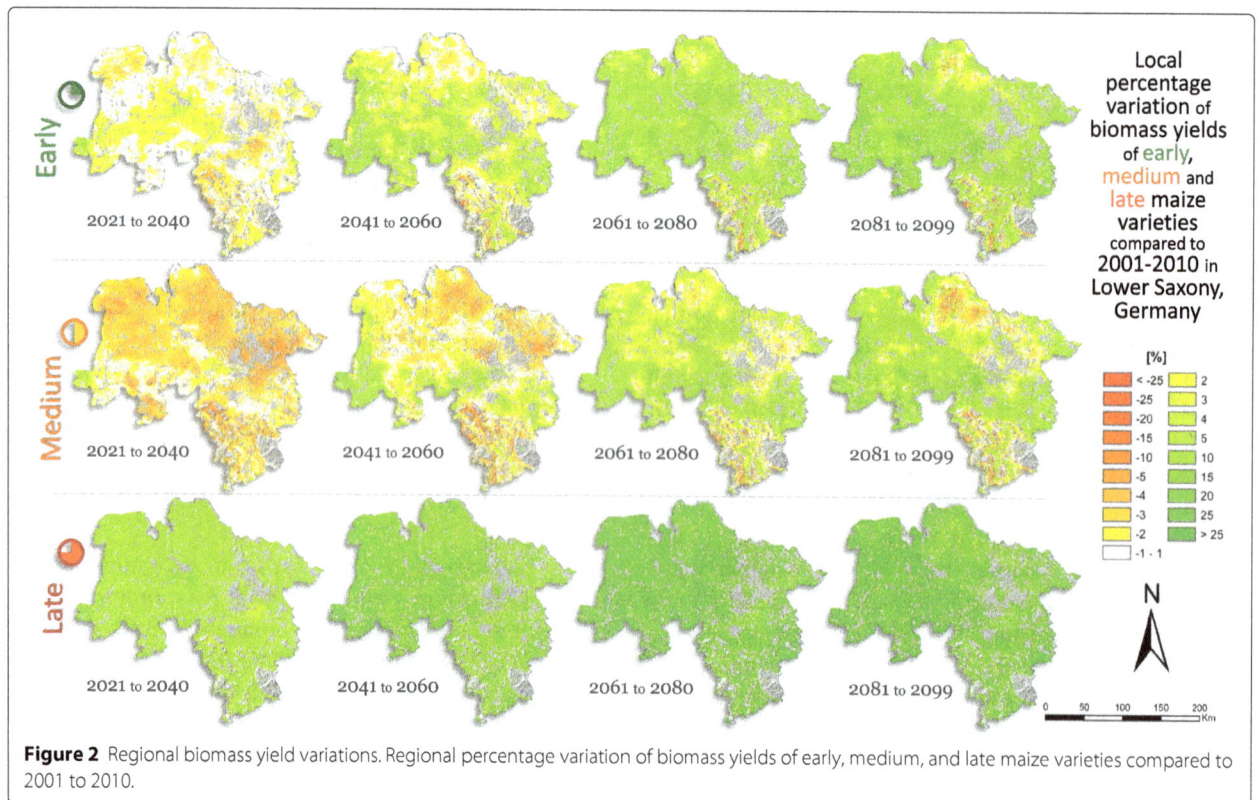

Figure 2 Regional biomass yield variations. Regional percentage variation of biomass yields of early, medium, and late maize varieties compared to 2001 to 2010.

applied with the results shown in Figure 3. On the left side, single parameters are tested against each other for their quantitative and qualitative input strength. For example, a positive correlation of +4 for medium maize and CO_2 indicates that a *rise* in atmospheric CO_2 concentrations by 1 ppm leads to a yield *increase* of 4 g/m^2. A negative correlation, as for example for summer temperature and medium variety, would, however, stand for a *decline* in yields by roughly 100 g/m^2 if summer temperatures would *rise* by 1°C.

The shown results are the mean output of all sampled sites. Though problems due to some autocorrelation were expected, the linear multivariate models performed quite well. For 2001 to 2099, the mean model p value over all sites was <0.001 for all varieties. For the late variety, this holds true not just for the mean but even if all sites are regarded individually. The early and medium varieties did, however, contain around 1,300 (1.5%) sites of less significance, which were, however, still within a margin of $0.01 > p > 0.001$. The models concerning the first half of the century, 2001 to 2050, show slightly worse results. Though the mean p value of all models is still <0.001, the number of models with higher p values increased. Even the late variety now did have around 1,500 sites exceeding this threshold, with early and medium variety on about 7,000 sites. Roughly a third of these exceptional sites have p values >0.05. In conclusion, the models are slightly better for the description of the long-term development than for the first half of the century.

For 2001 to 2099, two main influence variables are detected. Summer precipitation shows a strong positive correlation for all three varieties. As the amount of rain is expected to drastically decline throughout the century, this seems to be the main factor to limit future maize yields. On the other hand, atmospheric CO_2 concentrations have a comparable positive correlation and are thus possibly the main agent for a positive yield development. The amount of spring precipitation seems to be of higher importance for medium ($p = 0.09$) and especially early ($p = 0.04$) variety. Both do also show a negative connection with the rising summer temperatures (early $p = 0.09$, medium $p = 0.08$), at least to some degree. The late variety shows basically similar dependencies, however weaker. Instead, spring temperatures ($p < 0.01$) seem to be of much higher importance than for the other two varieties.

For 2001 to 2050, these indicators change only slightly. Still, summer precipitation and CO_2 concentrations remain the determining variables ($p < 0.001$). The late variety still shows some dependency towards spring temperatures ($p = 0.09$). For all three varieties, fall temperatures seem to be of higher importance in the first 50 years ($0.1 > p > 0.05$), whereas summer temperatures and spring precipitation have no apparent influence.

That these multivariate models are not entirely perfect becomes evident when, for example, winter precipitation and late maize are considered for 2001 to 2099. While not being *highly* significant, a certain connection between both variables is suggested. However, as winter months include December, January, and February, when no maize is grown, this also seems to be highly improbable. While the statistical model was believed to be reasonably good in determining the relative influence of each variable, there was a need to exclude variables that are not necessarily important.

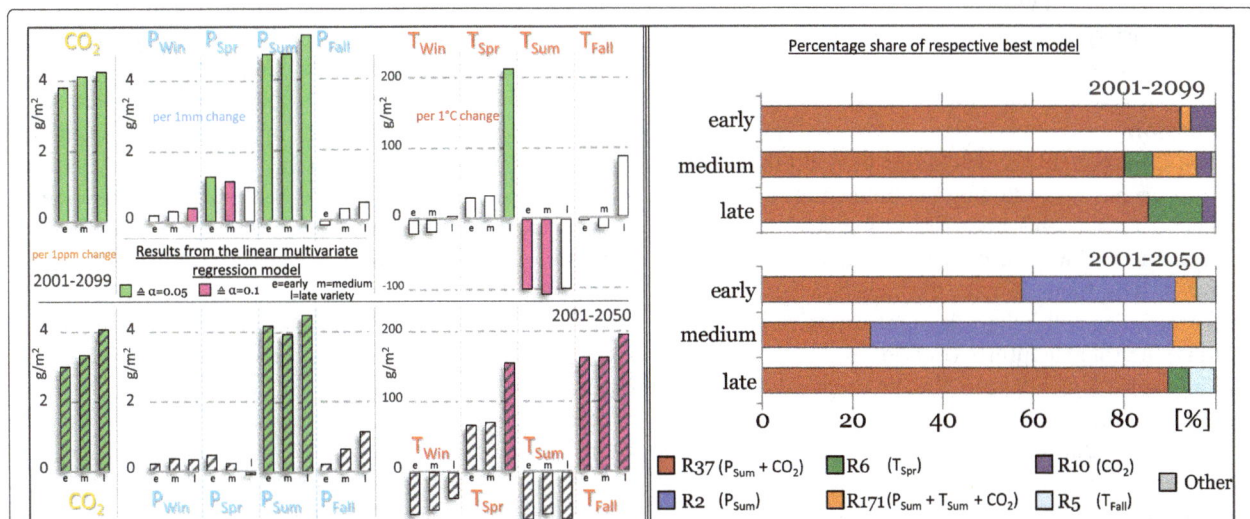

Figure 3 Results from the multivariate linear regression analysis. *Left side:* output of nine variables from the linear multivariate regression analysis for 2001 to 2099 (top) and 2001 to 2050 (bottom). For example, the topmost left bar for CO_2 is around the value 3.9, meaning that an increase of 1 ppm of CO_2 increases the annual yield by 3.9 g/m^2 - *Right side:* relative share of different linear (multivariate) models to the total number of models Rx represents the number of the 2047 possible runs through variable combination - P_x is precipitation by respective season, T_x for temperature.

Therefore, the next step was to identify the one linear multivariate regression model for each site that best describes its yield development, as shown on the right side in Figure 3. These results were largely in accordance with the results from the models with 11 input variables. Models in the following approach are numbered from R1 to R2047. R is here short for *Run* while the number indicates the variable combination. Higher numbers relate to more input variables. The relevant numbers are explained further in Figure 3.

For 2001 to 2099, the model best describing the yield development of all varieties was one containing only summer precipitation and CO_2 concentrations (R37); 92% of the early variety, 80% of the medium variety, and 85% of the late variety sites had this as the optimal model. The remaining sites of the early variety were best described by a model *only* containing CO_2 concentrations (R10). The same is partially true for the late variety, as 3% of the models show their best results when only including CO_2 (R10); however, models that only used spring temperatures (R6) accounted for the remaining 12%. This connection to spring temperatures was also identified in the models featuring all variables. The two runs R6 and R10 have a combined share of about 10% of the medium variety's remaining sites, while the other remaining 10% are a combination of summer precipitation, summer temperatures, and CO_2 concentration (R171).

For the period 2001 to 2050, the varieties did show a more differentiated picture. The late variety did still have R37 as the dominant model on 90% of its sites, while 5% were made up of R6 and another 5% of other not-further-distinguished models. The early variety had R37 on just 58% of its sites, 5% showing R171, and almost the entire rest of 34% from R2 with summer precipitation only. The medium variety had only 24% comprising of R37, 6% of R171, and a dominating 66% of R2.

Discussion

It should be noted that this study, like all other modeling approaches, is limited by the boundaries each model brings with itself. This study's results should thus be read as probable pathways if climatic variables do change as depicted. The focus lies on the question how different maize varieties will behave if input variables are altered within one probable future scenario.

While variations between climate models are well discussed and described, also for the climate model used in this study [22], Bassu et al. [23] took to crop models to estimate modeling spreads of maize yields between 23 crop models. In conclusion, they deemed the use of multi-model averages as being of merit as single results diverged markedly. Thus, it would be of interest to validate the results of this study by the use of different climate and crop models or their ensembles. As the selected approach

of this study is relatively time and resource consuming due to the large number of study sites, no alternative runs with other models have been conducted yet.

However, this study focuses on the differences of three varieties, a point that other studies often neglect [8] and that is not feasible with all crop models. If only studies that account for different varieties are regarded, the results from this study basically agree with the findings of, for example, Southworth et al. [17] in that the choice of variety will have a critical effect on how maize yields will develop under a future climate. It even agrees to the point that late varieties will show the most positive development which can be quite substantial with >25% in Lower Saxony towards the end of the century.

One reason for this beneficial development of late maize is clearly the fact that today's temperature sums in LS are not suitable for a full completion of its growing cycle. Temperature sums from 20th April to 15th October (minus 6°C temperature basis) vary today around $1,500$°C in LS and are therefore perfect for medium varieties but below optimum for late varieties. It seems that around 2030, when temperature sums have increased by about 100°C, the late variety can fully benefit from these temperatures. That the late variety disproportionately benefits from the generally rising temperatures is further supported by statistical analysis, as the late variety is the only one to show a substantial positive correlation to rising spring temperatures.

The future climatic conditions are, however, not entirely beneficial for the growth of late maize varieties. The main limiting factor, for all varieties, is the decline in summer precipitation. However, the time spent within these dry months in relation to the total growing time is shorter for the late variety as, for example, the medium one. Thus, the late variety can use the moister spring or fall conditions for a successful growth. Similarly, the early variety profits from the spring conditions while the medium variety would need more water during the summer months of which August will be the driest.

An adaption of sowing date could mitigate these negative effects to some degree. Some testing on single sites, however, suggests that the general yield development series of *late > early > medium* variety is not changed, though the absolute difference might. The influence of the sowing date on the results of this work are currently under evaluation.

In a related matter, critical development stages of maize, e.g., during flowering, where water shortage disproportionately restrains plant growth [24], were not sufficiently implemented in the used version of the crop model. This has changed in the current version and is under evaluation as well. While a shift in sowing dates is expected to be generally beneficial for maize yields, this increase in sensitivity is believed to have a rather detrimental effect.

While increasing temperatures are generally good for the late variety, temperatures during the growing season should not exceed optimum growing temperatures of 25°C to 30°C over longer time periods, as this would inhibit photosynthesis rates [25]. However, heat days will increase in LS throughout the century. The effects of this are visible in the multivariate regression models with an increasing negative impact of summer temperatures on yields.

In reality, this effect is expected to be even worse. The main reason is the use of a relatively smooth time series. This is firstly caused by the statistical nature of the climate model. The usage of mean values from ten climate model runs tends to eliminate extreme values. Secondly, the results of this downscaled climate data-set were 10-day values that were further combined into monthly averages. Temperature peaks were thus eliminated within the monthly means. The same is true for the monthly values of precipitation, as BioSTAR simply assumes that the monthly value is distributed evenly over each day of the month. This is clearly not the case in nature where a steady flow of water would be optimal for the plants' water supply. It will make some difference if 20 mm precipitates in 1 day followed by 9 dry days or if 2 mm for each of the 10 days is assumed. While a shortage in precipitation might be worse than elevated temperatures, it can be mitigated relatively easily by irrigation while the latter can hardly be opposed. As some areas in LS are already today under irrigation, it would be interesting to estimate probable changes in irrigation practices, meaning an estimation of the amount of water needed for optimal growth and taking the actually available amount of water into account.

That even relatively smooth precipitation series lead to varying yields becomes evident when the observed decadic variability of the yields is compared with the variability of precipitation. More precisely, the yields have their greatest variability in the decade 2051 to 2060, when summer precipitation has the greatest variability too. The same link can be found in the decade 2021 to 2030 but not for 2031 to 2050, as both decades have either high-yield or high-precipitation variability but not both. The underlying cause for the change in yield variability therefore seems to be more complex than a single dependency on summer precipitation but might well be a result of the models used.

All in all, the outlook for maize yields in LS can be described as good, especially considering that the choice of breed or variety can be quite beneficial. If absolute yields are considered today, the three varieties stand in a relation of 0.93:1:1.10 (early:medium:late, medium = 1) within the model if averaged over all modeled sites in LS. In the second half of the century, this will have shifted to a ratio 0.96:1:1.28, but can be as high as 0.96:1:1.37 for single years. The generally higher yields of the late variety

in combination with its all-out positive yield development will make it the number one choice for maize in the future.

Though this is generally positive, there exist some climatic circumstances that might negatively affect yield development that were not accounted for in this study. One is tropospheric ozone, as 30 ppb are sufficient to induce ozone intoxication in plants [26]. Since 1950, the concentration of tropospheric ozone has nearly doubled. Studies suggest that maize yields might be 2% to 5.5% higher today if this rise would not have happened [27]. It is, however, debatable if tropospheric ozone concentrations will further increase, at least in Europe, due to anthropogenic emission as CMIP 5 runs suggest [28].

A greater potential risk arises through common or invasive pests. Complicated interactions and feedbacks between climate, crop, and pests make concise predictions difficult [29]. However, as Fröhlich [13] points out, there is no expectation at all that the climatic change will lead to a reduction in infestation of any pest. In how far new cropping techniques or breeds will be able to counteract such problems is beyond the scope of this study.

Following the list of potentially negative effects of a climatic change, it seems almost surprising that the results from this study suggest quite the contrary: rising yields towards the century's end. The only variable contributing significantly towards rising yields is atmospheric CO_2; therefore, its actual future concentration will be crucial for maize yields.

Maize as a C_4 crop is not expected to profit from rising CO_2 through an elevated photosynthesis rate [30]. However, an increased water use efficiency is expected in C_3 as well as C_4 plants. This effect is accounted for by the crop model, resulting in a relatively linear decrease in the amount of water that is needed to produce the same amount of plant matter. The reduction is comparable for all three varieties and ranges between 25% to 30%.

In an environment where water is getting increasingly scarce, this is a desirable development. It appears that the negative impacts of summer temperature and precipitation are stronger until mid-century, especially for medium or early variety. The positive influence of CO_2 steadily increases to a point where the positive effects prevail and yields are rising.

That water saving through increased CO_2 concentrations can have such a strong effect is also pointed out by Taube and Herrmann [10], where grasslands profit from a rise even under increasing drought stress during summer months. This would be in line with Morgan et al. [31] who are emphasizing the importance of water saving through increased CO_2 in contrast to a direct fertilization effect.

CO_2 might still not be solely responsible for the rising yields. It undoubtedly plays a major role in doing so;

however, other factors that have not been included in the process evaluation might contribute as well. Mera et al. [32] included the effect of solar radiation in their research and found a non-linear contribution to yield development, however, not as prominent as changes in precipitation or water availability.

Conclusions

As could be shown, the changing climate will have a predominantly positive effect on the yield development of maize and its varieties in Lower Saxony. A stronger positive development is, however, not expected to set in before the second half of the 21st century.

The first half will be stagnant in yields for the early variety. In the last decades of the century, the yields will on average increase about 9%. The medium variety even shows a negative development in the first half that is later reversed. Towards the century's end, the yields then increase about 5% in comparison to today's yields. The late variety has the all-out best yield development, with an average increase of 25% for 2071 to 2099 and a strong positive trend beginning already around 2030. In addition to this above-average rise, the yields themselves are higher so that a transition of local agricultural practices towards the late variety is conceivable. The yield development of all yields is accompanied by an increase in yield variability during mid-century that seems to partially follow precipitation patterns.

Thus, the development will generally be positive in the long run, though the path for each variety diverges. As the varieties react in different ways to the changing annual pattern of temperature and precipitation, the results do indicate that the consideration of different varieties might also change the outcome of studies at different study sites. At any rate, the few other existing studies are hinting towards the same result [17,18]. Varieties with longer or shorter growing periods will have an advantage in areas where medium varieties are predominantly grown today.

Besides, for Lower Saxony or Germany in general, a decline in summer precipitation is not seen as an insurmountable obstacle for local agriculture, as there is no necessity for irrigation on most sites today and present water reserves would allow an expansion of irrigated areas at least to some degree. Intensive groundwater management will be a basis for this, as increasing winter precipitation could cover the water extraction during the summer months. New breeds and cropping techniques will also aid to further counteract the more negative effects of climate change, including the expansion of pests or hitherto unknown effects that might arise.

In conclusion, the maize yields in Lower Saxony will not suffer from long-lasting declines but will have a generally positive outlook over the course of the 21st century.

Methods

The basic approach used in this study was to use high-resolution climate data in combination with detailed soil information as the input for a crop model. All components involved are introduced in the following sections.

Area of interest: Lower Saxony

LS, with roughly 46,500 km^2 of land area, is the second largest of the 16 federal states of Germany, providing around 15% of the nations agricultural land [2]. Located to the north-west of Germany (Figure 4), the state lies in a transition zone between a more maritime (NW) towards a more continental climate (SE) [33] with an average annual temperature of around 9°C and a mean precipitation of 749 mm in the period of 1971 to 2000 [34].

Principally, LS consists of three distinguishable landscape structures: the coast, including the East Frisian Islands, the German North-Western Lowland (amounting for three quarters of LS' total land area) as well as a low-mountain range to its south, with the Harz as its most prominent representative [36]. The broad loess valleys to the south and especially the fertile "Börde" that fronts the low-mountain range to the north are the main cultivation areas for high-demand crops like winter wheat. The Lowland mainly consists of "Geest" land, Quaternary sediments that are particularly sandy to the north-east, with precipitation as low as 500 mm, making irrigation already necessary today on several sites. The west of LS is dominated by livestock farming with the coastal area predominantly used for grassland farming as high ground water levels prevent intensive use [37].

The regional differences manifest themselves in the average regional yields. In the period of 2003 to 2008, the average winter wheat yield south of Hanover was always above 8 t/ha, above 7 t/ha south of Oldenburg and generally below 7 t/ha in the north-east. Maize yields behave rather similarly, with dry maize silage (33% dry matter content) having the best yields to the south. The margin between the different parts of LS is, however, smaller for maize than for wheat and varies generally around 15 t/ha. As can be seen in Figure 4, the areas with the largest maize production coincide with areas where only little wheat is grown and where feed for livestock is in high demand.

The crop model

The crop model used in this study is a relatively new model called *BioSTAR*, developed at the Georg-August-University in Göttingen [38]. The model uses a CO_2-based crop development engine, thus taking a potential CO_2 fertilization effect into account.

The basic working principle uses temperature to determine the plants' development stages and a combination of temperature, solar radiation, and CO_2 concentration for the maximum photosynthesis rate. Both incrementally

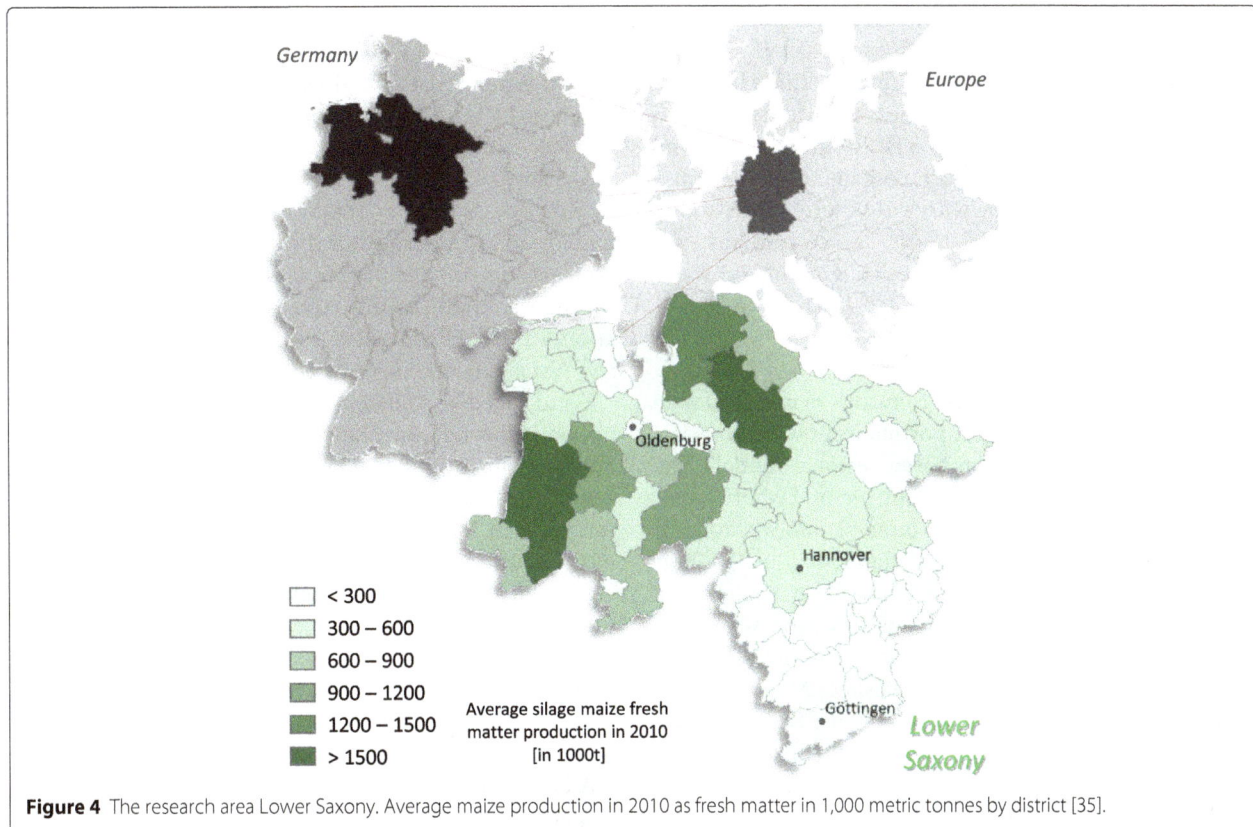

Figure 4 The research area Lower Saxony. Average maize production in 2010 as fresh matter in 1,000 metric tonnes by district [35].

build the plants' maximum possible biomass that is then recursively limited by precipitation, i.e., the soil water content. The model is suitable for large-scale as well as parcel size yield assessments. The philosophy behind it is an easy to use model with a robust output and a manageable amount of required input parameters.

The model was validated on sample sites in Lower Saxony with a general disagreement between actual and modeled yield of around 10%. The required climatic input variables are precipitation, temperature, atmospheric CO_2 concentration, solar radiation, relative air humidity, and wind-speed at 2-m altitude. In addition, information on the soil type is required. As the model was initially conceived as a tool for the estimation of bio-energy potentials, the maize crops only contain silage maize (no food maize). Furthermore, the three varieties do not consist of single breeds but represent an average of several early, medium, or late breeds. The breeds are already grown in Germany, though the late variety not yet in Lower Saxony due to temperature limitations. Breeds with a very high temperature demand, as grown today under a Mediterranean climate, are not included.

As a rather robust approach, the model leaves out some aspects that might well be of importance for a future yield development. Results in this study should thus be read as what would happen if nothing but the climatic input

variables would change. These neglected aspects include any technological advances, including any changes in farm management. Irrigation was not included in the modeling, whether for current or future yields, even if there do exist some areas today that are under irrigation. The sowing date was always the 115th day of the year and was not changed throughout the century. No extra fertilization was included and soil water content expected to be at 100% at the beginning of the growing season. No effects of a prior crop on a specific site are taken into account. The model either stopped on day 300 of a given year or when full maturity was attained, depending on what happened first.

At the time of the actual modeling, the BioSTAR model was still under active development. However, all crops had already been validated. This validation on sites in Lower Saxony was one reason for the choice of BioSTAR. The others are its applicability on a large number of sites and a differentiation of the three varieties.

Soil data

The soil data used in this study is part of the official digital soil survey map of LS in a resolution of 1:50,000, called BÜK50 [39]. This map was intersected with data from the CORINE land-use classification of 2005 for Lower Saxony to extract sites that are used for agricultural purposes only.

The result between the intersected soil and land-use map was a data-set of 91,014 sites with each used as a unique modeling area. The soil map contained codified information on the soil type and its thickness that were translated into the format required by BioSTAR. Fifteen 10-cm soil levels had to be identified, each containing the information on prevalent soil type with a 16th level representing everything below the initial 1.5 m. The crop model uses these information solely for the calculation of soil water content and flows.

Climate model and data

The climate data was derived from the regional climate model WETTREG, a German portmanteau word translating into "weather condition based regional model". The model uses a statistical downscaling method where large-scale atmospheric patterns are brought into a statistical relationship with local climate station data [40]. The initial link is created by using known measured data at these stations and globally gridded reanalysis data, with both ERA40 and NCEP/NCAR data using a k-means cluster approach. This link is then reestablished through GCM-derived gridded data, here from the ECHAM5 global climate model. For each large-scale weather pattern of the future, a pool of local station data is available that is then resampled several times to create the climate signal [41].

The actual climate model's name is WETTREG 2010, as the initial approach (today called WETTREG 2006) neglected weather patterns that are relatively rare today but will increasingly emerge in a future climate. Thus, two patterns were added to this latest version, significantly reducing the model bias in comparison to other climate models [42]. WETTREG 2010 was applied at 248 stations distributed throughout LS, whereas the mean of ten iterations at each station was used as the climate signal for the 21st century (A1B SRES scenario). Using spatial interpolation methodology, these point-based information were further upscaled to a grid of 100 × 100 m at the *Jülich Research Centre* through the CLINT *interpolation model* [43]. This resulted in a grid of 11,520,000 data points for each time step (with 10-day values amounting to 36 single steps per year) for temperature, precipitation, and potential evapotranspiration. The data was available for the years 1961 to 2100 with an additional data-set of interpolated measured station data from Germany's National Meteorological Service (DWD) for the years 1961 to 2005 for validation purposes. Both data-sets agreed reasonably well in temperature and precipitation (with WETTREG2010 showing a mean annual average bias of +0.02°C and −2.24% precipitation).

It should, however, be noted that this high-resolution downscaling cannot and does not improve the confidence of the initial climate projection. The main purpose of this downscaling is to capture regional differences within the

projected climatic change proposed by the ECHAM 5 GCM and the A1B scenario.

Furthermore, data on global radiation was taken from a run of ECHAM 5 in a global T31 grid of 48 × 96 that was calculated within the scope of the ENSEMBLES project [44]. The ECHAM 5 data was chosen for the purpose of data consistency as the WETTREG2010 data did also employ ECHAM 5 runs for the boundary conditions. The data-set was provided for the years 2001 until 2099, thus setting the limits for this study's timeframe. Global radiation was calculated as the sum of *surface net downward shortwave flux* and *surface net downward longwave flux*.

Wind speed was taken from official maps of LS of 2005 provided through the State Authority for Mining, Energy and Geology (LBEG) that uses the FAO approach for wind speed in a height of 2 m above grass. Typical wind speed ranges from 5 to 6 m/s at the coast to around 1 to 2 m/s in the south of LS. To present knowledge, no significant change in the wind speed pattern is anticipated for the future [45]; hence, the data was applied without further changes.

Relative air humidity was calculated backwards from the WETTREG2010 data on evapotranspiration, as this was derived through the Penman-Monteith approach.

All data was then intersected with the soil sites using the respective variable's mean value.

Climate change in Lower Saxony

Figure 5 gives a brief description of the average change of the climatic variables' temperature and precipitation in LS. The climatic comparison is done by 30-year intervals where 1971 to 2000 is used as present-day climate that might be seen as more current than the climatic normal period of 1961 to 1990 [46]. These intervals represent a near- (2011 to 2040), middle- (2041 to 2070), and long-term (2071 to 2100) climatic development.

There are no areas at any time that do show a stagnant or even decreasing temperature development. However, warming in spring is always below the annual average while the winter months are always above. Fall temperatures are slightly below annual average and summer months above, though both deviate less from the mean than spring and winter seasons do. The mean temperature increase is 0.95°C for near-, 2.30°C for middle-, and 3.40°C for long-term scenarios. The development is relatively uniform throughout LS with a slightly stronger (but still less than 0.5°C difference) development to the south-east.

The precipitation development is different in terms of being positive or negative depending on time and space. If only annual means are considered, almost no change in precipitation can be detected, although a moderate decline is visible. It, however, becomes increasingly obvious that the winter and summer seasons are drifting into

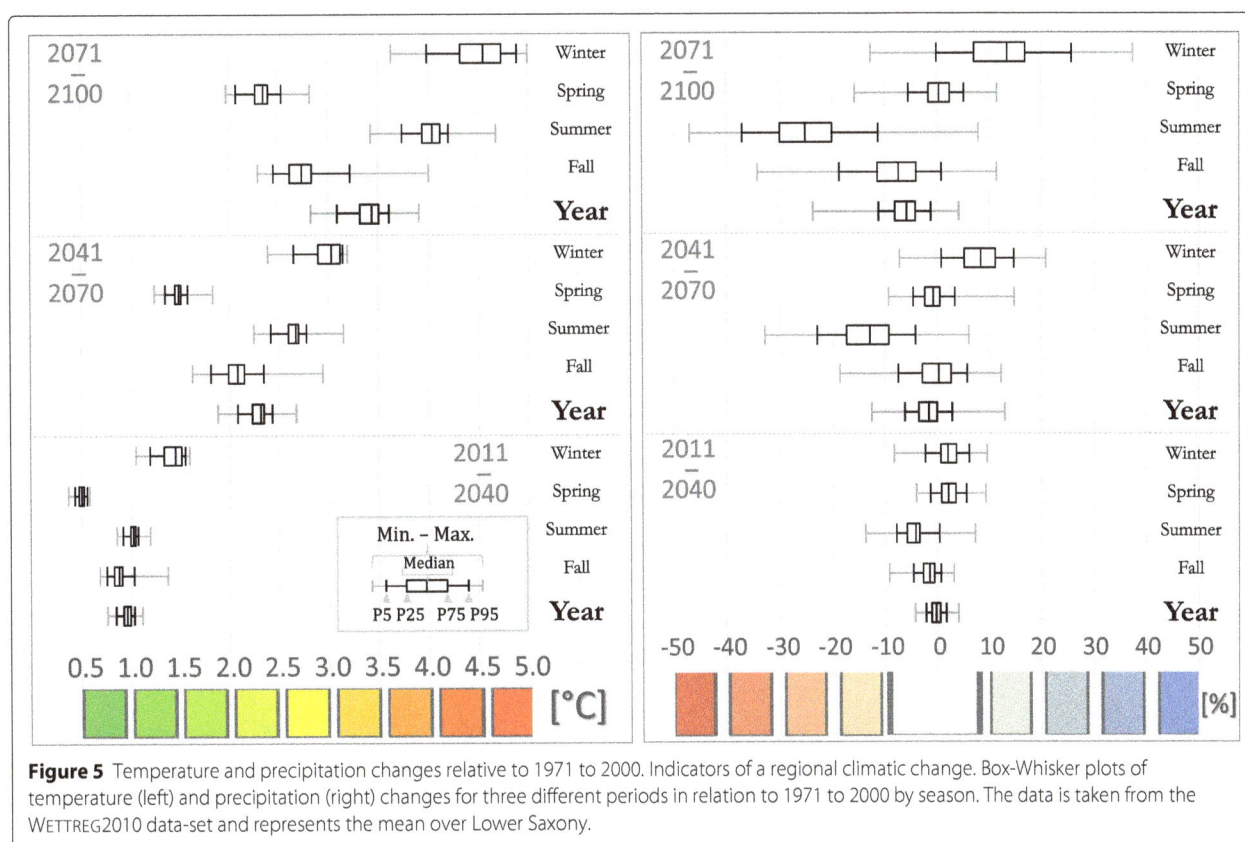

Figure 5 Temperature and precipitation changes relative to 1971 to 2000. Indicators of a regional climatic change. Box-Whisker plots of temperature (left) and precipitation (right) changes for three different periods in relation to 1971 to 2000 by season. The data is taken from the WETTREG2010 data-set and represents the mean over Lower Saxony.

opposite directions. While in the near future all seasonal differences remain in a window of more or less ±10%, these changes drastically amplify towards the end. The mean decline in precipitation is around −25% in the long-term perspective with some areas at a nearly −50% decrease. Winter increases are also substantial, but at around 15% towards the end of the century, they cannot fully counterbalance the summer losses.

In summary, all deviations from today's values will increase with passing time, fostering a local development towards a more winter rain climate that features increasingly hot and dry summers and mild wet winters.

Statistics

To account for extreme or unrealistic outliers, a two-way approach was devised for the original resulting data-set. At first, all sites with a biomass yield of 0 g/m^2 were excluded. This typically amounted to 706 sites that contain only bedrock in their soil levels. In a second step, all data below the 0.1 and above the 99.9 percentile were excluded, as values close to zero or unreasonably large yields were present. This proved to well eliminate outliers while preserving as much data as possible.

Basic statistics in this study include standard deviation, coefficient of variability (cov), linear regression models, and the coefficient of determination [47]. The time series

could well be described using linear regression models; however, tests with exponential, logarithmic, and second- and third-order polynomial and potential models did show about equal results.

The data was further explicitly tested for trends using a robust trend/noise (t/n) ratio, where the difference in yield from the years 2099 and 2001 was divided through the time series' standard deviation. A significant trend is assumed at a ratio of 1.96 or above, representing the $\alpha = 0.05$ level. As this test is often considered to be relatively weak, the non-parametric Mann-Kendall (MK) test was applied as well. A further advantage of MK is its ability to detect non-linear trends. Most statistics were applied for the time series of 2001 to 2030, 2001 to 2050, and 2001 to 2099.

Data comparing the first (2001 to 2050) and second (2051 to 2099) half of the century will sometimes give a ratio of 1st/2nd half. As the second half is here only 49 years long, only a ratio of 0.505 would mean that the value is equal for both halves.

To determine the climatic variables that significantly influence the yield development throughout the century, a multivariate regression model was used. In a first step, 11 variables were included in the model that was then run for all sites. These variables include, respectively, five temperature and precipitation values (annual, winter, spring, summer, fall mean) as well as atmospheric CO_2 concentration. This was done to get a general test of strength of all variables against each other at different sites. However, autocorrelation is very likely to occur, as at least temperature trends seem to be relatively equal across the five variables. Therefore, a best model approach was devised. Eleven variables can be assembled into 2,047 unique groups when their order is neglected. Each combination was treated as a new model and calculated on 3,740 randomly distributed sites. The multivariate model that explained the yield development best was then logged. If combinations gave equally good results, the first run, generally the one with less variables, was logged. This was done for the years 2001 to 2099 as well as 2001 to 2050 to identify possible changes in variable impact.

The statistics in this study have been calculated using *MS Excel 2010* and *Python* (v 2.7) with the addition of SciPy and NumPy [48], Pandas [49], and MatPlotLib [50]. The calculation of the multivariate regression models was done using *R* (v 2.15.2) and *rpy2* (v 2.3.0).

Abbreviations
LS: Lower Saxony, state in Germany; MK: Mann-Kendall test for trends; P_x: Precipitation sums of x (win = winter, spr = spring, sum = summer, fal = fall); Rx: Run-number of all possible variable combinations for a multivariate regression model; T_x: Temperature means of x (win = winter, spr = spring, sum = summer, fal = fall).

Competing interests
The authors declare that they have no competing interests.

Authors' contributions
JFD carried out the data analysis, graphical design and first draft of the study. MK took part in conceiving the study and participated in its design and coordination. Both authors contributed on specific aspects, read and approved the final manuscript.

Acknowledgements
We acknowledge support by the Open Access Publication Funds of the Göttingen University. We would like to thank the LBEG in Hannover for the provision of the climate data-set.

References
1. FAO. FAO online statistical database, Rome. 04.02.2014. 2014. http://faostat3.fao.org.
2. DeStatis. Statistical Yearbook of Germany 2013, 1st edn. Wiesbaden: Statistisches Bundesamt; 2013.
3. LWK. Homepage of the Chamber of Agricultre of Lower Saxony: homepage of the Chamber of Agricultre of Lower Saxony, Oldenburg. (10.03.2014). 2014. http://www.lwk-niedersachsen.de.
4. Hoeher G. Entwicklung Energiepflanzenanbau und Biogas in Niedersachsen, Berlin. (17.07.2013). 2007. www.ifeu.org/landwirtschaft/pdf/6_Hoeher_Maisanbau_Nieders.pdf.
5. NMELVL. Maisanbau: Mehr Vielfalt durch Alternativen und Blühstreifen. Hannover; 2012. http://www.ml.niedersachsen.de/download/78012.
6. LWK. Energiepflanzenfeldtage: Es muss nicht immer Mais sein, Oldenburg. (10.03.2014). 2014. http://www.lwk-niedersachsen.de/index.cfm/portal/2/nav/74/article/15172.html.
7. Wolf J, Diepen CA. Effects of climate change on grain maize yield potential in the European Community. Climatic Change. 1995;29(3):299–331.

8. Supit I, van Diepen CA, Wit AJWd, Wolf J, Kabat P, Baruth B, et al. Assessing climate change effects on European crop yields using the Crop Growth Monitoring System and a weather generator. Agric Forest Meteorology. 2012;164:96–111.
9. Meyer U, Hüther L, Manderscheid R, Weigel H-J, Lohölter M, Schenderlein A. Nutritional value of maize 2050 In: Schwarz FJ, editor. Optimierung des Futterwertes Von Mais und Maisprodukten. Landbauforschung/Sonderheft. Braunschweig: VTI; 2009. p. 107–14.
10. Taube F, Herrmann A. Relative benefit of maize and grass under conditions of climatic change In: Schwarz FJ, editor. Optimierung des Futterwertes Von Mais und Maisprodukten. Landbauforschung/Sonderheft. Braunschweig: VTI; 2009. p. 115–26.
11. Jacob D, Göttel H, Kotlarski S, Lorenz P, Sieck K. Klimaauswirkungen und Anpassung in Deutschland: Phase 1: Erstellung regionaler Klimaszenarien für Deutschland: Im Auftrag des Umweltbundesamtes, Dessau-Roßlau. Umweltbundesamt, Dessau-Roßlau, Climate Change (11/2008). 2008.
12. USF. Klimawandel und Landwirtschaft in Hessen: Mögliche Auswirkungen des Klimawandels auf landwirtschaftliche Erträge. INKLIM Baustein 2: Universität Kassel; 2005.
13. Fröhlich M. Klimawandel und Landwirtschaft: Auswirkungen der globalen Erwärmung auf die Entwicklung der Pflanzenproduktion in Nordrhein-Westfalen, Köln. Umwelt Landwirtschaft Natur-und Verbraucherschutz Landes Nordrhein-Westfalen: Des Ministerium für Klimaschutz; 2010.
14. Mirschel W, Wenkel K-O, Wieland R, Köstner B, Albert E, Luzi K. Auswirkungen des Klimawandels auf die Ertragsleistung ausgewählter landwirtschaftlicher Fruchtarten im Freistaat Sachsen, Müncheberg. Müncheberg: ZALF; 2008.
15. Mirschel W, Wieland R, Wenkel K-O, Guddat C, Michel H, Luzi K, et al. Regionaldifferenzierte Abschätzung der Auswirkungen des Klimawandels auf die Erträge von wichtigen Fruchtarten im Freistaat Thüringen mittels Ertragssimulation mit YIELDSTAT, Müncheberg. Müncheberg: ZALF; 2012.
16. Buttlar CV, Karpenstein-Machan M, Bauböck R. Klimafolgenmanagement Durch Klimaangepasste Anbaukonzepte Für Energiepflanzen in der Metropolregion Hannover-Braunschweig-Göttingen-Hildesheim. Stuttgart: Ibidem; 2013.
17. Southworth J, Randolph JC, Habeck M, Doering OC, Pfeifer RA, Rao DG, et al. Consequences of future climate change and changing climate variability on maize yields in the midwestern United States. Agric Ecosystems Environ. 2000;82(1-3):139–58.
18. Liu Z, Hubbard KG, Lin X, Yang X. Negative effects of climate warming on maize yield are reversed by the changing of sowing date and cultivar selection in Northeast China. Global Change Biol. 2013;(19):3481–92.
19. Wolf J, van Diepen C. Effects of climate change on silage maize production potential in the European Community. Agric Forest Meteorology. 1994;71(1-2):33–60.
20. Kwabiah AB. Growth and yield of sweet corn (Zea mays L,) cultivars in response to planting date and plastic mulch in a short-season environment. Scientia Horticulturae. 2004;102(2):147–66.
21. Meza FJ, Silva D, Vigil H. Climate change impacts on irrigated maize in Mediterranean climates: evaluation of double cropping as an emerging adaptation alternative. Agric Syst. 2008;98(1):21–30.
22. Kreienkamp F, Baumgart S, Spekat A, Enke W. Climate signals on the regional scale derived with a statistical method: relevance of the driving model's resolution. Atmosphere. 2011;2(2):129–45.
23. Bassu S, Brisson N, Durand J-L, Boote K, Lizaso J, Jones JW, et al. How do various maize crop models vary in their responses to climate change factors? Global Change Biol. 2014;20(7):2301–20.
24. Ehlers W. Wasser in Boden und Pflanze: Dynamik des Wasserhaushalts Als Grundlage Von Pflanzenwachstum und Ertrag. Stuttgart (Hohenheim): Ulmer (Eugen); 1996.
25. Endlicher W, (ed). 2007. Der Klimawandel: Einblicke, Rückblicke Und Ausblicke. Potsdam: Potsdam-Inst. für Klimafolgenforschung.
26. Long SP, Ainsworth EA, Leakey ADB, Nösberger J, Ort DR. Food for thought: lower-than-expected crop yield stimulation with rising CO2 concentrations. Science. 2006;312(5782):1918–21.
27. Avnery S, Mauzerall DL, Liu J, Horowitz LW. Global crop yield reductions due to surface ozone exposure: 1. Year 2000 crop production losses and economic damage. Atmos Environ. 2011;45(13):2284–96.
28. Fiore AM, Naik V, Spracklen DV, Steiner A, Unger N, Prather M, et al. Global air quality and climate. Chem Soc Rev. 2012;41(19):6663–83.

29. Schaller M, Weigel HJ. Analyse des Sachstands zu Auswirkungen von Klimaveränderungen auf die deutsche Landwirtschaft und Maßnahmen zur Anpassung, Braunschweig: FAL Agricultural Research, Sonderheft 316; 2007.

30. Lambers H, Pons TL, Chapin FS. Plant physiological ecology, 2edn. New York: Springer; 2008.

31. Morgan JA, Pataki DE, Körner C, Clark H, Del Grosso SJ, Grünzweig JM, et al. Water relations in grassland and desert ecosystems exposed to elevated atmospheric CO2. Oecologia. 2004;140(1):11–25.

32. Mera R, Niyogi D, Buol G, Wilkerson G, Semazzi F. Potential individual versus simultaneous climate change effects on soybean (C3) and maize (C4) crops: an agrotechnology model based study. Global Planet Change. 2006;54(1-2):163–82.

33. Seedorf HH, Meyer H-H. Historische Grundlagen und Naturräumliche Ausstattung. Neumünster: Wachholtz; 1992.

34. DWD. Website of the German Weather Service. 24.03.2014. 2014. http://www.dwd.de.

35. LSN. Statistical data of the Statistical Office of Lower Saxony, Hannover. (14.03.2014). 2014. http://www.nls.niedersachsen.de/.

36. Drachenfels OV. Überarbeitung der Naturräumlichen Regionen Niedersachsens. Informationsdienst Naturschutz Niedersachsen. 2010;30(4):249–52. (17.07.2013).

37. Heunisch C, Caspers G, Elbracht J, Langer A, Röhling H-G, Schwarz C, et al. Erdgeschichte von Niedersachsen: Geologie und Landschaftsentwicklung, Hannover. Landesamt für Bergbau, Energie und Geologie (LBEG), Hannover, Geoberichte. 2007;(6).

38. Bauböck R. GIS-gestütze Modellierung und Analyse von Agrar-Biomassepotentialen in Niedersachsen – Einführung in das Pflanzenmodell BioSTAR. Göttingen: PhD thesis, Georg-August-Universität. 2013.

39. Boess J, Gehrt E, Müller U, Ostmann U, Sbresny J, Steininger A. Erläuterungsheft zur Digitalen Nutzungsdifferenzierten Bodenkundlichen Übersichtskarte 1:50.000 (BÜK50n) von Niedersachsen, vol. 3. Hannover: Arbeitshefte Boden; 2004.

40. Enke W, Spekat A. Downscaling climate model outputs into local and regional weather elements by classification and regression. Climate Res. 1997;8:195–207.

41. Enke W, Deutschlander T, Schneider F, Kuchler W. Results of five regional climate studies applying a weather pattern based downscaling method to ECHAM4 climate simulation. Meteorologische Zeitschrift. 2005;14(2): 247–57.

42. Kreienkamp F, Spekat A, Enke W. Weiterentwicklung von WETTREG bezüglich neuartiger Wetterlagen. Climate & Environment Consulting Potsdam GmbH. 2010.

43. Müller U, Engel N, Heidt L, Schäfer W, Kunkel R, Wendland F, et al. Klimawandel und Bodenwasserhaushalt, Hannover. Landesamt für Bergbau, Energie und Geologie (LBEG), Hannover, Geoberichte. 2012;(20).

44. Roeckner E. ENSEMBLES STREAM2 ECHAM5C-MPI-OM SRA1B run1: World Data Center for Climate: CERA-DB ENSEMBLES2_MPEH5C_SRA1B_1_MM. (17.07.2013). 2009. http://cera-www.dkrz.de/WDCC/ui/Compact.jsp?acronym=ENSEMBLES2_MPEH5C_SRA1B_1_MM.

45. NMUEK. Empfehlung für eine niedersächsische Klimaanpassungsstrategie. 01.08.2013. 2012. http://www.umwelt.niedersachsen.de/klimaschutz/aktuelles/107128.html.

46. WMO. Homepage der WMO: working together in weather, climate and water. 21.06.2011. 2011. http://www.wmo.int.

47. Schönwiese C-D. Praktische Statistik Für Meteorologen und Geowissenschaftler, 4edn. Berlin [u.a.]: Borntraeger; 2006.

48. Jones E, Oliphant T, Peterson, Pearu, et al. SciPy: open source scientific tools for Python. (17.07.2013). 2001–. http://www.scipy.org/.

49. Pandas. pandas: Python Data Analysis Library. (17.07.2013). 2012. http://pandas.pydata.org/.

50. Hunter JD. Matplotlib: A 2D graphics environment. Comput Sci Eng. 2007;9(3):90–5.

15

Optimal test design for binary response data: the example of the fish embryo toxicity test

Nadia Keddig[1*], Sophia Schubert[1] and Werner Wosniok[2]

Abstract

Background: The fish embryo toxicity test (FET) is an established method in toxicology research for quantifying the risk potential of environmental contaminations and other substances. The typical results of the method are the half maximal effective concentration (EC_{50}) or the no observed effect concentration (NOEC). However, from an environmental perspective, it is most important to safely identify the concentration of the substance effect which lies above the effect under control condition (spontaneous effect). The common FET is not optimal to detect ECs for small target effects. This paper shows how to optimize the efficiency and consequently the benefit of the FET for small effects using an adequate experimental design. The approach presented here can be carried over to all test systems generating binary (yes/no) outcomes.

Results: The experimental design has three components in this context: determination of spontaneous response, sample size calculation, and dose allocation. A strategy for all three components is proposed from which a design is given including precision requirements and makes the most effective use of the experimental effort. This strategy amounts to expanding the usual FET guidelines of Organisation for Economic Co-operation and Development, German Institute for Standardization, or American Society for Testing and Materials by adding a planning step that adapts the test to the specific user's need.

Conclusions: For the practical calculation of an adapted design, a newly developed software is presented as R package *toxtestD*. It provides a user-friendly way of developing an optimal experimental design for the FET without in-depth statistical knowledge. The programme is suited for all experimental problems involving a binary outcome and a continuous concentration.

Keywords: Toxicity test planning; R statistic package; Zebrafish embryo; Spontaneous lethality; Sample size; Dose design

Background

Toxicity tests in ecotoxicology serve to detect and quantify toxic properties of chemical substances. Typically, a toxicity test is a laboratory test, which means that the experimenter chooses the test procedure, the number of subjects to test, the doses or concentrations to apply, as well as the appropriate way to quantify toxicity. Such quantifications are used to set thresholds for allowable concentrations in the environment. Choices in experimental design should ensure high quality of the results in terms of precise and unbiased toxicity quantification

and of statistical decisions with controlled error rates. The design should also make optimal use of the experimental effort. A proper experimental design is frequently demanded, but only a few publications on risk assessment deal with this aspect in detail. In this paper we use the fish embryo toxicity test (FET) as an example to demonstrate how to design a toxicity experiment attaining the required precision of results. We also include considerations on how to quantify toxicity using the FET results. The procedure proposed involves a four-parameter logistic dose-response model, which allows incorporating spontaneous effects as well as non-effects due to an insusceptible subpopulation. For the numerical operations of planning and analysis, we provide the R

* Correspondence: publication@gmx.net
[1]Institute of Fisheries Ecology, Thünen Institute (TI), Palmaille 9, 22767 Hamburg, Germany
Full list of author information is available at the end of the article

package *toxtestD*, whose reference manual can be downloaded from the CRAN homepage [1, 2].

In recent years, FET has predominantly replaced the fish acute toxicity test [3, 4]. Research projects like DanTox favour the embryos of the model organism *Danio rerio* to identify toxicity processes [5]. The classical version of the FET is established in research laboratories as well as in service laboratories [6–8].

A core component of the FET is exposing fertilized eggs, preferentially from zebrafish (*D. rerio*), in an early stage of cell division to an aquatic compound, which is charged with harmful substances. Responses to the tested substance can be death, coagulation, lethal or sublethal malformations, or teratogenic effects. The presence of effects is examined after 48 or 96 h post fertilization (hpf) [9, 4]. This test setup is used because early life stages are more sensitive than the adult life stage. In addition, early life stage tests operate faster than tests on full-grown parental fishes [10]. Following the norm of the German Institute of Standardization (DIN), ten fertilized, normally developed eggs per concentration and a negative control should be tested [4]. The Organisation for Economic Co-operation and Development (OECD) guideline recommends 20 eggs per test concentration and positive control, respectively, and 24 eggs per

negative control [9]. Both guidelines accept up to 10 % spontaneous deaths among negative controls [4, 9].

Effect quantification

Effect quantification means expressing the toxicity of a substance by a single number. The full information about the relation between concentration and toxicity (effect) is described by the concentration-response relation (see example in Fig. 1a). A major concept of effect quantification is the no observed effect concentration (NOEC). It is the result of comparing observed effects in treated groups to the effects observed in the control group. The other major effect quantification is the effective concentration value (EC_{xx}). It denotes the concentration which causes an effect of xx %. Depending on its application, EC_{xx} has been varyingly labelled as effective dose (ED_{xx}), lethal concentration/dose (LC_{xx}/LD_{xx}), benchmark dose (BMD), or virtual safe dose (VSD, for very small xx) (OECD 2013 [11]). Both concepts differ clearly with regard to their properties and interpretation.

NOEC, the controversial legacy

As stated by the guidance for the implementation of REACH (Registration, Evaluation, Authorisation and

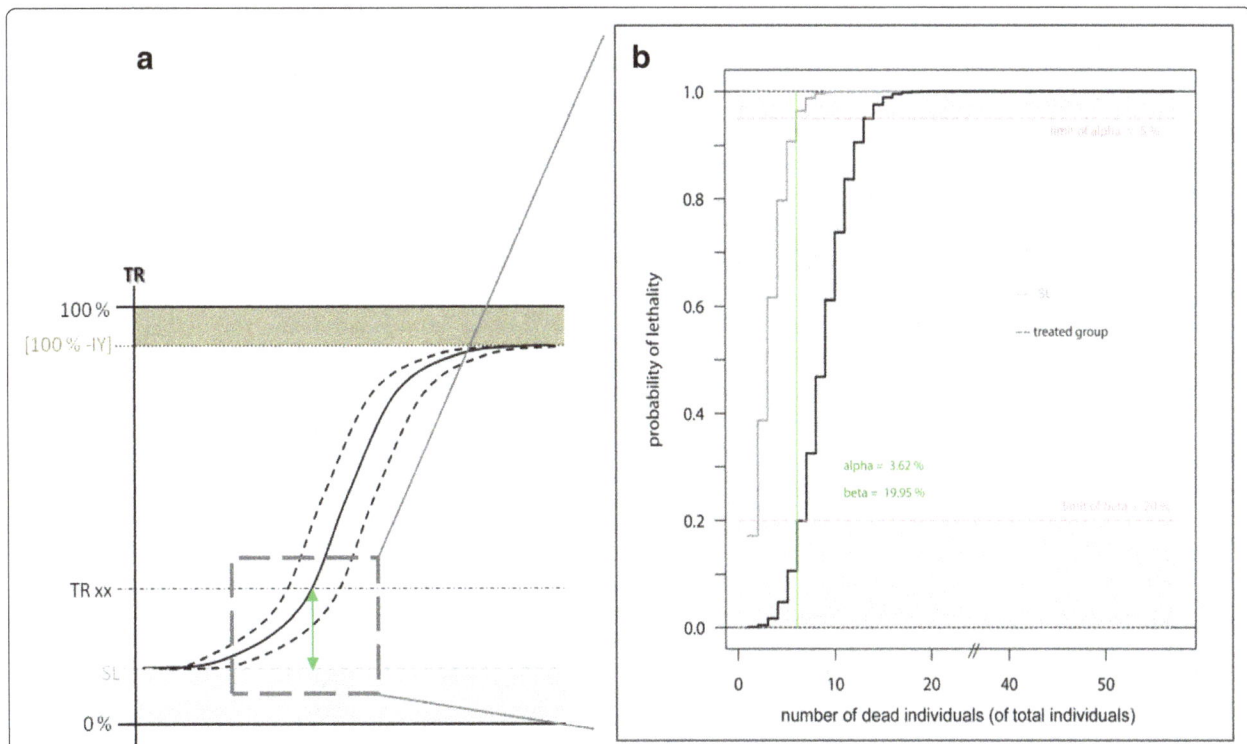

Fig. 1 Sample size calculation. Logistic distribution of a concentration-response relationship (*line*) with a confidence interval (*dotted lines* around). **a** Difference between SL and TR xx at the dose-response curve (*green arrow*) marks the basic distance for calculation of the sample size. The confidence interval of SL is not shown. **b** Calculation of the sample size with the two different distributions of SL (*light grey line*) and a treated group (*dark grey line*). Under the restrictions of alpha and beta (*red marked areas*) will the optimal number of individuals be estimated (green line). *IY* immunity, *SL* spontaneous lethality, *TR* total risk, *xx* target value

Restriction of Chemicals), the NOEC is *'the highest tested concentration for which there is no statistical significant difference of effect when compared to the control group'* [12]. Similar to other statistical tests, the test sequence leading to the NOEC will detect a substance-related effect with a given safety only if it has a certain size. The detectable size and the probability of detection depend on sample size, the number of concentration points, and their allocation. Changing the sample size may shift the NOEC value over the whole range of tested concentrations, e.g. if a test sequence is repeated with the identical set of concentrations, but with a different number of replicates per concentration, the highest concentration tested (for few replicates) or the smallest concentration (for many replicates) or a concentration somewhere in between may result as NOEC [13]. The high importance of sample size becomes evident when looking at a simple example: if a concentration causes one of 20 organisms to show an effect, but only 10 organisms are tested, it is not very likely that the experimenter will see an effect at all. Generally, an existing effect may be accidentally missed due to a too small sample size, an unfortunate choice of concentrations or just by chance [14].

The abbreviation NOEC contains the wording of a 'no observed effect' [15], and the NOEC only indicates a concentration which could not be shown to cause a response. However, in some cases, NOEC seems to be misunderstood as indicating a concentration that produces generally 'no effect', particularly when no effect was observed in the actual experiment. But a true effect greater than zero may be undetected in an experiment simply because of a small sample size. It would be seen in an experiment with larger sample size. A power calculation would unveil this situation. In a NOEC analysis involving only few replicates, the detection of a small effect cannot be expected due to the small statistical power of a statistical test on binary effects with few replicates [14]. As a NOEC is typically reported without the circumstances of its genesis, a user cannot comprehend whether a high NOEC is either due to weak toxicity or to an experiment with few replicates [15]. Moreover, NOECs from two experiments with different concentration patterns and varying replicate numbers can hardly be compared. Guidelines try to establish a minimum of experimental standards; nevertheless, resulting NOECs fluctuate still between concentrations generating 10 and 30 % effect [16]. NOECs are therefore considered as highly problematic in the scientific discussion [17, 15, 18].

Parametric modelling

The EC_{xx} calculation relies on a parametric assumption about the concentration-response relation (concentration-response curve) underlying the data. EC_{xx} links the pre-specified effect level to the effective concentration [19].

Presumably, the most frequently used target effect value is the mean effective concentration (EC_{50}), which relates to a mean response of 50 % [20]. Different from NOEC, a confidence interval (CI) can be calculated for both the whole curve and for every EC_{xx} value. The width of the CI for EC_{xx} is affected by the number of replications and the concentration allocation pattern. This can be exploited to set up an optimal experimental design that makes best use of the experimental effort. The EC_{xx} concept is commonly preferred over NOEC because of its fewer problematic attributes [12], in particular, for the fact that the expected value of EC_{xx} does not depend on sample size and that a confidence interval can be given.

Target shift to small effect sizes for threshold calculations

The effects of much smaller sizes than 50 % need to be detected to determine concentrations acceptable for health level and environmental conservation. Therefore, the target of the experiment is shifted towards smaller effects. The detection of small effects is necessary for employment and environmental protection to define concentration thresholds that should not be exceeded in order to keep the amount of adverse effects (response) due to exposure below the tolerable level [21, 22]. All ECs should be calculated from a concentration-response curve fitted to observed response data. Approximate calculations of EC_{xx} for small xx lead to diverging results as a consequence if controversial safety or assessment factors become necessary to apply [23]. Concluding the EC_{xx} from EC_{50} is an unsafe operation, as the difference between EC_{xx} and EC_{50} depends on the slope of the concentration-response curve, which is unknown and cannot be concluded from EC_{50}. The NOEC is by definition neither related to the size of an effect nor to a concentration-response curve; therefore, no EC_{xx} can reasonably be concluded from the NOEC. Small responses just above zero are generally hard to detect by a statistical test and proving the probability of a detectable response at zero requires an infinitely large number of biological objects in the test. As an example, if the substance effect increases and consequently the effect rate rises by 0.000001 (= 0.0001 %), the experimental group must contain at least 1/0.000001 = 1,000,000 objects to make the expected increase in response (by only one object) visible. In reality, experiments are designed with much smaller sample sizes simply for logistic reasons. This means that only concentrations with an effect clearly above zero can be detected, whereas the exact meaning of 'clearly above zero' needs to be calculated during planning and design of the experiment. It depends on sample size as well as on concentration-response relation. The required effect size, in order to fulfill the aim of the experiment, needs to be assessed for each particular problem. It defines the tolerable level mentioned above. Typically, a substance effect in the range of 1 to 10 % is set as tolerable level. This directs the focus on effect concentrations like EC_{01}, EC_{05}, and EC_{10}.

Changes in experimental design as a first step

In the discussion about how thresholds should be derived, the danger of using an insufficient data set has been identified as a basic point [17]. Actual norms and guidelines are optimized in regard to economic advisements [24]. Reducing time and equipment-dependent costs (including the number of organisms) when estimating a concentration-effect relationship seems to be more honoured than safely protecting the environment [25]. An example for a questionable proceeding is designing an experiment with high concentrations causing high effects and extrapolating the obtained data to the low effect situation. As fewer biological objects are needed, this approach has the advantage of being easier and more cost-efficient than an experiment with low doses, in which a higher number of objects is needed to generate effects [26]. The extrapolation strategy increases the random error of the estimated EC [17]. An adjustment for low risk effects is not considered in the procedure of OECD guidelines, which is typically proposed only for the optimal determination of EC_{50}.

We recommend determining EC_{xx} for small xx by organizing the FET according to the purpose of detecting small effects and then to estimate EC_{xx} from a fitted concentration-response curve. When developing an optimal design for small effect detection, it should be recalled that the FET is used in laboratory experiments, which gives full control over the experimental conditions, i.e. the number of different concentrations, the concentrations themselves, and the number of biological objects per concentration. This freedom will be exploited when developing an optimal design. Only small modifications of the standard FET are necessary to adjust to the shifted target question. The main steps in designing a FET experiment are choosing appropriate effect quantification, followed by setting up an optimal plan for the sample size, the number of concentrations, and the concentration allocations. In this context, optimal means determining the concentration of interest with a given precision while using as few organisms as possible.

Results and discussion

Before developing an experimental design for a toxicity test, a decision must be made on how to quantify the toxic effect. Both concepts presented, the NOEC and the EC_{xx}, have their merits and disadvantages.

The interpretation of a NOEC without additional information is not statistically sound. The NOECs state that when comparing the response of a control group to that of a group exposed to the NOEC concentration, no significant difference in response could be found. This may have two reasons: either there was really no difference in responses, or there was a difference in responses that could not be detected by the test due to the (too small) group size. As the number of cases per group is typically neither reported nor generally standardized, the effect size that may have been undetected is unknown and cannot be calculated. Therefore, there is a danger of underestimating the effect potential when using the NOEC compared to an effect-based analysis [17]. EC_{xx} has a clear interpretation as it is always an estimate of the concentration which causes a response of xx %.

The major criticism regarding the EC_{xx} concept is the need of specifying a mathematical model for the concentration-response relation. Such a model is not needed for the NOEC. However, a library of standard concentration-response models exists, from which an appropriate problem-specific model can be selected. For the example of the FET, a binary four-parameter logistic model (see Fig. 1a and Appendix) has been found suitable [27–29]. The EC_{xx} concept does not rely on using the logistic model; it can be adapted to every other strictly monotone concentration-response model. Also, non-parametric approaches can be used [26].

Different from the NOEC approach, a CI can be calculated for EC_{xx} as a measure of precision. The width of the EC_{xx} confidence interval depends, among others, on the value of xx. In contrast to the NOEC, EC_{xx} itself does not depend on the design of the experiment, which makes interlaboratory comparisons of EC_{xx} more consistent than comparing NOEC values [14], even if different experimental designs are involved.

When setting up a design for an experiment to determine acceptable concentrations in health prevention and environmental conservation, the main insight is that concentrations of small effects such as EC_{01}, EC_{05}, EC_{10} are relevant. Designs optimized for detecting EC_{50} are not suitable, but it is straightforward to build a design optimized for any specified effect size xx. There is no way to do so if a NOEC is used as risk quantification, because the NOEC concept means to search for an effect of zero, not for an effect of size xx > 0. Considering the advantages and disadvantages of both risk quantification concepts, we concluded using the EC_{xx} concept. NOECs are still used and generated by other authors [18, 30] despite their adverse properties and the debate to abandon them, which has been ongoing for more than 30 years. NOECs are not generated in this package because of the described reasons above.

The procedure for designing an experiment according to the strategy outlined in the 'Methods' section is implemented in the open-source statistic software R as the software package *toxtestD*, which is described below. Power consideration is part of the package, as requested since quite some time [31].

The need of a good experimental design is a well-communicated issue, but only a few publications on the

FET made reference to this [32], as well as current publications to concentration-response relationships [33]. The chosen procedure affects the sample size and the selection of concentrations in the experiment. Sample size will be a balance between contrasting interests: a high precision of the estimated EC, which requires a high number of biological organisms in the test, and the ethical and the economic aspects, which require using few objects. Even though embryos are not considered to be living organisms and are therefore not protected by animal welfare regulations [34], they are animals by ethical considerations [35]. Both interests ask for avoiding experiments which are uninformative because of too few test organisms. Following the suggested design, the experiments should be organized such that the effect of interest can be detected with reasonable precision without involving more biological organisms than necessary. We explicitly recommend following the suggested steps. Especially, the first step should be designed as single experiment for determining the spontaneous lethality (SL). The SL is an indication for the health of the breed and describes the response rate under control conditions. It should be determined with precision because it serves as a baseline for subsequent calculations. As the health status of breed may depend on lab conditions, the design of the experiment may be lab-specific. This give a serious baseline for the further experiments, the detection of the group size per concentration, and the allocation of concentrations for the main experiment. The consideration of SL is precisely important in FET. With the general approach described in this paper, the user is free to choose the necessary adjustments depending on the purpose and object.

The methods implemented in the package apply not only for the analysis of FET data but also to all other dose-response analysis tasks involving a binary target quantity. In all these cases, test designs can be developed which include a properly defined EC_{xx} by specifying the risk type and a reasonable power by regarding the error types I and II (see 'Methods' section for more explanations).

Conclusions

The quality of biological test procedures like the FET relies on using it in an appropriate experimental design. Test results are used for risk assessment and risk management. It is highly desirable that underlying tests are run transparently, with a sufficient number of objects warranting small error rates. Ethical considerations require concomitantly that samples larger than required to attain the accepted error rates should be avoided. This paper discusses standard approaches of risk quantification, concluding that the effect-oriented EC_{xx} concept for effect quantification is more favourable than the test-oriented NOEC concept. We therefore propose the effect size-oriented approach adjusted to small target effects. We suggest a way to organize an experiment according to this conclusion. Realizing such an experimental design is facilitated by the R software package *toxtestD*, which has been introduced in this paper. It organizes the design process in three steps. Being an open-source product, it is available for everybody and allows designing proper experiments also for non-statisticians. The procedure will be specific for the target quantity to be determined with the user-required precision and safety for the quantities of interest.

Methods
Basic considerations for an optimal test design
Error types
For calculation of the optimal test conditions, it is necessary to consider different error types as mathematical principals such as significance, power, and precision. In general, considering two concentrations (i.e. negative control and a test concentration >0), the associated observed effects are random values from two different distributions with concentration-specific mean values (Fig. 2). If the test concentration has no effect, both distributions and their means are identical. If the test concentration has an effect, the associated mean is higher than the control group mean. The two distributions will in general show a certain overlap. If the test concentration in an experiment generates a result in the overlapping

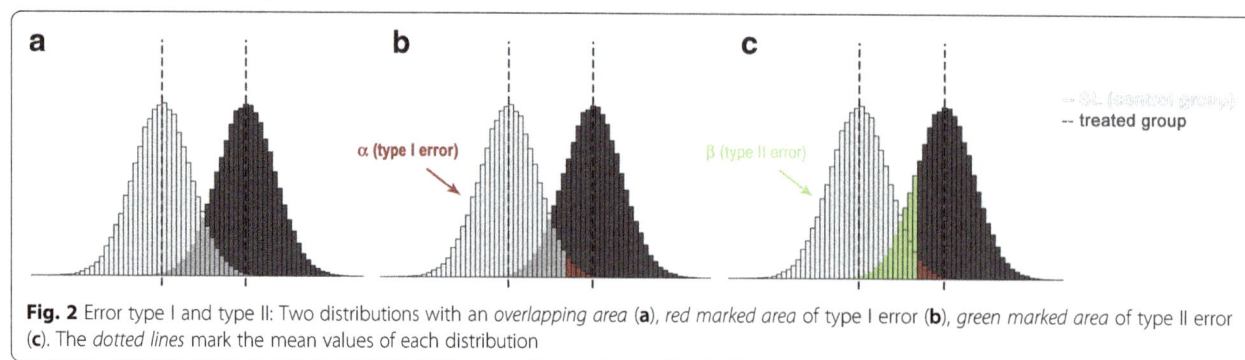

Fig. 2 Error type I and type II: Two distributions with an *overlapping area* (**a**), *red marked area* of type I error (**b**), *green marked area* of type II error (**c**). The *dotted lines* mark the mean values of each distribution

range, it cannot safely be concluded from this result that the test concentration has a systematic effect greater than the control group. The observed value has a considerable probability also under control conditions (Fig. 2a). This means that two different errors may occur when assessing response data from control and test concentrations. A type I error occurs if an effect is interpreted as a concentration effect although it is an effect of the negative control (Fig. 2b). The probability α of a type I error, also known as the level of significance (the p value), reflects the risk of the producer [36]. To keep the danger of a type I error small, the empirical significance level is computed in an actual statistical test procedure, and only if this probability is small (smaller than or equal to a pre-specified α), it is concluded that the test concentration had a systematically higher effect than the control condition. The default value for α is 0.05 or 5 %. The other error that might occur in a statistical test is that a systematically higher response from the test concentration is not recognized as such, so that an existing effect remains undetected (Fig. 2c). This is the type II error, the probability of which should also be restricted to a reasonable value. It reflects the risk of the consumer. A typical value for the accepted type II error is $\beta = 0.20$ or 20 %. The

complementary probability of the type II error (= $1 - \beta$), which is the probability of detecting a systematic difference between responses, is also known as the power or quality of the test. Consideration of the type II error seems to be much less common than considering the type I error [37], possibly because it demands an extra effort, but it is a constitutional component of experimental design. The probabilities of both errors depend (among others) on the group sizes involved. Increasing group sizes is the only way to reduce both error probabilities, and consequently, the crucial step in sample size planning is finding the minimal group sizes for which the accepted sizes of both errors are not exceeded (Fig. 1b). The distributions shown in Figs. 2 and 1b get narrower with increasing group size, which decreases the zone in which values from both groups overlap so that observations can more safely be attributed to one of the groups.

The present proposal uses $\alpha = 5$ % and $\beta = 20$ % as defaults for accepted errors. Both values are not fixed but are frequently used in experimental design. They may be changed, but the user should be careful when relaxing these defaults, as too liberal requirements make the test procedure ineffectual. It should be kept in mind that with a β close to 50 %, the user will declare a truly

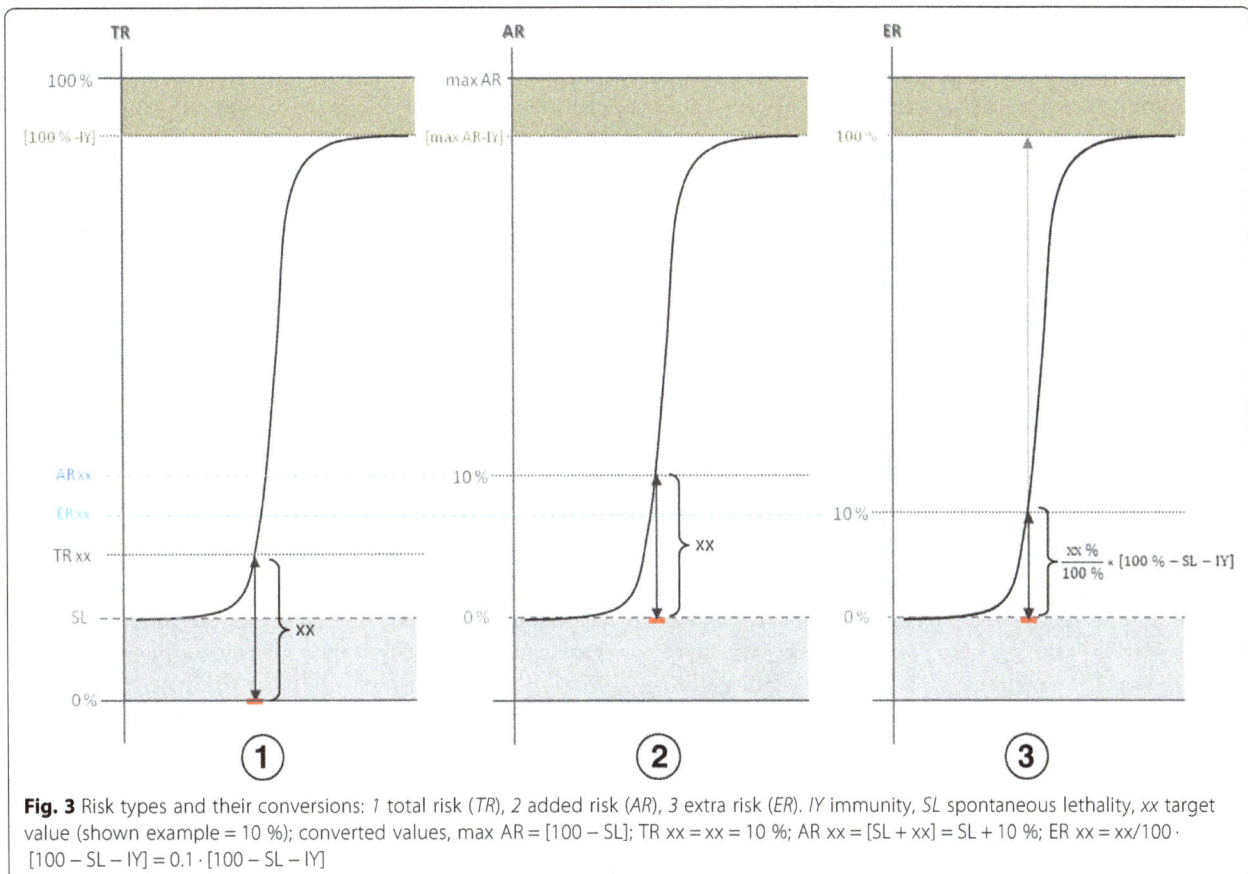

Fig. 3 Risk types and their conversions: *1* total risk (*TR*), *2* added risk (*AR*), *3* extra risk (*ER*). *IY* immunity, *SL* spontaneous lethality, *xx* target value (shown example = 10 %); converted values, max AR = [100 − SL]; TR xx = xx = 10 %; AR xx = [SL + xx] = SL + 10 %; ER xx = xx/100 · [100 % − SL − IY] = 0.1 · [100 % − SL − IY]

existing effect with a probability of 50 % as not existing. This situation is equivalent to tossing a coin to obtain the test result. Obviously, a power analysis compliant to the individual target is a fundamental component in experimental design [17].

Risk types

There are various ways of defining the response percentage (the xx part in EC_{xx}), differing by the way how the spontaneous and the immune level are incorporated. Three different common types of risks are considered here. They differ in the way of how the concentration-related increase in the response is expressed. They will further be named as risk types. Each risk type (Fig. 3) has its own interpretation of xx and its specific value for EC_{xx}, and it may also require its own specific experimental design. The risk types used here are extended versions of the US Environmental Protection Agency (EPA) definitions [38], whereas the extension consists in the additional consideration of immunity (IY) [28, 29]. Immunity describes the phenomenon that within a population, a subpopulation shows by chance no reaction at all. The EPA definitions result by setting IY = 0 %.

1. Total risk (TR): The total risk is the total response expressed as percentage of affected biological units among all treated units. Spontaneous lethality and immunity are ignored. Example: A desired xx = 10 % will force estimation of the concentration that generates an effect of 10 % (Fig. 3 (1)).

2. Added risk (AR): The reference frame is restricted below and above by spontaneous lethality and immunity. Only the response above the SL counts as an effect. Using AR, the total response associated with a target effect of size xx and a spontaneous lethality SL is xx + SL. Example: A desired AR of xx = 10 % and SL = 3.5 %, IY = 7 %, prompts estimating the concentration producing an effect of 10 % above the SL, equivalent to a total response of 13.5 %. The immunity parameter does generally not affect the EC_{xx} value but restricts its possible maximum to [100 % – SL – IY] (Fig. 3 (2)).

3. Extra risk (ER): The reference frame is the interval from SL to [100 % – IY]. Example: A desired ER with xx = 10 % and SL = 3.5 %, IY = 7 %, will force estimating the concentration which generates a total response of $[SL + 0.01 \cdot xx \cdot (100\ \% - SL - IY)]$ = 3.5 % + 0.01 · 10 · (100 % – 3.5 % – 7 %) = 12.45 % (Fig. 3 (3)).

The total response associated with a target effect of xx and using ER as risk type is smaller than or equal to the total response associated with the same xx, but with risk type AR.

Proposed software solution (*toxtestD*) for an optimal test design

To consider all our proposals, we implemented a package with a set of functions in R code [39], which do experimental design as outlined above in a user-friendly way. The R software is an open-source software. The package requires only a few inputs by the user. Sophisticated statistical understanding or modelling experiences are not necessary (though useful). Even though the concept is drafted for the fish embryo toxicity assay, it is possible to transfer the procedure in principle to all other toxicological questions, in which a continuous concentration equivalent generates a yes/no (binary) response per single study object. The package *toxtestD* should already be consulted during the planning phase of a test series. It contains the functions *spoD*, *setD*, and *doseD* which cover identification of the spontaneous lethality, the estimation of the necessary number of test organisms, and a concentration design according to the user's requirements. Examples for the application of all these functions are available after installation of the package by the command help (toxtestD) [40].

Determination of the spontaneous lethality by *spoD*

The first task when designing the experiment is calculating the sample size for determining the SL. The necessary sample size depends on the required precision of the estimated SL. It should be recalled that because SL will be calculated from data containing random variation, the resulting SL will also be a quantity with random error.

The function *spoD* offers two services. In the planning process, the total number of individuals or eggs to test is calculated, together with a proposal for partitioning the total data set into subgroups in order to identify the amount of biological variation in the separated tests. The calculations will be done for the denoted rate and additionally for the worst case in the interval given for the predicted SL. The optimal number of biological units is calculated by using an exact binomial test with α and β as specified. The previous mentioned random error can be quantified by a CI (see Appendix for calculation details), which contains the true value of SL (the response that would hold if no random fluctuation were present) with large probability, typically 95 %. We propose that as a default requirement, the limits of the CI for the SL should differ from the estimated SL by no more than ± 2.5 % (further on denoted by maxCI).

In the analysis process (initiated by setting analysis = TRUE), the spontaneous lethality together with its 95 % confidence interval and the biological variation are computed from the user's data. Biological variation becomes visible when comparing spontaneous rates from several experiments under control conditions. A χ^2 test is applied to check whether these rates vary according to binomial variation under the hypothesis of the same true spontaneous rate for all experiments. A significant result signals the presence of biological variation between experiments. If present, its standard deviation is reported. It is recommended to determine the spontaneous lethality very early under separate test conditions.

Specifications by the user

Subprocess planning

n: The maximal number (integer) of test organisms, with which the laboratory is willing to cope. Limiting the number is necessary to avoid non-essential calculations and thereby save computing time. The programme will invite the user to increase the number if the number is not high enough to estimate the SL with the given precision requirement.

SL.p: (optional SLmin, SLmax): To gain an optimal number of test objects for the determination of the true spontaneous lethality, the user needs at least a rough idea about the SL prior to test planning. This estimate is inserted in SL. It is possible to specify the SL either as single number or as an interval between 0 and 100 %. The maximum tolerated spontaneous lethality by OECD is 10 %. Datasets with higher SL should be discarded [9].

bio.sd.p (optional): The standard deviation of SL consists of normal random variation and a biological variation due to biological effects like season, daytime, or wellbeing. The default value of 2.008 % for bio.sd.p was determined from empirical data sets collected over 10 years by the Thünen Institute, Hamburg, Germany (U. Kammann and S. Schubert, personal communication) [41]. The value refers exclusively to dead eggs and lethal malformations (pursuant to the definition in DIN ENISO 15088) after 96 hpf in water in per cent [4, 9]. If not specified otherwise, this default will be used for determining the optimal number of partitions.

maxCI (optional): It is the maximally accepted absolute difference in per cent between mean SL and its confidence limits; default, 2.5 %.

print.result: If omitted, the result is written to a text file called '01_spontaneous lethality.txt' in the calling directory. If a file name is given in double quotes, the result is written to that file. Nothing is written if FALSE is chosen.

Subprocess analysis

analysis: The default value is FALSE, indicating that the function does planning. To analyse the own dataset, choose analysis = TRUE.

SLdataset: This is the R data frame containing the experimental data. It needs two columns titled 'n' and 'bearer'. In column n, the total number of observations of each single experimental run is listed. The column bearer comprises the number of organisms which are carriers (in the case of FET the counts of dead or lethal malformed eggs) within each single experimental run. Each row contains the outcome from one single experimental run.

Determination of the optimal number for each experimental run by *setD*

The second task should be the calculation of the optimal number per experimental run. The proposed calculations in *setD* involve a robustness consideration as it is done in the third task. We propose for every concentration in the FET a sample size such that a test for a concentration effect of size xx at the sought concentration EC_{xx} would detect this effect with a high (pre-specified) probability. Requiring a specified quality of the test leads to the necessary sample size per concentration. Two distributions will be estimated, one assuming SL as true response level, the other using SL + xx defined above (Fig. 1). The estimation of these two distributions bases on binomial probability. The default of the distance is xx % in respect to a posterior target of EC_{xx}. Additionally the distance depends on the reference frame. In consequence, it is necessary to choose the convenient risk type (see section 'Risk types'). The procedure increases the number of cases per concentration until the overlapping area of the two distributions corresponds to the specified levels of error types I and II (Fig. 1b).

Specifications by the user

nmax: Number (integer) of the maximum available number of organisms that can be tested in each treatment within an experimental run. The estimation of the optimal number will only start when this number is high enough to generate the response of at least one organism ($nmax \cdot p > 1$). If the chosen nmax is too small, a warning message is issued

SL.p: SL is calculated in per cent from own experimental data by the process *spoD*

immunity.p: A population of biological objects might contain a subpopulation which shows no reaction at all. Consequently, a reaction of 100 % will never be reached [42]. We call this effect immunity in our procedure. To account for this kind of non-response, the size of the immune subpopulation can be specified as percentage of the total population. The concentration-response curve then has [100 % – IY] as maximum. Please choose risk. type = 3 to include immunity in all calculations

risk.type: Please choose one of three risk types. Each type defines a specific reference frame for the concentration-response curve (for detailed information see the 'Risk types' section)

target.EC.p: The target response in per cent (e.g. 10 %, to calculate EC_{10}). Note that the interpretation of target.EC depends on the risk.type setting

plot: There are three possibilities:

plot = FALSE: no plots

plot = 'single': creates only one plot showing the two distr-ibutions under SL and under treated conditions with the optimal number of cases. Additionally, the real rates of error type one and two are given (see Fig. 1b). The special setting for risk.type is not included into this plot

plot = 'all': In addition to the single plot, this option provides an estimation for all possibilities of target values. This gives an impression which possibilities of detection exist under the chosen conditions. This option may need a lot of computer capacity and time. It should not be activated in general.

alpha.p & beta.p: See explanations in 'Error types' section

print.result: If omitted, the result is written to a text file called '02_sample size.txt' in the calling directory. If a file name is given in double quotes, the result is written to that file. Nothing is written if FALSE is chosen

Allocating concentration points by *doseD*

The third task, defining the concentration allocation for the main experiment, is guided by an idea of robustness similar to the second task. From a formal point of view, only as many concentration points as unknown parameters in the dose-response model are needed. It would however be unwise to involve only this minimum because it would not allow model checks. Concentration allocation needs at least a vague idea about position and scale of the concentration-response curve. Concentration-finding experiments, pilot studies, literature data, and similar sources are used to obtain these planning assumptions. Given an initial assumption, we propose the following concentration allocation strategy: calculate EC_{10}, EC_{50}, EC_{90} from the planning assumptions, assuming a logistic concentration-response relation and involving SL and IY, if the selected risk type requires so. The control concentration of zero and these three concentrations plus target concentrations, given by the EC_{xx} values of the experimenter's interest, constitute the concentration allocation pattern for the main experiment. Two-sided CIs with 95 and 99 % coverage probability will be calculated for the concentration-response curve. If more than four concentration points are chosen and there is an even number of free points, these will be allocated symmetrically around the chosen EC_{xx} value. If two free

points are available, these are located at the limits of the 95 % CI. If an even number of 4 or more free points is available, these are allocated equidistantly in the 99 % CI. For an odd number of free points, 1 point is located at EC_{xx} and the others are allocated according to the rule for an even number. Note that the 1 of EC_{10}, EC_{50}, EC_{90} can be used twice as an experimental concentration, if the user's target coincides with one of these. The strategy prefers low concentrations if several targets are specified by the user. This pattern is a robust strategy which focuses on the main interest of finding EC_{xx} but does not rely too strongly on the planning assumptions. If previous experience suggests that the planning assumptions are realistic, concentrations could be allocated more closely around the presumed EC_{xx}.

Specifications by the user

DP: The results from pre-tests must be given as a data frame with the columns 'name', 'organisms', 'death', 'concentration' and 'unit', which will be needed for the calculations of the dose scheme

immunity.p: immunity in per cent (see also settings of *spoD*)

SL.p: SL is calculated in percent from the user's experimental data by the function *spoD*

target.EC.p: effect in %, which is of special interest. It is possible to denote more than one target in the same calculation. For example: if EC_5 and EC_{10} are of special interest, then target. EC = *c* (5,10) may be chosen, and the dose points will be allocated around both targets

nconc: number of different concentrations the user is willing to test in each cycle

text: text = TRUE adds extended information in the plot

risk.type: Please choose one of three possible risk types. Each type defines another reference frame for concentration-response curve and target estimation (for detailed information see the 'Risk types' section). A plot for each risk type will be created separately

print.result: If omitted, the result is written to a text file called '03.dosestrategy.txt' in the calling directory. If a file name is given in double quotes, the result is written to that file. Nothing is written if FALSE is chosen.

Appendix

This appendix summarizes the main formulae that are proposed as part of the design procedure and are also incorporated in the R package.

The basic data in a concentration-response experiment is the number n of examined objects and the number r of responding objects. The empirical response rate is the ratio $p = r / n$. It is frequently expressed as a percentage by multiplying p by 100.

This appendix, however, uses only rates, not percentages, for easier notation.

A CI is a way to express how precisely the response rate p_{true} of the whole population is estimated by an empirical rate. A confidence interval CI = $(p_{\text{low}}, p_{\text{high}})$ for an empirical rate is obtained from observed n and r by [43]:

$$p_{\text{low}} = \frac{r \cdot F_{2(n-r+1),2r,1-\alpha/2}}{r + (n-r+1) \cdot F_{2(n-r+1),2r,1-\alpha/2}}$$

$$p_{\text{high}} = \frac{(r+1) \cdot F_{2(r+1),2(n-r),1-\alpha/2}}{n-r + (r+1) \cdot F_{2(r+1),2(n-r),1-\alpha/2}}.$$

The F terms in both equations are the quantiles of the F distribution with degrees of freedom and associated probability as given in the subscripts. The value of α controls the coverage probability. The calculated interval contains the value p_{true}, which holds for the whole population under study, with probability $1 - \alpha$. This statement must be understood in a strategic sense: If the actual experiment is replicated many times and the CI is calculated for each replicate, then the population rate p_{true} will be contained in $(1 - \alpha) \cdot 100$ % of the calculated CIs.

The equation for the CI makes use of the fact that the number r of responses has a binomial distribution, which means that the probability of observing r responses among n examined subjects is as follows:

$$\Pr(r, n, p_{\text{true}}) = \binom{n}{r} \cdot p_{\text{true}}^r \cdot (1 - p_{\text{true}})^{n-r}.$$

Different concentrations x_i of a substance generate their specific $p_{\text{true}}(x_i)$ values. A concentration-response (or dose-response) curve relates the probability p_{true} to the dose x involved. This relation will always be a non-linear one, because the concentration or dose may be any value ≥ 0 and the associated p_{true} must lie in the interval $[0,1]$. The present proposal uses the four-parameter logistic curve as concentration-response curve:

$$p_{\text{true}}(x) = \text{SL} + \frac{1 - \text{SL} - \text{IY}}{1 + e^{-(a+b \cdot x)}}.$$

The terms a and b in this equation control location and scale (slope) of the concentration-response relation. The values for these terms are estimated from I experiments with doses x_i and numbers (n_i, r_i) of examined and responding objects by a maximum likelihood approach by maximizing

$$\log \ell(a, b | n, r, x) = \sum_i^I n_i \cdot \log p_{\text{true}}(a, b | x) + (n_i - r_i) \cdot \log(1 - p_{\text{true}}(a, b | x)).$$

The optimization is done iteratively by a Newton–Raphson approach. EC_{xx} is obtained while using the estimates for (a, b) by

$$\text{EC}_{\text{xx}} = -\frac{1}{b} \left[\ln \left(\frac{1 - \text{SL} - \text{IY}}{\text{xx} - \text{SL}} - 1 \right) + a \right].$$

All calculations listed here are contained in the R package described.

Abbreviations
AR: Added risk; ASTM: ASTM International, known until 2001 as the American Society for Testing and Materials; BMD: Benchmark dose; DIN: German Institute for Standardization; EC: Effective concentration; ED: Effective dose; EPA: US Environmental Protection Agency; ER: Extra risk; FET: Fish embryo toxicity test; hpf: Hours post fertilization; IY: Immunity; LC: Lethal concentration; LD: Lethal dose; NOEC: No observed effect concentration; OECD: Organisation for Economic Co-operation and Development; REACH: Registration, Evaluation, Authorisation and Restriction of Chemicals; SL: Spontaneous lethality; TR: Total risk; VSD: Virtual safe dose.

Competing interests
The authors declare that they have no competing interests.

Authors' contributions
NK devised and realized the R package toxtestD and drafted the manuscript. SS has participated in the conception and design of the approach, performed all test runs (computer based as well as the FET), and has been involved in drafting the manuscript. WW has made substantial contributions to the conception and design of the R package and revised the manuscript critically for important intellectual content. All authors read and approved the final manuscript.

Acknowledgments
The authors thank Malte Damerau, Ulrike Kammann, Michael Haarich and Norbert Theobald for their thematic support. This study was incorporated into 'MERIT-MSFD: Methods for detection and assessment of risks for the marine ecosystem due to toxic contaminants in relation to implementation of the European Marine Strategy Framework Directive'. It was supported by a grant (grant number 10017012) from the German Federal Ministry of Transport and Digital Infrastructure (BMVI) and the German Maritime and Hydrographic Agency (BSH).

Author details
[1]Institute of Fisheries Ecology, Thünen Institute (TI), Palmaille 9, 22767 Hamburg, Germany. [2]Institute of Statistics, University of Bremen, Linzer Str. 4, 28334 Bremen, Germany.

References
1. r-project. http://cran.r-project.org/. 2015: Accessed: 18 Dec 2014.
2. Keddig N, Wosniok W. toxtestD package manual. http://cran.r-project.org/web/packages/toxtestD/toxtestD.pdf. 2014.
3. Lammer E, Carr GJ, Wendler K, Rawlings JM, Belanger SE, Braunbeck T. Is the fish embryo toxicity test (FET) with the zebrafish (Danio rerio) a potential alternative for the fish acute toxicity test? Comparative Biochemistry and Physiology Part C: Toxicology & Pharmacology. 2009;149(2):196–209. http://dx.doi.org/10.1016/j.cbpc.2008.11.006.
4. German Institute for Standardisation (DIN). Wasserbeschaffenheit – Bestimmung der akuten Toxizität von Abwasser auf Zebrafisch-Eier (Danio rerio) (ISO 15088:2007). DIN EN ISO 15088. 2009.

5. Keiter S, Peddinghaus S, Feiler U, von der Goltz B, Hafner C, Ho NY, et al. A novel joint research project using zebrafish (Danio rerio) to identify specific toxicity and molecular modes of action of sediment-bound pollutants. J Soils Sediments. 2010;10:714–7. doi:10.1007/s11368-010-0221-7.

6. Research Centre for Toxic Compounds in the Environment. http://www.recetox.muni.cz/index-en.php?pg=research-and-development–analyses-and-services–ecotoxicology. 2014: Accessed: 01 Apr 2014.

7. Fraunhofer-Gesellschaft. http://www.ime.fraunhofer.de/de/geschaeftsfelderAE/Verbleib_und_Wirkung_Agrochemikalien/Erweiterte_Standardtests.html#tabpanel-5. 2014: Accessed: 01 Apr 2014.

8. Microtest Laboratories. http://www.microtestlabs.com. 2014: Accessed: 1 Apr 2014.

9. Organisation for Economic Co-operation and Development. OECD guidelines for the testing of chemicals - fish embryo acute toxicity (FET) test. 2013;236 (adopted 26 July 2013).

10. Hutchinson TH, Solbe J, Kloepper-Sams PJ. Analysis of the ecetoc aquatic toxicity (EAT) database III—comparative toxicity of chemical substances to different life stages of aquatic organisms. Chemosphere. 1998;36(1):129–42. http://dx.doi.org/10.1016/S0045-6535(97)10025-X.

11. Crump KS. Calculation of benchmark doses from continuous data. Risk Anal. 1995;15(1):79–89.

12. European Chemicals Agency. Guidance on information requirements and chemical safety assessment Chapter R.10: Characterisation of dose [concentration]-response for environment. 2008. http://echa.europa.eu/documents/10162/13632/information_requirements_r10_en.pdf.

13. Van Der Hoeven N. How to measure no effect. Part III: statistical aspects of NOEC, ECx and NEC estimates. Environmetrics. 1997;8(3):255–61. doi:10.1002/(SICI)1099-095X(199705)8:3<255::AID-ENV246>3.0.CO;2-P.

14. Chapman PM, Caldwell RS, Chapman PF. A warning: NOECs are inappropriate for regulatory use. Environ Toxicol Chem. 1996;15:77–9.

15. Crane M, Newman MC. What level of effect is a no observed effect? Environ Toxicol Chem. 2000;19(2):516–9. doi:10.1002/etc.5620190234.

16. Moore DRJ, Caux P-Y. Estimating low toxic effects. Environ Toxicol Chem. 1997;16(4):794–801. doi:10.1002/etc.5620160425.

17. Warne MSJ, van Dam R. NOEC and LOEC data should no longer be generated or used. Australas J Ecotoxicol. 2008;14:1–5.

18. Landis WG, Chapman PM. Well past time to stop using NOELs and LOELs. Integr Environ Assess Manag. 2011;7(4):vi–vii. doi:10.1002/ieam.249.

19. Grasso P. Essentials of pathology for toxicologists. Taylor & Francis Inc., London, New York 2002.

20. van Ewijk PH, Hoekstra JA. Calculation of the EC50 and its confidence interval when subtoxic stimulus is present. Ecotoxicol Environ Saf. 1993;25(1):25–32. http://dx.doi.org/10.1006/eesa.1993.1003.

21. European Union. Directive 2008/56/EG - Marine strategy framework directive. In: Union OJotE, editor.2008.

22. Federal Institute for Occupational Safety and Health, (BAuA). Hazardous substances ordinance (Gefahrstoffverordnung – GefStoffV). 2010; updated. 2013.

23. Institute for Health and Consumer Protection. Technical guidance document on risk assessment. 2003.

24. European Union. Regulation (EC) No 1907/2006 of the European Parliament and of the council. 2006.

25. Fraysse B, Mons R, Garric J. Development of a zebrafish 4-day embryo-larval bioassay to assess toxicity of chemicals. Ecotoxicology and Environmental Safety. 2006;63(2):253–67. http://dx.doi.org/10.1016/j.ecoenv.2004.10.015.

26. Piegorsch WW, Xiong H, Bhattacharya RN, Lin L. Benchmark dose analysis via nonparametric regression modeling. Risk analysis: an official publication of the Society for Risk Analysis. 2013:1–17. doi:10.1111/risa.12066.

27. Kammann U, Vobach M, Wosniok W, Schäffer A, Telscher A. Acute toxicity of 353-nonylphenol and its metabolites for zebrafish embryos. Environ Sci Pollut Res. 2009;16(2):227–31.

28. Ratkowsky DA, Reedy TJ. Choosing near-linear parameters in the four-parameter logistic model for radioligand and related assays. Biometrics. 1986;42(3):575–82. doi:10.2307/2531207.

29. DeLean A, Munson PJ, Rodbard D. Simultaneous analysis of families of sigmoidal curves: application to bioassay, radioligand assay, and physiological dose-response curves. Am J Physiol Gastrointest Liver Physiol. 1978;235(2):G97–G102.

30. Carlsson G, Patring J, Kreuger J, Norrgren L, Oskarsson A. Toxicity of 15 veterinary pharmaceuticals in zebrafish (Danio rerio) embryo. Aquat Toxicol. 2013;126(0):30–41. http://dx.doi.org/10.1016/j.aquatox.2012.10.008.

31. Hayes JP. The positive approach to negative results in toxicology studies. Ecotoxicol Environ Saf. 1987;14(1):73–7. http://dx.doi.org/10.1016/0147-6513(87)90085-6.

32. Wedekind C, von Siebenthal B, Gingold R. The weaker points of fish acute toxicity tests and how tests on embryos can solve some issues. Environ Pollut. 2007;148(2):385–9. http://dx.doi.org/10.1016/j.envpol.2006.11.022.

33. Kent M, Buchner C, Barton C, Tanguay R. Toxicity of chlorine to zebrafish embryos. Dis Aquat Org. 2014;107(3):235–40. doi:10.3354/dao02683.

34. European Union. Directive 2010/63/EU of the European Parliament and of the council on the protection of animals used for scientific purposes. In: Union OJotE, editor.2010. p. 33–79.

35. Braunbeck T, Böttcher M, Hollert H, Kosmehl T, Lammer E, Leist E, et al. Towards an alternative for the acute fish LC (50) test in chemical assessment: the fish embryo toxicity test goes multi-species—an update. Altex. 2004;22(2):87–102.

36. Gad SC. Statistics and experimental design for toxicologists and pharmacologists. Taylor & Francis Group, LLC, Boca Raton, London, New York, Singapore; 2005.

37. Fairweather PG. Statistical power and design requirements for environmental monitoring. Mar Freshw Res. 1991;42(5):555–67.

38. United States Environmental Protection Agency (U.S. EPA). EPA's approach for assessing the risks associated with chronic exposure to carcinogens—Integrated Risk Information System (IRIS): http://www.epa.gov/iris/carcino.htm. 1992: Accessed 10 Sept 2014.

39. R Core Team. R: A language and environment for statistical computing. R Foundation for Statistical Computing. 2014;Vienna, Austria. http://www.R-project.org/

40. Keddig N, Wosniok W. toxtestD package—experimental design for binary toxicity tests (with examples). 2014. http://cran.r-project.org/web/packages/toxtestD/index.html.

41. Kammann U, Vobach M, Wosniok W. Toxic effects of brominated indoles and phenols on zebrafish embryos. Arch Environ Contam Toxicol. 2006;51(1):97–102. doi:10.1007/s00244-005-0152-2.

42. Dinse G, An EM. Algorithm for fitting a four-parameter logistic model to binary dose-response data. JABES. 2011;16(2):221–32. doi:10.1007/s13253-010-0045-3.

43. Johnson NL, Kotz S. Distributions in statistics: discrete distributions. Wiley Interscience; New York; Brisbane 1969.

16

Test strategy for assessing the risks of nanomaterials in the environment considering general regulatory procedures

Kerstin Hund-Rinke[1*], Monika Herrchen[1], Karsten Schlich[1], Kathrin Schwirn[2] and Doris Völker[2]

Abstract

Background: Engineered nanomaterials (ENMs) are marketed as a substance or mixtures and are additionally used due to their active agent properties in products such as pesticides or biocides, for which specific regulations apply. Currently, there are no specific testing strategies for environmental fate and effects of ENMs within the different regulations. An environmental test and risk assessment strategy for ENMs have been developed considering the general principles of chemical assessment.

Results: The test strategy has been developed based on the knowledge of national and international discussions. It also takes into account the conclusions made by the OECD WPMN which held an expert meeting in January 2013. For the test strategy development, both conventional and alternative endpoints were discussed and environmental fate and effects were addressed separately.

Conclusion: A tiered scheme as commonly used in the context of precautionary environmental risk assessment was suggested including the use of mathematical models and trigger values to either stop the procedure or proceed to the next tier. There are still several gaps which have to be filled, especially with respect to fate, to develop the test strategy further. The test strategy features a general approach. It is not specified to fulfil the information requirements of certain legislation (e.g. plant protection act, biocide regulation, REACH). However, the adaption of single elements of the strategy to the specific needs of certain legislation will provide a valuable contribution in relation to the testing of nanomaterials.

Keywords: Assessment, Test strategy, Nanomaterials, Ecotoxicology, Fate, Environment

Background

Engineered nanomaterials (ENMs) are marketed as a substance or mixtures and additionally used due to their active agent properties in products like, e.g. pesticides or biocides, for which substance and product specific regulations apply [1, 2]. Currently, there are no nano-specific information requirements for environmental fate and effects of ENMs within the different regulations. In several projects under the EU's Seventh Framework Programme for Research (FP7), test strategies to assess human and environmental risk are discussed and the topics to be further investigated listed. An excellent example is the project ITS-NANO (ITS: Intelligent Testing Strategy) which has delivered a detailed, stakeholder driven and flexible research prioritization (or strategy) tool, which identifies specific research needs, suggests connections between areas, and frames this in a time perspective [3]. In projects as MARINA or SUN, tools and strategies for a risk assessment of manufactured nanomaterials are being developed. Usually, the established procedure for testing and risk assessment of conventional chemicals is taken into account in these projects. Additional endpoints (such as biomarkers), test systems (such as more generation tests), and multiple application as well as ageing of ENMs in the respective test medium are discussed. The EU NanoSafety Cluster

*Correspondence: kerstin.hund-rinke@ime.fraunhofer.de
[1] Fraunhofer Institute for Molecular Biology and Applied Ecology IME, Auf dem Aberg 1, 57392 Schmallenberg, Germany
Full list of author information is available at the end of the article

was initiated by the DG RTD NMP to maximise the synergies between the EU projects addressing all aspects of nanosafety including toxicology, ecotoxicology, exposure assessment, mechanisms of interaction, risk assessment and standardisation (http://www.nanosafetycluster.eu/).

The aim of this study was to develop a test and risk assessment strategy for ENMs which specifically addresses environmental fate and effects (Fig. 1). For both of these, precise test systems and strategies of data collection, and evaluation are provided. To our knowledge, the level of detail and comprehensiveness, and the resulting recommendations exceed published approaches.

The test strategy has been developed based on a literature review and on the knowledge of national and international discussions, after comparison with the proposals presented by the Reach Implementation Project on Nanomaterials 2 (RIP-oN) and by German competent authorities for REACH (regulation concerning the registration, evaluation, authorisation and restriction of

chemicals) and CLP (regulation on classification, labelling and packaging of substances and mixtures). It also takes into account the conclusions agreed at the OECD WPMN (OECD working party on manufactured nanomaterials) which held an expert meeting in January 2013 on the suitability of test guidelines for environmental fate and ecotoxicity [4]. The literature review was performed in 2012, with additions in 2013 and 2014 with the aim to present an overview of the state of the art. It was not intended to provide a compilation of all available references. In the following, the main steps of the test strategy and results which emerged from this conceptual work are presented. The entire report is available at http://www.umweltbundesamt.de/publikationen/integrative-test-strategy-for-the-environmental.

Results and discussion
Overview on the test strategy

The presented approach (Fig. 2) is a life-cycle oriented one, and thus considers all stages along the life of the ENMs. In particular, these are: production, transport and distribution to the user, use, and waste management. Further, transport stages might occur, e.g. the transport of the used ENMs to an incineration plant. For each single stage, it has to be considered whether there is a potential for the ENMs to be released into the environment. Furthermore, with respect to each single stage, the initial environmental compartment in which the ENMs are expected to be released into has to be identified. In the test strategy, we consider the compartments: water, sediment and soil. If the release potential is negligible, this particular life-cycle stage needs no further consideration. It has to be noted that the definition of "negligible" and "non-negligible" with regard to this test strategy has still to be discussed.

At the beginning, the durability of the tested ENMs in the initial compartment should be screened (tier 0). The term "durability" means that the ENMs keep their status as a nanomaterial. For that screening, any information about the ENM properties and their possible loss (e.g. by rapid dissolution) is indispensable. Such information should be available (at least to some extent), e.g. from the manufacturer collecting it in the course of product design and development. In the case, that low durability is determined, meaning that the ENM rapidly loses its status of being a nanomaterial, the formed chemicals can be tested and assessed as conventional chemicals. In the case that medium to high durability is determined, the first tier of the assessment scheme is initiated. It has to be noted that any trigger value to differentiate between "high", "medium" and "low" durability has still to be discussed. We expect that for example metals and metal oxides will belong to the medium and high durability group, whereas

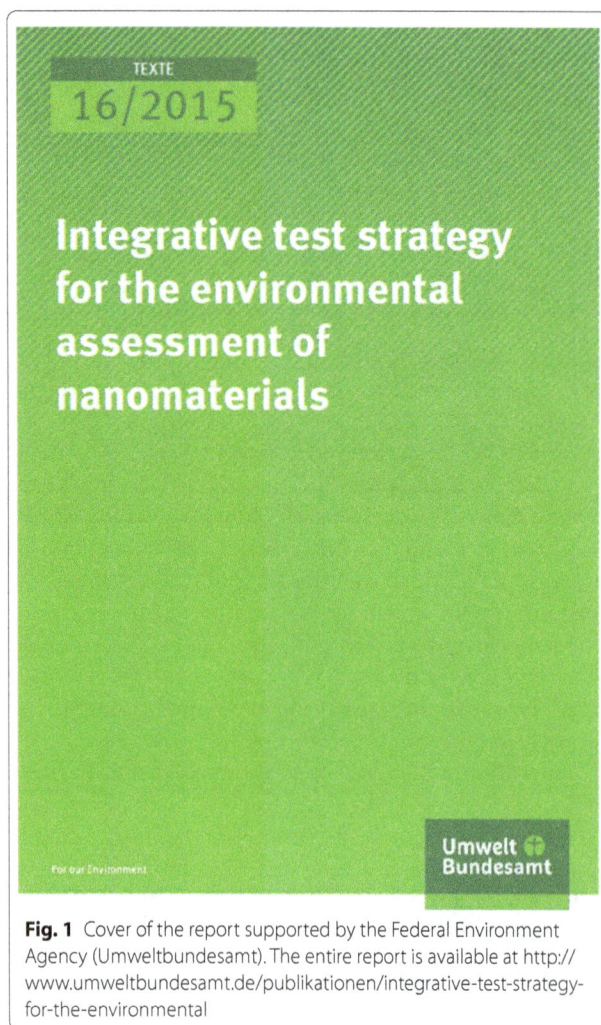

Fig. 1 Cover of the report supported by the Federal Environment Agency (Umweltbundesamt). The entire report is available at http://www.umweltbundesamt.de/publikationen/integrative-test-strategy-for-the-environmental

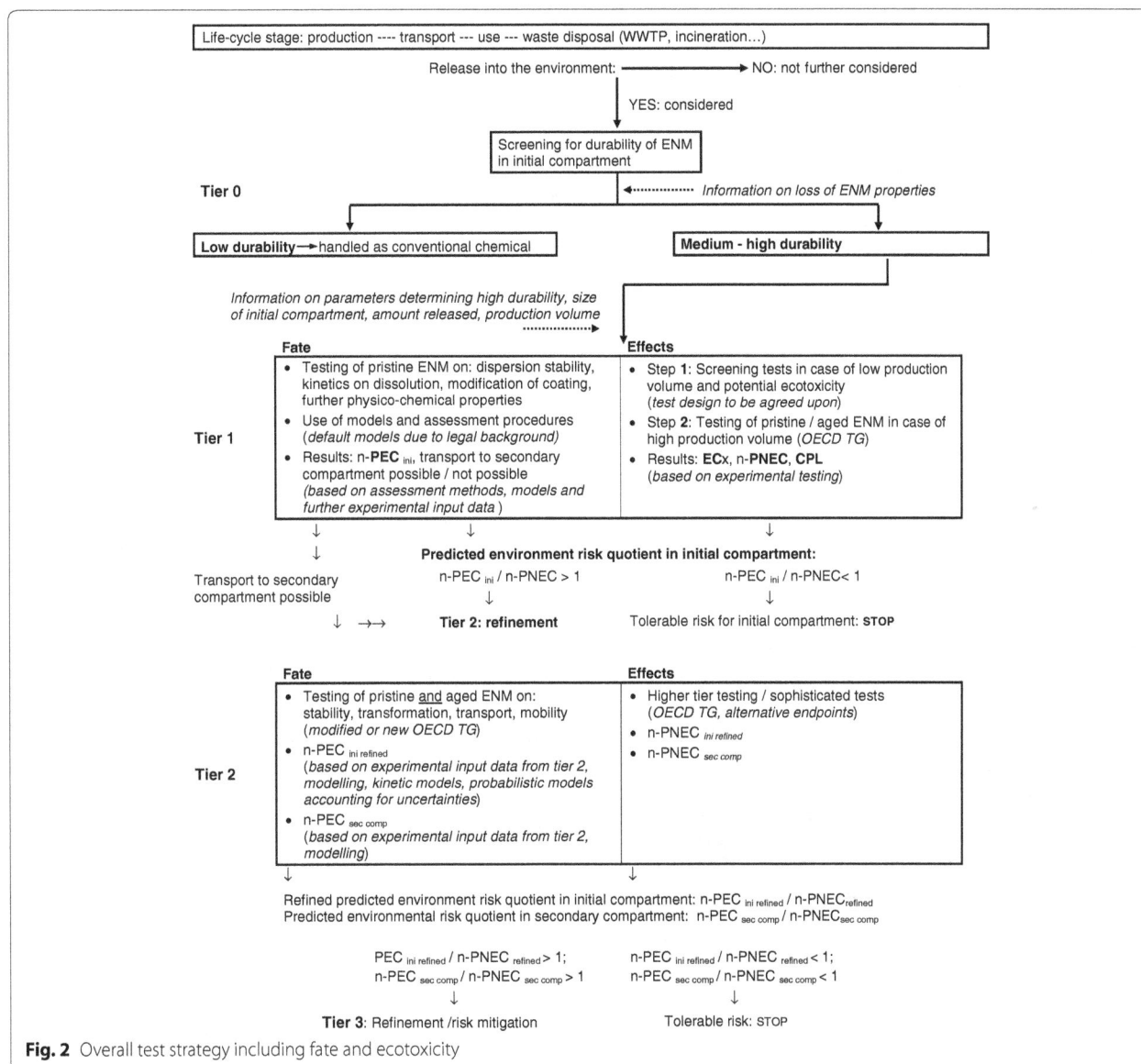

Fig. 2 Overall test strategy including fate and ecotoxicity

most of the "nano pesticides" will be allocated to the low durability group.

For tier 1, both a fate and effect assessment have to be performed. The assessments result in a predicted environmental concentration in the initial compartment (n-PEC$_{ini}$) and a predicted no effect concentration (n-PNEC). The prefix "n" is used to characterise the PEC and PNEC as concentrations for "nanomaterials".

The deduction of n-PEC$_{ini}$ needs the information on experimental physico-chemical characteristics as well as preliminary data on environmental behaviour of the ENMs, i.e. information on the agglomeration behaviour, stability of the coating, and alteration of the ENMs, e.g. by oxidation or dissolution. For information on some of

these endpoints, modified or even newly developed test guidelines and guidances, e.g. on agglomeration behaviour or dissolution rate, will be necessary. The deduction of n-PEC$_{ini}$ also needs information on the production volume as well as on the amount of the ENMs released in every life-cycle stage. Furthermore, it requires a specification of the volume of the initial compartment, e.g. the definition of a local or regional scenario. Finally, the definition of default models which is already applied for the exposure assessment of conventional chemicals, plant protection products and biocides is also considered for the derivation of n-PEC$_{ini}$.

The effect assessment resulting in n-PNEC for low production volume nanomaterials, which have non-toxic

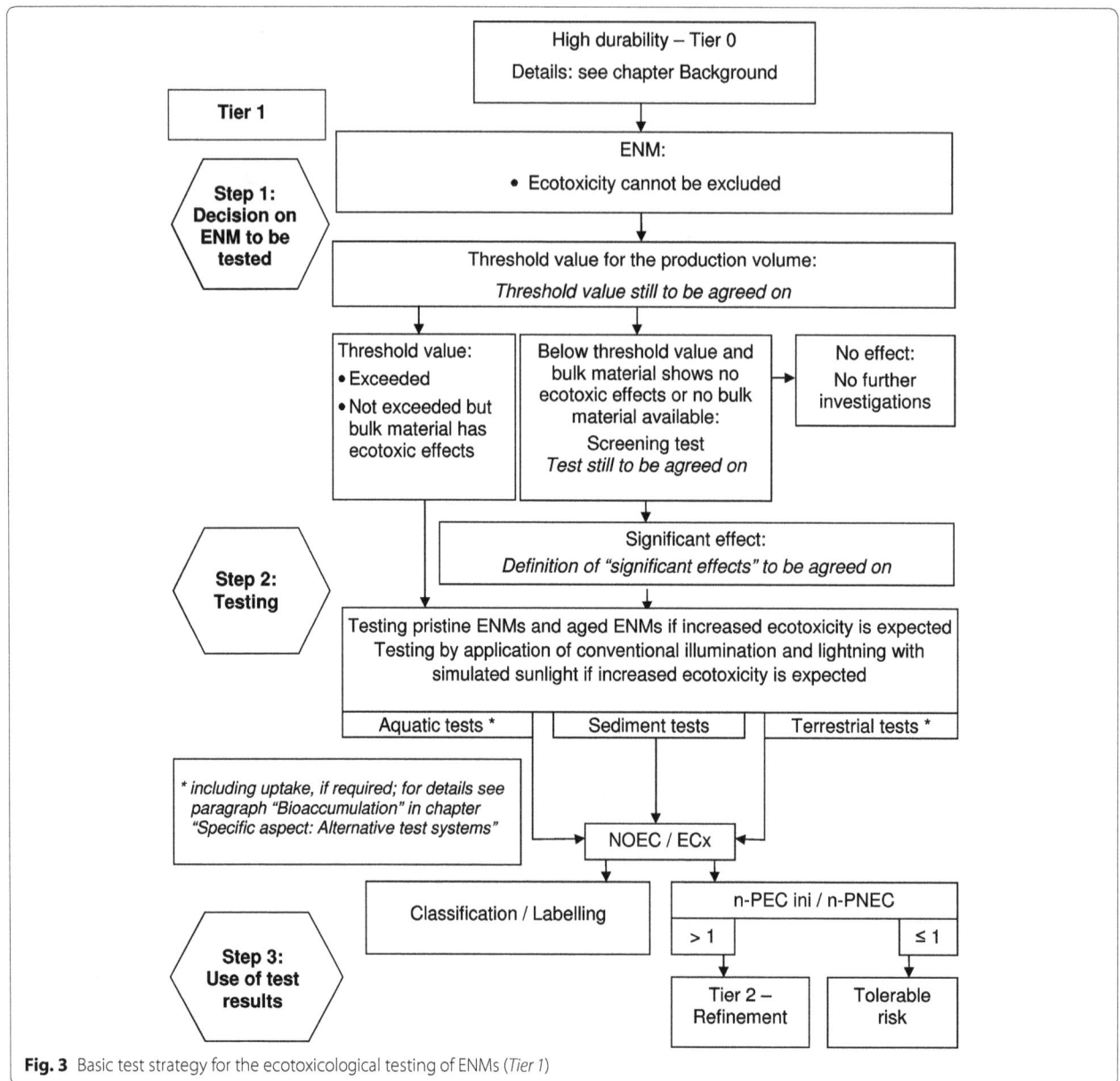

Fig. 3 Basic test strategy for the ecotoxicological testing of ENMs (*Tier 1*)

non-nano counterparts, is determined by screening tests representing the respective initial compartment (named as "step 1" in the scheme). For all other ENMs, OECD test guidelines suitable for the testing of ENMs are used (named as "step 2" in the scheme). Thus, the effect testing at tier 1 comprises two different levels of complexity. It has to be noted that trigger values for "low" and "high" production volume still have to be discussed. Furthermore, the test design of the screening tests needs mutual consent. NOEC values or ECx values are the outcomes of any of the experimental testing. Using assessment factors/uncertainty factors, well-known from the risk assessment of conventional chemicals, a predicted no effect

concentration (n-PNEC) can be derived. Besides n-PNEC values, a classification and product labelling (CPL) on the basis of the effect concentrations is conceivable.

Comparable to conventional chemicals, a risk quotient (n-PEC$_{ini}$/n-PNEC) can be derived. In case it is below 1, a tolerable risk for the initial compartment can be assumed. No further sophisticated risk assessment for the initial compartment is needed. In case it is above 1, the risk for the initial compartment might not be negligible and, thus, a refinement at tier 2 is needed.

Regardless of the risk quotient for the initial compartment, a possible ENM transport to a secondary compartment—e.g. the transport from the aqueous phase to

Test strategy for assessing the risks of nanomaterials in the environment...

179

the sediment and transport within the sediment—needs further consideration. The transport potential will be assessed on the basis of physico-chemical data, the preliminary tests on environmental behaviour, size, and size distribution rather than on complex fate tests. If transport to a secondary compartment is expected, this compartment has also to be addressed by a risk assessment at tier 2.

The refined n-PEC-assessment at tier 2 comprises two aspects: on the one hand, a refinement for the initial compartment (n-PEC$_{\text{ini refined}}$), and on the other hand an assessment for a second compartment if the ENM might be transported into it (n-PEC$_{\text{sec comp}}$). Both need experimental input data and modelling. The experimental fate testing at tier 2 will also need data from modified or even newly developed test guidelines and guidances. Furthermore, testing at tier 2 will also consider environmental behaviour of ENMs altered in the test system. n-PEC$_{\text{ini refined}}$ and n-PEC$_{\text{sec comp}}$ are assessed by the use of kinetic models. That accounts for the fact that environmental fate processes of ENMs are kinetic processes but not equilibrium processes as they are for conventional chemicals [5]. Furthermore, it is advisable, at least on the current state of knowledge, to use probabilistic models in order to account for the uncertainties of the model input parameters.

The refined n-PNEC-assessment at tier 2 comprises a higher tier testing, i.e. the use of more sophisticated tests such as water/sediment studies or aquatic mesocosm/terrestrial microcosm studies or the use of alternative endpoints whose appropriateness for utilisation in a refined PNEC assessment still needs to be investigated. In the case of a likely exposure of a secondary compartment, appropriate effect tests have to be performed. The testing and the use of assessment factors result in n-PNEC$_{\text{refined}}$.

Tier 2 yields a refined risk quotient for the initial compartment (n-PEC$_{\text{ini refined}}$/n-PNEC$_{\text{refined}}$) and, in the case of a likely exposure of a secondary compartment, in a risk quotient for that compartment (n-PEC$_{\text{sec comp}}$/n-PNEC$_{\text{sec comp}}$). As at tier 1, the trigger of 1 is used to either "STOP" or to proceed to a further tier. Tier 3 might comprise an additional even more sophisticated test refinement or measures for risk mitigation.

Effect assessment

The basic test strategy on effect assessment comprises three phases: (I) decision on the ENMs to be tested, (II) comprehensive testing and (III) use of test results (Fig. 3).

General aspects

ENMs to be tested In the first step of the test strategy, it has to be decided whether or not the nanospecific test procedures presented here have to be followed. Under following two conditions, it appears acceptable to waive testing:

- ENMs featuring physicochemical parameters well known for indicating nontoxic potential in every environmental compartment (water, sediment, soil) without any doubt.
- There is proof that a direct or indirect exposure of the environment can be excluded.

For regulatory testing, the decision on waiving of testing can be proposed by the registrant and has to be justified. The final decision on acceptance of the waiving is decided by the regulatory body. For chemical substances under REACH information requirement is triggered by their production volume. Four different ranges of production volumes per year are agreed on for which ecotoxicity data are required: 1 to <10 t/a; 10 to <100 t/a; 100 to <1000 t/a; 1000 t/a and more). Chemicals with a high production volume have to be tested more comprehensively than chemicals with a lower production volume. For chemicals with low production volumes, only aquatic tests are requested whereas for chemicals with high production volume additionally terrestrial and sediment tests are required. In the presented test strategy for ENMs, only one threshold value based on one, rather low production volume is intended. This value still has to be defined. ENMs with production volumes exceeding this threshold value have to be comprehensively tested addressing endpoints in the three compartments: water, sediment and soil. Also ENMs with a lower production volume than the threshold can be toxic to the environment. Therefore, two scenarios have to be considered. For low volume ENMs with bulk material of known ecotoxicity, comprehensive investigation has to be performed. For low volume ENMs where no bulk material is available, the bulk material shows no ecotoxicity, or ecotoxicity of the ENMs is expected due to their physico-chemical properties, a screening test has to be performed. If toxic effects are detected in the screening test, comprehensive testing has to be performed.

In summary, according to the presented test strategy testing is requested under the following conditions:

Comprehensive testing for:

- ENMs with production volume above the threshold value
- ENMs with production volumes below the threshold value and ecotoxic bulk material
- ENMs with production volumes below the threshold value and effects seen on the screening test

Screening test for:

- ENMs with production volume below the threshold value and bulk material is not ecotoxic or no bulk material is available or ecotoxic effect expected based on physical–chemical properties

If ENMs, due to their application, are covered by specific legislation where production volumes are not considered (e.g. ENMs used as biocides or pesticides), the aspect "production volume" is neglected in the presented test strategy. Instead, these ENMs are investigated starting with step 2 of tier 1 in order to comply with the requirements of the respective legislation, but also to consider nanospecific aspects as proposed within this test strategy (e.g. selection of tests: chronic tests instead of acute tests).

Screening test The screening test and the interpretation of the results still have to be discussed. The screening test should be used as a tool for identifying hazardous ENMs of low production volume whose corresponding non-nano counterparts give no hint on ecotoxicological potentials. Therefore, the test needs to have high sensitivity; the indicator function concerning effects on populations is less important in this case. The criteria for such tests should be (1) easy to perform and low work load, (2) short test duration, (3) sensitivity comparable to the sensitivity of the standardised endpoints to avoid too many "false" positive or "false" negative results. In several publications, high-throughput assays are described [6, 7]. Besides testing, also modelling might be a useful alternative. By the use of models toxic properties of ENMs such as oxidative stress potential of oxide, ENMs may be predictable [8, 9]. The suitability of such methods and procedures for the initial examination has to be investigated and the most reliable procedures need to be further developed. If the results indicate considerable toxicity, these ENMs enter step 2 of tier 1 (comprehensive testing).

Comprehensive testing Comprehensive testing is required for ENMs with production volumes exceeding the threshold value, for low volume ENMs which show toxicity in the screening test, and for ENMs which are covered by specific legislations where production volumes are not considered.

So far, there are considerable knowledge gaps with regard to the sensitivity of aquatic tests in comparison to terrestrial tests. It cannot be excluded that terrestrial tests are of comparable sensitivity or even more sensitive than aquatic tests as differences in the exposure concentration of the investigated ENMs between soil and aquatic tests are expected. In aquatic tests, the exposure and availability can change due to agglomeration and, depending on the test conditions, sedimentation. Sedimentation will result in increased exposure for sediment organisms. In contrast, in soil the exposure concentration of the organisms is not expected to change dramatically based on agglomeration. Therefore, the situation of exposure and availability and its changes are expected to be considerable differences between these three compartments and thus the test strategy includes a test programme which considers all three compartments, namely surface water, sediment, and soil.

Since it is assumed that ENMs preferentially enter the sediment compartment via the water phase, a test on sediment organisms performed using spiked water seems more appropriate to simulate the primary exposure scenario. Due to movement of the sediment organisms, sedimented ENMs can be incorporated into the sediment and spiking of sediment simulates the secondary exposure scenario. ENMs can be subjected to alterations of their physical–chemical characteristics in environmental media over time which in turn influences behaviour, bioavailability and toxicity. Furthermore, biodegradability of most ENMs is limited due to their inorganic condition and persistence is expected. Therefore, tests with longer incubation periods are preferred. Regarding the standardised test systems, the following test programme for ENMs is considered for the test strategy:

a. Aquatic tests
   Daphnids: OECD TG (test guideline) 211 [10]; algae: OECD TG 201 [11]; fish: OECD TG 210 "Fish, Early-life Stage Toxicity Test" [12]

b. Sediment test
   Chironomids: OECD 218, 219 [13, 14] (spiked sediment and spiked water phase) or lumbriculus: OECD TG 225 [15] (so far, a TG for the Lumbriculus test using spiked water is not available and a development is recommended)

c. Terrestrial tests
   Microflora: OECD TG 216 [16] using an inorganic nitrogen source instead of an organic one; earthworms: OECD TG 222 [17]; plants: OECD TG 208 [18]
   (Explanation for inorganic nitrogen source in a test according to OECD TG 216 [16]: Based on a recent study, it is anticipated that released ions tend to sorb to the additional organic nitrogen source, thus reducing their bioavailability. As a consequence, the use of an inorganic nitrogen source or a test on potential ammonium oxidation according to ISO Guideline 15685 [19, 20] can resolve this limitation.)

In this context, the quality control of the experiments and the validity of the test results have to be emphasised as addressed by Rösslein et al. [21] for tests in microtiter plates. Subjects such as homogeneity of spiking, sedimentation and concentration of ENMs over time, reactions with components of the test media and photoreactivity of the ENMs have to be considered and appropriate controls for ecotoxicological tests with organisms and complex test designs have to be established.

The need for revising current OECD test guidelines and the development of new ones was discussed by experts from science, industry and regulatory bodies at an OECD workshop on ecotoxicology and environmental fate of ENMs in 2013. An overview on the discussions and recommendations is given in Kühnel and Nickel [4]. The main subjects which have to be considered in the adaptation of the ecotoxicological test guidelines are spiking of terrestrial and aquatic test systems and the exposure of organisms in aquatic systems. For ENMs available as a powder, it has to be decided whether application via stock suspension (wet application) or via powder (dry application) is recommended. For aquatic tests, important issues were discussed at a workshop and recently published [22].

The metric to be used for the calculation of the toxicity is still being discussed. Besides mass, also size/surface area of ENMs and particle number may be suitable. To allow comparability with the results obtained with conventional substances, results should be presented on a mass basis. In addition, physical–chemical characterisation of the ENMs and the methods used for the determination should be reported. If required, the results can be recalculated using the metric "surface area" or "particle number". However, the reliability of a recalculation depends on the available information on particle size distribution of the respective ENM).

In several terrestrial and aquatic tests with various ENMs, a plateau with a maximum effect below 100 % is observed instead of concentration–effect relationships with a maximum of 100 % effect [23, 24]. The background of these observations is not yet systematically investigated but it is assumed that limitations in bioavailability and exposure are responsible. Therefore, a limit test with several test concentrations instead of only one test concentration is preferred to obtain information about the dose–response relationship.

The test conditions described in the test guidelines usually do not support photocatalytic activity. Simulated sunlight can increase ecotoxicity of photocatalytic active ENMs and possibly also of further ENM types [25, 26]. Aquatic tests should be performed according to the guidelines and additionally with simulated sunlight for photocatalytic ENMs. The tests with conventional lighting are recommended to address the unspecific properties of these ENMs in the absence of photoinduction and to link the results to results obtained by applying the test guidelines. The most sensitive result, independent of the illumination conditions, should be used for the assessment of hazard of ENMs. In addition, knowledge has to be improved with respect to illumination-dependent ecotoxicity of ENMs which are not specifically designed to feature photocatalytic activity but whose properties or behaviour are influenced by illumination [27].

Use of test results The test results can be used to describe the ecotoxicological properties of ENMs. Additionally, classification and labelling as well as an initial environmental hazard and risk assessment can be performed. For each purpose, only the relevant test results, as required in the respective regulation, need to be used. With respect to ecotoxicity, classification and labelling should address the most endangered environmental compartment. Currently, only the aquatic compartment is considered in classification and labelling and guidance for the other compartments has to be developed if required.

For the characterisation of the hazard with respect to risk assessment, PNEC values are required. For conventional chemicals, uncertainty in hazard can be considered using assessment factors [28]. Currently, there are no indications that assessment factors differing from the existing ones are needed for ENMs. For risk assessment, the PNEC values have to be compared with environmental concentrations (PEC). The topic of risk assessment is addressed in the chapters "Overview on the test strategy" and "Risk assessment approaches".

Specific aspect: alternative test systems in the test strategy for the assessment of ENMs

The appropriateness of the OECD test guidelines as well as other guidelines for nanomaterials has been reviewed and it is generally accepted that most endpoints are adequate and relevant also for ENMs [29]. Some modifications of the test procedures are required [4] and currently for some of the OECD test guidelines nanospecific guidance is drafted as well as some new OECD test guidelines are being developed.

Besides the application of the standardised test methods, alternative test methods and endpoints for the assessment of ENMs are published. So far, it is not clarified whether these endpoints provide additional information within the framework of regulation justifying the integration in the test strategy for ENMs. A literature review on alternative test methods such as behaviour, nutritional performance, indicator for oxygen stress, haematology, histology, genotoxicity, cytotoxicity, neurotoxicity, immunotoxicity, bioaccumulation and bio-diversity was performed in the project and the following conclusions were drawn:

1. The conventional endpoints used for hazard assessment are selected with respect to the protection of populations and cover parameters such as reproduction, mortality, growth. Effects on individuals are not considered. The results on alternative parameters reviewed in the literature usually address less complex and sub lethal reactions (e.g. determination of specific enzymes or gene activities) at a level of a single or some individuals, often resulting in an increased sensitivity. It is not always obvious whether an effect detected by a sensitive additional endpoint (e.g. indicators for oxygen stress) has an impact on the population level or indicates a compensation measure of the organism. Based on the literature review, it can be concluded that the advantage of considering alternative endpoints as additional input for a regulatory hazard assessment specific for ENMs is limited so far. Nevertheless, every additional parameter can provide additional information on ecotoxicity of ENMs and can support the assessment. In any case, in research, alternative endpoints play a major role by increasing the knowledge on the mode of action of ENMs.

2. There are some specific effects which are not detected with the conventional endpoints but which might have an impact on the population level and as such might be of relevance for assessing the hazard of ENMs for regulatory purposes.

   a. Immunotoxicity/genotoxicity
      The knowledge of the significance of effects on immunotoxicity and genotoxicity caused by ENMs in vitro and on the population level should be improved. Furthermore, the results have to be compared with the results obtained within the scope of studies on human toxicology. Based on this information, it can be decided whether these parameters are a suitable addition to the ecotoxicological test strategy.

   b. Bioaccumulation
      Bioaccumulation is actually considered for fate and behaviour aspects. However, in the present study, it was taken into consideration as an alternative endpoint delivering additional information on ecotoxicity. So far, the knowledge on physicochemical parameters indicating accumulation of ENMs is limited. Generally, the determination of bioaccumulation needs to account for the fact that uptake and distribution

processes of ENMs are kinetically driven. Thus, to obtain initial information on the accumulation potential and uptake of ENMs, a pragmatic screening procedure is to determine the ENM concentration in suitable test organisms (e.g. terrestrial and aquatic oligochaetes, daphnids, fish embryos and plants) at the end of the incubation period in an ecotoxicological test. If more detailed results are required, specific studies on bioaccumulation can be performed taking into account the discussions of the OECD expert meeting [4] and, once available, specific guidance on the accumulation of ENMs. Furthermore, this screening procedure for accumulation can be used to identify physicochemical parameters indicating bioaccumulation.

   c. Multi-generation tests
      It can be assumed that the effects become more pronounced if multi-generation tests are performed. Additionally, recovery studies can provide relevant information [30]. Even though the experimental effort is quite high, the consideration of multi-generation tests and recovery studies may result in a higher significance of the hazard assessment. However, uncertainty with respect to the significance of the assessment based on data of conventionally applied acute and chronic toxicity tests is considered, e.g. by assessment factors. Furthermore, it is assumed that multi-generation tests and recovery studies feature additional information specific not only for ENMs but also for conventional chemicals. There is no reason to consider such test approaches for only one group of chemicals. Nevertheless, knowledge on long-term effects should be improved to adapt the test strategy if necessary.

   d. Further test organisms
      ENMs agglomerate in aquatic systems, and increased concentrations in the sediment are expected [31]. The standardised test organisms *Chironomus riparius* and *Lumbriculus variegatus* develop in the sediment. It cannot be excluded that organisms living and grazing on the sediment as well as floated submerged, aquatic macrophytes are exposed to a higher extent compared to the standard test organisms if spiking of the water phase is performed. It is recommended that the sensitivity of potential suitable organisms (sediment organisms, aquatic macrophytes) and of the standard test organisms (*C.riparius*, *L. variegatus*, *Lemna minor*) are compared to decide on the suitability of further

test organisms and the potential replacement of traditionally applied organisms for the testing of ENMs.

e.   Behavioural tests

In the reviewed literature, behavioural tests appeared to be quite sensitive. However, so far, the information on the applicability on a wide range of ENMs is limited. To extend the knowledge, the behavioural test with earthworms [32] was studied in more detail in the experimental section of this project. It became obvious that the avoidance test with its short incubation period can provide important information on ecotoxicity and ageing of ENMs. However, a general utilisation within the test strategy for regulatory purposes, i.e. as a screening test, is not recommended since false-negative assessments cannot be excluded.

Fate assessment

General

The basic test strategy for fate endpoints of ENMs needed for their exposure assessment comprises three tiers: (0) screening for durability of ENMs in the initial compartment and thus a decision on the ENMs to be tested, (I) tier 1 to determine n-PEC $_{ini}$ and transport to secondary compartments (II) tier 2 to refine the results of tier 1. Tier 0 already has been presented in sufficient detail in the introduction and is not addressed furthermore. The determination of n-PEC$_{ini}$, transport to other compartments and the PEC refinement are based on experimental fate data. Examples of needs and challenges in tier 1 and tier 2 testing are presented in the following chapters and the test strategy considers the environmental compartments water, sediment, and soil. For the development of the test strategy on fate, a comprehensive literature review was performed. The aim of this evaluation was to summarise and analyse endpoints on fate and behaviour as well as the corresponding test methods for their importance and appropriateness to be implemented into a test strategy on the fate of ENMs. The detailed literature evaluation including results on single ENMs investigated using specific test methods is available in the report (http://www.umweltbundesamt.de/publikationen/integrative-test-strategy-for-the-environmental).

Tier 1 testing

The n-PEC$_{ini}$ is determined on the basis of information on the stability as a dispersion or emulsion, stability of the organic coating, and modification of the ENM, e.g. by oxidation, dissolution/solubility rate, size and size distribution. This information is also used to elucidate whether or not transport to secondary compartments is possible

which triggers refinement in tier 2. In addition, default models currently applied for the different compartments are used to deduce the n-PEC$_{ini}$.

Testing of fate endpoints of ENMs has to take into account that environmental fate processes of ENMs are mainly kinetically driven and include homo- and hetero-agglomeration (with suspended organic matter or biota) as well as transport processes like sedimentation in aquatic media [5]. Thus, it is commonly accepted [33–35] that guidelines which are based on partitioning processes are not suitable for ENMs, since employing partitioning coefficients to describe the behaviour of ENMs in the different environmental compartments will inevitably lead to misinterpretations of ENM distribution.

Since stability of ENMs in the environment is strongly influenced by the composition of the surrounding compartment, this indicates that for the determination of the n-PEC $_{ini}$ interaction with media compartments has to be considered (e.g. pH, NOM, ionic strength) [36–39].

Stability in the sense of biodegradation is measured based on the oxidation of organic carbon (e.g. BOD determination). However, these tests are expected to be applicable in rare cases only since most of the known ENMs are of inorganic nature. Thus, alternative approaches are needed to describe the general transformation of ENMs in the environment. The presented test strategy suggests that these approaches are covered by endpoints like agglomeration, dissolution or transformation upon ageing. In addition, transformation of ENMs based on the biological, chemical or physical loss of the coatings needs to be considered.

Based on the literature evaluation and discussions of the scientific and regulatory communities [4], it became obvious that for some of the mentioned endpoints like agglomeration and dissolution new or modified test guidelines are needed (e.g. [40–44]).

Tier 2 testing

In the case of an n-PEC$_{ini}$/n-PNEC-ratio of >1 risk for the considered compartment or transport to a secondary compartment cannot be excluded, a PEC refinement is needed. The refinement requires further experimental fate data as input for a more sophisticated modelling for exposure assessment. In the sense of the presented test strategy, most important endpoints to be considered include further (a) biotic transformation/degradation, mobility and transport in porous media, and sorption to soil, sediment and sludge of the pristine and aged ENM. The more complex the considered environmental matrix, the more environmental parameters interact with ENMs which themselves are of a complex nature. As a consequence thereof, the experimental test design needs to reflect this: the more complex the tested compartment, tier 2 considers experimental setups stronger mimicking the representative environmental compartment [45]. This can in particular be achieved using soil column experiments, e.g. as described in OECD 312 or by even more complex laboratory test systems such as model waste water treatment plants [46, 47] or fresh water mesocosms [33]. Most importantly, various techniques, in particular analytical techniques, should be combined to obtain a comprehensive and reliable picture of the ENM mobility, e.g. in porous media. It has to be considered that most experimental setups are likely to affect the form in which ENMs occur and might yield a result that is not representative of the behaviour under realistic environmental conditions. Furthermore, ENM properties like shape, crystal structure and surface properties influence mobility and transport as well as sorption/desorption to soil, sediment, and sludge and, therefore, have to be taken into account [48] within assessing fate in tier 2. It has to be noted that the concept of sorption is based on distribution coefficients and is of major importance for the description of solutes transport in soil. However, ENM association with soil is a non-equilibrium process, as it is also in other environmental compartments. Existing test methods appropriate for conventional chemicals, i.e. OECD TG 106, will generate misleading results [48, 49]. Alternative endpoints need to be employed to describe major processes influencing mobility and transport including agglomeration, deposition and re-mobilisation [48]. As already mentioned, for some of these endpoints new or modified test guidelines are needed.

PEC assessment, PEC models

Environmental fate processes of ENMs which are mostly influenced by aggregation, transformation and sedimentation are non-equilibrium but kinetic processes. ENMs do not reach thermodynamic equilibrium but are present in the environment as suspensions of different stability [5, 50]. Thus, conventional distribution models based on equilibrium processes such as the fugacity models developed by Mackay [51] are not applicable. ENM fate models have to be designed and evaluated which are capable of incorporating the environmental complexity to predict realistic environmental concentrations of ENMs. Furthermore, the use of kinetic models is essential in PEC assessment [4, 47].

Quite often reliable data are missing, e.g. on the quantity of emissions into the environment during production and usage. This situation can be overcome to some extent using probabilistic density functions [52].

Risk assessment approaches

Hazard and fate data are two essential parts of the analysis of the environmental risk of ENMs. Only a few references of the literature review conducted in this study deal with the risk assessment of ENMs. These comprise:

- Comparison of risk assessment of conventional substances and risk assessment of ENMs.
- Dealing with uncertainties and limited input information.
- Integration of ENM alteration and transformation in the risk assessment

Uncertainties regarding the potential impacts and risks associated with ENMs were discussed by Adam [34]. The authors combined life-cycle assessment (LCA) and risk assessment approaches. Because high uncertainties remain concerning the fate and effects of ENMs probabilistic approaches are needed, a Bayesian network was used. Nowack et al. [53] concluded that the risk due to ENMs cannot be determined exclusively for pristine ENMs, but has to consider alterations and transformation in the environment. Thus, the presented test strategy risk assessment considers information on pristine as well as aged ENMs as, based on the durability of the ENMs in the environment, alterations of the physical–chemical characteristics of ENMs are likely to occur and important to consider.

Conclusion

A test strategy is presented taking nanospecific aspects into account. The strategy for ecotoxicology is already more concrete than for environmental fate and is intended as a starting point for further discussions. There are still several gaps, such as

- Threshold values for the production volume (Fig. 1, step 1)
- Identification of suitable screening tests for substances with production volumes below the threshold value (Fig. 1, step 1)
- Trigger value for the screening tests to differentiate between "significant" and "not significant" effects (Fig. 1, step 1)
- Sensitivity of aquatic tests compared to terrestrial tests; in this context, the research gaps listed for aquatic tests [22] and the spiking methods for terrestrial tests (dry spiking vs. wet spiking) have to be considered (see "Comprehensive testing")
- Illumination-dependent ecotoxicity of ENMs not specifically designed to feature photocatalytic activity (see "Comprehensive testing")
- Further information on mode of action of ENMs to improve risk assessment (see "Comprehensive testing")

- Further information on specific effects currently not included in risk assessment (e.g. immunotoxicity, genotoxicity, multi-generation tests, necessity of further test organisms such as sediment organisms living and grazing on the sediment as well as aquatic macrophytes) (see "Alternative test systems")
- Current tests for fate endpoints are based on equilibrium situations. For fate testing, the test guidelines have to be modified to address the fact of non-equilibrium situations (e.g. OECD TG 106 adsorption/desorption) (see "PEC assessment, PEC models").

These gaps have to be filled in the near future to develop the test strategy further. The test strategy features a general approach to test and assess fate and effects of NMs. It features a first attempt to systematically test and assess effects and fate of ENMs in the environment. It has to be noted that the strategy is not yet developed sufficiently specified to fulfil the information requirements of certain legislation (e.g. plant protection act, biocide regulation, REACH). However, the adaption of single elements of the strategy to the specific needs of certain legislation will make a valuable contribution for the adjustment to the testing of nanomaterials.

Materials

The test strategy has been developed based on published literature, the knowledge of national and international discussions, after comparison with proposals presented by the European Commission and by German Federal Authorities. It also takes into account the conclusions made by the OECD WPMN which held an expert meeting in January 2013 [5].

To select appropriate parameters, test design and test methods for the test strategy on ENMs, recent literature was compiled for ecotoxicology and environmental fate-related key words. The following key words were applied and combined:

Nanoparticles, nanomaterials, ecotoxicology, nano, titanium dioxid, ecotoxicology, silver, soil, terrestrial, aquatic, toxicity, solubility, dissolution, release, partitioning, adsorption, desorption, sorption, sedimentation, transport, mobility, distribution, stability, hydrolysis, degradation, transformation, bioaccumulation, bioavailability, fate, PEC assessment, PEC models, PEC modelling, risk assessment.

The substances silver and titanium dioxide were specifically selected as much ecotoxicological work is done for these two types of nanomaterials.

Authors' contributions

KHR and KS (Fraunhofer IME) developed the test strategy with focus on ecotoxicity and MH with focus on fate. DV and KS (UBA) contributed significantly by comprehensive discussions on both ecotoxicity and fate. Every author contributed to draft the publication. All authors read and approved the final manuscript.

Author details

[1] Fraunhofer Institute for Molecular Biology and Applied Ecology IME, Auf dem Aberg 1, 57392 Schmallenberg, Germany. [2] German Federal Environment Agency (UBA), Wörlitzer Platz 1, 06844 Dessau, Germany.

Compliance with ethical guidelines

Competing interests

The authors declare that they have no competing interests.

References

1. Kah M, Hofmann T (2014) Nanopesticide research: current trends and future priorities. Environ Int 63:224–235
2. Windler L, Height M, Nowack B (2013) Comparative evaluation of antimicrobials for textile applications. Environ Int 53:62–73
3. Stone V, Pozzi-Mucelli S, Tran L, Aschberger K, Sabella S, Vogel U, Poland C, Balharry D, Fernandes T, Gottardo S et al (2014) ITS-NANO—prioritising nanosafety research to develop a stakeholder driven intelligent testing strategy. Part Fibre Toxicol 11:9. doi:10.1186/1743-8977-11-9
4. Kühnel D, Nickel C (2014) The OECD expert meeting on ecotoxicology and environmental fate—towards the development of improved OECD guidelines for the testing of nanomaterials. Sci Total Environ 472:347–353
5. Praetorius A, Tufenkji N, Goss K-U, Scheringer M, von der Kammer F, Elimeleche M (2014) The road to nowhere: equilibrium partition coefficients for nanoparticles. Environ Sci Nano 1:317–323
6. Tong T, Shereef A, Wu J, Binh CTT, Kelly JJ, Gaillard J-F, Gray KA (2013) Effects of material morphology on the phototoxicity of nano-TiO_2 to bacteria. Sci Total Environ 47:12486–12495
7. George S, Xia T, Rallo R, Zhao Y, Ji Z, Lin S, Wang X, Zhang H, France B, Schoenfeld D et al (2011) Use of a high-throughput screening approach coupled with in vivo zebrafish embryo screening to develop hazard ranking for engineered nanomaterials. ACS Nano 5:1805–1817
8. Burello E, Worth AP (2011) A theoretical framework for predicting the oxidative stress potential of oxide nanoparticles. Nanotoxicol 5:228–235
9. Zhang H, Ji Z, Xia T, Meng H, Low-Kam C, Liu R, Pokhrel S, Lin S, Wang X, Liao Y-P et al (2012) Use of metal oxide nanoparticle band gap to develop a predictive paradigm for oxidative stress and acute pulmonary inflammation. ACS Nano 6:4349–4368
10. OECD (2008) Test No. 211: *Daphnia magna* reproduction test. In: OECD Guidelines for the testing of chemicals, section 2. OECD Publishing, Paris
11. OECD (2011) Test No. 201: freshwater alga and cyanobacteria, growth inhibition test. In: OECD Guidelines for the testing of chemicals, section 2. OECD Publishing, Paris
12. OECD (2013) Test No. 210: fish, early-life stage toxicity test. In: OECD Guidelines for the testing of chemicals, section 2. OECD Publishing, Paris
13. OECD (2004) Test No. 218: sediment-water chironomid toxicity using spiked sediment. In: OECD Guidelines for the testing of chemicals, section 2. OECD Publishing, Paris
14. OECD (2004) Test No. 219: sediment-water chironomid toxicity using spiked water. In: OECD Guidelines for the testing of chemicals, section 2, OECD Publishing, Paris
15. OECD (2007) Test No. 225: sediment-water lumbriculus toxicity test using spiked sediment. In: OECD guidelines for the testing of chemicals, section 2. OECD Publishing, Paris
16. OECD (2000) Test No. 216: soil microorganisms: nitrogen transformation test. In: OECD Guidelines for the testing of chemicals, section 2. OECD Publishing, Paris
17. OECD (2004) Test No. 222: earthworm reproduction test (*Eisenia fetida/Eisenia andrei*). In: OECD guidelines for the testing of chemicals, section 2. OECD Publishing, Paris
18. OECD (2006) Test No. 208: terrestrial plant test: seedling emergence and seedling growth test. In: OECD Guidelines for the testing of chemicals, section 2. OECD Publishing, Paris

19. Hund-Rinke K, Schlich K (2014) The potential benefits and limitations of different test procedures to determine the effects of Ag nanomaterials and AgNO$_3$ on microbial nitrogen transformation in soil. Environ Sci Eur 26:28

20. ISO Guideline 15685 (2004) Soil quality—determination of potential nitrification and inhibition of nitrification—rapid test by ammonium oxidation. Genf, Schweiz: International Organization for Standardization

21. Rösslein M, Elliott JT, Salit M, Petersen EJ, Hirsch C, Krug HF, Wick P (2015) Use of cause-and-effect analysis to design a high-quality nanocytotoxicology assay. Chem Res Toxicol 28:21–30

22. Petersen EJ, Diamond S, Kennedy AJ, Goss G, Ho K, Lead JR, Hanna SK, Hartmann N, Hund-Rinke K, Mader B et al (2015) Adapting OECD aquatic toxicity tests for use with manufactured nanomaterials: key issues and consensus recommendations. Environ Sci Technol 49:9532–9547

23. Hund-Rinke K, Klawonn T (2013) Investigation of widely used nanomaterials (TiO2, Ag) and gold nanoparticles in standardised ecotoxicological tests. Texte | 29/2013. Umweltbundesamt. http://www.umweltbundesamt.de/en/publikationen/investigation-of-widely-used-nanomaterials-tio2-ag

24. Scott-Fordsmand JJ, Krogh PH, Schaefer M, Johansen A (2008) The toxicity testing of double-walled nanotubes-contaminated food to *Eisenia veneta* earthworms. Ecotoxicol Environ Saf 71:616–619

25. Adams LK, Lyon DY, Alvarez PJ (2006) Comparative eco-toxicity of nanoscale TiO$_2$, SiO$_2$ and ZnO water suspensions. Water Res 40:3527–3532

26. Bundschuh M, Zubrod JP, Englert D, Seitz F, Rosenfeldt RR, Schulz R (2011) Effects of nano-TiO$_2$ in combination with ambient UV-irradiation on a leaf shredding amphipod. Chemosphere 85:1563–1567

27. George S, Gardner H, Seng EK, Chang H, Wang C, Yu Fang CH, Richards M, Valiyaveettil S, Chan WK (2014) Differential effect of solar light in increasing the toxicity of silver and titanium dioxide nanoparticles to a fish cell line and zebrafish embryos. Envion Sci Technol 48(11):6374–6382

28. ECHA (2008) Guidance on information requirements and chemical safety assessment: chapter R.10: characterisation of dose [concentration]-response for environment. European Chemicals Agency ed. Helsinki, Finnland

29. Hankin SM, Peters SAK, Poland CA, Foss Hansen S, Holmqvist J, Ross BL, Varet J, Aitken RJ (2011) Specific advice on fulfilling information requirements for nanomaterials under REACH (RIP-oN2). Final Project Report. European Commission, Brussels

30. Völker C, Boedicker C, Daubenthaler J, Oetken M, Oehlmann J (2013) Comparative toxicity assessment of nanosilver on three Daphnia species in acute, chronic and multi-generation experiments. PLoS One 8(10):e75026

31. Gottschalk F, Sonderer T, Scholz RW, Nowack B (2009) Modeled environmental concentrations of engineered nanomaterials (TiO$_2$, ZnO, Ag, CNT, fullerenes) for different regions. Environ Sci Technol 43:9216–9222

32. ISO Guideline 17512-1 (2008) Soil quality—avoidance test for determining the quality of soils and effects of chemicals on behaviour—Part 1: test with earthworms (*Eisenia fetida* and *Eisenia andrei*). Genf, Schweiz: International Organization for Standardization

33. Lowry GV, Espinasse BP, Badireddy AR, Richardson CJ, Reinsch BC, Bryant LD, Bone AJ, Deonarine A, Chae S, Therezien M et al (2012) Long-term transformation and fate of manufactured Ag nanoparticles in a simulated large scale freshwater emergent wetland. Environ Sci Technol 46:7027–7036

34. Adam V, Quaranta G, Lawniczak S (2014) LCA-RA combined approach by using a Bayesian model: example of the aquatic ecotoxicity impact/risk of the nanoTiO$_2$ production. In: Science across bridges, borders and boundaries: Abstract Book, SETAC Europe 24th Annual Meeting. Basle, May 11–15

35. Kiser MA, Ladner DA, Hristovski KD, Westerhoff PK (2012) Nanomaterial transformation and association with fresh and freeze-dried wastewater activated sludge: implications for testing protocol and environmental fate. Environ Sci Technol 46:7046–7053

36. Dong H, Guan X, Lo IMC (2012) Fate of As(V)-treated nano zero-valent iron: determination of arsenic desorption potential under varying environmental conditions by phosphate extraction. Water Res 46:4071–4080

37. Neale PA, Malley EO, Jamting AK, Herrmann J, Escher BI (2014) Assessing the fate and effect of engineered nanomaterials in reference and wastewater derived organic matter. In: Science across Bridges, Borders and Boundaries: Abstract Book, SETAC Europe 24th Annual Meeting. Basle, May 11–15

38. Chowdhury I, Cwiertny DM, Walker SL (2012) Combined factors influencing the aggregation and deposition of nano-TiO2 in the presence of humic acid and bacteria. Environ Sci Technol 46:6968–6976

39. Ottofuelling S, Von der Kammer F, Hofmann T (2011) Commercial titanium dioxide nanoparticles in both natural and synthetic water: comprehensive multidimensional testing and prediction of aggregation behavior. Environ Sci Technol 45:10045–10052

40. Jafvert CT, Kulkarni PP (2008) Buckminsterfullerene's (C(60)) octanol-water partition coefficient (K(ow)) and aqueous solubility. Environ Sci Technol 42:5945–5950

41. Pakrashi S, Dalai S, Sneha RB, Chandrasekaran N, Mukherjee A (2012) A temporal study on fate of Al$_2$O$_3$ nanoparticles in a fresh water microcosm at environmentally relevant low concentrations. Ecotox Environ Safe 84:70–77

42. Unrine JM, Colman BP, Bone AJ, Gondikas AP, Matson CW (2012) Biotic and abiotic interactions in aquatic microcosms determine fate and toxicity of Ag nanoparticles. Part 1. Aggregation and dissolution. Environ Sci Technol 46:6915–6924

43. Ma R, Levard C, Marinakos SM, Cheng Y, Liu J, Michel FM, Brown GE Jr, Lowry GV (2012) Size-controlled dissolution of organic-coated silver nanoparticles. Environ Sci Technol 46:752–759

44. Aschberger K, Micheletti C, Sokull-Kluettgen B, Christensen FM (2011) Analysis of currently available data for characterising the risk of engineered nanomaterials to the environment and human health - lessons learned from four case studies. Environ Int 37:1143–1156

45. Auffan M, Pedeutour M, Rose J, Masion A, Ziarelli F, Borschneck D, Chaneac C, Botta C, Chaurand P, Labille J, Bottero J-Y (2010) Structural degradation at the surface of a TiO$_2$-based nanomaterial used in cosmetics. Environ Sci Technol 44:2689–2694

46. Wang Y, Westerhoff P, Hristovski KD (2012) Fate and biological effects of silver, titanium dioxide, and C-60 (fullerene) nanomaterials during simulated wastewater treatment processes. J Hazard Mater 201:16–22

47. Jarvie HP, Al-Obaidi H, King SM, Bowes MJ, Lawrence MJ, Drake AF, Green MA, Dobson PJ (2009) Fate of silica nanoparticles in simulated primary wastewater treatment. Environ Sci Technol 43:8622–8628

48. Kah M, Beulke S, Tiede K, Hofmann T (2013) Nanopesticides: state of knowledge, environmental fate, and exposure modelling. Environ Sci Technol 43:1823–1867

49. Forouzangohar M, Kookana RS (2011) Sorption of nano-C-60 clusters in soil: hydrophilic or hydrophobic interactions? J Environ Monitor 13:1190–1194

50. Sani-Kast NN, Praetorius A, Labille J, Ollivier P, Scheringer M, Hungerbuehler K (2014) Environmental fate models for engineered nanoparticles—simulating realistic conditions in a complex natural river system. In: Science across Bridges, Borders and Boundaries: Abstract Book, SETAC Europe 24th Annual Meeting. Basle, May 11–15

51. Mackay D (2001) Multimedia environmental models: the fugacity approach, 2nd edn. Lewis Publishers, Chelsea

52. Gottschalk F, Nowack B (2011) The release of engineered nanomaterials to the environment. J Environ Monit 13:1145–1155

53. Nowack B, Ranville JF, Diamond S, Gallego-Urrea JA, Metcalfe C, Rose J, Horne N, Koelmans AA, Klaine SJ (2012) Potential scenarios for nanomaterial release and subsequent alteration in the environment. Environ Toxicol Chem 31:50–59

17

The value of zebrafish as an integrative model in effect-directed analysis - a review

Carolina Di Paolo[1*], Thomas-Benjamin Seiler[1], Steffen Keiter[1,2], Meng Hu[3], Melis Muz[3], Werner Brack[3] and Henner Hollert[1,4,5,6]

Abstract

Bioassays play a central role in effect-directed analysis (EDA), and their selection and application have to consider rather specific aspects of this approach. Meanwhile, bioassays with zebrafish, an established model organism in different research areas, are increasingly being utilized in EDA. Aiming to contribute for the optimal application of zebrafish bioassays in EDA, this review provides a critical overview of previous EDA investigations that applied zebrafish bioassays, discusses the potential contribution of such methods for EDA and proposes strategies to improve future studies. Over the last 10 years, zebrafish bioassays have guided EDA of natural products and environmental samples. The great majority of studies performed bioassays with embryos and early larvae, which allowed small-scale and low-volume experimental setups, minimized sample use and reduced workload. Biotesting strategies applied zebrafish bioassays as either the only method guiding EDA or instead integrated into multiple bioassay approaches. Furthermore, tiered biotesting applied zebrafish methods in both screening phase as well as for further investigations. For dosing, most of the studies performed solvent exchange of extracts and fractions to dimethyl sulfoxide (DMSO) as carrier. However, high DMSO concentrations were required for the testing of complex matrix extracts, indicating that future studies might benefit from the evaluation of alternative carrier solvents or passive dosing. Surprisingly, only a few studies reported the evaluation of process blanks, indicating a need to improve and standardize methods for blank preparation and biotesting. Regarding evaluated endpoints, while acute toxicity brought limited information, the assessment of specific endpoints was of strong value for bioactivity identification. Therefore, the bioassay specificity and sensitivity to identify the investigated bioactivity are important criteria in EDA. Additionally, it might be necessary to characterize the most adequate exposure windows and assessment setups for bioactivity identification. Finally, a great advantage of zebrafish bioassays in EDA of environmental samples is the availability of mechanism- and endpoint-specific methods for the identification of important classes of contaminants. The evaluation of mechanism-specific endpoints in EDA is considered to be a promising strategy to facilitate the integration of EDA into weight-of-evidence approaches, ultimately contributing for the identification of environmental contaminants causing bioassay and ecological effects.

Keywords: Effect-directed analysis; Bioassay-guided fractionation; Zebrafish; Embryo; Larva; Bioassay; In vitro; In vivo

Review

Introduction

Zebrafish is a model vertebrate organism broadly applied in biological sciences, being one of the most important organisms that is used in different research areas as genetics, developmental biology and ecotoxicology [1]. More recently, its versatility has also been recognized by chemists, which provides an opportunity to enhance interdisciplinary studies involving biology and chemistry [2] as in effect-directed analysis (EDA) [3].

Bioassays in EDA

EDA, bioassay-guided fractionation and similar approaches are testing procedures applied to identify the individual bioactive compounds contained in highly complex matrices, such as natural products and environmental samples. Bioassays play a central role in EDA since biological activity directs the chemical fractionation and analysis steps as well as the testing strategy.

* Correspondence: carolina.dipaolo@bio5.rwth-aachen.de
[1]Department of Ecosystem Analysis, Institute for Environmental Research, ABBt - Aachen Biology and Biotechnology, RWTH Aachen University, Worringerweg 1, Aachen 52074, Germany
Full list of author information is available at the end of the article

Since fractionation of the sample is required to reduce the complexity of the original mixture, bioassays are needed to identify the active fractions and to guide further fractionation steps. Target and non-target chemical analyses are applied to select candidates and identify bioactive substances. Bioassays again play an important role in the confirmation phase, for biotesting of the pure substance identified as the bioactive compound [3-5].

Therefore, bioassay selection for EDA studies has to consider aspects that are rather specific to this application. For accurate identification of bioactive fractions, the bioassays should present high sensitivity and low internal test variability and be able to detect different chemicals that address similar endpoints or modes of action. Furthermore, due to limited sample amounts and large numbers of fractions to be tested, high-throughput low-volume bioassays are required [5].

Thus, *in vitro* bioassays are often selected for EDA studies; however, certain bioactivities require the organ or organism level for their proper identification, as for compounds in which metabolism plays an important role by interfering with formation or transformation of bioactive metabolites and bioaccumulation profiles [6]. These are the cases when bioassays with zebrafish early-life stages are considered to be of great value since they combine the organism-level endpoints with advantages of the *in vitro* format. Furthermore, biotesting strategies integrating organism-based and *in vitro* bioassays are expected to cover a broad range of bioeffects and related toxicants. The resulting diagnostic power strongly supports the identification of specific toxicants in EDA case studies [7].

Zebrafish model and bioassays in EDA

The zebrafish *Danio rerio* exhibits characteristics that make it a very attractive research model, including small size, ease of culture, high fecundity, rapid development, external fertilization and development, and transparency of the embryo. Bioassays with zebrafish embryos and larvae have further advantages that fit very well to EDA requirements. While these tests are relevant to evaluate acute [8] and chronic [9] effects in later life stages, the experimental setup exhibits several *in vitro* test characteristics, including a reduced volume of sample for testing and potential for high-throughput applications. Experiments with early life stages often do not require animal test authorization, and no external feeding is needed by embryos and larvae [1].

The zebrafish success as a model organism is in great part due to the work of pioneer scientists between the late 1960s and mid-1990s, as George Streisinger, who established the first zebrafish models and performed pioneer works on its genetics and developmental biology [10-12]; Charles Kimmel's descriptions of the cellular fate map [13] and the stages of development [14] in

embryos; and Christiane Nüsslein-Volhard, who performed a large-scale mutant screen to identify genes for vertebrate development control [15,16]. Following these ground-breaking studies, there was evident increase in the use of zebrafish in research [17], resulting in the sequencing of its genome [18], extensive information on its genetics, genomics, phenotypic and developmental biology [19], and the establishment of thousands of wild-type and transgenic zebrafish lines [20,21].

Importantly, zebrafish embryos and early larvae might be used to replace or refine experiments with adult fish, being increasingly applied in ecotoxicology to evaluate the toxicity of chemicals, plant protection products, biocides, pharmaceuticals, wastewater effluents and various aqueous environmental samples, and to assess sediment toxicity [1,22-24]. Recently, zebrafish embryo toxicity assays have been integrated in biotest batteries in environmental monitoring programmes, as the Joint Danube Survey [25] and the working group on bioassays of the NORMAN network [26]. Fish bioassays also play an important role in the implementation of the European Water Framework Directive (WFD) since they provide data for the derivation of environmental quality standards (EQS) and might represent a sensitive taxon for substances with specific modes of action [27]. Besides, biotests with fish are also included among recommended bioeffect-based tools for environmental assessment in the context of the WFD and the European Marine Strategy Framework Directive (MSFD) [7]. Consequently, current EU projects are investigating the contribution of zebrafish bioassays for water quality assessment and EDA of environmental samples, with focus on specific modes of action, mechanism-specific endpoints and adverse-outcome pathways [28,29]. Such initiatives are supported by the proposal that EDA contributes as an additional line of evidence in weight-of-evidence frameworks, such as the triad approach, ultimately leading to the identification of the contaminants responsible for the toxic effects observed in bioassays and the environment [30,31].

Context and objectives of this review

This review was developed in the context of the Marie Curie Initial Training Network 'EDA-EMERGE - Novel tools in effect-directed analysis for identifying and monitoring emerging toxicants on a European scale', funded by the European Commission within the Seventh Framework Programme for Research [28]. The literature review aimed to provide an overview of previous EDA investigations that applied bioassays with zebrafish, critically evaluating their objectives, methods, biotesting strategy and outcomes; discuss the potential contribution of further zebrafish bioassays for EDA; and propose strategies that might help optimizing the integration of such biotools into future EDA studies investigating environmental samples.

In order to meet these objectives, the literature was searched using the online tools Thomson Reuters Web of Science (WoS), ScienceDirect (SD) and Google Scholar (GS). In WoS, the terms were searched by topic (searching the fields Title, Abstract, Author Keywords and Keywords Plus® per record) in all databases, and in SD and GS, the terms were searched in all fields. The searches were done for publications in all years, except where indicated. The zebrafish terms used for search or filtration were a combination of 'zebrafish' or 'zebra fish' or *Danio rerio*. The EDA and life stage search terms are detailed below.

Zebrafish potential for EDA application
Zebrafish in EDA-relevant research areas
The application of the zebrafish model in EDA-related research areas was verified by search in WoS for the zebrafish terms as keywords in topic/all databases, followed by classification per research area (Figure 1). The search period was limited to between 2004 and 2014, to be in agreement with the publication years of EDA studies evaluated in this review. Outcomes are in good agreement with a recent review that applied much more sophisticated search strategy [17], indicating the usefulness of WoS for a first evaluation of research areas. Among the research areas strongly related to EDA, toxicology (8.9%) and pharmacology (9.0%) were each referred by circa 9% of the publications, while environmental sciences and ecology (3.0%) and chemistry (1.9%) were referred by a lower percentage. The prevalent research fields addressed by more than 20% of publications were mostly those that traditionally apply zebrafish, as genetics and heredity (40.6%), biochemistry and molecular biology (33.9%), developmental biology (30.7%), and zoology (24.6%)

Life stages referred to by research studies
The use of different life stages in studies with zebrafish was estimated by search for the zebrafish terms filtered by the life stage terms 'embryo*', 'larva*', 'juvenile*', 'adult*' and combinations of those. Again, the search period was limited to the publication years of reviewed EDA studies (2004 to 2014). As illustrated in Figure 2, more than half of the studies with zebrafish refer to embryos (52.6%) and almost one fourth of these mentioned also either adults or larvae, corresponding to 4.0% and 7.4% of total publications, respectively. The occurrence of studies mentioning adults (13.4%) and larvae (11.9%) was also representative, while circa 1% only referred to juvenile life stages.

EDA studies integrating zebrafish bioassays
Due to the heterogeneous nomenclature found in the literature, different EDA terms as listed in Weller 2012 [4] plus the term 'fractionation' were searched in quotation

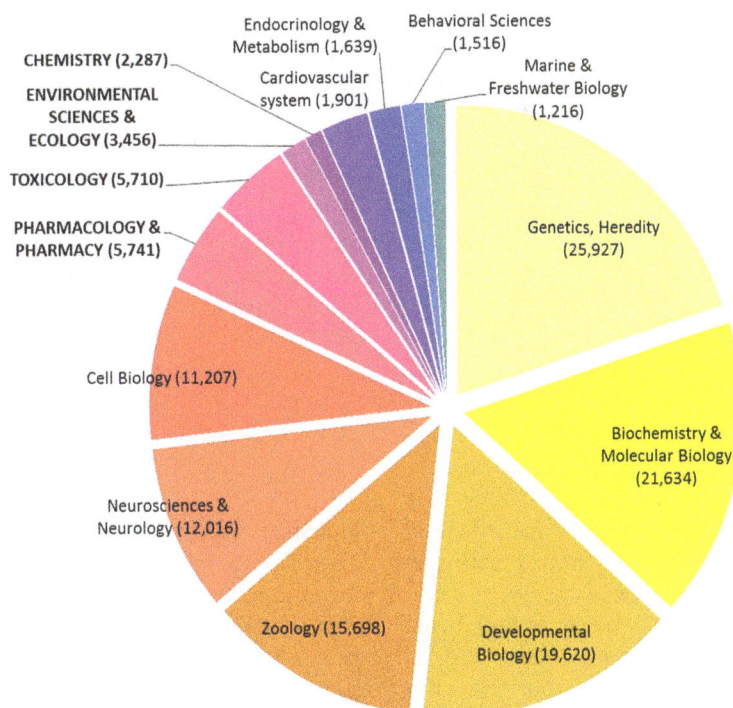

Figure 1 Records for the zebrafish terms, filtered by publication period (2004 to 2014), classified according to research area. Total number of records per period: 63,851. Search done in October 2014 (Web of Science).

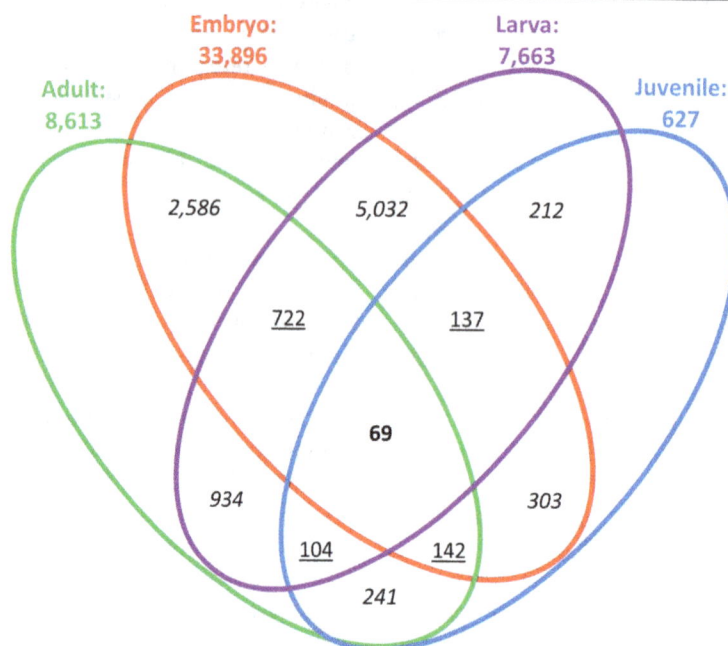

Figure 2 Records for the zebrafish and life stage terms, filtered by publication period (2004 to 2014). The life stage terms are 'embryo*', 'larva*', 'juvenile*', 'adult*' or combinations of these. Total number of records per period: 63,851. Search done in October 2014 (Web of Science).

marks, using the different search tools. After filtering by the zebrafish terms mentioned above, resulting publications were screened for confirmation of the searched content. Review papers, or studies that did not include the EDA procedure or zebrafish bioassays, were excluded. Two studies that followed procedures for toxicity identification evaluation (TIE) [32,33] instead of EDA were included since the similarities between both approaches [5] make them relevant for this review. In total, 29 publications were found (Table 1), which were carefully evaluated for research area, objective, investigated matrix, bioassay endpoint and setup, biotesting strategy, and study outcomes.

Research areas and investigated matrices
Two main fields were prevalent among EDA studies using zebrafish bioassays: drug discovery from natural products and environmental toxicology (Figure 3). Natural product studies aimed to identify bioactive compounds for pharmacological applications, investigating mostly plant extracts [19,34-45] but also extracts of bacteria [46], cyanobacteria and algae [47], seaweed [48] and marine organisms [49]. Environmental toxicology studies aimed to identify the toxic compounds in various environmental samples, including marine and fluvial sediments [50-52], soil [53], cyanobacteria and algae [54,55], industrial effluent [33], rubber tyre leachates [32], oil sand process waters [56,57] and river pore water [58]. Finally, fish skin extracts were investigated in a behavioural sciences study [59].

Prevalent life stages and exposure setups
EDA studies applied mostly bioassays with early embryos and larvae, following exposure to chemical extracts and fractions in multiwell-plates, often with exposure of several individuals in the same well (Table 2). Zebrafish up to 5 days post fertilization (dpf) were the life stages mostly applied, except for experiments that extended the assays up to 6 to 7 dpf [34,49,53,56,57] or a few studies with adults [33,47,59]. Environmental toxicology studies for the most part performed exposure not only in 24-well plates (200 µL to 2 mL per embryo or larva) but also in crystallization dishes, scintillation vials or beakers (450 µL to 5 mL per embryo or larva, 40 to 300 mL per adult), while natural product studies were performed exclusively in multiwell-plate setup (<100 to 250 µL per embryo or larva). The exposure of several individuals in the same well or vessel was observed for most of the studies, reflecting the need to reduce workload for EDA biotesting.

Biotesting strategy
The EDA investigations guided only by zebrafish bioassays followed either a single test setup (e.g. [36,53,55-57]) or a combination of methods (e.g. [44,52]) to evaluate endpoints in zebrafish. Other studies applied methods with additional experimental models, mostly cell-based (e.g. [19,45,51]) but also bacteria [46,50] and rodent [39] assays. When the application of multiple biossays aimed to evaluate distinct bioactivities, the tests were mostly performed

Table 1 Records for the EDA terms after filtering and after confirming that studies performed EDA

Search terms	WoS, all databases by topic (n)	WoS, filtered by zebrafish terms (n)	WoS, confirmed content (n)	Confirmed papers in WoS + SD + GS (n)
Bioassay(–)guided fractionation	5,134	14	8	9
Effect(–)directed analysis	189	14	3	5
Bioassay(–)guided isolation	454	9	7	8
Bioassay(–)directed fractionation	464	8	0	1
Toxicity(–)identification evaluation	346	2	1	2
Bioactivity(–)screening	92	3	2	2
Activity(–)guided fractionation	641	2	1	1
Bioactivity(–)guided fractionation	492	1	0	2
Bio(–)guided fractionation	136	1	1	1
Fractionation	232,733	84	15	21
Total[a]				29

The sum of papers from WoS, ScienceDirect (SD) and Google Scholar (GS) after confirmation of content is also presented. Search done in October 2014. [a]The total number differs from the sum since some studies resulted in more than one search.

in parallel. For instance, bioassay batteries evaluated the occurrence of different toxicity mechanisms [50,51] or effects on different trophic levels in the two TIE studies [32,33]. When instead the aim of multiple methods was to analyse different aspects of the same bioactivity or toxicity mechanism, there was the prevalence of tiered approach biotesting [19,39,45]. When applied in screening phase, zebrafish bioassays aimed to identify active fractions by organism-level endpoints, which were later further investigated by additional methods with zebrafish [52] or with other experimental models [19,45]. As an example, zebrafish bioassays were applied to screen extracts and fractions for anti-angiogenic effects, followed by further

investigations on human cells and transgenic zebrafish embryos [19]. On the other hand, zebrafish bioassays applied only as secondary tests aimed mostly at the confirmation of bioeffect occurrence at the organism level [49] or to evaluate the occurrence of acute toxicity in fish [46,47].

Use of solvents in bioassays

In biotesting, solvents were used for transference of samples into exposure vessels or as carriers. The first approach was applied using acetonitrile [55] or ethanol [54], including also solvent control conditions, and proceeding to solvent evaporation before adding exposure media. The use of solvents as carriers in bioassays showed the prevalence of

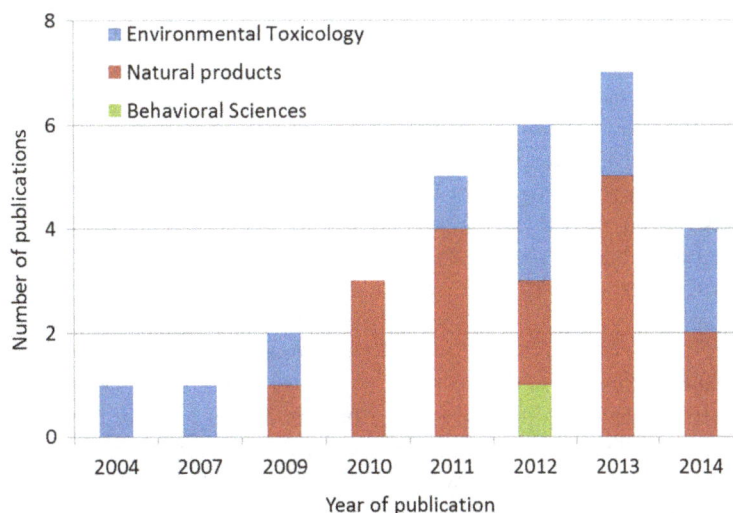

Figure 3 EDA studies applying zebrafish bioassays for the investigation of natural products and environmental samples. Number of studies that were evaluated in the literature review identified per main research area and year of publication (Web of Science, ScienceDirect and Google Scholar, October 2014).

Table 2 Exposure setup for studies on natural products or environmental toxicology

Exposure format	Fish per well or vessel (n)	Natural products Medium volume per fish and respective reference	Environmental toxicology Medium volume per fish and respective reference
96-well plates	1	100 µL [19,30,34,39,42]	100 µL [50]
	4	<100 µL [45]	-
24-well plates	1	-	2 mL [32]
	5	n.a. [46]	200 µL [51,54,55]
	10	100 to 250 µL [38,40,43]	-
	15	<100 µL [41]	-
	25	<100 µL [35]	-
12-well plates	10	n.a. [48,49]	-
6-well plates	25 to 30	n.a. [36,37]	-
Beakers, dishes, vials	5 to 40	40 mL[b] [47]	0.45 to 5 mL[a] [53,56-58], 300 to 1,000 mL[b] [33,59]

n.a. means information not available; [a]bioassay with embryos and larvae, [b]bioassays with adults.

dimethyl sulfoxide (DMSO), in concentrations ranging from 0.01% [35], 0.1% [31,45], 0.2% [26], 0.5% [46,47], 1% [33,48-50] up to 2% [51]. In addition, ethanol was also used as a carrier by a few studies, in concentrations of 0.001% for experiments with larvae [40,41] or 0.435% for experiments with zebrafish adults [42].

It is relevant to mention that the different procedures for extraction, cleanup, pre-concentration and fractionation of samples, already extensively reviewed elsewhere [3,4,60], also involve the use of different solvents and chemicals. Criteria for the use of solvents in EDA studies are the following: low or lack of toxicity in the biotest, the capacity of the solvent to dissolve complex extracts and fractions and the possibility to use the solvent in chemical analysis. The latter is the precondition to make sure that the chemical mixture tested in the bioassay resembles the mixture evaluated in chemical analysis. While DMSO excellently meets the first criterion, it is less suitable for dissolving complex mixtures when compared to other possible alternatives, and it completely fails the criterion related to the use in chemical analysis. Thus, the investigation of other solvents as possible carriers for exposure in zebrafish embryo testing might help to reduce possible artefacts during solvent exchange to DMSO. The evaluation of process blanks in order to exclude artefact toxicity is crucial for successful EDA and will be discussed below.

Positive/negative controls and biotesting of blanks

Positive control conditions that were specific to the evaluated endpoints were often described. For that, there was exposure of the zebrafish to compounds known to cause specific effects such as anti-convulsant activity [39], glucose uptake [45], pro-angiogenesis [36,37], anti-angiogenesis [19,44], or estrogenic effects [52,56]. Regarding negative control conditions, most of the studies

reported the testing of solvent controls in the same concentration as for the respective sample testing [19,32,34,35,38-40,43,44,47,49-51,53,55]. Some studies have additionally evaluated a medium only condition in addition to the solvent control [36,37,42,54,56,57].

The preparation and biotesting of blanks was described only in few of the evaluated EDA studies and in the two TIE studies. In the EDA studies, there was submission of the respective solvents [51,53,56,57] or of HPLC-grade water [58] through the same or part of the procedures that were applied to samples (i.e. sample preparation, extraction, fractionation). The TIE studies described blank preparation by treatment of milliQ water [32] or 0.1 M KCl solution [33] in the same way as samples for all procedures. In all these studies, the prepared blanks were evaluated in bioassays in the same way as done for samples and fractions. Another strategy was the use of a fraction that showed to be negative for the evaluated effect as a blank condition [45]. The exchange between elution solvents and DMSO was identified as a critical step since solvent traces might interfere with bioassays; therefore, blank testing was suggested to always be performed [51].

Investigated endpoints and studies outcomes

The specificity and sensitivity of bioassays and endpoints in identifying the bioactivity or adverse effects in fractions were considered to be a key issue for the relevance of zebrafish bioassays in EDA. Therefore, it is recommended to identify the critical aspects for endpoint assessment, to optimize bioassays accordingly and to demonstrate the validity of the bioassay by testing known bioactive compounds [61]. The endpoints and bioassays described in the different studies are summarized in Table 3 and discussed in the context of respective study objectives and outcomes.

Table 3 Organism-level effects and assessed endpoints, according to the research area of studies and the investigated sample matrices

Organism-level effect	Endpoint	Research area	Matrix and reference
Acute toxicity	Lethality and acute toxicity (respiratory rate, heart rate, movement) endpoints	Environmental toxicology	Sediment extracts [50], industrial effluents [33], rubber tyre leachates [32], oil sand process waters [57]
		Natural products	Cyanobacteria and algae extracts [47], seaweed hydrolysates [48]
Acute and developmental toxicity	Lethal and morphological endpoints	Environmental toxicology	Sediment extracts [51], soil extracts [53], microalgae extracts [54], cyanobacteria extracts [55], river pore water extract [58]
		Natural products	Bacteria extracts [46]
Developmental toxicity	Phenotypic endpoints	Natural products	Plant extracts [41]
Anti- or pro-angiogenesis	ISV formation and/or function in wild-type or *fli1:EGFP* zebrafish	Natural products	Plant extracts [19,36-38,42-44]
Anti-angiogenesis and anti-inflammatory	ISV outgrowth in *fli1:EGFP*, leukocyte migration after tail transection	Natural products	Plant extracts [40]
Glucose uptake	Uptake of fluorescein-tagged glucose bioprobe	Natural products	Plant extracts [45]
Lipid storage modulation	Uptake and metabolism of fluorescent fatty acid	Natural products	Heterofibrin molecules from *Spongia* sp. [49]
Antioxidant effects	ROS generation, cell death	Natural products	Plant extracts [35]
Estrogenicity	GFP induction in *tg(cyp19a1b-GFP)*	Environmental toxicology	Sediment extracts [52]
	vtg1 gene expession by qPCR	Environmental toxicology	Oil sand process waters [56]
Anti-convulsant	Locomotor activity, electrographic activity and epileptiform discharges	Natural products	Plant extracts [39]
	Inhibition of pentylenetetrazol-induced seizure activity, WISH for brain c-fos expression	Natural products	Plant extracts [34]
Fear behaviour	Alarm response, olfactory bulb activation in *Ta1:GCaMP2*	Behavioural sciences	Skin extracts [59]

ISV, intersegmental vessels; *fli1:EGFP*, transgenic zebrafish line with expression of enhanced green fluorescent protein marker in endothelial cells of vasculature; ROS, reactive oxygen species; GFP, green fluorescent protein; WISH, whole-mount *in situ* hybridization; *Ta1:GCaMP2*, transgenic zebrafish line with expression of the calcium indicator GCaM.

Acute toxicity and lethality

Bioactive sediment fractions [51] and components partially responsible for toxicity in oil sand process water fractions [57] have been identified by acute toxicity bioassays. The two TIE studies reported inconsistent acute toxicity of industrial effluents [33] and rubber tyre leachates [32]. One study investigating seaweed hydrolysates evaluated *in vivo* toxic potential through acute toxicity testing [48].

It may be summarized that EDA studies that focused on acute toxicity and lethality had only modest success in determining active compounds. These are unspecific responses that might occur due to exposure to very broad range of compounds; therefore, fractionation typically results in the distribution of toxicity over many different fractions. However, also other unrelated factors might have been involved, as for example, the high complexity of investigated matrices in the reviewed studies. Nevertheless, acute toxicity testing might be a powerful tool in TIE,

when applied to evaluate highly contaminated sites with acute toxicity caused by compounds that are well characterized [62].

Teratogenesis and developmental toxicity

Assessment of teratogenesis and developmental effects was done in studies that identified the bioactive compounds from microalgae, cyanobacteria and plant [41,54,55], river pore water [58] and developmental toxicants in soil [53]. Most studies evaluated traditionally assessed morphological endpoints, while one investigation of plant fractions focused on ectopic tail formation [41]. One study identified embryotoxicity in sediment extracts but not in respective fractions, which was attributed to losses of active compound or of synergistic effect during fractionation [50].

An aspect shared by the successful studies was the meticulous experimental characterization of the original matrices and respective fractions regarding their teratogenic effects and developmental toxicity potential. For instance, there was the determination of the optimal exposure period to identify a phenotype of interest that caused minimal acute toxicity [41]. Characteristic phenotypical effects were identified for specific fractions [53], also on a dose-dependent manner [54,55]. Two studies also investigated if additive or synergistic effects occurred between different fractions [55] or between aryl hydrocarbon receptor (AhR) agonists by co-exposure to a CYP1A inhibitor [58].

Angiogenesis modulation

Bioassays investigating pro- and anti-angiogenesis modulation by different bioactive plants were the most frequent studies in natural products. To this end, studies applied wild-type zebrafish [30,42,43] or the transgenic *fli1:EGFP* [63] zebrafish line [19,36-38,40], in which the zebrafish *fli1* promoter drives the expression of enhanced green fluorescent protein in blood vessels. In wild-type zebrafish, staining of the vessels was applied to facilitate scoring [30,43], while the transgenic line allowed *in vivo* observation of the embryonic vasculature. Selected endpoints evaluated specific cellular-morphological phenotypes, as intersegmental vessel formation. In these assays, the exposure start and duration were set to the most sensitive developmental windows related to the assessed endpoints.

All of the evaluated studies were successful in identifying at least one bioactive compound causing angiogenesis modulation, indicating that the identification of highly specific endpoints on the organism level might be a good requirement for the efficient use of zebrafish bioassays in EDA. The use of transgenic zebrafish lines is also considered to be a great asset for studies that evaluate specific morphological effects since it can facilitate endpoint observation and increase sensitivity of bioassays.

Energy uptake and storage

EDA was successful in identifying known and novel insulin-mimetic compounds in plants [45] with the contribution of zebrafish bioassays to characterize glucose uptake modulation. The study applied fluorescein-tagged glucose bioprobes and measured fluorescence by microscopy imaging and microplate reader, obtaining dose- and time-dependent responses. Another study applied a fluorescent fatty acid analogue to evaluate fatty acid storage modulation in zebrafish embryos by extracts from marine sponge [49]. In this case, the characterization of effects was done by extraction of zebrafish lipids followed by thin-layer chromatography. These studies demonstrated that the use of fluorescent bioprobes is a good tool to evaluate effects on the uptake and storage capacity of zebrafish, allowing not only for qualitative but also quantitative analysis of effects.

Antioxidant effects

Zebrafish embryos were integrated into an EDA study that identified and purified aloe vera polysaccharide with protective effects against oxidative stress [35]. Tests with zebrafish bioassays provided valuable information on organism-level responses regarding the generation of reactive oxygen species and oxidative stress-induced cell death, which were observed in a dose response manner.

Estrogenicity assessment by gene expression

Estrogenic effects were investigated in extracts and fractions of oil sand process waters by vitellogenin gene expression (*vtg1*) through quantitative polymerase chain reaction (qPCR) in zebrafish early larvae [56]. Estrogenicity was also assessed by the use of transgenic zebrafish embryos that exhibit green fluorescence protein expression in response to aromatase (*cyp19a1b*) gene induction, with confirmation of results by qPCR [52].

Gene expression analysis by qPCR showed to be a useful EDA endpoint in zebrafish embryos and larvae when background information allows the selection of specific biomarker genes for the studied effect, as for estrogenicity. The evaluation of sets of genes by qPCR is considered to be a promising strategy for endocrine disruption investigation, when following optimized experimental design regarding exposure intervals and evaluated zebrafish developmental stages [64]. The transgenic zebrafish embryos were also considered to be experimental models compatible with EDA, and their integration in future studies is expected to be facilitated by automated image analysis procedures [52].

Neuroactivity and behaviour

EDA was applied to identify anticonvulsant compounds present in plants, by co-exposure of evaluated samples with a convulsant compound, followed by the analysis of larvae total locomotor activity. Effects were assessed with video-tracking and software analysis and with electroencephalogram recording analysis [39]. Another EDA study of plant neuroactivity applied similar bioassays, in combination with larvae whole-mount *in situ* hybridization to assess increased brain *c-fos* gene expression as an indicator of seizure onset and brain damage [34]. Both studies identified anticonvulsant compounds, demonstrating the usefulness of the zebrafish model to identify neuroactivity in EDA. Also for neuroactivity and behaviour, the assessment of specific endpoints and setting the bioassay accordingly demonstrated to be an effective EDA biotesting approach.

The identification of neuroactive compound extracts of a mixture of red algae and cyanobacteria was investigated by a biotest battery including *in vitro* and organism-level methods [47]. Bioassays with zebrafish adults aimed at evaluating the neurotoxic potential of the matrix. However, evaluated endpoints were non-specific acute toxicity and mortality, which provided only minor contribution to the overall study outcomes.

The bioactive compounds responsible for fear behaviour response in fish were investigated by the exposure of zebrafish adults to fish skin extracts and fractions [59]. Video tracking was used to quantify alarm behaviour by measuring swimming speed and vertical position. The study identified the bioactive compound and proposed a new class of odorants that trigger alarm behaviour in fish. This study required the development of experimental setup and endpoint assessment that were specific to the evaluated behavioural alteration, confirming the importance of this step also for behavioural assessment.

Summary and discussion

Over the last 10 years, EDA studies guided by zebrafish bioassays have successfully identified bioactive or toxic compounds present in diverse biological matrices or environmental samples. Embryos and early larvae were the prevalent zebrafish life stages in these studies, with exposure being done in multiwell-plates, often with several individuals in the same well. In consequence, the sample volume for biotesting was minimized and the workload was reduced, which are important aspects in EDA. Zebrafish bioassays showed also versatility in terms of biotesting strategy, being applied alone or as a part of biotest batteries and in both screening phase as well as for further investigation of active fractions in tiered biotesting.

In spite of its limited capacity to dissolve complex matrix extracts, DMSO was the main carrier solvent applied in zebrafish bioassays. As a result, it was used in concentrations up to two orders of magnitude higher than the recommended for single compound biotesting (0.01%) [65]. Additionally, DMSO is not suitable for chemical analysis, which restricts the characterization of samples evaluated in biotesting. Therefore, the investigation of alternative carrier solvents would be an asset for zebrafish bioassays in EDA. Passive dosing methods are also promising options, as recently done in EDA investigation of sediments through the use of silicone rods for dosing of extracts and fractions in algae bioassay [66]. In fact, a loaded polymer silicone cast has successfully been integrated in zebrafish embryo assay for dosing of polycyclic aromatic hydrocarbons [67].

The EDA procedures for sample extraction, cleanup, pre-concentration and fractionation involve the use of different solvents and chemicals. Nevertheless, while most of the studies evaluated solvent and medium control conditions, the investigation of process blanks in bioassays was reported only by a small number of studies. In addition, methods for blank preparation varied considerably between these studies. Since the biotesting of process blanks is crucial for effective EDA, there is a

need to improve and standardize the procedures for their preparation and biotesting in future studies.

Most of the successful EDA studies applied specific and sensitive bioassays evaluating molecular, morphological or behavioural endpoints. Some studies optimized bioassays by identifying the most adequate exposure windows and assessment setups to maximize the specific endpoint response and minimize the interference of acute toxicity [41]. Further improvements might be achieved by advancing methods for the analysis of endpoints. For instance, the automated analysis of morphological phenotypes in transgenic or wild-type zebrafish would reduce workload and increase reliability in EDA [52]. Also, EDA of environmental samples would benefit from the analysis of bioassay results in correlation with previously characterized responses to specific classes of pollutants. That is the case of gene expression analysis of biomarker genes for specific mechanisms and modes of action. When analysed in correlation with respective gene modulation by known classes of compounds [68], biomarker gene responses might indicate the presence of certain classes of chemicals [69]. Similarly, EDA studies evaluating behavioural phenotypes to identify neuroactivity might rely in the near future on databases of behavioural profiles for different classes of compounds [70,71].

Such outcomes support the idea that EDA investigations of toxic environmental samples would benefit of the application of endpoint- and mechanism-specific methods with zebrafish. That is in fact a great advantage since mechanism-specific toxicity methods with zebrafish are broadly developed, as for AhR-mediated toxicity [72], genotoxicity [73] and neurotoxicity [74]. Furthermore, EDA guided by such zebrafish bioassays could integrate broader environmental assessment strategies, complementing effect-based approaches [7] and weight-of-evidence frameworks [30]. In this way, EDA would support the identification of contaminants causing bioassay and ecological effects, and the clarification of links between ecosystem functioning and the responses at different biological levels [30,62]. Finally, the evaluation of toxic aquatic contaminants through EDA guided by zebrafish bioassays might improve the protection of water bodies in the context of the European WFD and MSFD [7,27]. In conclusion, endpoint- and mechanism-specific zebrafish bioassays are considered of great relevance not only for guiding EDA studies but also for integrating EDA into environmental assessment and monitoring, ultimately contributing for environmental quality improvement [7,75].

Conclusions

Zebrafish bioassays have successfully guided different EDA studies; however, further method developments are still needed. Alternative dosing procedures should be investigated, and process blank preparation and biotesting should be standardized. Endpoint- and mechanism-specific bioassays with embryos and larvae are considered to be the most promising zebrafish biotests for future EDA of environmental samples. When integrated into broader environmental assessment strategies, EDA guided by specific zebrafish bioassays might support the identification of compounds causing bioassay and ecological effects, ultimately contributing for environmental quality improvement.

Competing interests

The authors declare that they have no competing interests.

Authors' contributions

CDP was responsible for the general design of the review and wrote the first draft of the manuscript. The other authors - TBS, SK, MH, MM, WB and HH - contributed with specific information concerning their respective expertise. All authors helped revise the draft of the manuscript. All authors read and approved the final manuscript.

Acknowledgements

The EDA-EMERGE project is supported by the EU Seventh Framework Programme (FP7-PEOPLE-2011-ITN) under the grant agreement number 290100.

Author details

[1]Department of Ecosystem Analysis, Institute for Environmental Research, ABBt - Aachen Biology and Biotechnology, RWTH Aachen University, Worringerweg 1, Aachen 52074, Germany. [2]Man-Technology-Environment Research Centre, School of Science and Technology, Örebro University, SE-701 82 Örebro, Sweden. [3]UFZ-Helmholtz Centre for Environmental Research, Permoserstrasse 15, Leipzig 04318, Germany. [4]College of Resources and Environmental Science, Chongqing University, 1 Tiansheng Road, Beibei, Chongqing 400715, China. [5]College of Environmental Science and Engineering and State Key Laboratory of Pollution Control and Resource Reuse, Tongji University, 1239 Siping Road, Shanghai, China. [6]State Key Laboratory of Pollution Control and Resource Reuse, School of the Environment, Nanjing University, Nanjing, China.

References

1. Strähle U, Scholz S, Geisler R, Greiner P, Hollert H, Rastegar S, et al. Zebrafish embryos as an alternative to animal experiments—a commentary on the definition of the onset of protected life stages in animal welfare regulations. Reprod Toxicol. 2012;33:128–32.
2. Basu S, Sachidanandan C. Zebrafish: a multifaceted tool for chemical biologists. Chem Rev. 2013;113:7952–80.
3. Brack W. Effect-directed analysis: a promising tool for the identification of organic toxicants in complex mixtures? Anal Bioanal Chem. 2003;377:397–407.
4. Weller MG. A unifying review of bioassay-guided fractionation, effect-directed analysis and related techniques. Sensors. 2012;12:9181–209.
5. Burgess RM, Ho KT, Brack W, Lamoree M. Effects-directed analysis (EDA) and toxicity identification evaluation (TIE): complementary but different approaches for diagnosing causes of environmental toxicity. Environ Toxicol Chem. 2013;32:1935–45.
6. Di Paolo C, Gandhi N, Bhavsar SP, Van den Heuvel-Greve M, Koelmans AA. Black carbon inclusive multichemical modeling of PBDE and PCB biomagnification and transformation in estuarine food webs. Environ Sci Technol. 2010;44:7548–54.
7. Wernersson A S CM, Maggi C, Tusil P, Soldan P, James A, Sanchez W, et al. Technical report on aquatic effect-based monitoring tools technical report 2014–077 EU Commission. doi:102779/7260. 2014.
8. Knobel M, Busser FJ, Rico-Rico A, Kramer NI, Hermens JL, Hafner C, et al. Predicting adult fish acute lethality with the zebrafish embryo: relevance of test duration, endpoints, compound properties, and exposure concentration analysis. Environ Sci Technol. 2012;46:9690–700.
9. Volz DC, Belanger S, Embry M, Padilla S, Sanderson H, Schirmer K, et al. Adverse outcome pathways during early fish development: a conceptual framework for identification of chemical screening and prioritization strategies. Toxicol Sci. 2011;123:349–58.
10. George Streisinger. Biographical Memoirs V.68: 353-362. Washington, DC: The National Academies Press; 1995.
11. Streisinger G, Walker C, Dower N, Knauber D, Singer F. Production of clones of homozygous diploid zebra fish (Brachydanio rerio). Nature. 1981;291:293–6.
12. Streisinger G, Coale F, Taggart C, Walker C, Grunwald DJ. Clonal origins of cells in the pigmented retina of the zebrafish eye. Dev Biol. 1989;131:60–9.
13. Kimmel CB, Warga RM, Schilling TF. Origin and organization of the zebrafish fate map. Development. 1990;108:581–94.
14. Kimmel CB, Ballard WW, Kimmel SR, Ullmann B, Schilling TF. Stages of embryonic development of the zebrafish. Dev Dyn. 1995;203:253–310.
15. Nüsslein-Volhard C. The zebrafish issue of development. Development. 2012;139:4099–103.
16. Nüsslein-Volhard C. The identification of genes controlling development in flies and fishes (Nobel Lecture). Angewandte Chemie Int Ed Eng. 1996;35:2176–87.
17. Kinth P, Mahesh G, Panwar Y. Mapping of zebrafish research: a global outlook. Zebrafish. 2013;10:510–7.
18. Howe K, Clark MD, Torroja CF, Torrance J, Berthelot C, Muffato M, et al. The zebrafish reference genome sequence and its relationship to the human genome. Nature. 2013;496:498–503.
19. Bradford Y, Conlin T, Dunn N, Fashena D, Frazer K, Howe DG, et al. ZFIN: enhancements and updates to the zebrafish model organism database. Nucleic Acids Res. 2010;39(1):D822–9.
20. Zebrafish International Resource Center - ZIRC [http://www.zebrafish.org]
21. EZRC - European Zebrafish Resource Center [http://www.ezrc.kit.edu]
22. Braunbeck T, Boettcher M, Hollert H, Kosmehl T, Lammer E, Leist E, et al. Towards an alternative for the acute fish LC(50) test in chemical assessment: the fish embryo toxicity test goes multi-species – an update. Altex. 2005;22:87–102.
23. Hallare A, Seiler T-B, Hollert H. The versatile, changing, and advancing roles of fish in sediment toxicity assessment—a review. J Soils Sediments. 2011;11:141–73.
24. EURL-ECVAM. Recommendation on the zebrafish embryo acute toxicity test method (ZFET) for acute aquatic toxicity testing. Report EUR 26710. 35p. In: Recommendation on the zebrafish embryo acute toxicity test method (ZFET) for acute aquatic toxicity testing. Report EUR 26710. 35p. 2014.
25. Joint Danube Survey [http://www.danubesurvey.org]
26. NORMAN Network [http://www.norman-network.net]
27. European Commission. Technical guidance for deriving environmental quality standard. Guidance document n°27, technical report 2011–055, European Communities, 203 p. In: Technical guidance for deriving environmental quality standard. Guidance document n°27, technical report 2011–055, European Communities, 203 p. 2011.
28. Brack W, Govender S, Schulze T, Krauss M, Hu M, Muz M, et al. EDA-EMERGE: an FP7 initial training network to equip the next generation of young scientists with the skills to address the complexity of environmental contamination with emerging pollutants. Environ Sci Eur. 2013;25:18.
29. Brack W, Altenburger R, Schüürmann G, Krauss M, López Herráez D, van Gils J, et al. The SOLUTIONS project: challenges and responses for present and future emerging pollutants in land and water resources management. Sci Total Environ. 2014;503–504:22–31.
30. Hecker M, Hollert H. Effect-directed analysis (EDA) in aquatic ecotoxicology: state of the art and future challenges. Environ Sci Pollut Res. 2009;16:607–13.
31. Chapman PM, Hollert H. Should the sediment quality triad become a tetrad, a pentad, or possibly even a hexad? J Soils Sediments. 2006;6:4–8.
32. Wik A, Nilsson E, Källqvist T, Tobiesen A, Dave G. Toxicity assessment of sequential leachates of tire powder using a battery of toxicity tests and toxicity identification evaluations. Chemosphere. 2009;77:922–7.
33. Fang Y-X, Ying G-G, Zhang L-J, Zhao J-L, Su H-C, Yang B, et al. Use of TIE techniques to characterize industrial effluents in the Pearl River Delta region. Ecotoxicol Environ Saf. 2012;76:143–52.
34. Buenafe OE, Orellana-Paucar A, Maes J, Huang H, Ying X, De Borggraeve W, et al. Tanshinone IIA exhibits anticonvulsant activity in zebrafish and mouse seizure models. ACS Chem Neurosci. 2013;4:1479–87.
35. Kang MC, Kim SY, Kim YT, Kim EA, Lee SH, Ko SC, et al. In vitro and in vivo antioxidant activities of polysaccharide purified from aloe vera (Aloe barbadensis) gel. Carbohydr Polym. 2014;99:365–71.
36. Liu C-L, Cheng L, Kwok H-F, Ko C-H, Lau T-W, Koon C-M, et al. Bioassay-guided isolation of norviburtinal from the root of Rehmannia glutinosa, exhibited angiogenesis effect in zebrafish embryo model. J Ethnopharmacol. 2011;137:1323–7.

37. Liu C-L, Kwok H-F, Cheng L, Ko C-H, Wong C-W, Wai Fong Ho T, et al. Molecular mechanisms of angiogenesis effect of active sub-fraction from root of Rehmannia glutinosa by zebrafish sprout angiogenesis-guided fractionation. J Ethnopharmacol. 2014;151:565–75.

38. Crawford AD, Liekens S, Kamuhabwa AR, Maes J, Munck S, Busson R, et al. Zebrafish bioassay-guided natural product discovery: isolation of angiogenesis inhibitors from East African medicinal plants. PLoS One. 2011;6:e14694.

39. Orellana-Paucar AM, Serruys A-SK, Afrikanova T, Maes J, De Borggraeve W, Alen J, et al. Anticonvulsant activity of bisabolene sesquiterpenoids of Curcuma longa in zebrafish and mouse seizure models. Epilepsy Behav. 2012;24:14–22.

40. Bohni N, Cordero-Maldonado ML, Maes J, Siverio-Mota D, Marcourt L, Munck S, et al. Integration of Microfractionation, qNMR and zebrafish screening for the in vivo bioassay-guided isolation and quantitative bioactivity analysis of natural products. PLoS One. 2013;8:e64006.

41. Gebruers E, Cordero-Maldonado ML, Gray AI, Clements C, Harvey AL, Edrada-Ebel R, et al. A phenotypic screen in zebrafish identifies a novel small-molecule inducer of ectopic tail formation suggestive of alterations in non-canonical Wnt/PCP signaling. PLoS One. 2013;8:e83293.

42. Krill D, Madden J, Huncik K, Moeller PD. Induced thyme product prevents VEGF-induced migration in human umbilical vein endothelial cells. Biochem Biophys Res Commun. 2010;403:275–81.

43. Han L, Yuan Y, Zhao L, He Q, Li Y, Chen X, et al. Tracking antiangiogenic components from Glycyrrhiza uralensis Fisch. based on zebrafish assays using high-speed countercurrent chromatography. J Sep Sci. 2012;35:1167–72.

44. He MF, Liu L, Ge W, Shaw PC, Jiang R, Wu LW, et al. Antiangiogenic activity of Tripterygium wilfordii and its terpenoids. J Ethnopharmacol. 2009;121:61–8.

45. Lee J, Jung D-W, Kim W-H, Um J-I, Yim S-H, Oh WK, et al. Development of a highly visual, simple, and rapid test for the discovery of novel insulin mimetics in living vertebrates. ACS Chem Biol. 2013;8:1803–14.

46. Dash S, Nogata Y, Zhou X, Zhang Y, Xu Y, Guo X, et al. Poly-ethers from Winogradskyella poriferorum: antifouling potential, time-course study of production and natural abundance. Bioresour Technol. 2011;102:7532–7.

47. Suyama TL, Cao Z, Murray TF, Gerwick WH. Ichthyotoxic brominated diphenyl ethers from a mixed assemblage of a red alga and cyanobacterium: structure clarification and biological properties. Toxicon. 2010;55:204–10.

48. Fitzgerald C, Gallagher E, O'Connor P, Prieto J, Mora-Soler L, Grealy M, et al. Development of a seaweed derived platelet activating factor acetylhydrolase (PAF-AH) inhibitory hydrolysate, synthesis of inhibitory peptides and assessment of their toxicity using the Zebrafish larvae assay. Peptides. 2013;50:119–24.

49. Rae J, Fontaine F, Salim AA, Lo HP, Capon RJ, Parton RG, et al. High-throughput screening of Australian marine organism extracts for bioactive molecules affecting the cellular storage of neutral lipids. PLoS One. 2011;6:e22868.

50. Higley E, Grund S, Jones P, Schulze T, Seiler T, Lubcke-von VU, et al. Endocrine disrupting, mutagenic, and teratogenic effects of upper Danube River sediments using effect-directed analysis. Environ Toxicol Chem/SETAC. 2012;31(5):1053–62. Epub 2012 Mar 2023.

51. Kammann U, Biselli S, Hühnerfuss H, Reineke N, Theobald N, Vobach M, et al. Genotoxic and teratogenic potential of marine sediment extracts investigated with comet assay and zebrafish test. Environ Pollut. 2004;132:279–87.

52. Fetter E, Krauss M, Brion F, Kah O, Scholz S, Brack W. Effect-directed analysis for estrogenic compounds in a fluvial sediment sample using transgenic cyp19a1b-GFP zebrafish embryos. Aquat Toxicol. 2014;154:221–9.

53. Legler J, van Velzen M, Cenijn PH, Houtman CJ, Lamoree MH, Wegener JW. Effect-directed analysis of municipal landfill soil reveals novel developmental toxicants in the zebrafish Danio rerio. Environ Sci Technol. 2011;45:8552–8.

54. Berry JP, Gantar M, Gibbs PDL, Schmale MC. The zebrafish (Danio rerio) embryo as a model system for identification and characterization of developmental toxins from marine and freshwater microalgae. Comp Biochem PhysiolPart C Toxicol Pharmacol. 2007;145:61–72.

55. Jaja-Chimedza A, Gantar M, Gibbs PDL, Schmale MC, Berry JP. Polymethoxy-1-alkenes from aphanizomenon ovalisporum inhibit vertebrate development in the zebrafish (Danio rerio) embryo model. Mar Drugs. 2012;10:2322–36.

56. Reinardy HC, Scarlett AG, Henry TB, West CE, Hewitt LM, Frank RA, et al. Aromatic naphthenic acids in oil sands process-affected water, resolved by GCxGC-MS, only weakly induce the gene for vitellogenin production in zebrafish (Danio rerio) larvae. Environ Sci Technol. 2013;47:6614–20.

57. Scarlett AG, Reinardy HC, Henry TB, West CE, Frank RA, Hewitt LM, et al. Acute toxicity of aromatic and non-aromatic fractions of naphthenic acids extracted from oil sands process-affected water to larval zebrafish. Chemosphere. 2013;93:415–20.

58. Fang M, Getzinger GJ, Cooper EM, Clark BW, Garner LVT, Giulio RTD, et al. Effect-directed analysis of Elizabeth river pore water: developmental toxicity in zebrafish (Danio rerio). Environ Toxicol Chem. 2014:n/a-n/a.

59. Mathuru Ajay S, Kibat C, Cheong Wei F, Shui G, Wenk Markus R, Friedrich Rainer W, et al. Chondroitin fragments are odorants that trigger fear behavior in fish. Curr Biol. 2012;22:538–44.

60. Brack W, editor. Effect-directed analysis of complex environmental contamination, Handbook of environmental chemistry series 15. Berlin, Heidelberg: Springer; 2011. p. 345.

61. Cordero-Maldonado ML, Siverio-Mota D, Vicet-Muro L, Wilches-Arizábala IM, Esguerra CV, de Witte PAM, et al. Optimization and pharmacological validation of a leukocyte migration assay in zebrafish larvae for the rapid In Vivo bioactivity analysis of anti-inflammatory secondary metabolites. PLoS One. 2013;8:e75404.

62. Connon RE, Geist J, Werner I. Effect-based tools for monitoring and predicting the ecotoxicological effects of chemicals in the aquatic environment. Sensors. 2012;12:12741–71.

63. Lawson ND, Weinstein BM. In vivo imaging of embryonic vascular development using transgenic zebrafish. Dev Biol. 2002;248:307–18.

64. Schiller V, Wichmann A, Kriehuber R, Muth-Kohne E, Giesy JP, Hecker M, et al. Studying the effects of genistein on gene expression of fish embryos as an alternative testing approach for endocrine disruption. Comp Biochem Physiol Toxicol Pharmacol. 2013;157:41–53.

65. OECD. Test No. 236: Fish Embryo Acute Toxicity (FET) Test. Paris: OECD Publishing; 2013.

66. Bandow N, Altenburger R, Streck G, Brack W. Effect-directed analysis of contaminated sediments with partition-based dosing using green algae cell multiplication inhibition. Environ Sci Technol. 2009;43:7343–9.

67. Seiler TB, Best N, Fernqvist MM, Hercht H, Smith KE, Braunbeck T, et al. PAH toxicity at aqueous solubility in the fish embryo test with Danio rerio using passive dosing. Chemosphere. 2014;112:77–84.

68. Kosmehl T, Otte JC, Yang L, Legradi J, Bluhm K, Zinsmeister C, et al. A combined DNA-microarray and mechanism-specific toxicity approach with zebrafish embryos to investigate the pollution of river sediments. Reprod Toxicol. 2012;33:245–53.

69. Keiter S, Peddinghaus S, Feiler U, von der Goltz B, Hafner C, Ho N, et al. DanTox—a novel joint research project using zebrafish (Danio rerio) to identify specific toxicity and molecular modes of action of sediment-bound pollutants. J Soils Sediments. 2010;10:714–7.

70. Ali S, Champagne DL, Richardson MK. Behavioral profiling of zebrafish embryos exposed to a panel of 60 water-soluble compounds. Behav Brain Res. 2012;228:272–83.

71. Kokel D, Bryan J, Laggner C, White R, Cheung CY, Mateus R, et al. Rapid behavior-based identification of neuroactive small molecules in the zebrafish. Nat Chem Biol. 2010;6:231–7.

72. Schiwy S, Braunig J, Alert H, Hollert H, Keiter SH. A novel contact assay for testing aryl hydrocarbon receptor (AhR)-mediated toxicity of chemicals and whole sediments in zebrafish (Danio rerio) embryos. Environ Sci Pollut Res Int. 2014.

73. Kosmehl T, Hallare AV, Reifferscheid G, Manz W, Braunbeck T, Hollert H. A novel contact assay for testing genotoxicity of chemicals and whole sediments in zebrafish embryos. Environ Toxicol Chem/SETAC. 2006;25:2097–106.

74. de Esch C, Slieker R, Wolterbeek A, Woutersen R, de Groot D. Zebrafish as potential model for developmental neurotoxicity testing: a mini review. Neurotoxicol Teratol. 2012;34:545–53.

75. Brack W, Klamer HJ, Lopez De Alda M, Barcelo D. Effect-directed analysis of key toxicants in European river basins a review. Environ Sci Pollut Res Int. 2007;14:30–8.

18

Naturally toxic: natural substances used in personal care products

Ursula Klaschka

Abstract

Background: Nature offers an incredible diversity of chemical compounds that boast a wide array of physiological effects. Many natural substances are employed in personal care products. Which of these natural substances are hazardous ingredients? How do European legal instruments regulate natural substances with toxic effects?

Results: 1,358 natural substances appear in the 'International Nomenclature of Cosmetic Ingredients' (INCI list, 'inventory …… of ingredients employed in cosmetic products') [Commission Decision 96/335/EC], most of them are herbal products, others are of animal, fungal, or bacterial origin. Out of these, 655 natural substances are enrolled in the EU database for classification and labeling, with 56% classified as hazardous chemicals, 38% classified due to their hazards to human health (35% due to their effects on skin and eyes), and 21% due to their hazards to the environment. 53 natural substances in the INCI list are classified as carcinogens, mutagens, and substances toxic to reproduction. Many classifications are not in line with expectations from experience, such as severe classifications of substances derived from some basic food plants or lacking classification of known medical plants or plants with sensitizing potential.
Classification and labeling is a trigger for the registration requirements according to REACH. It must be assumed that there are more substances that should undergo the REACH process among the 703 natural substances that do not turn up in the C&L inventory.

Conclusions: Many natural substances used in personal care products have toxic properties. The interdisciplinary compilation and analysis of regulatory instruments concerning natural substances revealed some inconsistencies which need further analysis and urgent correction to ensure prudent handling in consumer products.

Keywords: Classification and labeling; CLP regulation; Cosmetics; Cosmetics regulation; Hazardous substances; INCI; Natural substances; Personal care products; REACH

Background

Substances obtained from plants, animals, or other organisms have always enjoyed great popularity as ingredients in conventional personal care products or in natural cosmetics. Which natural substances employed in personal care products are hazardous ingredients? How are they regulated?

Natural substances are substances which occur in nature. There are various definitions, for example according to REACH (Art. 3 (39)) [1] 'substances which occur in nature: means a naturally occurring substance as such, unprocessed or processed only by manual, mechanical or gravitational means, by dissolution in water, by flotation, by extraction with water, by steam distillation, or by heating solely to remove water or which is extracted from air by any means'. In this study, a subset of natural substances originating from defined organisms which are applied in personal care products is considered. Most natural substances are complex mixtures of compounds belonging to various chemical substance classes, e.g. alkaloids, lipids, peptides, phenolics, sugars, terpenes [2-4]. The compositions of these mixtures derived from a certain species can vary. For example the composition of natural substances derived from a plant depends on the conditions of the plant growth like climate or soil parameters, on the harvest time, on production methods, or on storage conditions [5] and of course on the parts of the plants which were used. Even morphologically identical plants can

Correspondence: klaschka@hs-ulm.de
University of Applied Sciences Ulm, Prittwitzstr 10, D-89075 Ulm, Germany

contain different chemical secondary metabolites, so called chemotypes [6].

Many natural compounds are known to be physiologically active in the organisms of origin executing special physiological roles, such as storage or defense. Many are secondary metabolites and present in only minor concentrations in the organism of origin [2]. Their physicochemical and their physiological effects are as manifold as their chemical structures. In most cases, the mechanism of action is not known, but effects, such as anticancer, antimicrobial, antifungal, molluscidal, insecticidal, effects on skeletal muscles, blood pressure action, and gastroprotective action, have been observed [2,3,6,7]. These potent effects can be used to the advantage of human health (e.g. some natural pharmaceuticals, (e. g. glycosides from *Digitalis* sp. like digitoxin help against heart trouble) or to the disadvantage of the consumer (e.g. elevated doses of digitoxin lead to intoxications) [2].

The focus of this study was on natural substances derived from plants, animals, mushrooms, or bacteria which are listed in the 'Inventory ...... of ingredients employed in cosmetic products' (INCI list [8]). The aim was to analyze the European regulatory requirements for these substances with a special emphasis on classification and labeling according to the CLP regulation [9].

According to the CLP regulation, manufacturers and importers have to notify the classification and labeling to the European Chemicals Agency (ECHA) regardless of the production volumes. The CLP regulation provides for two approaches: on one hand, the harmonized classification which applies predominantly on effects of high concern, like carcinogenicity, germ cell mutagenicity, toxicity for reproduction as well as on substances which are persistent, bioaccumulative, and toxic, and which leads to partial classifications. The harmonized classifications are the minimum mandatory classifications listed in Annex VI of the CLP regulation. On the other hand, self-classification for all other hazardous properties is effectuated by the manufacturer, importer, or downstream user. Notifiers need to make sure that their self-classifications are correct: They need to consider all accessible existing information, evaluate their reliability, relevance, and adequacy in an expert judgment with a weight of evidence approach and use the evaluated data for self-classification [9-11].

Results and discussion
Natural substances used in personal care products
The international INCI list contains more than 16,000 ingredients [12], and the published European list [8] that was used in this study consists of around 8,500 substances. 1,358 ingredients specified in the INCI list are natural substances derived from organisms (according to the selection criteria as described in the 'Methods' section) (Table 1).

Out of these, 655 appear in the C&L inventory [13], which is the official European database that contains classification and labeling information on notified and registered substances received from manufacturers and importers.

The INCI list [8] was scanned for natural substances derived from natural organisms according to the criteria described in the methods. 1,358 natural substances were found. Out of these, 655 natural substances appear in the C&L inventory with 369 being classified as hazardous substances.

CAS and EC numbers are a prerequisite for the unequivocal retrieval of the classification and labeling of a definite substance in the C&L inventory. It was found that some substances appear in the INCI list with identical EC and CAS numbers, even when they have different origins and names (e.g. *Betula alba* bark extract and *Betula alba* leaf extract, both with CAS number 84012-15-7 and EC number 281-660-9, or *Citrus aurantium amara* flower distillate and *Citrus aurantium amara* peel extract both with 72968-50-4 and 277-143-2). It is apparent that these 'substances' should be differing chemical mixtures. Similar inconsistencies were found a number of times but could not be investigated further in this study.

Table 2 illustrates that these natural substances are mainly produced from higher plants, but some are also animal products or originate from other organisms. This table also shows the number of notifiers who submitted the same classification and labeling information for a certain substance in the C&L inventory. In case a substance was classified differently by various notifier groups, only the clear majority was considered here. The number of notifiers ranges between one to more than 1,000 e.g. *Ricinus communis* oil (1,469 notifiers) and Xanthan gum (polysaccharide secreted from the bacterium *Xanthomonas campestris*) (1,357 notifiers). 208 herbal natural substances are notified by more than 500 notifiers. This indicates the dissemination of a substance but does not allow to estimate amounts produced per year (see also the section concerning REACH). It must be emphasized that substances itemized in the INCI list can be applied not only in personal care products but also in other products, e.g. food [2-4] or medicine [3,4,14]. Therefore, the number of notifiers does not need to correlate to the quantitative importance of a certain natural substance in the cosmetic market. However, the numbers show the abundance of companies involved in the classification and labeling of these natural substances.

Natural substances from animals are, for example, amniotic fluid, aorta extract, *Bombyx* lipida, brain extract, bubulum oil (oil from the feet of cattle), equus extract, heart extract, liver extract, marrow extract, maromota oil, or serum. Less than 500 and more than 100 companies

Table 1 Overview of the number of substances analyzed in the study

	Number of substances
INCI list	More than 8,500
Natural substances	1,358
Natural substances in the C&L inventory	655 (100%)
Natural substances classified	369 (56%)
Natural substances classified (effects on human health)	257 (39%)
Natural substances classified (effects on environment)	185 (28%)
Natural substances classified (effects on human health and environment)	177 (27%)
Natural substances classified (physical effects, effects on human health and environment)	94 (14%)

notified e.g. shellac, cera alba (beeswax), albumen (egg white), colostrum, lac powder, ovum, royal jelly.

The majority of animal or human material substances are not notified by any manufacturer or importer (e.g. embryo extract, neural extract, pellis lipida (animals' skin lipids), spleen extract, spinal cord extract, sus extract (extract from the skin of the pig), thymus extract (thymus glands extract), or umbilical extract.

Algae that are used for the production of natural substances are e.g. *Caulerpa taxifolia, Chlorella vulgaris, Chondrus crispus* (Irish moss or carrageen moss), *Corallina officinalis, Delesseria sanguinea, Fucus serratus,* *Laminaria digitata, Laminaria japonica, Palmaria palmate* (dulse), and *Ulva lactuca* (sea lettuce). The extract of the brown algae *Fucus vesiculosus* (bladderwrack algae) has the greatest number of notifiers (123).

Most mushrooms entries are *Saccharomyces* ferment, and the yeast extract (faex extract) has the largest number of notifiers (115 notifiers).

Among the bacteria that are used for natural substances in personal care products are the following genera, e.g. *Aspergillus, Lactobacillus, Lactococcus,* and *Streptococcus.*

The lichen extracts are produced from *Evernia prunastri, Usnea barbata* (tree moss), and *Cetraria Islandica* (Iceland moss).

Table 3 shows examples of plant species from the INCI list which are used also in other applications but personal care or which are wild herbs. Some are widely used aromatic herbs, e.g. *Laurus nobilis* (laurel), *Myristica fragrans* (nutmeg), *Piper nigrum* (pepper), and *Zingiber officinale* (ginger). Others are food plants, e.g. *Allium cepa* (onion), *Arachis hypogaea* (peanut), *Citrus aurantium* (bitter orange), *Coffea arabica* (coffee), *Daucus carota* (carrot), *Glycine soja* (soybean), *Theobroma cacao* (cacao), *Triticum vulgare* (wheat), *Vitis vinifera* (grape), and *Zea mays* (corn). There are other plants known for their use as phytopharmaceuticals, e.g. *Arnica montana* (arnica), *Calendula officinalis* (calendula), *Chamomilla recutita* (matricaria, chamomile), *Valeriana officinalis* (valerian), and *Viscum album* (mistletoe). *Nicotiana tabacum* (tobacco) is also among the plant species used, as well as many wild or

Table 2 Numbers of natural substances in the INCI list [8] originating from various organism genera

Organism group	Number of substances in the INCI list	Number of substances which were notified in the C&L inventory by					Not in the C&L inventory
		>1,000 notifiers	500 to 1,000 notifiers	100 to 499 notifiers	10 to 99 notifiers	1 to 9 notifiers	
Higher plants	1,189	1[a]	207	50	201	148	582
Animals and humans[b]	84	0	0	16[c]	6	10	52[d]
Mushrooms[e]	34	0	0	1[f]	0	0	33
Algae[g]	31	0	0	1[h]	3	6	21
Bacteria[i]	17	1[j]	0	0	0	0	16
Lichens[k]	3	0	1	0	1	1	0
Total	1,358	2	208	68	212	165	703

Grouped according to the number of notifiers in the C&L inventory [13].
[a]*Ricinus communis* oil.
[b]In several cases the species was not named for mammalian substances.
[c]For example, shellac, cera alba (beeswax), albumen (eggwhite), colostrum, lac powder, ovum, royal jelly.
[d]No notifiers, e.g.: embryo extract neural extract, pellis lipida (animals' skin lipids), royal jelly extract, spleen extract, spinal cord extract, sus extract (extract from the skin of the pig), thymus extract (thymus glands extract), umbilical extract.
[e]Most entries are *Saccharomyces* ferment.
[f]Faex extract (yeast extract).
[g]For example: *Caulerpa taxifolia, Chlorella vulgaris, Chondrus crispus* (Irish moss or carrageen moss), *Corallina officinalis, Delesseria sanguinea, Fucus serratus, Laminaria digitata, Laminaria japonica, Palmaria palmate* (dulse), *Ulva lactuca* (sea lettuce).
[h]*Fucus vesiculosus* extract (bladderwrack algae).
[i]For example, the following genera: *Aspergillus, Lactobacillus, Lactococcus, Streptococcus.*
[j]Xanthan gum (polysaccharide secreted by the bacterium *Xanthomonas campestris*).
[k]Extracts from *Usnea barbata* (tree moss), *Cetraria Islandica* (Iceland moss), and *Evernia prunastri.*

Table 3 Examples of herbal natural substances in the INCI list [8]

Aromatic herbs	*Borago officinalis* (borage), *Laurus nobilis* (laurel), *Myristica fragrans* (nutmeg), *Piper nigrum* (pepper), *Prunus amygdalus amara* (bitter almond), *Rosmarinus officinalis* (rosemary), *Zingiber officinale* (ginger)
Food	*Ananas sativum* (pineapple), *Allium cepa* (onion), *Arachis hypogaea* (peanut), *Asparagus officinalis* (asparagus), *Beta vulgaris* (beetroot), *Citrus aurantium* (bitter orange), *Cocos nucifera* (coconut), *Coffea arabica* (coffee), *Daucus carota* (carrot), *Foeniculum vulgare* (fennel), *Fragaria vesca* (strawberry), *Glycine soja* (Soybean), *Juglans regia* (walnut), *Lactuca sativa* (lettuce), *Musa sapientium* (banana), *Oryza sativa* (rize), *Secale cereale* (rye), *Spinacia oleracea* (spinach), *Theobroma cacao* (cacao), *Triticum vulgare* (wheat), *Vitis vinifera* (grape), *Zea mays* (corn)
Officinal plants	*Achillea millefolium* (yarrow), *Aloe barbadensis* (aloe), *Arnica montana* (arnica), *Calendula officinalis* (calendula), *Chamomilla recutita* (matricaria, chamomile), *Eucalyptus globulus* (eucalyptus), *Echinacea spec.* (coneflower), *Mentha piperita* (peppermint), *Salvia officinalis* (sage), *Valeriana officinalis* (valerian), *Viscum album* (mistletoe)
Other agricultural crop plants	*Nicotiana tabacum* (tobacco)
Wild or garden plants	*Acer pseudoplatanus* (harewood, great maple), *Bellis perennis* (daisy), *Berberis vulgaris* (barberry), *Betula alba* (birch), *Buxus sempervirens* (boxwood), *Castanea sativa* (chestnut), *Crocus sativus* (saffron crocus), *Equisetum arvense* (horsetail), *Ginkgo biloba* (maidenhair tree), *Hedera helix* (ivy), *Helianthus annuus* (sunflower), *Lupinus albus* (lupin), *Nasturtium officinale* (watercress), *Nerium oleander* (oleander), *Papaver rhoeas* (corn poppy), *Petasites hybridus* (butterbur), *Ranunculus ficaria* (pilewort), *Sambucus nigra* (elder), *Syringa vulgaris* (lilac), *Taraxacum officinale* (dandelion), *Trifolium pratense* (clover), *Tussilago farfara* (coltsfoot), *Urtica dioica* (stinging nettle)

Many natural substances in the INCI list derived from plants are known from other usages than cosmetic applications. Some popular plants are specified here with their respective common English names in brackets as examples.

garden plants, e.g. *Buxus sempervirens* (boxwood), *Castanea sativa* (chestnut), *Crocus sativus* (saffron crocus), *Equisetum arvense* (horsetail), *Hedera helix* (ivy), *Lupinus albus* (lupin), *Nerium oleander* (oleander), *Papaver rhoeas* (corn poppy), *Syringa vulgaris* (lilac), *Taraxacum officinale* (dandelion), *Tussilago farfara* (coltsfoot), and *Urtica dioica* (stinging nettle).

The INCI list indicates the possible functions in cosmetic products. Some natural substances are 'antiseborrheic', 'emollient', 'keratolytic', 'refreshing', 'skin conditioning', 'skin protecting', 'smoothing', 'soothing', 'solvent', or 'tonic'. Only one 'colorant' (*Beta vulgaris/Beta vulgaris* extract) and one 'hair dye' (*Lawsonia inermis* extract, which is the extract of the henna plant) is among them. Some are 'antimicrobials', but no one has the function 'preservative' (Note: There are many non-natural compounds listed in the INCI with the function 'preservative', e.g. benzylalcohol or benzylparaben.). For *Usnea barbata* extract, the function 'deodorant' and for ten other natural substances, e.g. *Citrus aurantium bergamia* oil, *Mentha viridis* oil, *Prunus amygdalus amara* oil, *Ruta graveolens* oil, and *Sambucus nigra* oil, 'masking' was the only function indicated. Some natural substances are 'deodorant' or 'masking' in addition to other functions, e.g. 'tonic'. Other cosmetic ingredients used due to their odor are listed in the INCI list part II 'Perfume and aromatic raw materials' [15] and not in the INCI list analyzed here. It was found that some substances with different INCI names, different origins, obviously different chemical compositions, and also different functions in a product were itemized with identical EC and CAS numbers in the INCI list, e.g. *Avena sativa* bran extract (abrasive), *Avena sativa* extract (emollient), and *Avena sativa* meal extract (soothing), all three with the CAS 84012-26-0 and the EC 281-672-4, or *Citrus aurantium dulcis* flower oil (astringent and tonic), *Citrus aurantium dulcis* flower water (skin conditioning), *Citrus aurantium dulcis* peel cera (emollient and skin conditioning), *Citrus aurantium dulcis* peel extract (emollient, skin conditioning), and *Citrus aurantium dulcis* seed extract (skin conditioning), all five with the CAS 8028-48-6 and the EC 232-433-8.

Legal restrictions
Classifications according to the CLP regulation
Out of the 1,358 natural substances (Table 1), 703 natural compounds could not be found in the C&L inventory [13]. There are several reasons why a substance does not show up there: the classification was not notified yet, the substance might not have been scrutinized for its hazardous properties yet, or the substance is no longer produced, imported, or used by any company. In general, natural substances are self-classified by producers or importers. None of the natural substances appear in Annex VI of the CLP regulation, which is the list of harmonized classification and labeling of hazardous substances, whereas several single components of natural substances turn up there, such as e.g. nicotine, strychnine, atropine, digitoxin, papaverine, or aconitine.

Some notifier numbers appear repeatedly which might indicate the size of consortia that submitted the classifications to the ECHA: e.g. 14 substances were classified by 226 manufacturers or importers, 30 by 500, 45 by 747, and 26 by 748 manufacturers. Many natural substances are classified differently by various notifier groups. If the classifications differ between notifiers, they should make an effort to find a joint solution [9, Art. 41] which has apparently not been done for many natural substances. The differences could be the consequences of slightly different compositions or contaminations of the natural substances, of different data bases used, of the application of calculation methods instead of measured test data, or of different classification approaches due to precaution or strategic reasons. It is also possible that some classifications are not correct. If the self-classifications of various companies differ, the classification which the clear majority of companies agreed upon is used in this study. It was not the purpose here to validate whether the majority is right or not. In most cases, the majority is clear. However, there were also cases, where one consortium of notifiers did not classify or label at all, whereas the other consortium of a similar size did. For example, *Brassica nigra* extract was not classified by 29 notifiers, whereas 28 notifiers classified it with H226, H301, H311, H315, H319, H330, H335, H400, and H410. Several substances were classified by only one company and not classified by many others.

Table 1 shows that 369 natural substances that appear in the C&L inventory are classified as hazardous substances. 257 substances are classified due to their hazardous effects on human health, and 185 substances are classified due to their hazardous effects on the environment. Eight substances are classified only because of their effects on the environment (e.g. tall oil H412 (352 notifiers)); 177 substances are classified due to their hazardous effects on human health and the environment, e.g. *Picea excelsa* extract H304, H315, H317, H319, H400, and H410 (500 notifiers) or *Lavandula hybrida* extract H304, H315, H317, H373, and H411 (748 notifiers).

136 substances were classified due to physical effects only (H225, H226) and 94 substances due to their physical, health, and environmental hazards, e.g. *Rosa damascena* extract H226, H315, H317, H318, H341, H351, and H412 (352 notifiers); *Melaleuca alternifolia* oil H226, H302, H304, H315, H317, H319, and H411 (981 notifiers).

Table 4 is a compilation of H phrases found for natural substances in the C&L inventory. The H-phrases due to physical properties (H226 and H225) may be mainly due to the solvents used for extraction or other preparation methods of the natural products. No natural substance is 'fatal if swallowed', but 45 are 'toxic' or 'harmful if swallowed'.

226 (35%) of the substances are assigned H phrases concerning effects on skin and eyes. Several natural products are called 'skin conditioning', 'skin protecting', or 'tonic' in the INCI list, even if they are classified due to their hazardous properties for skin or eyes (H310, H311, H312, H314, H315, H317, H318), e.g., *Nicotiana tabacum* extract (H312), *Pelargonium graveolens* oil (H304, H315, H317, H318, H412), *Ribes nigrum* extract (H304, H317, H411), *Rosa centifolia* water (H315, H317, H319, H341, H351), *Rosa damascena* distillate (H226, H315, H317, H318, H341, H351, H412), *Ruta graveolens* extract (H317, H411), *Thuya occidentalis* extract (H226, H301, H304, H317, H411), *Vanilla planifolia* extract (H317), *Verbena*

officinals extract (H304, H315, H317, H411), and *Vetiveria zizanoides* extract (H315, H317, H319).

Many natural substances produced from food plants are classified due to hazardous properties, e.g. *Carum carvi* extract (H226, H302, H304, H315, H317, H400, H410), *Carum petroselinum* extract (H226, H304, H317, H400, H410), *Daucus carota* extract (H226, H304, H317, H319, H411), *Daucus carota* juice (H226, H304, H317, H319, H411), *Rosmarinus officinale* extract (H226, H304, H317, H373, H411), *Triticum vulgare* extract (H317), and *Foeniculum vulgare* oil (H226, H304, H315, H317, H319, H341, H351, H371, H400, H410).

Many plants are generally known for their ingredients of high sensitization potential [2], such as *Achillea millefolium, Arachis hypogaea, Arnica montana, Ginkgo biloba, Glycine max, Helianthus annuus, Inula helenium,* and *Ricinus communis.* 21 entries of natural substances in the INCI list originate from these plants, but only *Achillea millefolium* oil is classified due to its sensitization potential (H317).

Seven natural substances are classified as 'fatal', 'toxic', or 'harmful if inhaled'. Three notifiers classify a natural substance as 'May cause allergy or asthma symptoms or breathing difficulties if inhaled'. Eight substances are classified as 'May cause respiratory irritation', e.g. *Quillaia saponaria* extract (783 notifiers).

130 substances are classified with H304 'May be fatal if swallowed and enters airways', e.g. *Picea excelsa* extract and *Pinus sylvestris* leaf extract.

Carcinogens, mutagens, and substances toxic to reproduction (CMR substances) should be prohibited in cosmetic products according to Preamble (32) of the Cosmetics Regulation 1223/2009 [16], but there are exceptions for these substances where their 'use has been found safe by the SCCS' (Scientific Committee for Consumer Safety) [17]. Natural substances are classified 53 times with H341, H350, H351, H360, and H361 (H phrases indicating CMR substances), several of them with more than 500 notifiers.

Six natural substances are classified with H371, among them is *Lawsonia inermis* extract (the extract of the henna plant), classified also with H317. The active ingredient of the hair dye in the henna plant is 2-hydroxy-1,4-naphthoquinone, also called lawsone, CAS 83-72-7 EC 201-496-3 which is classified by the majority of notifiers (23) with H315, H319, and H335.

Many medicinal plants contain relevant physiological active ingredients. Some natural substances used for personal care products are produced from medicinal plants and are classified, e.g. *Chamomilla recutita* extract (H304, H315, H317, H412), *Chenopodium ambrosioides* extract (H301, H304, H311, H315, H317, H411), and *Valeriana officinalis* extract (H304, H317, H319, H400, H410), whereas others are not classified, e.g. *Arnica*

Table 4 Classifications of natural substances from the INCI list [8] according to the C&L inventory [13]

	(1)	(2)	(3)
H225	Highly flammable liquid and vapor	39	*Acacia catechu* (747), *Pyrus malus* extract (747)
H226	Flammable liquid and vapor	223	*Allium sativum* extract (747), *Pinus pinaster* extract (501)
H300	Fatal if swallowed	0	-
H301	Toxic if swallowed	9	*Allium sativum* extract (747), *Artemisia absinthium* extract (500), *Tanacetum vulgare* extract (128)
H302	Harmful if swallowed	36	*Prunus amygdalus amara* extract (847), *Prunus amygdalus amara* oil (22), *Sassafras officinale* extract (27), *Viola odorata* extract (746)
H304	May be fatal if swallowed and enters airways	130	*Pelargonium graveolens* oil (814), *Picea excelsa* extract (500), *Pinus sylvestris* leaf extract (502)
H310	Fatal in contact with skin	1	*Nicotiana tabacum* extract (27)
H311	Toxic in contact with skin	3	*Brassica nigra* extract (28), *Chenopodium ambrosioides* extract (69), *Piper methysticum* extract (7)
H312	Harmful in contact with skin	7	*Cinnamomum cassia* extract (834), *Cinnamomum zeylandicum* extract (247), *Prunus amygdalus amara* oil (23)
H314	Causes severe skin burns and eye damage (skin corrosive)	4	*Origanum majorana* extract (747), *Satureia hortensis* extract (748), *Thymus serpillum* extract (129), *Thymus vulgaris* extract (750)
H315	Causes skin irritation	137	*Chamomilla recutita* extract (772), *Citrus aurantium dulcis* oil (109)
H317	May cause an allergic skin reaction	198	*Citrus medica limonum* oil (127), *Valeriana officinalis* extract (502)
H318	Causes serious eye damage	17	*Melissa officinalis* extract (128), *Rosa centifolia* oil (89)
H319	Causes serious eye irritation	58	*Rosa centifolia* water (69), *Valeriana officinalis* extract (502)
H330	Fatal if inhaled	2	*Brassica nigra* extract (28), *Tropaeolum majus* extract (13)
H331	Toxic if inhaled	1	*Piper methysticum* extract (7)
H332	Harmful if inhaled	4	*Artemisia vulgaris* extract (128), *Piper methysticum* extract (7), *Polianthes tuberosa* extract (748), *Prunus amygdalus amara* oil (23)
H334	May cause allergy or asthma symptoms or breathing difficulties if inhaled	3	pollen extract (1), *Tabebuia impetiginosa* bark extract (1), *Tabebuia impetiginosa* leaf extract (1)
H335	May cause respiratory irritation	8	*Carum petroselinum* seed oil (23), *Quillaia saponaria* extract (783)
H336	May cause drowsiness and dizziness	0	-
H340	May cause genetic defects	0	-
H341	Suspected of causing genetic defects	24	*Artemisia dracunculus* extract (748), *Illicium verum* oil (748)
H350	May cause cancer	4	*Cinnamomum ceylandicum* extract (500), *Myristica fragrans* extract (500), *Myristica fragrans* oil (61), *Sassafras officinale* extract (27)

Table 4 Classifications of natural substances from the INCI list [8] according to the C&L inventory [13] *(Continued)*

H351	Suspected of causing cancer	23	*Artemisia dracunculus* extract (748), *Crocus sativus* extract (58), *Cymbopogon nardus* oil (747), *Foeniculum vulgare* extract (500), *Illicium verum* oil (747), *Laurus nobilis* extract (747), *Levisticum officinale* extract (749), *Ocimum basilicum* extract (500), *Pimpinella anisum* extract (747), *Rosa damascena* distillate (748)
H360	May damage fertility or the unborn child	1	*Urtica urens* extract (747)
H361	Suspected of damaging fertility or the unborn child	1	*Melia azadirachta* seed oil (1)
H362	May cause harm to breast-fed children	0	-
H370	Causes damage to organs	0	-
H371	May cause damage to organs	6	*Artemisia vulgaris* extract (128), *Foeniculum vulgare* oil (65), *Laurus nobilis* oil (64), *Lawsonia inermis* extract (1), *Ocimum basilicum* oil (84), *Rosmarinus officinalis* oil (89)
H373	May cause damage to organs through prolonged or repeated exposure	5	*Artemisia absinthium* extract (500), *Lavandula hybrida* extract (748), *Rosmarinus officinalis* extract (748), *Salvia officinalis* extract (500), *Salvia officinalis* water (500)
H400	Very toxic to aquatic life	74	*Apium graveolens* extract (501), *Valeriana officinalis* extract (502)
H410	Very toxic to aquatic life with long lasting effects	77	*Carum carvi* extract (504), *Citrus grandis* peel extract (575)
H411	Toxic to aquatic life with long lasting effects	63	*Coriandrum sativum* extract (766), *Daucus carota* extract (747)
H412	Harmful to aquatic life with long lasting effects	41	*Humulus lupulus* extract (747), *Illicium verum* oil (747)
H413	May cause long lasting harmful effects to aquatic life	5	*Origanum vulgare* extract (67), *Thymus vulgaris* oil (66), *Tropaeolum majus* extract (13)

Column (1) are the substance names in the INCI list.
Column (2) are the CAS and EC numbers in the INCI list or in the REACH registrations and the production tonnage bands as given in the REACH registrations.
Column (3) are the H-phrases given by the clear majority in the C&L inventory with the number of notifiers in brackets and the H-phrases as given in the REACH registration dossier. The number of notifiers in the C&L inventory which classify a substances in the same way as in the REACH dossier is also listed.

montana extract, *Calendula officinalis*, or *Ricinus communis* oil.

H phrases due to hazardous properties for the environment are attributed 185 times to natural substances, with 184 for long lasting effects (H410, H411, H412, and H413), i.e. that these substances are not readily degradable.

REACH

The following paragraphs in the REACH regulation [1] concern natural substances used in personal care products:

In the introduction, it says (35) 'The Member States, the Agency and all interested parties should take full account of the results of the RIPs, (REACH Implementation Projects (added by the author)) in particular with regard to the registration of substances which occur in nature (Table 5).

There are exemptions from the obligation to register in accordance with Art. 2(7b) e.g. 'Substances occurring in nature other than those listed under paragraph 7 (which are inorganic substances or petrochemicals (added by the author)), if they are not chemically modified, unless they meet the criteria for classification as dangerous according to Directive 67/548/EEC.' The natural substances in the INCI list that are classified as hazardous substances should therefore be registered according to REACH. 19 natural substances in the INCI list are registered according to REACH with annual production tonnage bands up to 100,000 to 1,000,000 t/a (Table 5). As substances used as medicinal products, food and feeding stuff do not need to be registered, evaluated, and authorized according to REACH Art. 2(5), production volumes indicated in Table 5 cover the uses in other applications, e.g. in personal care products. The

Table 5 Natural substances in the INCI list that are registered according to REACH in descending order of the production volumes

(1)	(2)	(3)
Tall oil	8002-26-4/232-304-6 100,000 to 1,000,000 t/a	H412 (352) (not classified in the REACH dossier and also not classified by 200 notifiers)
Citrus aurantium dulcis extract, *Citrus aurantium dulcis* flower extract, *Citrus aurantium dulcis* flower oil, *Citrus aurantium dulcis* flower water, *Citrus aurantium dulcis* peel cera	8028-48-6/232-433-8 > 10,000 t/a	H225, H304, H315, H317, H400, H410 (500) H226, H304, H315, H317, H411 (REACH dossier and 62 notifiers)
Eucalyptus globulus extract	84625-32-1/283-406-2 1.000 to 10.000 t/a	H226, H304, H317, H411 (905) H226, H304, H315, H317, H411 (REACH dossier and 97 notifiers)
Citrus medica limonum extract, *Citrus medica limonum* juice, *Citrus medica limonum* juice extract, *Citrus medica limonum* juice powder, *Citrus medica limonum* peel extract	84929-31-7/284-515-8 100 to 1,000 t/a	H226, H304, H315, H317, H400, H410 (507) H226, H304, H315, H317, H410 (REACH dossier and 96 notifiers)
Mentha arvensis extract	90063-97-1/290-058-5 100 to 1,000 t/a	H315, H317, H411 (891) H302, H315, H317, H319 H401, H411 (REACH dossier and 6 notifiers)
Mentha piperita extract, *Mentha piperita* water	84082-70-2/282-015-4 100 to 1,000 t/a	H315, H412 (785) H315, H317, H319, H402, H412 (REACH dossier and 4 notifiers)
Pogostemon cablin extract (Patchouli ext.)	84238-39-1/282-493-4 (INCI) 939-227-3 (REACH registration) 100 to 1,000 t/a	H304 (792) H304, H411 (REACH dossier and 1 notifier)
Myristica fragrans extract	84082-68-8/282-013-3 10 to 100 t/a	H226, H304, H315, H317, H341, H350 (500) H226, H304, H317, H341, H350, H410 (REACH dossier and 31 notifiers)
Myristica fragrans oil	8008-45-5 10 to 100 t/a	H226, H304, H317, H350, H400, H410 (61) H226, H304, H317, H341, H350, H400, H410 (only in the REACH dossier and by no notifier)
Eugenia caryophyllus extract	84961-50-2/284-638-7 (INCI) 904-912-8 (REACH registration) intermediate only	H304, H315, H317, H319 (749) (not classified in the REACH dossier, whereas all notifiers classify the substance)

Column (1) shows the substance names in the INCI list.
Column (2) shows the CAS and EC numbers in the INCI list or in the REACH registrations and the production tonnage bands as given in the REACH registrations.
Column (3) shows the H phrases given by the clear majority in the C&L inventory with the number of notifiers in brackets and the H phrases as given in the REACH registration dossier. The number of notifiers in the C&L inventory which classify substances in the same way as in the REACH dossier is also listed.

other 350 natural substances which are classified and labeled according to the C&L inventory have not been registered so far. Some of them might be registered in 2018, the registration deadline for production volumes of 1 to 100 t/a. Furthermore, it must be assumed that some of the 703 natural substances that do not turn up in the C&L inventory so far also will have to be registered according to REACH in future. Classifications in the REACH registration are not in line with the classifications by the majority of notifiers in the C&L inventory as shown in Table 5.

If substances registered according to REACH are classified differently by various notifiers, registrants should be participants in a substance information exchange forum (SIEF) with the aim to agree classification and labeling ([1] Art. 29(2)). As some natural substances are CMR substances, as shown above, they might be candidates for REACH Annex XIV (list of substances subject to authorization) ([1] Art. 57). As data on cosmetic ingredients need not be transmitted in the supply chain

([1] Art. 2(6b)), it is not foreseen that the final formulator receives the information about classification and labeling.

None of the natural substances considered in this study is in the list of substances of very high concern (SVHC).

According to the REACH regulation, the chemical safety report does not need to consider the risks to human health from the use of cosmetic products ([1] Chapter 1 Art. 14(5b)). This is valid for synthetic ingredients as well as for natural substances.

Cosmetic regulation

There are several restrictions for natural substances used in personal care products.

Annex II of the Cosmetic Regulation [16] is the 'List of substances prohibited in cosmetic products'. It lists 35 natural substances, most of which are prohibited because they are highly toxic, such as *Atropa belladonna* L. (deadly nightshade) and its preparations, Cantharides, *Cantharis vesicatoria* (Spanish fly (beetle)), *Chenopodium ambrosioides* L. (wormseed) (essential oil), *Claviceps purpurea* Tul. (ergot fungus), and its alkaloids and galenical preparations, *Conium maculatum* L. (poison hemlock) (fruit, powder, galenical preparations), *Prunus laurocerasus* L. (cherry laurel water), *Strychnos* species (e.g. *Strychnos nux-vomica*), and their galenical preparations, *Veratrum* spp. (false hellebore) and their preparations, and *Pyrethrum album* L. and its galenical preparations.

These natural substances are not listed in the INCI list but related substances produced by the same plants: For example, *Chenopodium ambrosioides* L. (essential oil) is prohibited, whereas *Chenopodium ambrosioides* L. (extract) is classified by 69 notifiers, and the INCI list indicates the function of 'skin conditioning'.

The oil from the seed of *Laurus nobilis L.* is prohibited in Annex II. However, *Laurus nobilis* extract of the leaves has the same CAS and EINECS numbers and was classified by 747 notifiers in the C&L inventory. (The oil from the fruits in the INCI list has a different CAS number and was classified by 64 notifiers in the C&L inventory).

Fig leaf absolute (*Ficus carica* L.) is prohibited when used as a fragrance ingredient, *Ficus carica* extract is classified by 74 notifiers in the C&L inventory, and the INCI list indicates the function 'skin conditioning'.

Verbena oil (*Lippia citriodora* Kunth.) is prohibited when used as a fragrance ingredient, *Lippia citriodora* Kunth. extract is classified by 128 notifiers in the C&L inventory, and the INCI list indicates the function 'tonic'.

Further restrictions exist for the origin of bovine products (e.g. adeps bovi, brain extract, marrow extract, or neural extract).

Annex III of the Cosmetic Regulation contains the 'List of substances which cosmetic products must not contain except subject to the restrictions laid down'. Oakmoss extract (*Evernia prunastri* extract) as a strong contact allergen is part of this list and must be named in the list of ingredients referred to in Art. 19(1)(g) when its concentration exceeds 0.001% in leave-on products and 0.01% in rinse-off products. Despite this restriction, there were 747 notifiers for oakmoss extract in the C&L inventory. According to the INCI list, oakmoss extract is used with the function of skin conditioning. Oakmoss is also listed in the INCI list part II (perfume and aromatic raw materials), together with treemoss extract (*Evernia furfuracea* lichen extract).

It is not evident to the author why the reference in the restriction of the cosmetic colorant *Beta vulgaris* (which has a strong red color) in the INCI list (IV 1) leads to a substance of green color.

Red list of endangered species

A dozen natural substances from the INCI list originate from plants that are on the red list of endangered species in Germany (class 2 or 3) [18], e.g. *Drosera rotundifolia* (sundew), *Orchis morio* (green-veined orchid), *Gentiana lutea* (great yellow gentian), *Arctostaphylos uva-ursi* (bearberry), or *Arnica montana* (mountain arnica). There were no notifiers of *Drosera rotundifolia* (sundew) and *Orchis morio*. Also, no notifier is entered for *Arnica montana*, although this plant is used frequently for pharmaceutical products. Some plants which are endangered in Germany and used for the production of natural substances may be cultivated (e.g. *Gentiana lutea*, *Arctostaphyllos uva-ursi*, and since recently also *Arnica montana*) or they may be collected in other countries (e.g. *Arnica montana*). *Gentiana lutea*, and *Arctostaphyllos uva-ursi* are notified by 747, respectively 25 notifiers. There are additional rare and protected plants, which are used for the production of personal care products, but which are itemized in the INCI list and not in the German red list. For example, alpine valerian (*Valeriana celtica*, German name: Speick) is used by the German Speick company for the production of soap. This company was granted an exception permit for the collection of *Valeriana celtica* in the Austrian Alps.

Conclusions

The interdisciplinary analysis and compilation of the various regulatory instruments for natural substances used in personal care products reveals the heterogeneity of their origins, the relevance of their hazardous properties, and the importance to clarify relevant regulatory inconsistencies.

The following conclusions could be drawn, with each conclusion leading to further questions.

Classifications

The classifications in the C&L inventory must be assumed to be correct. However, some classifications are difficult to comprehend as they are not in line with expectations from experience, and some basic properties of the classification process make the interpretation difficult.

- In many cases, some notifiers classified the substance, whereas others did not. Could the differences be due to differences of the biological material used? Or could they be due to pesticidal or other contaminations, impurities, solvent residues, or added stabilizers or preservatives? Or could they be due to oxidation or other reaction processes of the ingredients?
- At several occasions, natural substances originating from plants with known toxic or physiologically active ingredients are not classified (e.g *Achillea millefolium, Arachis hypogaea, Arnica montana, Ginkgo biloba, Glycine max, Helianthus annuus, Inula helenium, Ricinus communis*). Did the

preparation procedures destroy, separate or dilute the active ingredients sufficiently?

- The pharmaceutical properties, the composition, and the toxicity of natural substances have been studied for centuries and vast knowledge on natural compounds is gathered [2-4]. Have these indicators been always respected in the classification and labelling process as they should be [9-11]?
- Many natural substances do not (yet) appear in the C&L inventory. How many of them would need to be classified as hazardous chemicals?
- Many natural substances are classified due to their long lasting effects on the environment. Which components are responsible for these effects?
- Some natural substances derived from basic food plants are classified. Which chemical components are responsible for the respective classifications? For example, what makes *Allium sativum* extract (garlic extract) H301 'toxic if swallowed'? Which components are responsible that *Daucus carota* juice is classified with H226, H304, H317, H319, and H411?
- The classifications of the natural substances are self-classifications. How can a high quality standard be ensured?
- The C&L inventory is updated regularly, but the updated notifications are not flagged. This means working with an ever changing data set, which is especially annoying for a data compilation like in the present study that was effectuated over several months.
- Some identifications of natural substances produced by natural organisms are apparently inconsequential. Could the intense discussion about substance identity in the REACH process lead to improvement for the definition of natural substances in the INCI list?

REACH

Classification and labeling is not only an interesting information for manufacturers and consumers, but it is also a trigger for the registration requirements according to REACH.

- Are there more than the 19 natural substances (Table 5) in the INCI list that should be classified as hazardous substances and registered according to REACH?
- According to the REACH regulation, the chemical safety report does not need to consider the risks to human health from the use of cosmetic products. What would be the result if it did consider these risks?
- In some cases, effects on the environment are described in detail e.g. effects of the hair dye henna on the development of zebra fish [19]. Were such

studies considered in the classification procedure? What are the results of risk assessments of natural substances which are used in large amounts?

The cosmetic regulation does not ask for a different safety evaluation of natural compared to synthetic ingredients. Is it still justified that natural products receive a special treatment in the frame of REACH?

INCI list

The INCI list is a very valuable list, but several aspects need to be corrected in an update to come.

- Many natural substances, especially animal products, have no notifier. I assume that consumers consider several of the animal products as rather 'nauseating', e.g. testicular extract or sus extract, and would not want them on their body surface. Are they still relevant? If so, do they correspond to today's hygiene requirements?
- For several plants, the old taxonomic names are listed in the INCI list. Are there no notifiers for some of them because of the new taxonomic names?

Consumer protection

Toxic effects of natural substances used in personal care products have been known for a long time and have led to many regulations to protect consumers and the environment. However, there are some relevant inconsistencies which make it rather difficult to believe that products containing natural substances are always safe.

- Many natural substances used in personal care products are classified as hazardous substances. The vast majority of health risk phrases concerned the hazardous effects on the skin. Why are many natural substances catalogued in the INCI list with the function of 'skin conditioning' if they are classified as substances with hazardous effects for the skin?
- Many natural substances in the INCI list are classified with H-phrases indicating CMR substances. Where and in what amounts are these substances used?
- Personal or environmental risks posed by a substance in personal care products depend on exposure to a certain substance. Are the exposures of concern? What are the risks for the environment and for consumers?
- Many natural substances used in personal care products are classified as hazardous substances. How many consumers know the classifications of natural substances, especially as long as the products need not be classified [20]? How could consumers

be informed about the hazardous properties and the relevant exposure of natural substances in personal care products [21]?

- Many natural organisms are endangered or live in declining habitats. How could sustainable production of natural substances be guaranteed?

Nature offers an incredible diversity of chemical compounds with a wide array of physiological effects. Natural substances are not only used in personal care products, but some are used in food, as spices, or in medicine (Table 3) [2-4,14]. The borderline between these uses is not sharp. Many herbal substances used as phytopharmaceuticals have hazardous properties [4], and many ingredients in natural flavorings have toxic properties (like beta-asarone, coumarin, hydrogen cyanide, methyl eugenol, capsaicin, eucalyptol, hypericin, thujone, aloin/aloe-emodin, berberine, and pyrrolizidine alkaloids [2]). Therefore, it is not surprising that many natural substances used in personal care products are naturally toxic, too, and demand a prudent handling and sensible regulations.

Methods

As transparency is considered to be very important in the European chemical policy, the ambition of this study was to use publically available information only.

Substances were selected from the 'Inventory and a common nomenclature of ingredients employed in cosmetic products' [8,22] which originate clearly from a specific organism indicated by its scientific name. In most of these cases, the two partite scientific Latin name of the organism, the preparation (e.g. *Ricinus communis* oil), and a short description of the substance is given in the INCI list (e.g. *Ricinus communis* oil is the fixed oil obtained from the seeds of *Ricinus communis*, Euphorbiaceae. It consists primarily of the glycerides of the fatty acid ricinoleic) as well as the CAS and the EC number of the substance and the function in cosmetic products (for *Ricinus communis* oil: emollient/skin conditioning/moisturizing/smoothing/solvent). The natural substance can be 'total organism', 'tissue', 'extract', 'oil', 'water', 'juice', 'ferment', 'cera', 'powder', 'gum', 'root', 'butter', 'fruit', 'bran', or 'flour'. Fermented substances were considered, in case the respective fermenting microorganism was named. Entries were considered in case the species of origin was clear by the familiar name (e.g. cera alba is wax of the honey bee *Apis mellifera*).

The following entries were not considered further in the present study:

- Other natural substances that are applied in personal care products, but do not appear in the INCI list.

- Substances which underwent subsequent chemical reactions (e.g. hydrolyzation or lysis) (such as 'cocoyl hydrolyzed collagen, acid chlorides, coco, reaction products with protein hydrolyzates')
- Specific chemicals with a definite chemical structure (e.g. coco betaine)
- Substances with a common chemical structural property (such as cocoglycerides)
- Inorganic substances and substances on the basis of mineral oil
- Fragrances which are listed in the INCI list Section II 'Perfume and aromatic raw materials' [15]. This list contains approximately 2,750 substances; more than a quarter of them are natural substances according to the criteria of this study.

Classification and labeling of hazardous substances according to the European Regulation on Classification, Labelling and Packaging [9] can be found in the publically accessible C&L inventory database according to REACH Title XI [13]. Classification and labeling is an ongoing process, leading to a continuous update of the data. Data used here were collected between August 2013 and June 2014. CAS numbers or EC numbers were used preferentially in the C&L inventory to find the respective hazard classifications. Ingredients with neither a CAS nor an EC number in the INCI list were searched with their INCI names in the C&L inventory. For reasons of clarity and comprehensibility, only the H phrases are presented here, without the GHS symbols.

Competing interests

The author declares that she has no competing interests.

Acknowledgement

I thank Sabrina Hartman for collecting the data in the INCI list and the C&L inventory.

This article belongs to a series of contributions submitted from members of the 'Division of Environmental Chemistry and Ecotoxicology' of the 'German Chemical Society (GDCh)'.

References

1. Regulation (EC) No 1907/2006 concerning the Registration, Evaluation, Authorisation and Restriction of Chemicals (REACH). http://echa.europa.eu/de/regulations/reach/legislation.
2. Teuscher E, Lindequist U. Biogene Gifte (in German). Stuttgart: WBG; 2010.
3. Ramawat KG, Mérillon J-M. Natural products. Phytochemistry, Botany and Metabolism of Alkaloids, Phenolics and Terpenes. Berlin Heidelberg: Springer; 2013.
4. Wichtl M: Teedrogen und Phytopharmaka. Ein Handbuch für die Praxis auf wissenschaftlicher Grundlage (in German). Stuttgart: WBG; 2008.
5. Buchbauer G. Über biologische Wirkungen von Duftstoffen und ätherischen Ölen. Med Wochenschr. 2004;154(21–22):539–47.
6. Lahlou M. Essential oils and fragrance compounds: bioactivity and mechanisms of action. Flavour Fragrance J. 2004;19:159–65.
7. Polya G. Biochemical Targets of Plant Bioactive compounds. A Pharmacological Reference Guide to Sites of Action and Biological Effects. London, New York: Taylor and Francis; 2003.

8. Commission Decision: 96/335/EC of 8 May 1996 establishing an inventory and a common nomenclature of ingredients employed in cosmetic products. *Off J L* 1996, 1–684: 132, 01/06/1996. [http://eur-lex.europa.eu/LexUriServ/LexUriServ.do?uri=OJ:L:2006:097:0001:0528:EN:PDF]

9. Regulation No 1272/2008 of the European Parliament and of the Council of 16 December 2008 on classification, labelling and packaging of substances and mixtures, amending and repealing Directive 67/548/EEC and 1999/45/EC and amending Regulation (EC) No 1907/2006. [http://eur-lex.europa.eu/LexUriServ/LexUriServ.do?uri=OJ:L:2008:353:0001:1355:EN:PDF]

10. Marcello I, Giordano F, Costamagna FM. Information gathering for CLP classification. Ann Ist Super Sanita. 2011;47(2):132–9.

11. Schöning G. Classification & Labelling Inventory: role of ECHA and notification requirements. Ann Ist Super Sanita. 2011;47(2):140–5.

12. Chemical Inspection and Regulation service CIRS [http://www.cirs-reach.com/Cosmetic_Inventory/International_Nomenclature_of_Cosmetic_Ingredients_INCI.html]

13. C&L Inventory. [http://echa.europa.eu/web/guest/information-on-chemicals/cl-inventory-database]

14. Newman DJ, Gragg GM. Natural products as sources of new drugs over the 30 years from 1981 to 2010. J Nat Prod. 2012;75(3):311–35.

15. SCCNFP: Opinion Concerning the 1st Update of the Inventory of Ingredients Employed in Cosmetic Products. Section II: Perfume and Aromatic Materials. 2000. SCCNFP/0389/00/Final http://www.leffingwell.com/cosmetics/out131_en.pdf.

16. Regulation 1223/2009 of the European Parliament and of the Council on Cosmetic Products. [http://eur-lex.europa.eu/LexUriServ/LexUriServ.do?uri=CELEX:32009R1223:EN:NOT]

17. OPINION on Fragrance Allergens in Cosmetic Products SCCS/1459/11 [http://ec.europa.eu/health/scientific_committees/consumer_safety/docs/sccs_o_102.pdf]

18. Ludwig G, Schnittler M. Rote Liste der Pflanzen Deutschlands (in German). 1996. [http://www.bfn.de/fileadmin/MDB/documents/RoteListePflanzen.pdf.

19. Manjunatha B, Wei-bing P, Ke-chun L, Marigoudar SR, Xi-qiang C, Xi-min W, et al. The effects of henna (hair dye) on the embryonic development of zebrafish (*Danio rerio*). Env Sci & Pollut Res. 2014;21(17):10361–7.

20. Klaschka U. Dangerous cosmetics - criteria for classification, labelling and packaging (EC 1272/2008) applied to personal care products. Env Sci Europe. 2012;24:37.

21. Klaschka U, Rother H-A. "Read this and be safe!" Comparison of regulatory processes for communicating risks of personal care products to European and South African consumers Env Sci Europe. 2013;25:30.

22. Cosing http://ec.europa.eu/consumers/cosmetics/cosing/.

19

The potential benefits and limitations of different test procedures to determine the effects of Ag nanomaterials and AgNO₃ on microbial nitrogen transformation in soil

Kerstin Hund-Rinke[*] and Karsten Schlich

Abstract

Background: The procedure described in The Organisation for Economic Co-operation and Development (OECD) test guideline (TG) 216 is used to assess the effects of chemicals on microbial nitrogen transformation in soil, and the results are considered in regulatory risk assessments. We investigated the suitability of this method to characterize the effects of two silver nanomaterials and a soluble silver salt. We applied three different test procedures: (i) nitrogen transformation using the complex organic nitrogen source lucerne meal (OECD TG 216), (ii) nitrogen transformation using the inorganic nitrogen source $(NH_4)_2SO_4$ (following OECD TG 216), and (iii) ammonium oxidation (ISO 15685). The results were compared with substrate-induced respiration (OECD TG 217).

Results: The standard nitrogen transformation test using lucerne meal suggested that the test materials had no effect, whereas significant effects were identified with the other two test procedures. The absence of effects with lucerne meal probably reflected the sorption of silver ions to the additional organic nitrogen source, thus reducing its bioavailability, or blocking the silver nanomaterial oxidation sites by sorption of organic matter.

Conclusions: This common test used in the context of chemical registration is therefore unsuitable for the detection of potential effects caused by silver nanomaterials and soluble silver salts because it can yield false negative results. We instead recommend the use of an inorganic nitrogen source. The observed effects were not specific to nanomaterials. The time course of the effect in the nitrogen transformation test based on $(NH_4)_2SO_4$ and the potential ammonium oxidation test varied according to the test substance, indicating different kinetic behaviors of ion release.

Keywords: Silver nanomaterials; Nitrogen transformation; Ecotoxicity; Soil

Background

Microorganisms play an important role in the breakdown and transformation of organic matter in fertile soils with many species contributing to different aspects of soil fertility. Several guidelines are currently used to investigate the effects of chemicals on selected microbial activities, as outlined in chapter R7c of the ECHA guidance for the implementation of Regulation (EC) 1907/2006 - the Registration, Evaluation, Authorisation and Restriction of Chemicals (REACH) [1]. Two Organisation for Economic Co-operation and Development (OECD) test guidelines are available to detect chemical effects on soil microflora, i.e., the nitrogen and carbon transformation tests [2,3]. These are designed to determine the long-term adverse effects of substances on nitrogen or carbon transformation in aerobic soils over at least 28 days. According to the guidance listed above [1], the nitrogen transformation test is considered sufficient for most non-agrochemical substances. Further ISO-standard methodologies are available, but in the absence of corresponding OECD guidelines, these methods are used less often than the two microbial assays mentioned above. One of the ISO methods evaluates the potential activity of the

* Correspondence: kerstin.hund-rinke@ime.fraunhofer.de
Fraunhofer Institute for Molecular Biology and Applied Ecology IME, Auf dem Aberg 1, 57392 Schmallenberg, Germany

nitrifying population by measuring the accumulation of nitrite over a short incubation period of 6 h [4].

The quantity of nanosilver produced and utilized worldwide is estimated to be 55 tons per annum (median value), 5.5 tons per annum in Europe and in a range of 2.8 to 20 tons per annum in the USA [5]. Nanosilver is often applied in coatings to take advantage of its antimicrobial properties, but it is also found in paints, cleaning agents, consumer electronics, cosmetics, and textiles [6,7]. The antimicrobial mechanisms of silver nanomaterials have recently been reviewed [8] and may include the formation of pits on the cell surface, the generation of free radicals that damage the cell membrane, the release of silver ions - which affect Gram-positive and Gram-negative bacteria [7] - and the interaction between silver ions and the thiol and phosphate groups of enzymes and other organic molecules.

Several studies have already been carried out to monitor the potential adverse effects of nanosilver on microorganisms in the environment. One such study showed that nanosilver was toxic to bacteria in sewage sludge, especially nitrifying bacteria [9], but that the toxicity was strongly influenced by the ambient medium, e.g., the sulfur content [10]. Another study showed that a commercially-available spray containing nanosilver, which promotes the growth of plants according to the manufacturer, had a negative impact on microbial biomass in the soil and enhanced basal respiration [11]. The latter study did not state which test guidelines were used. However, another study in which tests were carried out in accordance with OECD and ISO guidelines showed that nanosilver applied to soil as pristine material or adsorbed to sewage sludge had a negative impact on nitrifying bacteria, microbial biomass, and soil exoenzymes [12], supporting earlier reports that nanosilver inhibits soil exoenzymes [13]. Studies have also been carried out to determine the effect of nanosilver on other soil organisms, although the observed effect concentrations were higher than those recorded for soil microbes [14-16].

Here we investigated whether the type of nitrogen source influences nitrite and nitrate accumulation caused by two silver nanomaterials differing in parameters such as size and shape, and by a soluble silver salt, over an incubation period of 28 days. We applied three different procedures addressing the nitrogen mineralization in the nitrogen cycle. Briefly, the nitrogen cycle consists of the following: (i) nitrogen mineralization with the individual steps ammonification (organic nitrogen $\rightarrow$ ammonium) and nitrification (ammonium $\rightarrow$ nitrite $\rightarrow$ nitrate), (ii) nitrogen immobilization and assimilation (nitrate $\rightarrow$ organic nitrogen; ammonium $\rightarrow$ organic nitrogen), (iii) denitrification (nitrite $\rightarrow N_2$), and (iv) nitrogen fixation ($N_2 \rightarrow$ organic nitrogen) [17]. At the beginning of the test, we added the complex insoluble organic nitrogen source

lucerne meal according to OECD test guideline (TG) 216 [2] (first approach), or the soluble inorganic nitrogen source $(NH_4)_2SO_4$ (second approach), and periodically measured the nitrate content/nitrate accumulation during the 28-day incubation period. The addition of the ammonium salt (second approach) was based on a former German test guideline for the investigation of pesticide side effects [18]. In a third approach, we incubated non-augmented soil and measured the transformation of $(NH_4)_2SO_4$ to nitrite in periodic subsamples [4]. This test has a short incubation period of 6 h and was applied 2 to 3 times during the 28-day incubation period. The three procedures all cover a 28-day exposure of the soil microflora to silver. The measured microbial activity, however, depends on the time point at which the nitrogen source is added. The test on nitrite accumulation provides information on the activity of the respective microorganisms at the time when samples are taken: the nitrogen source is added to a subsample, and microbial activity is determined. The test on nitrate accumulation monitors the microbial activity from test start up to the sampling day since the nitrogen source is added at the start of the 28-day incubation period. The effects on nitrite and nitrate accumulation were compared with the results of the carbon transformation test (soil-induced respiration) [3], in which short-term aerobic respiration after the addition of an easily degradable soluble carbon source (glucose) is indicated by oxygen depletion. Short-term respiration was determined 2 to 3 times during the 28-day incubation period. Analogous to the short-term test on nitrite accumulation, the short-term respiration test provides information on the microbial activity at the sampling time. The selected approaches allowed us to determine the suitability of the current procedure for the assessment for silver nanomaterials in the context of chemical regulation.

Results

Most experiments were repeated. The results were comparable. In the following, the results of one test for each experimental design are presented in detail in Tables 1 and 2 as well as in the Figures. The EC_{50} values of the second experiment are presented in Table 1 and proof the statements of the first experiments.

Nitrogen transformation test

The N-transformation test according to OECD TG 216 monitors the formation of nitrate from a nitrogen source (in our investigation: powdered lucerne or NH_4SO_4). The test was carried out in the presence of three test substances (Figures 1 and 2). An increase in the soil nitrate concentration due to microbial activity was observed. The nitrate concentration increased in the untreated control regardless of the nitrogen source, but there were clear

Table 1 EC$_{10}$ and EC$_{50}$ values for silver nanomaterials and AgNO$_3$ on day 28

Test parameters	Experiment 1		Experiment 2
	EC$_{10}$^{a} [mg/kg]	EC$_{50}$^{a} [mg/kg]	EC$_{50}$^{a} [mg/kg]
NM-300K^b			
Nitrate accumulation, lucerne	---	>15	>15
Nitrate accumulation, (NH$_4$)$_2$SO$_4$	0.9 [0.2 to 1.6]	7.3 [5.1 to 11.5]	6.5 [0.3 to 63.5]
Potential ammonium oxidation	0.5 [0.0 to 1.0]	1.6 [0.5 to 4.8]	3.7 [2.1 to 8.3]
Substrate-induced respiration	1.0 [n.d.]	5.0 [n.d.]	13 [n.d.]
NM-302^c			
Nitrate accumulation, lucerne	---	>100	>100
Nitrate accumulation, (NH$_4$)$_2$SO$_4$	5.6 [4.7 to 6.6]	44.6 [41.2 to 48.5]	Not determined
Potential ammonium oxidation	0.5 [n.d.]	6.0 [n.d.]	17.1 [14.3 to 20.6]
Substrate-induced respiration	2.1 [n.d.]	67.5 [n.d.]	74.7 [42.8 to 174.8]
AgNO$_3$^{b}			
Nitrate accumulation, lucerne	---	>15	Not determined
Nitrate accumulation, (NH$_4$)$_2$SO$_4$	0.7 [0.50 to 0.8]	2.8 [2.6 to 3.1]	Not determined
Potential ammonium oxidation	1.4 [1.2 to 1.6]	2.9 [2.7 to 3.2]	2.4 [n.d.]
Substrate-induced respiration	---	>15	>15

^{a}Confidence interval shown in square brackets; ^{b}five test concentrations; spacing factor of 3; ^{c}screening approach: three test concentrations; spacing factor of 10; n.d., not determinable.

differences in the observed effects of the test substances depending on which nitrogen source was applied.

At the end of the test, there was no statistically significant deviation in the nitrate concentration in the treated soil samples supplemented with the complex organic nitrogen source lucerne meal, although a short-term concentration-dependent effect was detected on day 7 in soils treated with either AgNO$_3$ or NM-300 K (Figure 1). In contrast, applying the inorganic nitrogen source (NH$_4$)$_2$SO$_4$ resulted in a concentration-dependent effect relationship throughout the 28-day incubation period (Figure 2). The highest test concentrations of NM-300 K (15 mg/kg) and NM-302 (100 mg/kg) caused a 65% and 70% decrease of the nitrate concentration by the

end of the test, and the highest concentration of AgNO$_3$ (15 mg/kg) caused a 100% inhibition. Lower test concentrations resulted in 0% to 71% decrease of the nitrate concentration compared to the control.

Potential ammonium oxidation test
The potential ammonium oxidation test according to ISO 15685 monitors the transformation of ammonium to nitrite. Concentration-dependent effects were also observed in the potential ammonium oxidation activity test (Figure 3). For technical reasons described in the 'Methods' section, the initial effect determination measurement was taken on day 1 rather than day 0 in this test, at which point it was already clear that both NM-300 K and AgNO$_3$ had a

Table 2 Experiment 1; EC$_{50}$ for nitrate accumulation and potential ammonium oxidation activity over the incubation period

Test substance	Sampling time	N-transformation with lucerne^a EC$_{50}$ [mg/kg]	N-transformation with (NH$_4$)$_2$SO$_4$^{a} EC$_{50}$ [mg/kg]	Potential ammonium oxidation^a EC$_{50}$ [mg/kg]	C-transformation EC$_{50}$ [mg/kg]
NM-300K^b	Day 7	10.1 [4.1 to 1,048]	22.4 [11.1 to 402.3]	2.3 [n.d.]	2.0 [n.d.]
	Day 21	>15	7.1 [4.9 to 11.4]	Not determined	Not determined
	Day 28	>15	7.3 [5.1 to 11.5]	1.6 [0.5 to 4.8]	5.0 [n.d.]
NM-302^c	Day 7	>100	190 [165 to 226]	>100	>100
	Day 21	>100	60.7 [51.6 to 71.6]	19.4 [17.4 to 21.8]	24.7 [n.d.]
	Day 28	>100	44.6 [41.2 to 48.5]	6.0 [n.d.]	67.5 [n.b.]
AgNO$_3$^{b}	Day 7	2.5 [1.1 to 5.8]	2.7 [n.d.]	3.8 [n.d.]	Approximately 15^d
	Day 21	33.9 [n.d.]	2.5 [1.5 to 2.6]	Not determined	Not determined
	Day 28	>15	2.8 [2.6 to 3.1]	2.9 [2.7 to 3.2]	>15

^{a}Confidence interval shown in square brackets; ^{b}five test concentrations; spacing factor of 3; ^{c}screening approach: three test concentrations; spacing factor of 10; ^{d}50% inhibition at highest test concentration (15 mg/kg). n.d., not determinable.

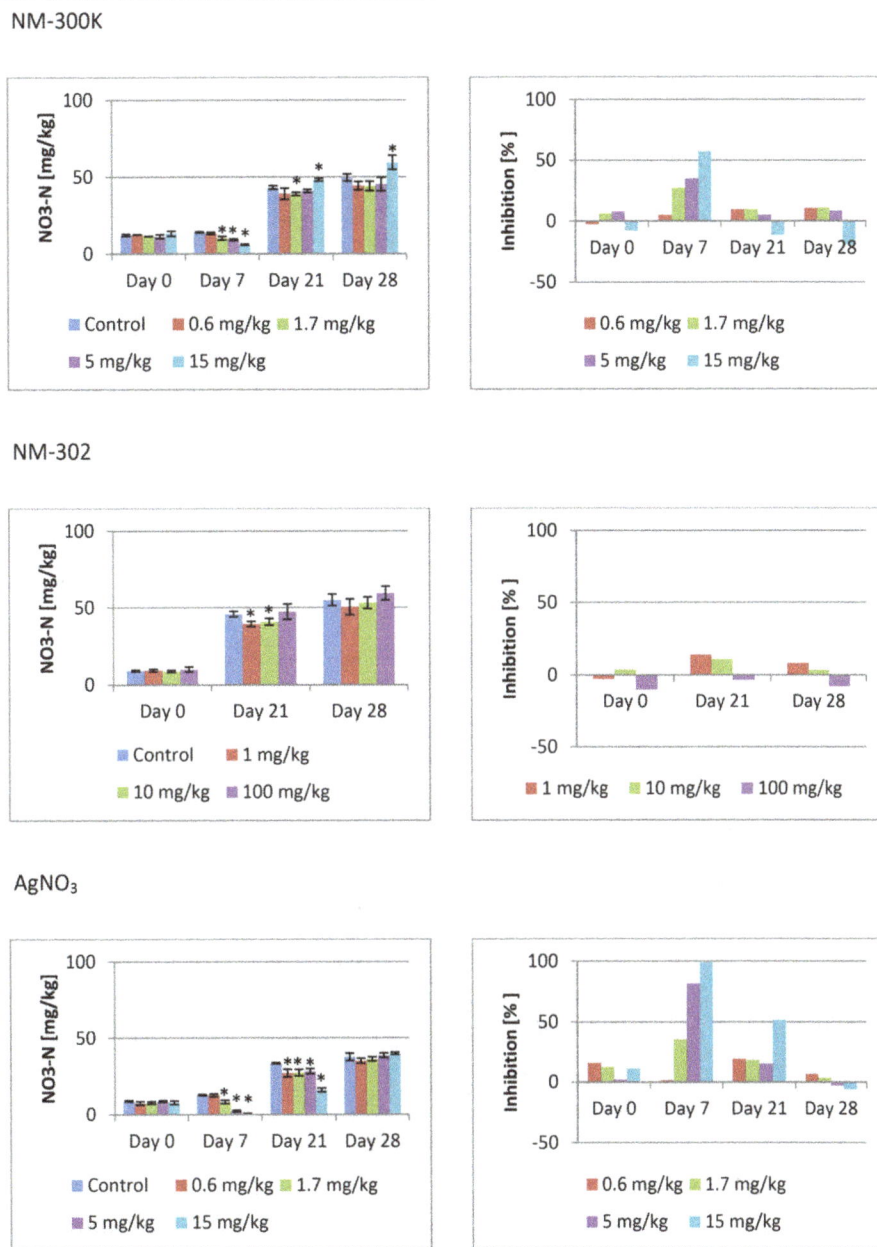

Figure 1 Effect of silver nanomaterials and AgNO₃ on nitrate concentration with powdered lucerne as nitrogen source [2]. The results for AgNO₃ were corrected by the nominal nitrate concentration added with the test substance. Left figure column: mean activity and standard deviation. Right figure column: percentual deviation. Asterisks indicate a statistically significant difference to controls ($p \leq 0.05$).

significant inhibitory effect. At the end of the test, the highest test concentrations of all three substances achieved nearly 100% inhibition, whereas lower test concentrations resulted in 0% to 94% inhibition.

Carbon transformation test

The test describes changes in size and activity of microbial communities responsible for carbon transformation. Carbon transformation is measured via glucose-induced respiration activity. Inhibition was also detected in the carbon transformation test, although we observed differences in the inhibition profile between the nanomaterials and AgNO₃ (Figure 4). Neither of the nanomaterials showed any effects at the start of the test, but a concentration-dependent effect was observed at the next measuring point followed by recovery by the end of the test. AgNO₃ had a significant inhibitory effect immediately after application (sampling time 3 h) and also at the 7-day

NM-300K

NM-302

AgNO$_3$

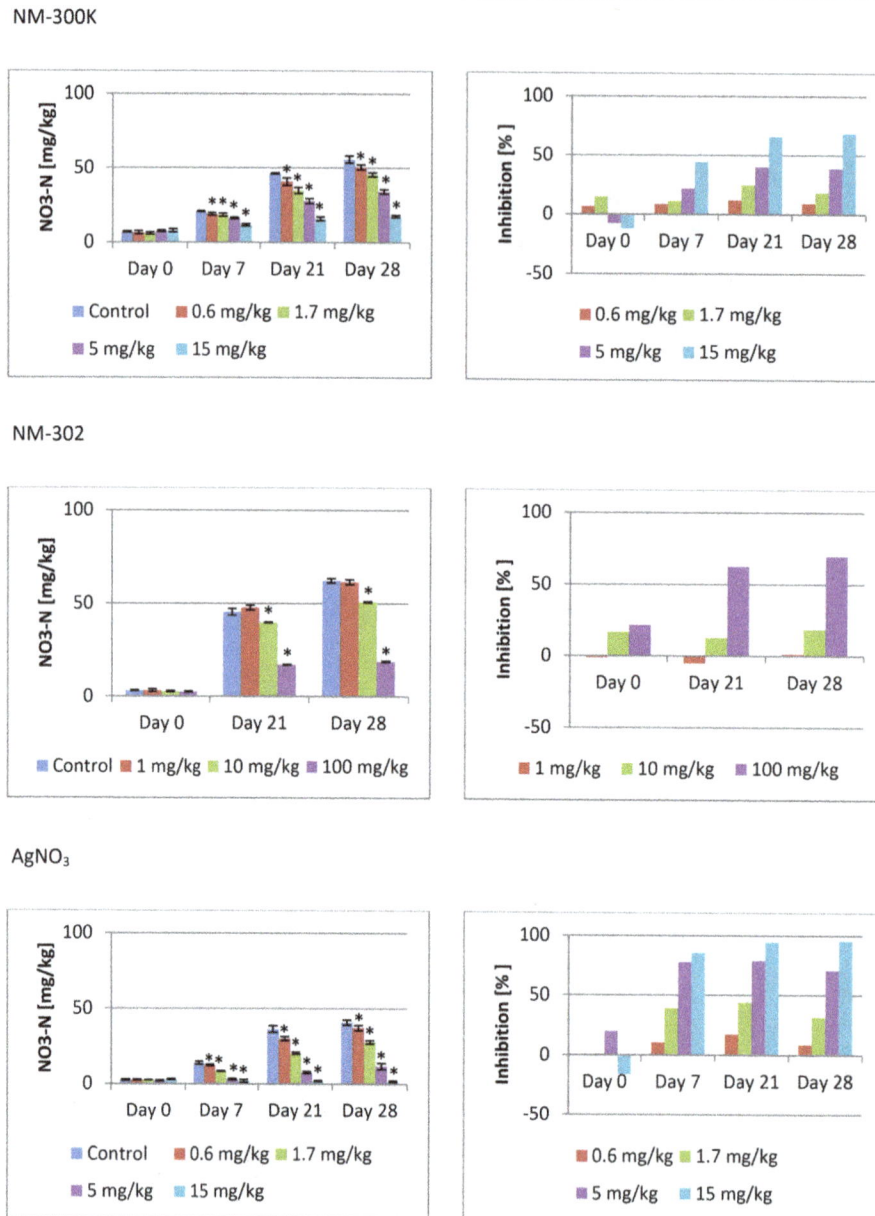

Figure 2 Effect of silver nanomaterials and AgNO$_3$ on nitrate concentration with (NH4)$_2$SO$_4$ as nitrogen source [2]. The results for AgNO$_3$ were corrected by the nominal nitrate concentration added with the test substance. Left figure column: mean activity and standard deviation. Right figure column: percentual deviation. Asterisks indicate a statistically significant difference to controls ($p \leq 0.05$).

measuring point albeit only at the highest test concentration. After 28 days, there was no statistically significant residual effect on respiration activity.

Time course of the effects

Table 1 presents the EC$_{10}$ and EC$_{50}$ values (experiment 1) and EC$_{50}$ values (experiment 2) for every test procedure and test substance on day 28. In the following the results of experiment 1 are presented in detail. The EC$_{50}$ values of the second experiment proof the statements of the first experiments. If only the results determined after

an incubation period of 28 days are considered, the toxicity of NM-300 K and AgNO$_3$ are comparable in terms of potential ammonium oxidation activity and nitrate accumulation when applying an inorganic soluble nitrogen source, but no inhibition of nitrate accumulation was observed when lucerne meal was used as the additional nitrogen source. In experiment 1, the EC values for NM-300 K and AgNO$_3$ for (i) nitrate formation using (NH$_4$)$_2$SO$_4$ as nitrogen source and (ii) the potential ammonium oxidation activity differ by a factor of less than 3 (e.g., EC$_{50}$ potential ammonium oxidation for NM-300 K

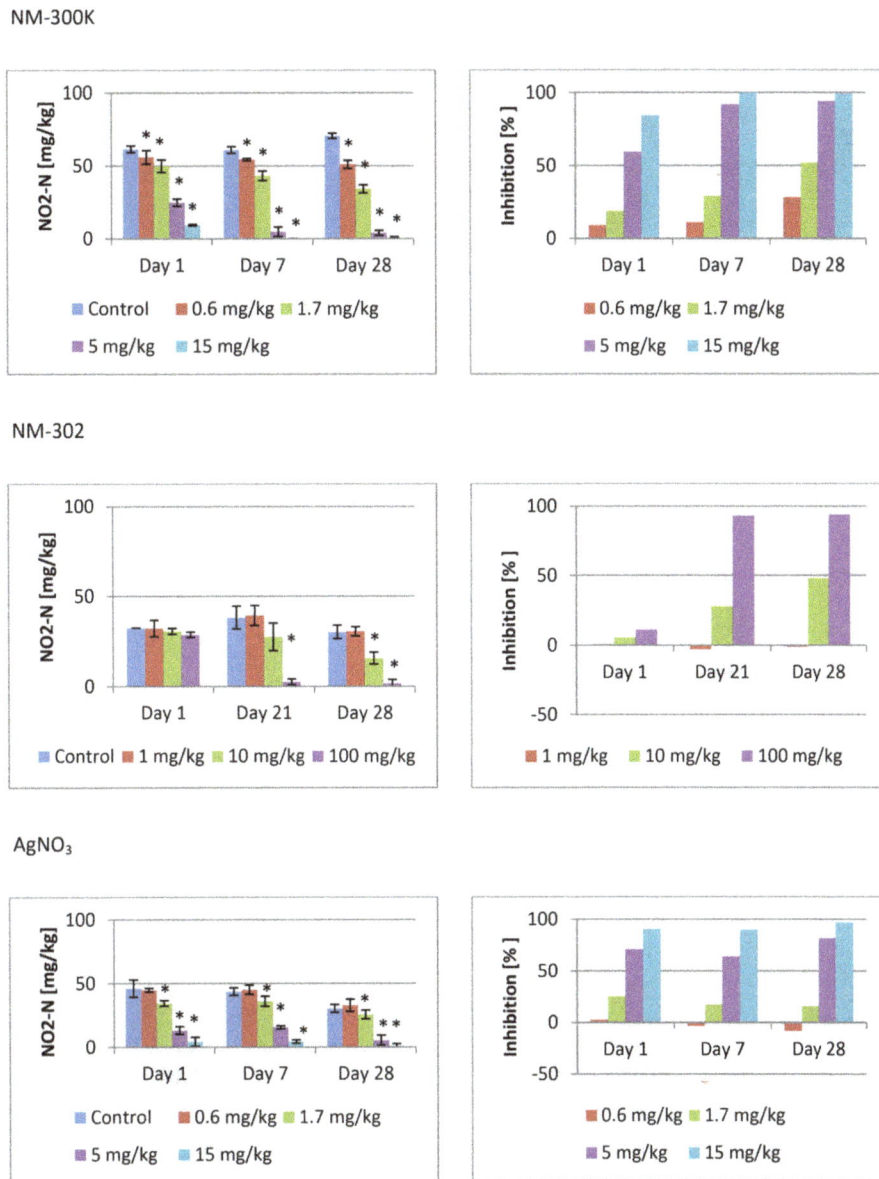

NM-300K

NM-302

AgNO$_3$

Figure 3 Effect of silver nanomaterials and AgNO$_3$ on potential ammonium oxidation according to ISO 15685 [4]. Left figure column: mean activity and standard deviation. Right figure column: percentual deviation. Asterisks indicate a statistically significant difference to controls ($p \leq 0.05$).

and AgNO$_3$ = 1.6 and 2.9 mg/kg, respectively; EC$_{50}$ nitrate accumulation for NM-300 K and AgNO$_3$ = 7.3 and 2.8 mg/kg, respectively). NM-302 appeared to be slightly less toxic than NM-300 K and AgNO$_3$ in terms of the effect on nitrate accumulation. The EC values differed at least by a factor of 6, but this was a screening approach carried out with three test concentrations and a spacing factor of 10, suggesting the EC values for NM-302 are less reliable. The toxicity of the three silver materials measured in terms of the effect on potential ammonium oxidation appeared to be comparable, with EC$_{50}$ values of 1.6 to 6 mg/kg. As stated above, even the highest test

concentrations of the test substances failed to inhibit nitrate accumulation when lucerne meal was used as the nitrogen source, thus no EC values were calculated for day 28.

The toxicity of the three silver materials measured in terms of the effect on substrate-induced respiration (carbon transformation) is also represented by the EC values for day 28. These values show no toxicity for AgNO$_3$, medium toxicity for NM-302 (EC$_{50}$ = 68 mg/kg), and relatively high toxicity for NM-300 K (EC$_{50}$ = 5 mg/kg).

Table 2 shows the time course of the EC$_{50}$ values of experiment 1 over the incubation period, revealing clear

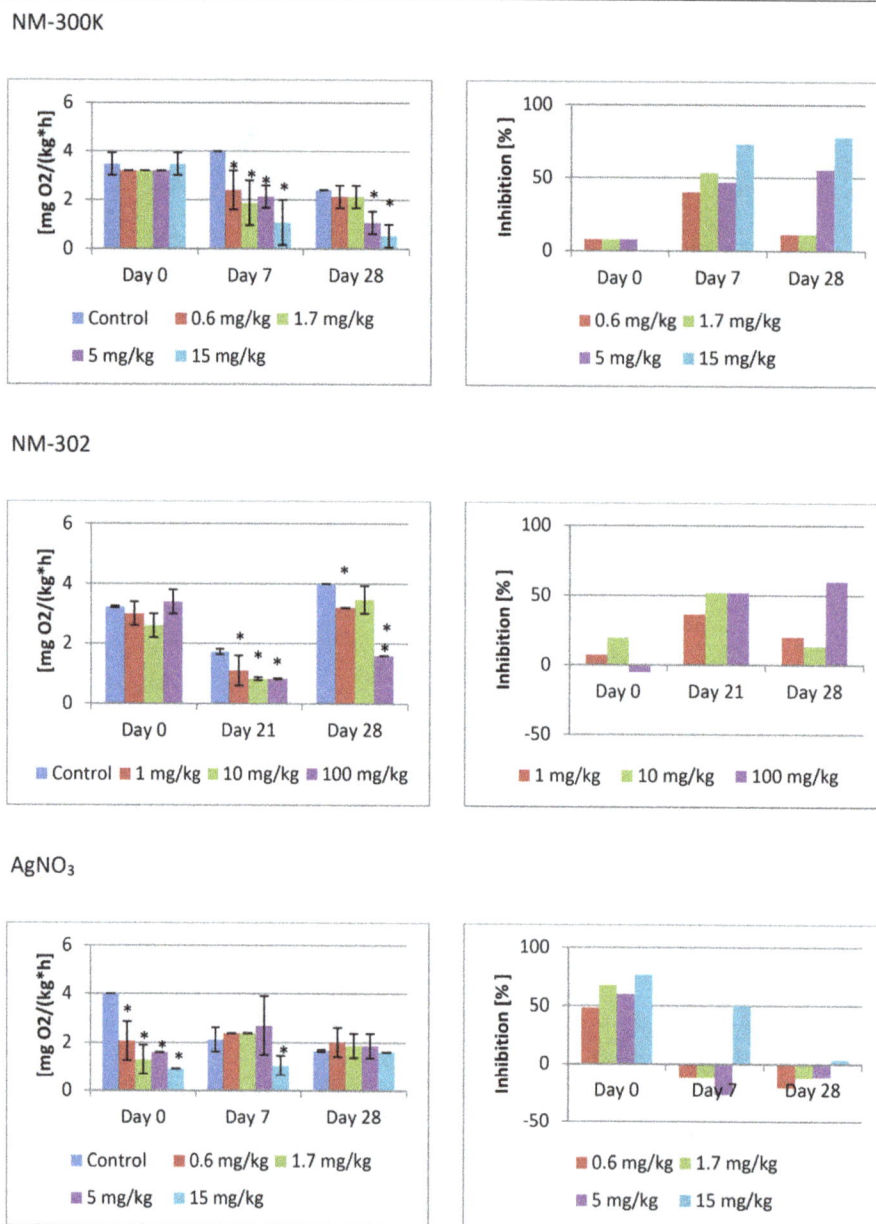

Figure 4 Effect of silver nanomaterials and AgNO₃ on glucose-induced respiration according to OECD TG 217 [3]. Left figure column: mean activity and standard deviation. Right figure column: percentual deviation. Asterisks indicate a statistically significant difference to controls ($p \leq 0.05$).

differences between the test substances and the test procedures. When lucerne meal was used as the nitrogen source, we observed either no effects or only short-term intermediate effects on nitrate accumulation with all three test substances. There was no difference in AgNO₃ toxicity on days 7 and 28 based on the nitrate concentration applying an inorganic nitrogen source (2.7 and 2.8 mg/kg, respectively), but the toxicity of both NM-302 and NM-300 K increased during the test. For NM-300 K, toxicity equilibrium was achieved during the incubation period with comparable EC₅₀ values recorded

on days 21 and 28 (7.1 and 7.3 mg/kg, respectively). The toxicity of NM-302 increased throughout the incubation period (EC₅₀ = 190 mg/kg on day 0 and 44.6 mg/kg on day 28).

The effects of AgNO₃ and NM-302 on potential ammonium oxidation activity were similar to the corresponding nitrate accumulation profiles, with comparable values for AgNO₃ on both testing dates (potential ammonium oxidation; 3.8 and 2.9 mg/kg) and a significant increase in the toxicity of NM-302 (from >100 mg/kg to 6 mg/kg). The toxicity of NM-300 K to potential ammonium oxidation was comparable at both measuring points (2.3 mg/kg on day 7 and 1.6 mg/kg on day 28).

The microflora involved in glucose-induced respiration recovered by the end of the test. The effect was more obvious with NM-300 K based on test concentrations (Figure 4) instead of EC_{50} values (Table 2). This nanomaterial caused about 50% inhibition on day 7 at test concentrations of 0.6, 1.7, and 5 mg/kg. On day 28 however, although there was still about 50% inhibition at the test concentration of 5 mg/kg, only 10% inhibition was recorded in soils treated with 0.6 and 1.7 mg/kg of the test substance (Figure 4).

Discussion

Comparison: nitrite and nitrate accumulation - carbon transformation

Tests according to OECD TGs 216 and 217 are used for the assessment of chemicals as required, e.g., in the context of regulation. OECD TG 216 describes a laboratory test that is used to investigate the long-term effects of chemicals on the microbial nitrogen transformation resulting in nitrate accumulation after a single exposure. The measured nitrate concentration reflects the sum of several microbial activities, such as ammonification, nitrification, denitrification, and formation of microbial biomass. Denitrification is expected to be of minor relevance due to the aerobic test conditions. Sieved soil is supplemented with powdered plant meal (e.g., lucerne meal). According to the OECD test guideline, for non-agrochemicals, the nitrate concentrations in treated and control samples are measured at the start of the test and after incubation for 28 days; for agrochemicals additional sampling points are requested. We applied this test to two silver nanomaterials and the soluble silver salt $AgNO_3$, and by considering only the start and end point, we found no evidence of toxicity towards soil-nitrifying bacteria. However, the same concentrations of test materials showed clear concentration-effect curves in the carbon transformation test (substrate-induced respiration) [3]. For strongly biocidal compounds, dose-related inhibition of substrate-induced respiration combined with a stimulated N-mineralization is typical [19]. In the following chapters, we discuss/consider the unusual sensitivity of both test systems (effects in C-transformation test; no effect in N-transformation test) when investigating Ag-NM and soluble $AgNO_3$.

Influence of nitrogen source (refer to 'Nitrogen transformation test' and 'Potential ammonium oxidation test' sections)

Usually, the effects of chemicals on nitrite and nitrate accumulation determined in tests using an inorganic nitrogen source are more obvious than in tests investigating the effects on soil respiration [20-22]. The different sensitivities of the endpoints may reflect the diversity of the microflora responsible for the different activities and hence their potential for recovery. This means that the ability to respire using easily degradable carbon sources such as glucose (the principle of the carbon transformation test) is more widespread than the ability to transform ammonium to nitrite and nitrate. When we replaced the powdered plant meal in the N-transformation test according to OECD TG 216 with $(NH_4)_2SO_4$ as a form of inorganic nitrogen, the same tests revealed clear concentration-effect curves for each test substance, and similar results were obtained when we tested the potential ammonium oxidation activity of the soil. These tests specifically monitored the transformation of ammonium to nitrate and ammonium to nitrite and did not require the introduction of complex organic matrices into the soil that might interfere with the bioavailability of the test substances. Indeed, organic materials such as sewage sludge have already been shown to interact with silver nanomaterials, with sorption rates greater than 90% for uncoated particles [23-25] and lower rates for functionalized particles [26]. Studies with biofilm communities and pure cultures have also shown that organic matter such as exopolysaccharides and humic acids can protect microorganisms and reduce the toxicity of silver nanomaterials [27-29]. The toxicity of ZnO nanomaterials towards *Folsomia candida* was found to be dependent on the soil pH and the proportion of organic matter [30]. These data, together with our results in the present investigation, suggest that silver nanomaterials and the ions released therefrom interfere with organic matter which reduces their long-term toxic effects on the sum of microbial activities resulting in nitrate accumulation in soil. The sorption of ions can also explain the missing effect of the soluble silver salt on day 28. The suggested mechanisms for the inhibition of silver nanomaterial dissolution are through surface adsorption of natural organic matter which in turn will block silver nanomaterial oxidation sites or released silver ions reducing back to silver through reaction with humic/fulvic acids [31]. This demonstrates that the ecotoxicity of such substances can be underestimated using the nitrogen transformation test recommended for the hazard and risk assessment of chemicals. Hazard assessment as required in the context of regulation can result in false negative conclusions. In the first step of a risk assessment, conservative conditions have to be simulated and refinements are foreseen only in case a risk cannot be excluded under these conditions [32]. As a consequence, for nanomaterials, an inorganic nitrogen source should be used in the first step. If a risk cannot be excluded, a complex organic nitrogen source could be considered, which reflects more natural conditions.

The toxicity of silver nanomaterials among other things depends on the release of free ions [33,34]. The impact of

soluble metal ions in aquatic and terrestrial ecotoxicological tests is usually demonstrated by comparing the effect of nanomaterials against the corresponding soluble
salts [16,29,35]. For aquatic environments, comparable results are achieved if the test is based on ion concentrations
because dissolution is the principal exposure mechanism
[33,36]. However, both dissolution and sorption must be
considered when testing soils. For example, the toxicity of
Zn^{2+} in F. candida exceeds that of Zn nanomaterials if the
calculations are based on the total zinc concentration
[37-39], but is less than that of zinc nanomaterials if calculations are based on the pore water Zn^{2+} concentration
[39]. However, different results were observed when testing the toxicity of silver in earthworms: there was no
significant difference in toxicity based on the total concentration of silver nanomaterials/soluble silver salts
and based on the amount of free Ag^+ [16]. This discre-
pancy has yet to be explained, but it may depend on
nanomaterial-specific factors (differences between zinc
and silver, size of nanomaterials, and velocity of ion release), differences in the time of ion determination (directly after spiking [39]; at test end (56 days) [16]), or
differences in species habitat: F. candida lives in air-filled
soil pores and earthworms in bulk soil. We did not measure the pore water Ag^+ concentration in our investigation,
but based on the total silver concentration, our results
suggest there is no significant difference in the toxicity of
soluble silver salts and NM-300 K at the end of the test.

Time course of effects
Although there was only a negligible difference in the effect of the test substances on the sum of microbial activities resulting in a nitrate accumulation after 28 days,
the effect profile changed during the test in different
ways for each substance. $AgNO_3$ showed immediate toxicity, whereas the small spherical nanoparticles showed
the rapid development of toxicity followed by toxicity
equilibrium and the larger rod-shaped nanomaterials became toxic more slowly but toxicity was still increasing
at the end of the test. These differences may depend on
the surface area of the nanomaterials, with the smaller
particles having a larger surface area and thus a higher
release of ions and a more rapid effect. These effect profiles may indicate the stability of the nanomaterials, the
dynamic release of ions, and/or the overall toxicity, but
this cannot be addressed specifically without further experiments involving prolonged incubation periods that
achieve equilibrium of toxicity, thus allowing various relationships such as velocity and duration or shape and
velocity to be characterized in more detail. An influence
of the morphology of NM has also been demonstrated
in other studies [40,41].

The toxicity of each test substance was greater when
measured as an impact on potential ammonium oxidation

[4] than when measured as an impact on nitrate accumulation using $(NH_4)_2SO_4$ as the nitrogen source in the
N-transformation test [2]. There are several differences
between the tests that may explain this phenomenon.
Only the transformation step from ammonium to nitrite
is investigated in the first test, whereas the whole process
from ammonium to nitrate is covered by the second, i.e.,
that further to the transformation of ammonium to nitrate
this test includes additional transformation steps of the nitrogen cycle (release of ammonium from organic nitrogen
sources and nitrogen immobilization). The potential ammonium oxidation test provides information on the current activity of the respective microorganisms at the
sampling time point, whereas the effect on nitrogen transformation is based on the total nitrate concentration per
test concentration and sampling day.

Conclusions
The procedure described in OECD TG 216 is often used
to assess chemical effects on the soil microflora responsible for nitrate accumulation under the framework
of the REACH regulation and the resp. Guidance Document [1]. We found that the test is unsuitable for the
detection of effects caused by silver nanomaterials and
soluble silver salts, suggesting that the deficiency relates
not to the testing of nanomaterials but substances that
release soluble ions in significant amounts that are the
basis of the observed toxicity. We therefore recommended an incubation period of the soil of at least 28 days
according to OECD TG 216 and the use of an inorganic
nitrogen source instead of organic nitrogen or the periodical measurement of potential ammonium oxidation
instead of nitrogen transformation [4]. Finally the effect
development time course in the nitrogen transformation
test with $(NH_4)_2SO_4$ and the potential ammonium oxidation activity test may provide a useful indicator for the
stability of silver nanomaterials, although additional experiments are required to characterize this relationship
in more detail.

Methods
Test soil
The experiments were carried out using the reference
soil RefeSol 01A [42] (sieved ≤2 mm) which is a loamy,
medium-acidic and lightly humic sand (Table 3). RefeSol
soils were selected as reference soils by the German
Federal Environment Agency (Umweltbundesamt, UBA)
and are known to be suitable for testing the influence of
substances on the habitat function of soils (bioavailability, effects on organisms). RefeSol 01A matches the
properties stated in various OECD terrestrial ecotoxicological guidelines (e.g., tests with plants and soil microflora). The soils were sampled in the field and stored
in high-grade stainless steel basins with drainage and

Table 3 Physicochemical properties of RefeSol 01A

Physicochemical properties	RefeSol 01A
pH	5.7
C_{org} [%]	0.93
N_{total} [mg/kg]	882
CEC_{eff} [mmolc/kg]	37.9
Sand [%]	71
Silt [%]	24
Clay [%]	5
WHC_{max} [ml/kg]	227

CEC_{eff}, effective cation exchange capacity; WHC_{max}, maximum water-holding capacity.

ground contact at the Fraunhofer IME in Schmallenberg. Red clover was sown on the stored soils to preserve microbial activity. No pesticides were used. Soil was sampled 1 to 4 weeks before the test. If the soil was too wet for sieving, it was dried at room temperature to 20% to 30% of the maximum water-holding capacity (WHC_{max}) with periodic turning to avoid surface drying. If the tests did not start immediately after sieving, the soil was stored in the dark at 4°C under aerobic conditions [43].

Test substances

We used one silver nanomaterial with the code NM-300 K and a silver material with the code NM-302 containing nanostructures but also larger particles. Both materials had been selected for the OECD Sponsorship Programme on nanomaterials and are referred to as nanomaterials [44]. NM-300 K is a colloidal silver dispersion with a nominal silver concentration of 10% (w/w). The primary particles of the nanomaterial have a size of approximately 20 nm, measured on TEM images. The general morphology of the primary subunits of the NM is nearly equiaxed and rounded, or slightly elongated. Their suggested 3D structure is spherical or slightly ellipsoidal (determinations performed by EM-service of CODA-CERVA, Brussels, Belgium). NM-300 K is a mixture of a stabilizing agent (NM-300 K DIS) comprising 4% (w/w) each of polyoxyethylene glycerol trioleate and polyoxyethylene sorbitan monolaurate (Tween-20) and silver nanomaterials. NM-302 comprises silver nanomaterial (purity ≥99%, thickness 100 to 200 nm, and length 0.6 to 12 μm) dispersed in an aqueous solution containing (according to the manufacturer) silver nanowires (8.6% w/w), polyvinylpyrrolidone (<1% w/w), acrylic/acrylate copolymer (<2% w/w), and polycarboxylate ether (<2% w/w). The general morphology of the primary subunits of the NM is longitudinal or rounded, but slightly elongated. Their suggested 3D structures are rod-like or poly-angular (determinations performed by EM-service of CODA-CERVA, Belgium). AgNO$_3$ was purchased from Merck, Darmstadt, Germany.

Spiking of soil and incubation

The procedure used to mix the test materials with the soil has been described [45]. All the values below refer to the concentration of silver (regardless of the test substance) and the dry weight of the soil. Both NM-300 K and AgNO$_3$ were tested at concentrations of 0.56, 1.67, 5.0, and 15.0 mg/kg. NM-302 was tested at concentrations of 1, 10, and 100 mg/kg. NM-300 K and NM-302 consisted of silver nanomaterials in dispersant. For NM-300 K, the pure dispersant was tested in amounts comparable to the amounts present in the NM-300 K applications. For NM-302, only the highest concentration of dispersant was investigated. Neither of the applied tests on nitrogen transformation, potential ammonium oxidation, and on carbon transformation revealed statistically significant differences between the untreated control and soil spiked with dispersant (data not shown).

For the carbon transformation [3] and potential ammonium oxidation activity [4] tests, the soil was incubated without nutrients. For the nitrogen transformation test with soil microflora [2], sieved and spiked soil was supplemented with 5 g powdered plant material (lucerne meal) per kg of dry soil or with 1 g (NH$_4$)$_2$SO$_4$ per kg of dry soil, corresponding to 212 mg nitrogen per kg. The (NH$_4$)$_2$SO$_4$ was added in a 20-mL aqueous solution.

The soil was then adjusted to 50% ± 5% WHC_{max}, transferred to glass vessels (500 g for the nitrification and ammonium oxidation activity tests and 1,500 g for the carbon transformation tests), and incubated under aerobic conditions at 20°C ± 2°C for 4 weeks. Every 7 days, evaporated water was replaced. The incubation corresponded to the procedure described in OECD TGs 216 and 217. According to the OECD test guidelines sampling times for non-agrochemicals are mandatory on days 0 (3 h after spiking) and 28. Additional samples were analyzed on days 7 and/or 21. The ammonium oxidation activity was determined over a period of 6 h on day 1 instead of day 0 because spiking, incubation for 3 h, and testing is not possible within 1 day.

Ecotoxicological determinations

Nitrate levels were measured as an indicator of the soil nitrification activity after mixing with lucerne meal or (NH$_4$)$_2$SO$_4$ as a nitrogen source. Nitrate was extracted by shaking soil samples (20 g dm) with 0.1 M KCl (100 mL minus the water content of the soil sample) for 60 min at 150 rpm. Three replicates for the control and every test concentration were sampled. The mixtures were filtered and nitrate in the filtrate was measured using a Spectroquant® NOVA 400 spectrophotometer (Merck, Darmstadt, Germany) at 370 nm immediately after preparation. For the determinations we used a test kit (nitrate cell test; Spectroquant 1.14542.0001; measuring range 0.5 to 18.0 mg/L NO$_3$-N; Merck, Darmstadt, Germany). The determination is based on the reaction of nitrate ions with benzoic acid derivates in the presence of sulfuric acid resulting in a red nitro compound that is determined photometrically. Commercially available nitrate standard solutions CRM (Spectroquant, 2.5 and 15 mg/L; Merck, Darmstadt, Germany) were used for checking the procedure. Soil extracts with nitrate concentrations exceeding the measuring range were diluted with deionized water.

The potential ammonium oxidation activity [4] was measured in a slurry of 25 g dry soil matter in 100 mL mineral test medium (0.56 mM KH_2PO_4, 1.44 mM K_2HPO_4, 5 mM $NaClO_3$, 1.50 mM $(NH_4)_2SO_4$). The slurries were incubated on an orbital shaker at 25°C ± 2°C, and 10-mL samples were removed after 2 and 6 h. The samples were mixed with 10 mL 4 mol/L KCl, and after filtration, the nitrite levels in the filtrate were measured with sulfanilamide and N-(1-naphthyl)-ethylene-diamine dihydrochloride giving a red-violet diazo dye. The absorbance of the diazo dye was measured using a Cary 300 Scan UV–VIS spectrophotometer (Varian Deutschland GmbH, Darmstadt, Germany) at 530 nm. $NaNO_2$ was used for the calibration curve (0.01 to 1.0 NO_2-N/mL).

Substrate-induced respiration [3] was measured in three 500-mL Erlenmeyer flasks per treatment filled with 100 g dry soil matter and 400 mg glucose, mixed homogenously. The vessels were incubated in darkness at 20°C ± 1°C for 24 h. Respiration was measured during incubation with an OxiTop Sensomat system (Aqualytic/ WTW, Weilheim, Germany). Oxygen was consumed during respiration and the resulting low pressure was measured. The CO_2 generated was bound to KOH to prevent interference with the measurement. The microbial respiration rate was calculated over a linear time scale.

Statistical analysis

In the ecotoxicological tests, probit analysis was used to estimate the EC_{10} and EC_{50} values and the dose–response curves. The significance of differences between treatments in the long-term tests was determined using Student's t-test. Statistical analysis was carried out using ToxRat® Pro v2.10 software (Alsdorf, Germany) for ecotoxicity response analysis [46].

Competing interests
The authors declare that they have no competing interests.

Authors' contributions
KHR and KS designed the studies, headed the test performance, and drafted the manuscript. Both authors read and approved the final manuscript.

Acknowledgements
The authors would like to thank the MARINA Framework 7 project for providing the funding for this study and the EM-service of CODA-CERVA, Belgium, for the characterization of the nanomaterials.

References

1. ECHA: *Guidance for the Implementation of REACH - Guidance on Information Requirements and Chemical Safety Assessment. Chapter R.7c: Endpoint Specific Guidance*. Helsinki: European Chemicals Agency; 2012 [http://echa.europa. eu/documents/10162/13632/information_requirements_r7c_en.pdf]
2. OECD: *Guideline 216: OECD Guideline for the Testing of Chemicals. Soil Microorganisms: Nitrogen Transformation Test*. Paris: Organisation for Economic Co-operation and Development; 2000.
3. OECD: *Guideline 217: OECD Guideline for the Testing of Chemicals. Soil Microorganisms: Carbon Transformation Test*. Paris: Organisation for Economic Co-operation and Development; 2000.
4. ISO 15685: *Soil Quality - Determination of Potential Nitrification and Inhibition of Nitrification - Rapid Test by Ammonium Oxidation*. Geneva: International Organization for Standardization; 2012.
5. Piccinno F, Gottschalk F, Seeger S, Nowack B: **Industrial production quantities and uses of ten engineered nanomaterials in Europe and the world.** *J Nanopart Res* 2012, **14**(1109: 1-1109):11.
6. Chaloupka K, Malam Y, Seifalian AM: **Nanosilver as a new generation of nanoproduct in biomedical applications.** *Trends Biotechnol* 2010, **28**:580–588.
7. Yuan G, Cranston R: **Recent advances in antimicrobial treatments of textiles.** *Text Res J* 2008, **78**:60–72.
8. Prabhu S, Poulose E: **Silver nanoparticles: mechanism of antimicrobial action, synthesis, medical applications, and toxicity effects.** *Int Nano Lett* 2012, **2**(32:1-32):10.
9. Choi O, Hu Z: **Size dependent and reactive oxygen species related nanosilver toxicity to nitrifying bacteria.** *Environ Sci Technol* 2008, **42**:4583–4588.
10. Choi O, Clevenger TE, Deng B, Surampalli RY, Ross L Jr, Hu Z: **Role of sulfide and ligand strength in controlling nanosilver toxicity.** *Water Res* 2009, **43**:1879–1886.
11. Hänsch M, Emmerling C: **Effects of silver nanoparticles on the microbiota and enzyme activity in soil.** *J Plant Nutr Soil Sc* 2010, **173**:554–558.
12. Schlich K, Klawonn T, Terytze K, Hund-Rinke K: **Hazard assessment of a silver nanoparticle in soil applied via sewage sludge.** *Environ Sci Eu* 2013, **25**:17.
13. Shin YJ, Kwak JI, An YJ: **Evidence for the inhibitory effects of silver nanoparticles on the activities of soil exoenzymes.** *Chemosphere* 2012, **88**:524–529.
14. Shoults-Wilson WA, Reinsch BC, Tsyusko OV, Bertsch PM, Lowry GV, Unrine JM: **Role of particle size and soil type in toxicity of silver nanoparticles to earthworms.** *Soil Sci Soc Am J* 2011, **75**:365–377.
15. Yang Y, Chen Q, Wall JD, Hu Z: **Potential nanosilver impact on anaerobic digestion at moderate silver concentrations.** *Water Res* 2012, **46**:1176–1184.
16. Schlich K, Klawonn T, Terytze K, Hund-Rinke K: **Effects of silver nanoparticles and silver nitrate in the earthworm reproduction test.** *Environ Toxicol Chem* 2013, **32**:181–188.
17. van Beelen P, Doelman P: **Significance and application of microbial toxicity tests in assessing ecotoxicological risk of contaminants in soil and sediment.** *Chemosphere* 1997, **34**(3):455–499.
18. Biologische Bundesanstalt für Land- und Forstwirtschaft: Auswirkungen auf die Aktivität der Bodenmikroflora. *Richtlinien für die amtliche Prüfung von Pflanzenschutzmitteln Teil VI*, 1-1:1990.
19. Malkomes H-P: **Microbiological-ecotoxicological soil investigations of two herbicidal fatty acid preparations used with high dosages in weed control.** *UWSF - Z Umweltchem Ökotox* 2006, **18**:13–20.
20. Hund-Rinke K, Simon M: **Terrestrial ecotoxicity of eight chemicals in a systematic approach.** *J Soils Sediments* 2005, **5**:59–65.
21. Agnihotri VP: **Persistence of captan and its effects on microflora, respiration, and nitrification of a forest nursery soil.** *Can J Microbiol* 1971, **17**:377–383.
22. Remde A, Hund K: **Response of soil autotrophic nitrification and soil respiration to chemical pollution in long-term experiments.** *Chemosphere* 1994, **29**:391–404.
23. Wang Y, Westerhoff P, Hristovski KD: **Fate and biological effects of silver, titanium dioxide, and C60 (fullerene) nanomaterials during**

simulated wastewater treatment processes. *J Hazard Mater* 2012, **201–202**:16–22.

24. Tiede K, Boxall ABA, Wang X, Gore D, Tiede D, Baxter M, David H, Tear SP, Lewis J: **Application of hydrodynamic chromatography-ICP-MS to investigate the fate of silver nanoparticles in activated sludge.** *J Anal At Spectrom* 2010, **25**:1149–1154.

25. Kiser MA, Ryu H, Jang H, Hristovski K, Westerhoff P: **Biosorption of nanoparticles to heterotrophic wastewater biomass.** *Water Res* 2010, **44**:4105–4114.

26. Kiser MA, Ladner DA, Hristovski KD, Westerhoff PK: **Nanomaterial transformation and association with fresh and freeze-dried wastewater activated sludge: implications for testing protocol and environmental fate.** *Environ Sci Technol* 2012, **46**:7046–7053.

27. Sheng Z, Liu Y: **Effects of silver nanoparticles on wastewater biofilms.** *Water Res* 2011, **45**:6039–6050.

28. Wirth SM, Lowry G, Tilton RD: **Natural organic matter alters biofilm tolerance to silver nanoparticles and dissolved silver.** *Environ Sci Technol* 2012, **46**:12687–12696.

29. J-y R, Sim SJ, Yi J, Park K, Chung KH, Ryu D-y, Choi J: **Ecotoxicity of silver nanoparticles on the soil nematode *Caenorhabditis elegans* using functional ecotoxicogenomics.** *Environ Sci Technol* 2009, **43**:3933–3940.

30. Waalewijn-Kool PL, Rupp S, Lofts S, Svendsen C, van Gestel CAM: **Effects of soil properties on the toxicity of ZnO nanoparticles to *Folsomia candida* in a comparsion of four natural soils.** In *Ecotoxicological Assessment of ZnO Nanoparticles to Folsomia candida.* Edited by Waalewijn-Kool PL. The Netherlands: PhD thesis, VU University Amsterdam, Department of Ecological Science; 2013.

31. Liu J, Hurt R: **Ion release kinetics and particle persistence in aqueous nano-silver colloids.** *Environ Sci Technol* 2010, **44**:2169–2175.

32. ECHA: *Guidance for the Implementation of REACH - Guidance on Information Requirements and Chemical Safety Assessment. Chapter R.19: Uncertainty Analysis.* Helsinki: European Chemicals Agency; 2012 [http://echa.europa.eu/documents/10162/13632/information_requirements_r19_en.pdf]

33. Kim J, Kim S, Lee S: **Differentiation of the toxicities of silver nanoparticles and silver ions to the Japanese medaka (*Oryzias latipes*) and the cladoceran *Daphnia magna*.** *Nanotoxicology* 2011, **5**:208–214.

34. Yin L, Cheng Y, Espinasse B, Colman BP, Auffan M, Wiesner M, Rose J, Liu J, Bernhardt ES: **More than the ions: the effects of silver nanoparticles on *Lolium multiflorum*.** *Environ Sci Technol* 2011, **45**:2360–2367.

35. Asghari S, Johari S, Lee J, Kim Y, Jeon Y, Choi H, Moon M, Yu I: **Toxicity of various silver nanoparticles compared to silver ions in *Daphnia magna*.** *J Nanobiotechnol* 2012, **10**(14):1–14. 11.

36. Lee Y-J, Kim J, Oh J, Bae S, Lee S, Hong IS, Kim S-H: **Ion-release kinetics and ecotoxicity effects of silver nanoparticles.** *Environ Toxicol Chem* 2012, **31**:155–159.

37. Kool PL, Ortiz MD, van Gestel CAM: **Chronic toxicity of ZnO nanoparticles, non-nano ZnO and ZnCl$_2$ to *Folsomia candida* (Collembola) in relation to bioavailability in soil.** *Environ Pollut* 2011, **159**:2713–2719.

38. Waalewijn-Kool P, Diez Ortiz M, Gestel CM: **Effect of different spiking procedures on the distribution and toxicity of ZnO nanoparticles in soil.** *Ecotoxicology* 2012, **21**:1797–1804.

39. Waalewijn-Kool PL, Ortiz MD, Lofts S, van Gestel CAM: **The effect of pH on the toxicity of zinc oxide nanoparticles to *Folsomia candida* in amended field soil.** *Environ Toxicol Chem* 2013, **32**:2349–2355.

40. Chithrani BD, Ghazani AA, Chan WCW: **Determining the size and shape dependence of gold nanoparticle uptake into mammalian cells.** *Nano Lett* 2006, **6**:662–668.

41. Pal S, Tak YK, Song JM: **Does the antibacterial activity of silver nanoparticles depend on the shape of the nanoparticle? A study of the Gram-negative bacterium *Escherichia coli*.** *Appl Environ Microbiol* 2007, **73**:1712–1720.

42. **Refesol.** [http://www.refesol.de/english/]

43. ISO 18512: *Soil Quality - Guidance on Long and Short Term Storage of Soil Samples.* Geneva: International Organization for Standardization; 2007.

44. OECD (WPMN): *Safety of Manufactured Nanomaterials - Sponsorship Programme for the Testing of Manufactured Nanomaterials*, OECD's Working Party on Manufactured Nanomaterials (WPMN). 2007

[http://www.oecd.org/science/nanosafety/sponsorshipprogramme forthetestingofmanufacturednanomaterials.htm]

45. Hund-Rinke K, Schlich K, Klawonn T: **Influence of application techniques on the ecotoxicological effects of nanomaterials in soil.** *Environ Sci Eur* 2012, **24**(12):1–30. 12.

46. **ToxRat Professional - Software for Ecotoxicity Response Analysis.** Alsdorf, Germany: Online Version 2.10. ToxRat® Solutions GmbH.

20

A decision-analytic approach to predict state regulation of hydraulic fracturing

Igor Linkov[1*], Benjamin Trump[1], David Jin[1,4], Marcin Mazurczak[2,3] and Miranda Schreurs[2]

Abstract

Background: The development of horizontal drilling and hydraulic fracturing methods has dramatically increased the potential for the extraction of previously unrecoverable natural gas. Nonetheless, the potential risks and hazards associated with such technologies are not without controversy and are compounded by frequently changing information and an uncertain landscape of international politics and laws. Where each nation has its own energy policies and laws, predicting how a state with natural gas reserves that require hydraulic fracturing will regulate the industry is of paramount importance for potential developers and extractors. We present a method for predicting hydraulic fracturing decisions using multiple-criteria decision analysis. The case study evaluates the decisions of five hypothetical countries with differing political, social, environmental, and economic priorities, choosing among four policy alternatives: open hydraulic fracturing, limited hydraulic fracturing, completely banned hydraulic fracturing, and a cap and trade program.

Results: The result is a model that identifies the preferred policy alternative for each archetypal country and demonstrates the sensitivity the decision to particular metrics. Armed with such information, observers can predict each country's likely decisions related to natural gas exploration as more data become available or political situations change.

Conclusions: Decision analysis provides a method to manage uncertainty and address forecasting concerns where rich and objective data may be lacking. For the case of hydraulic fracturing, the various political pressures and extreme uncertainty regarding the technology's risks and benefits serve as a prime platform to demonstrate how decision analysis can be used to predict future behaviors.

Keywords: Hydraulic fracturing; Multi criteria decision analysis; Policy alternatives; Energy policy

Background

Horizontal drilling and high-volume hydraulic fracturing, collectively known as 'fracking,' opened up the possibility for new natural gas extraction across the globe. While not a novel technology, fracking has taken off in the USA and Canada in regions with substantial yet traditionally difficult to harvest shale gas [1]. Despite the potential for economic benefits, the use of such technology introduces risks to humans and the environment [2]. These risks are compounded over the extended time period by which wells tapping rich natural gas deposits remain in operation [3]. Uncertainty regarding the likelihood and consequences of harmful events has driven many local, regional, and national governments to issue warnings, regulations, and moratoriums on the industry [4,5]. Natural gas companies are challenged to identify rich deposits that can be extracted as cheaply and efficiently as possible with the longest expected payoff, while at the same time hedging against the likelihood that industry regulations will change. As such, the ability of a company to predict future state behavior in regulating a highly uncertain practice is critical to its long-term survival and success.

In this paper, we present a decision tool that simulates state behavior with regard to the regulation of hydraulic fracturing. This tool takes into account a variety of factors ranging from drill site profitability to the public perception of hydraulic fracturing as an acceptable practice, a realistic enumeration of the various pressures facing government legislators currently considering how to regulate the industry. We make particular use of decision analysis to

* Correspondence: Igor.Linkov@usace.army.mil
[1]US Army Engineer Research and Development Center, 696 Virginia Rd, Concord, MA 01742, USA
Full list of author information is available at the end of the article

aggregate a variety of quantitative and qualitative state inputs, which is ultimately used to produce a prediction regarding state action towards hydraulic fracturing. While this demonstration is a fictional representation of state action by making use of archetypal countries and hypothetical data, users could input their own data to construct a more realistic set of predictions for a chosen set of states.

Results and discussion

The overall scores displayed in Figure 1 represent the weighted summation of value functions applied to each alternative for each archetype. The main result is a ranked list of policy alternatives for each archetypal country. The policy alternative with the highest score is the most preferred option, while the alternative with the lowest

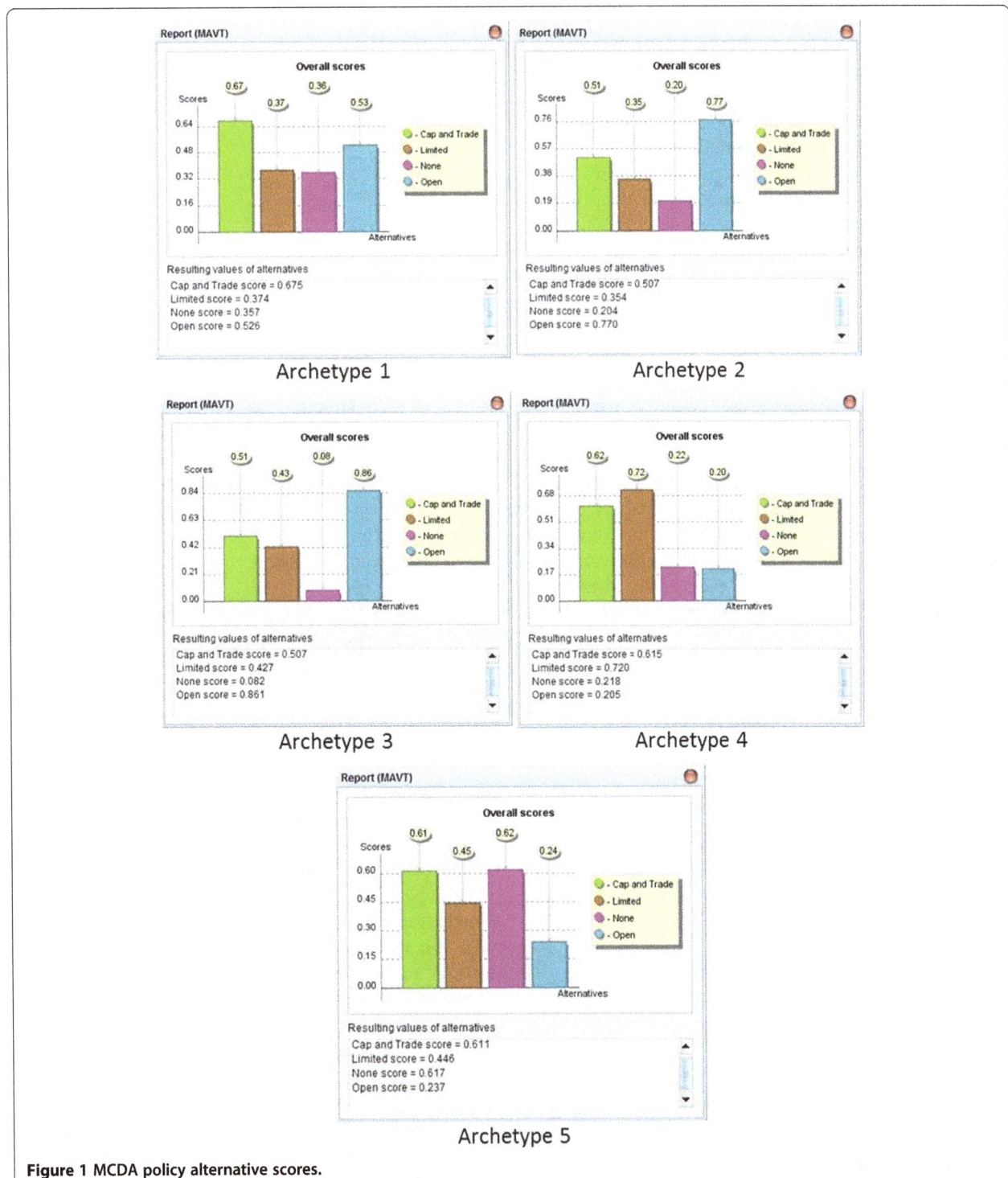

Figure 1 MCDA policy alternative scores.

score is the least preferred option. Each archetypal country has a current optimal policy solution - although some are more resolute than others. The model output (Figure 1) for each country indicates a ranking of the policy options based upon their relative optimality, indicating where a policy option is clearly favored (e.g., open hydraulic fracturing in country 3), or where results are less clear and require a more thorough sensitivity analysis, such as with country 5. Ultimately, each country's policy options serve as a mathematical reflection of the political, environmental, economic, and social pressures facing national lawmakers. For example, within country 1 (developed capitalist democracy), rich gas reserves coupled with strong fuel consumption reduce preference for the option of banning hydraulic fracturing altogether, yet existing environmental concerns (having the highest criterion weight) also reduce preference for only limited regulation. Consequently, a cap and trade system for gas well development rises as a politically and economically acceptable alternative.

Due to the strong tendency of preferences to shift within public policy, understanding how changes in the elements of the hydraulic fracturing decision model impact each country's preferred policy alternative is critical to reducing uncertainty. Sensitivity analysis is helpful to analyze how a shift in a specific criterion or sub-criterion's weight affects the overall decision. Figures 2 and 3 display the sensitivity analysis output for the 'Environment' criterion of archetype 5 and the 'Economics' criterion of archetype 4, respectively. The vertical axes show the value of each policy alternative as a function of the weight of the criterion by percent

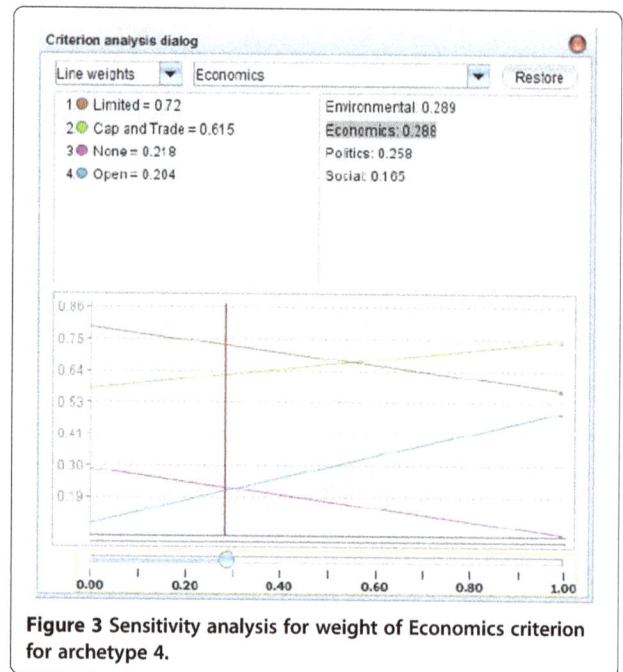

Figure 3 Sensitivity analysis for weight of Economics criterion for archetype 4.

on the horizontal axes. Figure 2 shows the preference score for each alterative as a function of the weight applied to the Environment criterion in country 5. The intersections illustrate the critical points where the preferred alternative changes from the current policy alternative to a different one. Here, even a minor decline in the importance of Environment relative to the other criteria could contribute to a shift from 'no hydraulic fracturing' to 'cap and trade'. As such, this weighting scheme is considered 'sensitive' and should be considered alongside any expected developments in national priorities. For archetype 4, the weight of the Economics criterion must increase from 29% to over 55% in order for the preferred policy alternative to transition from 'limited' to 'cap and trade,' indicating that this archetypal country is relatively insensitive to economic criteria. Such an analysis can be conducted on all decision criteria for all countries in order to anticipate future shifts in priorities and preferences and determine the point where a different policy option may become a better choice.

Conclusions

With this hypothetical case study, we demonstrate how decision analysis can be used to predict the behavior of governments in anticipation of hydraulic fracturing policy. Such a tool could prove valuable not only to drilling companies whose livelihoods depend upon understanding and predicting drilling regulation but also to academics and researchers seeking to gain greater understanding and transparency of how different political pressures impact high-level decision making. This form of decision aid is particularly helpful to

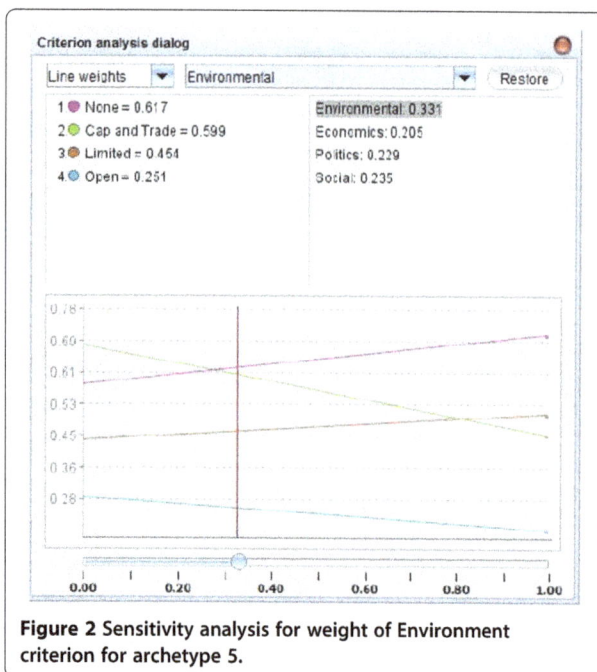

Figure 2 Sensitivity analysis for weight of Environment criterion for archetype 5.

better understand the breadth of issues and concerns facing policymakers along with the uncertainty regarding how future developments may affect major policy decisions.

Though a skilled analyst is required to perform these computations, decision analysis (multi-criteria decision analysis (MCDA) in particular) can supplement traditional political science research in forecasting behavior on issues with subjective, disparate, or highly uncertain data. The archetypal examples above demonstrate how a variety of differing inputs may be aggregated and quantified in a manner where an analyst can meaningfully compare a host of policy options. In this case, organizations interested in hydraulic fracturing can use available information to predict to what degree where countries with gas deposits will regulate drilling. Such a decision tool could in turn reduce the uncertainty and confusion surrounding the variety of inputs to consider and guide organizational decision making over a span of time. Further sensitivity analysis can help describe how changes in available data or public sentiment might affect the preferred policy alternative of governing bodies. Such knowledge can help decision makers adapt quickly as new information becomes available.

Decision analysis is not perfect - it does rely upon expert elicitation to acquire traditionally qualitative information. However, it provides a method to manage uncertainty and address forecasting concerns where rich and objective data may be lacking. For the case of hydraulic fracturing, the various political pressures and extreme uncertainty regarding the technology's risks and benefits serve as a prime platform to demonstrate how decision analysis can be used to predict future behaviors.

Methods

Multi-criteria decision analysis (MCDA) is a method for decision structuring that permits the use of both quantitative and qualitative data sources with high uncertainty or subjectivity [6]. Specifically, MCDA helps aggregate the impact of various unrelated inputs into a ranked list of quantitative results in a transparent process [7]. One type of qualitative data is stakeholder options, which in this study includes subjective metrics of a country's willingness to accept certain hydraulic fracturing risks in order to acquire the perceived economic benefits. The Decision Evaluation for Complex Risk Network Systems (DECERNS) software [6,8] is used to incorporate stakeholder opinions with available data on the risks of unconventional drilling methods to determine to what degree each factor influences a state's behavior towards hydraulic fracturing.

The MCDA prediction model requires the construction of a value hierarchy, a tree that represents the major factors and policy solutions which influence a stakeholder's decision on a given issue [9]. These factors include the main criteria which pressure lawmakers and policymakers

to regulate in a certain fashion. The first branch of value hierarchy development identifies the overarching criteria, including 'Political,' 'Environmental,' 'Social,' and 'Economic' factors. Further refined, these factors are broken down into individual elements which each represent a specific factor lawmakers must consider when forming hydraulic fracturing policy. The list discussed below is not exhaustive and can be expanded upon to meet the needs of a real world scenario.

Political factors relate to the partisan behaviors of a state's citizens and governing officials. State policy, influenced by public opinion, can shift quickly when the risks and benefits of hydraulic fracturing are relatively uncertain. One proposed element is the goal of *energy independence* (Figure 4), which considers the amount of energy that could be produced domestically, relative to current import levels, if hydraulic fracturing were to be officially sanctioned. Specifically, *energy independence* incorporates a state's degree of desire to reduce reliance on foreign energy. Another element is *legislative leaning*, the proportion of a country's legislature that is favorable towards hydraulic fracturing. *Legislative leaning* takes into account the general political affiliations of policymakers, under the belief that certain affiliations are prone to support or oppose hydraulic fracturing compared to others. A third example element for the Political factor includes *environmental consciousness*, which serves as a measure of the knowledge of environmental issues, especially those posed by hydraulic fracturing, of politicians and the general public.

Social factors include those that may have an effect upon a government's ability and likelihood to regulate a potentially risky and uncertain activity such as hydraulic fracturing. Three elements - the degree of public *trust in government*, the *national attitudes towards hydraulic fracturing*, and the existing rate of *underemployment* within gas-rich regions - are used as measures of the public's willingness to accept the risks of hydraulic fracturing in order to gain economic benefits (Figure 4). For example, it is assumed that the higher the level of underemployment within a gas-rich area, the more likely residents are to pressure their lawmakers to approve the further development of hydraulic fracturing sites.

Environmental factors focus explicitly upon the perceived risks that hydraulic fracturing activities pose to human and environmental health. As a more objective category, three elements that are analyzed here are the degree to which *existing environmental legislation* prevents/limits gas drilling, the *availability of other energy sources* (preferably renewable) within state borders, and the *availability of water* to be used in the hydraulic fracturing process (Figure 4). Of these factors, the availability of a relatively close and plentiful water supply is crucial to the drilling process, as millions

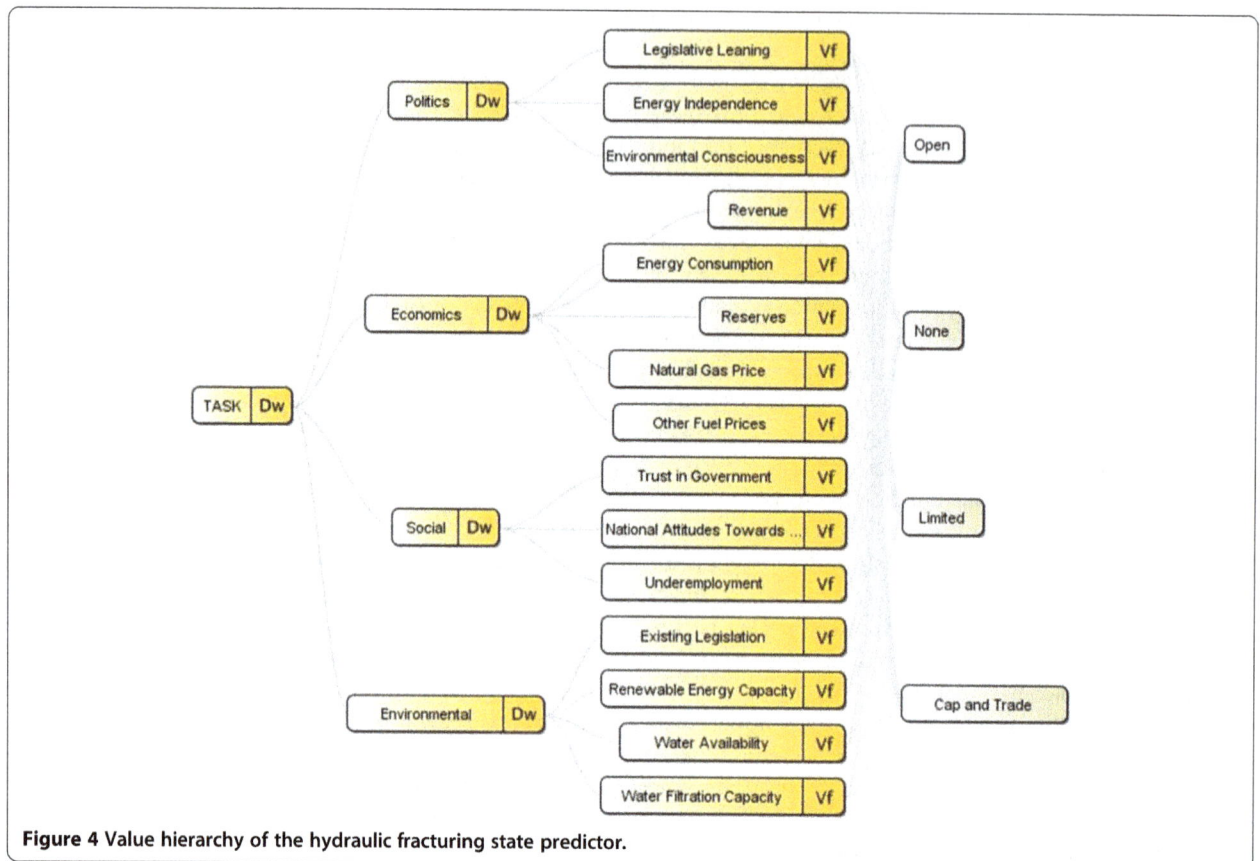

Figure 4 Value hierarchy of the hydraulic fracturing state predictor.

of gallons of water are required by the fracturing technology. Without access to water, and the ability to safely dispose of it after use, unconventional drilling becomes strategically difficult to carry out or approve from a policymaking perspective.

Lastly, economic factors consider the direct benefits and potential financial limitations of hydraulic fracturing for a particular state. Benefits include an estimate of the expected revenue from hydraulic fracturing based upon the availability of gas resources and cost of compressing and transporting the extracted gas. Potential limitations to consider include the trend in prices for other energy sources, which will have a direct effect upon the demand and price for natural gas. Though not universally true, this factor will be a strong influence on most states' decisions, as such natural gas serves as a method by which a state could procure millions of dollars in revenues that were previously unobtainable.

Ultimately, the development of the major decision criteria and their associated individual elements attempt to capture the pressures facing governmental officials required to act upon an industry where they have imperfect information. The value hierarchy for this policy problem is then expanded to include four potential policy outcomes: a full moratorium on hydraulic fracturing, a partial/regional/conditional moratorium on hydraulic

fracturing, the limitation of drilling via a 'cap and trade' system, and a full and open allowance for all drilling (Figure 4). Other policy options certainly exist; however, for this case study, these options are chosen to show how the government officials' decisions are further complicated by the presence of multiple policy outcomes. MCDA can be used to quantify the impact of each criterion on each of the policy alternatives. The quantified values serve as an approximate measure of the degree of pressure each sub-criterion expresses for lawmakers to regulate in a given way.

With the value hierarchy developed, a relative weighting scheme of the criteria would then be elicited from a selection of expert and stakeholder interviews. This qualitative information would be acquired using surveys or content analysis of such experts' feedback, although for this case we utilize hypothetical weights. In general, weights are constructed in a manner where all branches of a criterion are normalized relative to their perceived importance. All major criteria (the four noted in this case) are additionally normalized across each other. Qualitative and quantitative scores are converted into a maximizing scale for each alternative. Mathematically, scores will aggregate 'up the tree,' where results will combine into a single risk attitude score for each individual alternative per country. For the final risk scores, higher

Table 1 Characteristics of archetypal countries

Country archetype	Political structure	Economic potential	Environmental situation	Social trust in government
1 Developed democratic	Capitalist democratic	High energy demand; significant gas reserves	Some environmental protection	Moderate
2 Former communist social democratic	Socialist democratic	High energy demand; moderate gas reserves	Limited environmental protection in favor of industrialization	Limited
3 Developing communist	Communist – centrally planned	Growing energy demand; significant gas reserves	Limited environmental protection in favor of industrialization	Limited
4 Developing third-world social state	Socialist democratic	Energy diversification; some natural gas potential	Substantial environmental protection	Moderate
5 Developed social democratic	Socialist democratic	High energy demand; limited tight gas reserves	Substantial environmental protection	High

scores correspond to less acceptance of risk (more negative risk attitude) regarding hydraulic fracturing. States with higher scores will see more restrictive policies regarding unconventional drilling as the preferred alternatives. The alternatives are subjectively ranked regarding their degree of regulatory restriction from least restrictive to most regulated as open, cap and trade, limited, and closed/no available drilling. Equation 1 gives a mathematical representation of the additive nature of the decision model:

$$V(a) = w_1(V_1(a_1)) + \ldots + w_m(V_m(a_m)) \qquad (1)$$

The alternatives tree is evaluated from the bottom up. At any particular level of interest, a value or utility function, V, is applied to the aggregate alternative scores, a, and is adjusted using the weight, w, given by the stakeholder for that criterion, m. MCDA produces a ranked list of alternatives for each individual country. While the model will indicate the preferred alternative given available information under existing circumstances, it is highly likely that *future* economic, environmental, political, and social needs will shift due to a variety of factors. As such, it is crucial to understand how easily preference may shift in this model from one alternative to another. Sensitivity analysis is used to relax certain conditions and incrementally shift criteria weights (holding all other element constant) to identify the threshold at which the preferred alternative changes. If a significant change in criteria weights or score inputs required in order to change the preferred alternative, the alternative is said to be insensitive to the parameters. If optimality is perceived to shift easily or often, these results are determined to be sensitive to the criteria considered, and policymakers will need to take into account which alternatives may offer the best utility over time instead of simply the currently preferred alternative.

Case study: predicting hydraulic fracturing policy with archetypal examples

This case study is designed to demonstrate how MCDA may be used to predict future state behavior regarding the regulation of unconventional drilling activities. Five archetypal countries were created to represent different combinations of social and economic pressures (Table 1). Specifically, each nation is modeled with unique environmental, economic, social, and political factors that policymakers must consider as they decide how to regulate hydraulic fracturing. All are considered to have some volume of natural gas that could be acquired only through the process of hydraulic fracturing.

In this limited case study, it is already apparent that the factors noted in Table 1 are highly varied and difficult to collectively assess for an individual country. Using MCDA, this information is quantified to for use as scores that can then be aggregated with sub-criteria weights. For now, the MCDA predictor is focused upon the four policy alternatives discussed above, although more can be integrated for additional analysis. The DECERNS software tool was used to evaluate the model using the multi-attribute value theory method (Figure 1).

Abbreviation
MCDA: Multi-criteria decision analysis.

Competing interests
The authors declare that they have no competing interests.

Authors' contributions
IL conceived the idea and the general framework. BT and DJ drafted and revised the manuscript together. BT developed the conceptual model with significant assistance from DJ, MM, and MS. MM and MS helped develop the model and archetypes. All authors contributed to the editing of the paper. All authors read and approved the final manuscript.

Acknowledgements
The authors would like to thank Cate Fox-Lent for discussion and help in preparing this paper. The idea of this paper originated while Igor Linkov participated within the Science Fellow Program at the U.S. embassy in Berlin; some of his views within this article are derived from discussion and experience working with the U.S. Embassy and Department of State personnel engaged in the program. Permission was granted by the USACE Chief of Engineers to publish this material. The views and opinions expressed in this paper are those of the individual authors and not those of the US Army, or other sponsor organizations.

Author details
[1]US Army Engineer Research and Development Center, 696 Virginia Rd, Concord, MA 01742, USA. [2]Free University of Berlin, Kaiserswerther Straße 16-18, Berlin 14195, Germany. [3]Wrocław University of Technology, Wybrzeże Stanisława Wyspiańskiego 27, Wrocław 50-370, Poland. [4]Massachusetts Institute of Technology, 77 Massachusetts Avenue, Cambridge, MA 02139, USA.

References
1. Schmidt CW: **Blind rush? Shale gas boom proceeds amid human health questions.** *Environ Health Perspect* 2011, **119:**a348.
2. Wiseman H: **Untested waters: the rise of hydraulic fracturing in oil and gas production and the need to revisit regulation.** *Fordham Environ Law Rev* 2009, **20:**115.
3. Ridley M: *The Shale Gas Shock.* London: The Global Warming Policy Foundation; 2011.
4. Daniel AJ, Bohm B, Layne M: *Hydraulic Fracturing Considerations for Natural Gas Wells of the Marcellus Shale.* Cincinnati: Groundwater Protection Council Annual Forum; 2008.
5. Esch M: *NY 'Fracking' Ban: Governor David Paterson Orders Natural Gas Hydraulic Fracturing Moratorium for Seven Months in New York.* New York: Associated Press; 2010.
6. Linkov I, Moberg E: *Multi-Criteria Decision Analysis: Environmental Applications and Case Studies.* Boca Raton: CRC Press; 2012.
7. Linkov I, Rosoff H, Valverde LJ, Bates ME, Trump B, Friedman D, Evans J, Keisler J: **Civilian Response Corps force review: the application of multi-criteria decision analysis to prioritize skills required for future diplomatic missions.** *J Mult Criteria Decis Anal* 2012, **19:**155–168.
8. Sullivan T, Yatsalo B, Grebenkov A, Linkov I: **Decision Evaluation for Complex Risk Network Systems (DECERNS) Software Tool.** In *Decision Support Systems for Risk-Based Management of Contaminated Sites.* Edited by Marcomini A, Walter Suter G II, Critto A. New York: Springer; 2009:257–274.
9. Linkov I, Trump B, Pabon N, Keisler J, Collier Z, Scriven J: *A Decision Analytical Approach for Department of Defense Acquisition Risk Management.* Arlington: Military Operations Research; 2013.

21

Post-ozonation in a municipal wastewater treatment plant improves water quality in the receiving stream

Roman Ashauer*[iD]

Abstract

Background: Removal of organic micropollutants from wastewater by post-ozonation has been investigated in a municipal wastewater treatment plant (WWTP) temporarily upgraded with full-scale ozonation, followed by sand filtration, as an additional treatment step of the secondary effluent. Here, the SPEAR (species at risk) indicator was used to analyse macroinvertebrate abundance data that were collected in the receiving stream before, during and after ozonation to investigate whether ozonation improved the water quality.

Results: The SPEAR values indicate a better water quality downstream the WWTP during ozonation. With ozonation the relative abundance of vulnerable macroinvertebrates in the stream receiving the treated wastewater increases from 18 % (CI 15–21 %) to 30 % (CI 28–32 %). This increase of 12 % (CI 8–16 %) indicates improved ecological quality of the stream and shifts classification according to the Water Framework Directive from poor to moderate.

Conclusions: The SPEAR concept, originally developed to indicate pesticide stress, also appears to indicate toxic stress by a mixture of various micropollutants including pharmaceuticals, personal care products and pesticides. The responsiveness of the SPEAR indicator means that those macroinvertebrates that are vulnerable to pesticide pollution are also vulnerable to micropollutants from WWTPs. The change in the macroinvertebrate community downstream the WWTP indicates that toxicity by pollutants decreased by more than one order of magnitude during ozonation. Ozonation followed by sand filtration has favourable impacts on the composition of the macroinvertebrate community and can improve the water quality in the receiving stream.

Keywords: Treatment of wastewater, Good biological status, Stream macroinvertebrates, Trait-based ecological risk assessment, Micropollutant removal, Biodiversity

Background

Micropollutants, for example, pharmaceuticals, personal care products or biocides, are discharged with municipal wastewater and may be hazardous to the environment [1–3]. Ozonation is one of the techniques suggested for tertiary treatment to remove micropollutants from wastewater [1, 4, 5], but the ecotoxicological consequences of wastewater ozonation are ambiguous [6]. Formation of toxic by-products through ozonation is possible, although these can be eliminated in subsequent sand filtration [7, 8].

Removal of organic micropollutants from wastewater by post-ozonation has recently been investigated in a municipal wastewater treatment plant [9, 10]. The wastewater treatment plant (WWTP) Wüeri in Regensdorf, Switzerland was upgraded with ozonation as an additional treatment step of the secondary effluent. Ozonation followed by sand filtration was shown to remove most of the micropollutants [9]. Of those compounds that were detected in the secondary effluent, 17 compounds were reduced by more than 90 % during ozonation, another 17 compounds between 50 and 90 % and four compounds were reduced by less than 50 % [9]. A

*Correspondence: Roman.Ashauer@york.ac.uk
Environment Department, University of York, Heslington, York YO10 5DD, UK

complementary study using an in vitro mode-of-action-based bioassay battery also demonstrated that ozonation reduced the toxicity of the mixture of micropollutants in the effluent in this experiment [11]. The bioassay battery used enriched samples and measured mode-of-action specific toxicity. The treatment efficiency of the ozonation step was 65 and 76 % for non-specific toxicity in the bacterium *Vibrio fischeri* and the algae *Pseudokirchneriella subcapitata*, respectively, 86 % for inhibition of photosystem II in algae, 86 % for estrogenicity, 60 % for inhibition of acetylcholinesterase and complete removal of genotoxicity [11]. Consistent with chemical analysis, micropollutants which are readily oxidised by ozonation, e.g. those causing estrogenicity, showed greatest reduction of toxicity [11]. Furthermore, another complementary study using fish early life stage toxicity tests (FELST) [8] found that the ozonation step led to reduced growth and development in the FELST, although post-treatment with sand filtration eliminated such toxic effects. Altogether these three studies showed reduced micropollutant loads and toxicity in the wastewater after ozonation together with sand filtration compared to conventionally treated wastewater.

The ultimate aim of upgrading WWTPs, for example, with ozonation followed by sand filtration, is to improve the water quality in the receiving stream. Hence, I investigated if ozonation of wastewater improved the water quality in the Furtbach, the stream into which the WWTP in Regensdorf discharges its effluent. My objective is to use the macroinvertebrate data that were collected as part of the full-scale ozonation experiment [8, 9, 11] and investigate whether ozonation followed by sand filtration improved the water quality as indicated by the abundance of vulnerable macroinvertebrates. The macroinvertebrate data were analysed using the SPEAR (species at risk) indicator. Specifically, I asked: How much did the proportion of vulnerable species in the receiving stream's macroinvertebrate community change, when the wastewater was ozonated?

Results and discussion

According to the classification of Beketov et al. [12] the macroinvertebrates indicate a poor biological status of the stream upstream and downstream of the WWTP without ozonation (Fig. 1). The poor quality of the upstream sites can be, at least partially, explained by pollution upstream of the WWTP [13]. Ozonation increases the abundance of vulnerable macroinvertebrates in the stream receiving the treated wastewater from 18 % (CI 15–21 %) to 30 % (CI 28–32 %). This increase of 12 % (CI 8–16 %) indicates improved ecological quality of the stream and shifts classification according to the WFD from poor to moderate [12].

Other researchers have found that WWTP effluents change the assemblages of macroinvertebrates in receiving stream mesocosms [14], although they attributed the effect to increased nutrients and reduced dissolved oxygen. Here, ozonation followed by sand filtration increases the relative abundance of vulnerable species present downstream of the WWTP and leads to an improved water quality classification. Ozonation even appears to improve the water quality downstream of the WWTP compared to upstream (Fig. 1). This seems plausible given the large relative volume that the wastewater contributes to the stream, although the low replication within this study and the large number of possible confounding factors requires further research on this aspect.

It is noteworthy that the effect of the ozonation treatment can be detected in the stream macroinvertebrate composition after only 8 and 16 months. The number of $SPEAR_{pesticides}$ values in each group was small and consisted of data from different locations and sampling dates (spring and autumn), all of which can be assumed to increase variability in the macroinvertebrate community composition. The raw data of this analysis, i.e. taxa lists and abundances, are given in the Additional file 1.

Another finding is that the SPEAR concept, originally developed to indicate pesticide stress, also appears to indicate toxic stress by a mixture of various micropollutants including pharmaceuticals, personal care products and pesticides. The responsiveness of the SPEAR indicator, also known as $SPEAR_{pesticides}$, does not necessarily mean that the stressors are pesticides; rather it means that those macroinvertebrates that are vulnerable to pesticide pollution are also vulnerable to pollution by micropollutants from WWTPs. An improvement of the ecological status in the receiving stream due to the additional ozonation step followed by sand filtration as indicated by SPEAR is consistent with the reduction of overall micropollutant load found by chemical analysis [9] and monitoring with bioassays [11]. A differentiation of the effects of various micropollutants was not possible with SPEAR. Chemical analysis of the receiving stream water before and during ozonation also confirmed the reduction of micropollutant loads, for example, the concentrations of carbamazepine, diclofenac, clarithromycin and sulfamethoxazole were reduced by ozonation from 0.51 µg/L to below 3 ng/L, 0.41 µg/L to below 10 ng/L, 0.12 µg/L to below 3 ng/L and 0.12 µg/L to below 6 ng/L, respectively [9, 15]. Furthermore, feeding trials with leaf discs conditioned in wastewater from the same WWTP as studied here showed that *Gammarus fossarum* preferred the leaf discs that were conditioned in wastewater treated with high doses of ozone over those leaf discs that were conditioned in untreated wastewater [6] and in situ feeding rate trials showed that ozonation increases detritus processing in the stream [16].

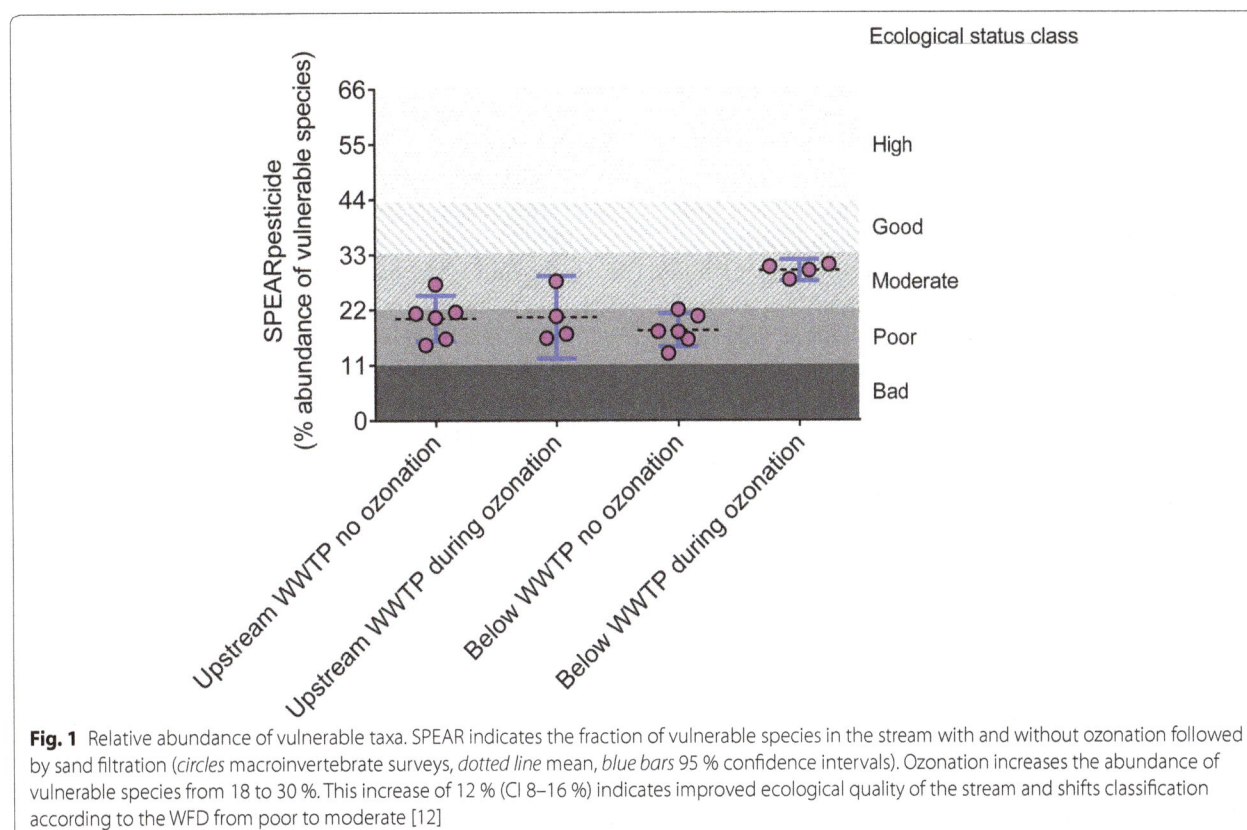

Fig. 1 Relative abundance of vulnerable taxa. SPEAR indicates the fraction of vulnerable species in the stream with and without ozonation followed by sand filtration (*circles* macroinvertebrate surveys, *dotted line* mean, *blue bars* 95 % confidence intervals). Ozonation increases the abundance of vulnerable species from 18 to 30 %. This increase of 12 % (CI 8–16 %) indicates improved ecological quality of the stream and shifts classification according to the WFD from poor to moderate [12]

Analysis of the macroinvertebrate community on the receiving water bodies downstream of WWTPs can clearly contribute to answer the question whether post-treatment technologies help achieve water quality goals, in particular when existing knowledge about vulnerability of species is built into the data analysis as ,for example, with the SPEAR concept. Not all WWTPs contribute as much water volume to the stream as the one studied here, thus future studies may need to increase their power by larger sample size and improved design [17, 18].

The impact of WWTP effluents on stream macroinvertebrate assemblages has been documented before [19, 20]. The clear effects of upgrading the WWTP with ozonation, more specifically the increase in vulnerable species in downstream samples from 18 to 30 %, would correspond to a reduction of toxicant loads by approximately 1.5 toxic units (*Daphnia magna*) according to the regressions in [21]. In other words the change in SPEAR in the macroinvertebrate community downstream the WWTP indicates that toxicity by pollutants decreased by more than one order of magnitude during ozonation.

Conclusion

The previously reported reduction in chemical loads and reduced toxicity measured by an in vitro bio-test battery during ozonation followed by sand filtration in the WWTP has favourable impacts on the composition of the macroinvertebrate community and water quality in the receiving stream.

Methods

Study site

The WWTP discharges into a small stream (Furtbach) with a catchment of 12 km^2 consisting of 24 % forest, 42 % agriculture and 29 % urban use (5 % other uses). The Furtbach stream has an average slope of 0.1 %, holds water all year round and the substrate consists mostly of large (fist to nut size) to small (nut to pea size) gravel with 10–20 % sand and 10 % or less silt at all four macroinvertebrate sampling sites [22]. The Furtbach originates from a small lake approximately 5 km upstream of the WWTP and discharges into the river Limmat approximately 9 km downstream. More details about the sampling site characteristics can be found in [22].

The WWTP approximately doubles the discharge in the stream (WWTP treats 5500 m^3 d^{-1} on average under dry conditions, WWTP discharge ranges from 30 to 120 L s^{-1} and constitutes ca. 60 % of the water in the stream under dry weather conditions [9]). The WWTP consists of primary sedimentation, activated sludge treatment (nitrifying and denitrifying) and secondary clarification followed by sand filtration. The ozonation step was added after the secondary clarifier and before the sand filtration. The ozonation reactor had a retention time of 8–15 min during dry conditions and 3 min during storm water (which was judged not sufficient under stormwater conditions [9]). Ozonation followed by sand filtration was in operation, with short breaks, from July 2007 until the end of October 2008 [15]. Then the ozonation equipment was removed from the WWTP. Ozone dose and residence time in the reactor varied between 357 and 1157 g_{O3}/kg_{DOC} and 4–10 min, respectively [11] and were regulated based on online measurements of dissolved organic carbon. More details on the treatment processes, chemistry of the wastewater and operation of the ozonation can be found in [9] and [15], including a wide range of additional parameters measured.

Various sources of pollution upstream of the WWTP exist, for example, several storm water overflow channels discharge in the stream (Fig. 2), in June 2007 there was a contamination incident with an unspecified fungicide [15] and chemical analysis of the upstream water in June 2007 found several pesticides or biocides and their biotransformation products, as well as some pharmaceuticals in the ng/L range [13].

Macroinvertebrate data

There were three sampling sites upstream and three downstream of the WWTP (see Table 1; Fig. 2). Before, during and after ozonation followed by sand filtration was installed at the wastewater treatment plant, the macroinvertebrates in the receiving stream were sampled, identified to species or family level (according to [23]) and abundances recorded [22]. There were two sampling dates before (5 October 2006, 26 February 2007), two during the period with ozonation followed by sand filtration (26 February 2008, 20 October 2008) and two sampling dates after the ozonation treatment was dismantled (10 March 2014, 14 October 2014). Thus, the ozonation was in operation already for 8 and 16 months when the macroinvertebrates were sampled to measure effects of the ozonation treatment in 2008 and the ozonation treatment had been dismantled for over 5 years before the sampling in 2014. The 2006–2008 macroinvertebrate data were collected by Aqua-Plus, Zug, Switzerland [22] on behalf of AWEL (Amt für Abfall, Wasser, Energie und Luft; Zürich, Switzerland). The macroinvertebrate data from the year 2014 were provided by AWEL (Patrick Steinmann, pers. comm.). More details and raw data can be found in [15, 22], as well as on the website of AWEL (http://www.gewaesserqualitaet.zh.ch).

The SPEAR indicator and micropollutants

The SPEAR concept was developed as a tool to reveal impacts on stream macroinvertebrate communities related to chemical stress by pesticides [21, 24]. Species are classified according to their vulnerability into species at risk and species not at risk. Vulnerability classification takes into account ecological and physiological traits of the species, more specifically the toxicological sensitivity to organic pollutants including pesticides [25] as the only physiological trait, as well as the generation time, migration ability and time of emergence as ecological traits [21, 24]. Although some methodical aspects of the SPEAR approach such as the sensitivity ranking relative to *D. magna* and the neglect of mode-of-action specific sensitivity differences can be criticised [26], SPEAR values were shown to correlate with pesticide contamination in several catchments throughout Europe [24, 27], also when family-level data were used [12]. The approach taken here, using the SPEAR concept, assumes that the species traits that make SPEAR indicative of chemical stress by pesticides are also defining the vulnerability of macroinvertebrates to micropollutants present in WWTP effluent.

Calculation of species at risk (SPEAR)

The 2006 to 2008 taxa lists and their abundance [22] were entered into the SPEAR web calculator (http://www.systemecology.eu/SPEAR/index.php, accessed 7 June 2010), whereas the 2014 macroinvertebrate data were analysed using the SPEAR calculator v0.9.0 (http://www.systemecology.eu/spearcalc/, accessed 6 November 2015, using family-level taxa, default traits and no recovery areas). Four taxa [Ostracoda (1 entry), Collembola (3 entries), *Gordius aquaticus* (4 entries), Podura (1 entry)] were deleted because they could not be found in the SPEAR database. In the web calculator Central Europe was selected as region, $SPEAR_{pesticides}$ was calculated and absence or presence of recovery areas was not assessed (no values assigned). The explanatory power of the SPEAR indicator does not suffer significantly if family-level data are used instead of species-level data [12], the taxonomic resolution of the macroinvertebrate data used here is sufficient.

The relative abundance of species at risk (vulnerable species) is calculated as [27]

$$\% SPEAR = \frac{\sum_{i=1}^{n} \log(x_i + 1) \times y}{\sum_{i=1}^{n} \log(x_i + 1)}, \quad (1)$$

where n is the number of taxa, x_i is the abundance of taxon i and y is 1 if the taxon i is classified as species at risk (vulnerable), otherwise y is 0.

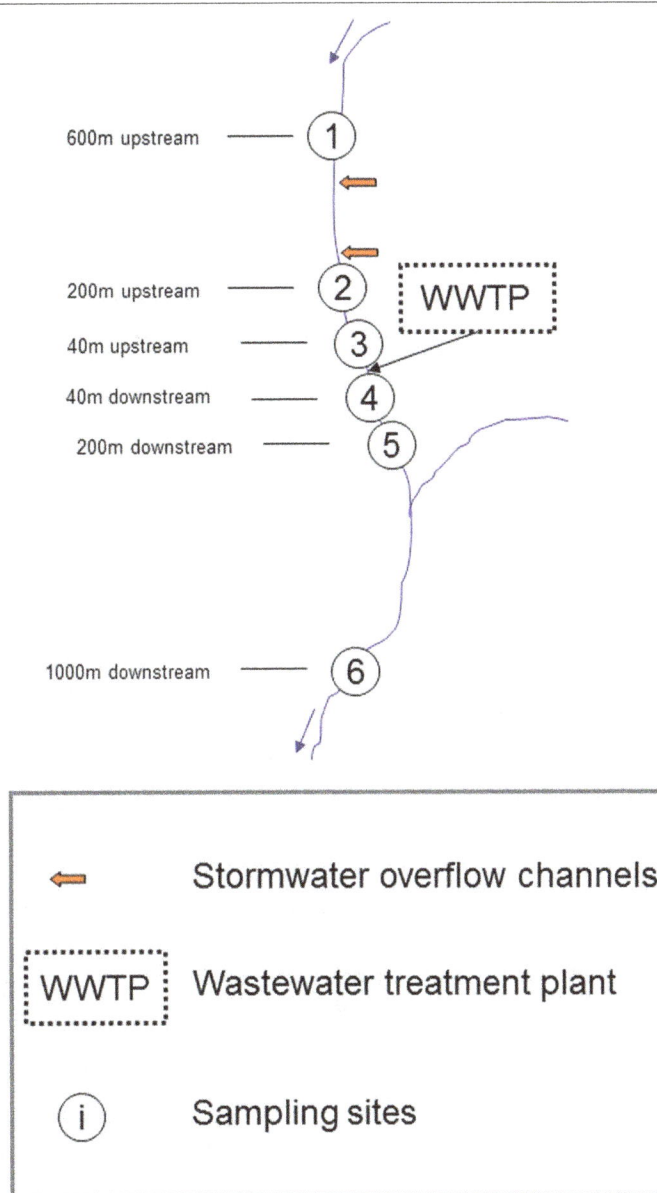

Fig. 2 Location of the WWTP and the sampling sites at the stream Furtbach

Statistical analysis

The SPEAR data can be grouped into four groups A, B, C and D (Table 1, Fig. 1) to better illustrate the analysis. These groups are (A) without ozonation upstream the WWTP, (B) during ozonation upstream the WWTP, (C) without ozonation below the WWTP and (D) with ozonation below the WWTP (Table 1). As the macroinvertebrate samples consist of only few replicates I followed recent developments in statistical reasoning and calculated the confidence interval (CI) of the difference that ozonation makes [28, 29]. Thus, I answered one question with the analysis of the SPEAR data: How much of a difference did ozonation make for the proportion of vulnerable species in the macroinvertebrate community in the receiving stream? The difference was calculated assuming normally distributed errors and equal variances and was carried out using GraphPad Prism version 6.03 (http://www.graphpad.com). All CIs are 95 % confidence intervals.

My analysis assumes that there is no effect of season and distance to the WWTP in the SPEAR data. Alternatively one can carry out a paired analysis, which reduces the number of data points to four in each group and results in larger confidence intervals. However, the

Table 1 Sampling sites, dates and SPEAR values

Ozonation	Sampling date	Sampling site (site number in brackets, see Fig. 1)	Coordinates[a]	SPEAR (% relative abundance of species at risk)[b]
Group A				
No	5 October 2006	600 m upstream of WWTP (1)	2'676'819/1'256'070	16.20
No	5 October 2006	200 m upstream of WWTP (2)	2'676'457/1'256'202	21.27
No	26 February 2007	600 m upstream of WWTP (1)	2'676'098/1'256'159	21.52
No	26 February 2007	200 m upstream of WWTP (2)	2'675'322/1'256'133	27.15
No	10 March 2014	40 m upstream of WWTP (3)	2'676'296/1'256'225	20.42
No	14 October 2014	40 m upstream of WWTP (3)	2'676'296/1'256'225	15.03
Mean (95 % confidence intervals)				20.27 (15.72, 24.81)
Group B				
Yes	20 October 2008	600 m upstream of WWTP (1)	2'676'819/1'256'070	16.38
Yes	20 October 2008	200 m upstream of WWTP (2)	2'676'457/1'256'202	27.75
Yes	26 February 2008	600 m upstream of WWTP (1)	2'676'098/1'256'159	17.20
Yes	26 February 2008	200 m upstream of WWTP (2)	2'675'322/1'256'133	20.70
Mean (95 % confidence intervals)				20.51 (12.27, 28.75)
Group C				
No	5 October 2006	200 m downstream of WWTP (5)	2'676'819/1'256'070	13.31
No	5 October 2006	1000 m downstream of WWTP (6)	2'676'457/1'256'202	20.71
No	26 February 2007	200 m downstream of WWTP (5)	2'676'098/1'256'159	16.07
No	26 February 2007	1000 m downstream of WWTP (6)	2'675'322/1'256'133	17.58
No	10 March 2014	40 m downstream of WWTP (4)	2'676'211/1'256'205	22.03
No	14 October 2014	40 m downstream of WWTP (4)	2'676'211/1'256'205	17.64
Mean (95 % confidence intervals)				17.89 (14.59, 21.19)
Group D				
Yes	20 October 2008	200 m downstream of WWTP (5)	2'676'819/1'256'070	30.57
Yes	20 October 2008	1000 m downstream of WWTP (6)	2'676'457/1'256'202	29.84
Yes	26 February 2008	200 m downstream of WWTP (5)	2'676'098/1'256'159	31.01
Yes	26 February 2008	1000 m downstream of WWTP (6)	2'675'322/1'256'133	28.01
Mean (95 % confidence intervals)				29.86 (27.75, 31.96)

[a] Geographic coordinates: North/East

[b] See Eq. (1), indicator also known as $\text{SPEAR}_{pesticides}$

results are very similar to those given above: ozonation increases the abundance of vulnerable macroinvertebrates downstream the WWTP from 17 % (CI 12–22 %) to 30 % (CI 28–32 %). This increase of 13 % (CI 7–19 %) also shifts classification according to the WFD from poor to moderate [12].

Abbreviations

WWTP: wastewater treatment plant; SPEAR: species at risk ($\text{SPEAR}_{pesticides}$); FELST: fish early life stage toxicity test; AWEL: Amt für Abfall, Wasser, Energie und Luft; Zürich, Switzerland (local authority).

Authors' information

RA works as senior lecturer in the Environment Department at the University of York. His area of research is ecotoxicology and environmental risk assessment of chemicals. From 2007 to 2012 he worked at Eawag, the Swiss Federal Institute of Aquatic Science and Technology which conducted the full-scale ozonation experiment, but RA was not involved in that.

Acknowledgements

I thank Patrick Steinmann, Pius Niederhauser, Nele Schuwirth, Matthias Liess and Mikhail Beketov for advice and discussions, as well as Cornelia Kienle, Juliane Hollender and Beate Escher for comments on an earlier version of this manuscript. The macroinvertebrate data were kindly provided by AWEL (Amt für Abfall, Wasser, Energie und Luft; Abteilung Gewässerschutz; Zürich). During the early stages of this study (2010-2012) the author was funded by the Swiss Federal Office for the Environment (Grants 09.033.PJ/I362-1602 and 09.0012. PJ).

Competing interests

The author declares that he has no competing interests.

References

1. Ternes T (2007) The occurrence of micopollutants in the aquatic environment: a new challenge for water management. Water Sci Technol 55(12):327–332
2. Schwarzenbach RP, Escher BI, Fenner K, Hofstetter TB, Johnson CA, von Gunten U et al (2006) The challenge of micropollutants in aquatic systems. Science 313(5790):1072–1077
3. Ternes TA, Joss A, Siegrist H (2004) Scrutinizing pharmaceuticals and personal care products in wastewater treatment. Environ Sci Technol 38(20):392A–399A
4. Paraskeva P, Graham NJD (2002) Ozonation of municipal wastewater effluents. Water Environ Res 74(6):569–581
5. Huber MM, Göbel A, Joss A, Hermann N, Löffler D, McArdell CS et al (2005) Oxidation of pharmaceuticals during ozonation of municipal wastewater effluents: a pilot study. Environ Sci Technol 39(11):4290–4299
6. Bundschuh M, Gessner MO, Fink G, Ternes TA, Sögding C, Schulz R (2011) Ecotoxicological evaluation of wastewater ozonation based on detritus-detritivore interactions. Chemosphere 82(3):355–361
7. Stalter D, Magdeburg A, Oehlmann J (2010) Comparative toxicity assessment of ozone and activated carbon treated sewage effluents using an in vivo test battery. Water Res 44(8):2610–2620
8. Stalter D, Magdeburg A, Weil M, Knacker T, Oehlmann J (2010) Toxication or detoxication? In vivo toxicity assessment of ozonation as advanced wastewater treatment with the rainbow trout. Water Res 44(2):439–448
9. Hollender J, Zimmermann SG, Koepke S, Krauss M, McArdell CS, Ort C et al (2009) Elimination of organic micropollutants in a municipal wastewater treatment plant upgraded with a full-scale post-ozonation followed by sand filtration. Environ Sci Technol 43(20):7862–7869
10. Zimmermann SG, Wittenwiler M, Hollender J, Krauss M, Ort C, Siegrist H et al (2011) Kinetic assessment and modeling of an ozonation step for full-scale municipal wastewater treatment: micropollutant oxidation, by-product formation and disinfection. Water Res 45(2):605–617
11. Escher BI, Bramaz N, Ort C (2009) JEM Spotlight: monitoring the treatment efficiency of a full scale ozonation on a sewage treatment plant with a mode-of-action based test battery. J Environ Monit 11(10):1836–1846
12. Beketov MA, Foit K, Schaefer RB, Schriever CA, Sacchi A, Capri E et al (2009) SPEAR indicates pesticide effects in streams—Comparative use of species–and family-level biomonitoring data. Environ Pollut 157(6):1841–1848
13. Singer H, Jaus S, Hanke I, Lück A, Hollender J, Alder AC (2010) Determination of biocides and pesticides by on-line solid phase extraction coupled with mass spectrometry and their behaviour in wastewater and surface water. Environ Pollut 158(10):3054–3064

14. Grantham TE, Canedo-Argüelles M, Perrée I, Rieradevall M, Prat N (2012) A mesocosm approach for detecting stream invertebrate community responses to treated wastewater effluent. Environ Pollut 160:95–102
15. Abegglen C, Escher BI, Hollender J, Koepke S, Ort C, Peter A et al (2009) Ozonation of treated effluent—final report pilot plant Regensdorf. Swiss Federal Office for the Environment 2009, Bern (report in German with English summary)
16. Bundschuh M, Pierstorf R, Schreiber WH, Schulz R (2011) Positive effects of wastewater ozonation displayed by in situ bioassays in the receiving stream. Environ Sci Technol 45(8):3774–3780. doi:10.1021/es104195h
17. Underwood AJ (1994) On beyond BACI: sampling designs that might reliably detect environmental disturbances. Ecol Appl 4(1):3–15
18. Stewart-Oaten A, Bence JR, Osenberg CW (1992) Assessing effects of unreplicated perturbations: no simple solutions. Ecology 73(4):1396–1404
19. Dyer SD, Wang X (2002) A comparison of stream biological responses to discharge from wastewater treatment plants in high and low population density areas. Environ Toxicol Chem 21(5):1065–1075
20. Jin SR, Sang DK, Nam IC, An KG (2007) Ecological health assessments based on whole effluent toxicity tests and the index of biological integrity in temperate streams influenced by wastewater treatment plant effluents. Environ Toxicol Chem 26(9):2010–2018
21. Liess M, Schaefer RB, Schriever CA (2008) The footprint of pesticide stress in communities-species traits reveal community effects of toxicants. Sci Total Environ 406(3):484–490
22. AquaPlus (2009) Furtbach (ZH): biologische untersuchungen oberhalb und unterhalb der ARA Regensdorf—äusserer aspekt, pflanzlicher bewuchs, kieselalgen und zoobenthos. untersuchungen der jahre 2006, 2007 und 2008. Kurzbericht zürich: AquaPlus, commisioned by Baudirektion Kanton Zürich, AWEL, Abteilung Gewässerschutz
23. BUWAL (2005) Methoden zur untersuchung und Beurteilung der Fliessgewässer. Makrozoobenthos Stufe F (flächendeckend)
24. Liess M, Von Der Ohe PC (2005) Analyzing effects of pesticides on invertebrate communities in streams. Environ Toxicol Chem 24(4):954–965
25. Wogram J, Liess M (2001) Rank ordering of macroinvertebrate species sensitivity to toxic compounds by comparison with that of Daphnia magna. Bull Environ Contam Toxicol 67(3):360–367
26. Rubach MN, Baird DJ, Van Den Brink P (2010) A new method for ranking mode-specific sensitivity of freshwater arthropods to insecticides and its relationship to biological traits. Environ Toxicol Chem 29(2):476–487
27. Schaefer RB, Caquet T, Siimes K, Mueller R, Lagadic L, Liess M (2007) Effects of pesticides on community structure and ecosystem functions in agricultural streams of three biogeographical regions in Europe. Sci Total Environ 382(2–3):272–285
28. Cumming G (2014) The new statistics: why and how. Psychol Sci 25(1):7–29. doi:10.1177/0956797613504966
29. Halsey LG, Curran-Everett D, Vowler SL, Drummond GB (2015) The fickle P value generates irreproducible results. Nat Methods 12(3):179–185. doi:10.1038/nmeth.3288

Permissions

The contributors of this book come from diverse backgrounds, making this book a truly international effort. This book will bring forth new frontiers with its revolutionizing research information and detailed analysis of the nascent developments around the world.

We would like to thank all the contributing authors for lending their expertise to make the book truly unique. They have played a crucial role in the development of this book. Without their invaluable contributions this book wouldn't have been possible. They have made vital efforts to compile up to date information on the varied aspects of this subject to make this book a valuable addition to the collection of many professionals and students.

This book was conceptualized with the vision of imparting up-to-date information and advanced data in this field. To ensure the same, a matchless editorial board was set up. Every individual on the board went through rigorous rounds of assessment to prove their worth. After which they invested a large part of their time researching and compiling the most relevant data for our readers.

The editorial board has been involved in producing this book since its inception. They have spent rigorous hours researching and exploring the diverse topics which have resulted in the successful publishing of this book. They have passed on their knowledge of decades through this book. To expedite this challenging task, the publisher supported the team at every step. A small team of assistant editors was also appointed to further simplify the editing procedure and attain best results for the readers.

Apart from the editorial board, the designing team has also invested a significant amount of their time in understanding the subject and creating the most relevant covers. They scrutinized every image to scout for the most suitable representation of the subject and create an appropriate cover for the book.

The publishing team has been an ardent support to the editorial, designing and production team. Their endless efforts to recruit the best for this project, has resulted in the accomplishment of this book. They are a veteran in the field of academics and their pool of knowledge is as vast as their experience in printing. Their expertise and guidance has proved useful at every step. Their uncompromising quality standards have made this book an exceptional effort. Their encouragement from time to time has been an inspiration for everyone.

The publisher and the editorial board hope that this book will prove to be a valuable piece of knowledge for researchers, students, practitioners and scholars across the globe.

List of Contributors

Roland Bauböck
Department of Cartography, GIS and Remote Sensing, Research Project 'BIS', University of Göttingen, Goldschmidtstraße 5, Göttingen 37077, Germany

Bernd Nowack, Nicole C Mueller, Harald F Krug and Peter Wick
Empa - Swiss Federal Laboratories for Materials Science and Technology, Lerchenfeldstrasse 5, St Gallen 9014, Switzerland

Caren Rauert
Section International Chemicals Management, Federal Environment Agency (UBA), Wörlitzer Platz 1, 06844 Dessau-Roßlau, Germany

Anton Friesen, Anja Kehrer, and Karen Willhaus
Section Biocides, Federal Environment Agency (UBA), Wörlitzer Platz 1, 06844 Dessau-Roßlau, Germany

Janina Wöltjen, Sabine Duquesne and Georgia Hermann
Section Plant Protection Products, Federal Environment Agency (UBA), Wörlitzer Platz 1, 06844 Dessau-Roßlau, Germany

Ulrich Jöhncke and Michael Neumann
Section Chemicals, Federal Environment Agency (UBA), Wörlitzer Platz 1, 06844 Dessau-Roßlau, Germany

Ines Prutz, Jens Schönfeld and Astrid Wiemann
Section Pharmaceuticals, Washing and Cleaning Agents, Federal Environment Agency (UBA), Wörlitzer Platz 1, 06844 Dessau-Roßlau, Germany

Axel Bergmann and Frank-Andreas Weber
IWW Water Centre, Department Water Resources Management, Moritzstrasse 26, Muelheim 45476, Germany

Georg Meiners and Frank Müller
ahu AG Wasser Boden Geomatik, Kirberichshofer Weg 6, Aachen 52066, Germany

Andreas Toschki
gaiac Research Institute for Ecosystem Analysis and Assessment, Kackerstr. 10, 52072 Aachen, Germany

Stephan Jänsch and Jörg Römbke
ECT Oekotoxikologie GmbH, Böttgerstr. 2-14, 65439 Flörsheim, Germany

Martina Roß-Nickoll
RWTH Aachen University, Institute for Environmental Research, Worringer Weg 1, 52074 Aachen, Germany

Wiebke Züghart
Federal Agency for Nature Conservation (BfN), Konstantinstr. 110, 53179 Bonn, Germany

J. Kowalczyk, H. Schafft and M. Lahrssen-Wiederholt
Federal Institute for Risk Assessment, Max-Dohrn-Str. 8-10, 10589 Berlin, Germany

S. Riede and G. Breves
Department Institute of Physiology, University of Veterinary Medicine Hannover, Foundation, Bischofsholer, Damm 15, 30173 Hannover, Germany

Roland Bauböck and Martin Kappas
Department of Cartography, GIS and Remote Sensing, Research Project 'BIS', University of Göttingen, Goldschmidtstraße 5, Göttingen 37077, Germany

Marianne Karpenstein-Machan
Interdisciplinary Centre of Sustainable Development, Research Project 'BIS', University of Göttingen, Goldschmidtstraße 1, Göttingen 37077, Germany

Gilles-Eric Séralini, Emilie Clair, Robin Mesnage, Steeve Gress, Nicolas Defarge and Joël Spiroux de Vendômois
Institute of Biology, EA 2608 and CRIIGEN and Risk Pole, MRSH-CNRS, Esplanade de la Paix, University of Caen, Caen, Cedex 14032, France

Manuela Malatesta
Department of Neurological, Neuropsychological, Morphological and Motor Sciences, University of Verona, Verona 37134, Italy

Didier Hennequin
Risk Pole, MRSH-CNRS, Esplanade de la Paix, University of Caen, Caen, Cedex 14032, France

Winfried Schröder, Gunther Schmidt and Simon Schönrock
Landscape Ecology, University of Vechta, P.O. Box 1553, Vechta 49364, Germany

Kathrin Schwirn, Lars Tietjen and Inga Beer
Federal Environment Agency, WoerlitzerPlatz 1, Dessau-Rosslau 06844, Germany

Ute Schoknecht
Division 4.1 Biodeterioration and Reference Organisms, BAM Federal Institute for Materials Research and Testing, Unter den Eichen 87, D-12205 Berlin, Germany

Ute Kalbe
Division 4.3, Contaminant Transfer and Environmental Technologies, BAM Federal Institute for Materials Research and Testing, Unter den Eichen 87, D-12205 Berlin, Germany

André van Zomeren
Department of Environmental Assessment, ECN, Westerduinweg 3, 1755 LE Petten, the Netherlands

Ole Hjelmar
Waste and Soil, DHI, Agern Allé 5, DK-2970 Hørsholm, Denmark

Sophia Schubert, Nadia Keddig, Reinhold Hanel and Ulrike Kammann
Institute of Fisheries Ecology, Thünen Institute (TI), Palmaille 9, 22767 Hamburg, Germany

Ann-Sofie Wernersson, Bengt Fjällborg and Tobias Porsbring
Swedish Agency for Marine and Water Management, Gullbergs Strandgata 15, Göteborg 404 39, Sweden

Mario Carere, Ines Lacchetti and Laura Mancini
Department of Environment and Primary Prevention, ISS-Italian Institute of Health, Viale Regina Elena, 299, 00161 Rome, Italy

Chiara Maggi, Antonella Ausili and Loredana Manfra
ISPRA - Institute for Environmental Protection and Research, Via Brancati 48, 00144 Rome, Italy

Petr Tusil and Premysl Soldan
T.G. Masaryk Water Research Institute, Podbabská 2582/30, 6, 160 00 Praha, Czech Republic

Alice James, Wilfried Sanchez and Valeria Dulio
National Institute for Industrial Environment and Risks INERIS, Rue Jacques Taffanel, 60550 Verneuil en Halatte, France

Katja Broeg
Baltic Eye, Östersjöcentrum, Stockholms Universitet, SE-106 91, Universitetsvägen 10 A, Stockholms, Sweden

Georg Reifferscheid and Sebastian Buchinger
Federal Institute of Hydrology, Am Mainzer Tor 1, 56068 Koblenz, Germany

Hannie Maas
Rijkswaterstaat,Water, Traffic and Environment, Zuiderwagenplein 2, 8224 Lelystad, The Netherlands

Esther Van Der Grinten
National Institute for Public Health and the Environment (RIVM), 3720, Postbus 1BA, Bilthoven, The Netherlands

Simon O'Toole
EPA, Johnstown Castle Estate, PO Box 3000, Wexford, Ireland

Laura Marziali and Stefano Polesello
IRSA-CNR, Via del Mulino 19, 20047 Brugherio, Italy

Karl Lilja, Maria Linderoth and Tove Lundeberg
Swedish Environmental Protection Agency, Naturvårdsverket, SE-106 48 Stockholm, Sweden

DG Joakim Larsson, Johan Bengtsson-Palme and Lars Förlin
University of Gothenburg, Guldhedsgatan 10, SE-405 30 Gothenburg, Sweden

Cornelia Kienle, Petra Kunz, Etienne Vermeirssen, Inge Werner and Robert Kase
Swiss Centre for Applied Ecotoxicology Eawag-EPFL, Überlandstrasse 133, CH-8600 Dübendorf, Switzerland

Craig D Robinson
Marine Scotland Science, 375 Victoria Road, AB11 9DB Aberdeen, Scotland

Brett Lyons and Ioanna Katsiadaki
Cefas, Pakefield Road, Lowestoft, Suffolk NR33 0HT, UK

Caroline Whalley
Defra, Area 3D, Nobel House, Smith Square, London SW1P 3JR, UK

Klaas den Haan
CONCAWE, Vorstlaan165, B-1160 Brussels, Belgium

Marlies Messiaen
Eurometaux, Avenue de Broqueville 12, B-1150 Brussels, Belgium

Helen Clayton
DG Environment - European Commission, Avenue de Beaulieu 9,B-1160 Brussels, Belgium

Teresa Lettieri, Raquel Negrão Carvalho and Bernd Manfred Gawlik
DG Joint Research Centre - European Commission, Via Enrico Fermi 2749, I-21027 Ispra, Italy

Henner Hollert and Carolina Di Paolo
Department of Ecosystem Analysis, Institute for Environmental Research, ABBt RWTH Aachen University, Worringerweg 1, 52074 Aachen, Germany

Werner Brack
Helmholtz Centre for Environmental Research-UFZ, Permoserstraße 15, 04318 Leipzig, Germany

Ulrike Kammann
Thünen-Institut of Fisheries-Ecology, Palmaille 9, 22767 Hamburg, Germany

Jan F Degener and Martin Kappas
Department of Geography, University of Göttingen, Goldschmidtstr.5, 37077 Göttingen, Germany

Nadia Keddig and Sophia Schubert
Institute of Fisheries Ecology, Thünen Institute (TI), Palmaille 9, 22767 Hamburg, Germany

Werner Wosniok
Institute of Statistics, University of Bremen, Linzer Str. 4, 28334 Bremen, Germany

Kerstin Hund-Rinke, Monika Herrchen and Karsten Schlich
Fraunhofer Institute for Molecular Biology and Applied Ecology IME, Auf dem Aberg 1, 57392 Schmallenberg, Germany

Kathrin Schwirn and Doris Völker
German Federal Environment Agency (UBA), Wörlitzer Platz 1, 06844 Dessau, Germany

Carolina Di Paolo, Thomas-Benjamin Seiler, Steffen Keiter and Henner Hollert
Department of Ecosystem Analysis, Institute for Environmental Research, ABBt - Aachen Biology and Biotechnology, RWTH Aachen University, Worringerweg 1, Aachen 52074, Germany

Steffen Keiter
Man-Technology-Environment Research Centre, School of Science and Technology, Örebro University, SE-701 82 Örebro, Sweden

Meng Hu, Melis Muz and Werner Brack
UFZ-Helmholtz Centre for Environmental Research, Permoserstrasse 15, Leipzig 04318, Germany

Henner Hollert
College of Resources and Environmental Science, Chongqing University, 1 Tiansheng Road, Beibei, Chongqing 400715, China
College of Environmental Science and Engineering and State Key Laboratory of Pollution Control and Resource Reuse, Tongji University, 1239 Siping Road, Shanghai, China

State Key Laboratory of Pollution Control and Resource Reuse, School of the Environment, Nanjing University, Nanjing, China

Ursula Klaschka
University of Applied Sciences Ulm, Prittwitzstr 10, D-89075 Ulm, Germany

Kerstin Hund-Rinke and Karsten Schlich
Fraunhofer Institute for Molecular Biology and Applied Ecology IME, Auf dem Aberg 1, 57392 Schmallenberg, Germany

Igor Linkov, Benjamin Trump and David Jin
US Army Engineer Research and Development Center, 696 Virginia Rd, Concord, MA 01742, USA

Marcin Mazurczak and Miranda Schreurs
Free University of Berlin, Kaiserswerther Straße 16-18, Berlin 14195, Germany

Marcin Mazurczak
Wrocław University of Technology, Wybrzeże Stanisława Wyspiańskiego 27, Wrocław 50-370, Poland

David Jin
Massachusetts Institute of Technology, 77 Massachusetts Avenue, Cambridge, MA 02139, USA

Roman Ashauer
Environment Department, University of York, Heslington, York YO10 5DD, UK